Biologically-Inspired Systems

Volume 9

Series Editor
Stanislav N. Gorb
Christian Albrecht University of Kiel, Kiel, Germany

Motto: Structure and function of biological systems as inspiration for technical developments

Throughout evolution, nature has constantly been called upon to act as an engineer in solving technical problems. Organisms have evolved an immense variety of shapes and structures from macro down to the nanoscale. Zoologists and botanists have collected a huge amount of information about the structure and functions of biological materials and systems. This information can be also utilized to mimic biological solutions in further technical developments. The most important feature of the evolution of biological systems is multiple origins of similar solutions in different lineages of living organisms. These examples should be the best candidates for biomimetics. This book series will deal with topics related to structure and function in biological systems and show how knowledge from biology can be used for technical developments in engineering and materials science. It is intended to accelerate interdisciplinary research on biological functional systems and to promote technical developments. Documenting of the advances in the field will be important for fellow scientists, students, public officials, and for the public in general. Each of the books in this series is expected to provide a comprehensive, authoritative synthesis of the topic.

More information about this series at http://www.springer.com/series/8430

Jan Knippers • Klaus G. Nickel • Thomas Speck
Editors

Biomimetic Research for Architecture and Building Construction

Biological Design and Integrative Structures

Springer

Editors
Jan Knippers
Institute of Building Structures
and Structural Design (ITKE)
University of Stuttgart
Stuttgart, Germany

Klaus G. Nickel
Applied Mineralogy (AMIN)
University of Tübingen
Tübingen, Germany

Thomas Speck
Plant Biomechanics Group, Botanic Garden,
Faculty of Biology
University of Freiburg
Freiburg, Germany

Freiburg Centre for Interactive Materials
and Bioinspired Technologies (FIT)
University of Freiburg
Freiburg, Germany

ISSN 2211-0593 ISSN 2211-0607 (electronic)
Biologically-Inspired Systems
ISBN 978-3-319-46372-8 ISBN 978-3-319-46374-2 (eBook)
DOI 10.1007/978-3-319-46374-2

Library of Congress Control Number: 2016960678

Printed on acid-free paper

This Springer imprint is published by Springer Nature
The registered company is Springer International Publishing AG
The registered company address is: Gewerbestrasse 11, 6330 Cham, Switzerland

Preface

This book comprises the first survey of the topics addressed by the Collaborative Research Centre SFB-TRR 141 'Biological Design and Integrative Structures – Analysis, Simulation and Implementation in Architecture', funded by the Deutsche Forschungsgemeinschaft since October 2014. The SFB-TRR 141 provides a collaborative framework for architects and engineers from the University of Stuttgart, biologists and physicists from the University of Freiburg and geoscientists, material scientists and evolutionary biologists from the University of Tübingen. The programme is based on the belief that biomimetic research has the potential to lead everyone involved to new findings far beyond his/her individual reach and the realm of his/her intrinsic field of research.

The contributions in this book summarize the presentations concerning the various research projects given at the first public conference of the SFB TRR 141 that took place during the initial stages of our collaborative research programme at the 'Staatliches Museum für Naturkunde, Stuttgart', in November 2015. Hence, the contributions to this book present topics, targets and the first interesting findings rather than the final results of the SFB TRR 141. In addition, the book comprises the contributions of some external speakers who were invited to the conference.

We wish to thank everyone who contributed to the conference and to the book but, even more than that, to the creative spirit of mutual inspiration that the SFB TRR 141 has developed, even at its earliest stages.

We are also grateful for the funding that is provided by the Deutsche Forschungsgemeinschaft and that has made all of this possible.

Stuttgart, Germany
Tübingen, Germany
Freiburg, Germany

Jan Knippers
Klaus G. Nickel
Thomas Speck

Acknowledgements

We would like to thank our reviewers whose comments have markedly improved the articles presented in the book.

Maria Abou Chakra, Max-Planck-Institut für Evolutionsbiologie, Plön
Joachim Bluhm, Universität Duisburg-Essen
Roland Ennos, University of Hull
Claudia Erbar, Universität Heidelberg
Claudia Fleck, Technische Universität Berlin
Maik Gude, Technische Universität Dresden
Michael Herdy, INPRO, Berlin
Marc-Thorsten Hütt, Jacobs University, Bremen
George Jeronimidis, University of Reading
Roland Kappel, Karlsruher Institut für Technologie
Cezar Khursigara, University of Guelph
Axel Kilian, Princeton University
Wilfried Konrad, Universität Tübingen
Johannes Kreißig, DGNB, Stuttgart
Wolfgang Maier, Universität Tübingen
Jürgen Meinhardt, Fraunhofer ISC, Würzburg
Göran Pohl, Pohl Architekten, Stuttgart
Wolfgang Rähle, Staatliches Museum für Naturkunde Stuttgart
Markus Rüggeberg, ETH Zürich
Michael Scheffler, Universität Magdeburg
Simon Schleicher, UC Berkeley
Thomas Stützel, Ruhr-Universität Bochum
Peter Supancic, Montanuniversität Leoben
Oliver Tessmann, Technische Universität Darmstadt
Georg Toepfer, Zentrum für Literatur- und Kulturforschung, Berlin
Georg Vrachliotis, Karlsruher Institut für Technologie
Peter Weiliang Xu, The University of Auckland

The editors are very grateful to Kristina Hajek, Manager of the SFB-TRR 141, for coordinating this book with utmost patience and commitment.

Structure and Function of Biological Systems as Inspiration for Technical Developments

Throughout evolution, organisms have evolved an immense variety of materials, structures and systems. This book series deals with topics related to structure-function relationships in diverse biological systems and shows how knowledge from biology can be used for technical developments (bio-inspiration, biomimetics).

Prof. Dr. Stanislav N. Gorb, Christian Albrecht University of Kiel, Kiel, Germany

Contents

Chapter 1
Biomimetic Research: A Dialogue Between the Disciplines

Jan Knippers, Thomas Speck, and Klaus G. Nickel

Recent advances in computational imaging, simulation and fabrication have opened a new era in biomimetics: local differentiation at various scales, the main feature of natural constructions enabling multiple networked working principles, can for the first time not only be analysed, but to a certain extent also be transferred to the scale of building construction. The exploration and biomimetic transfer of the vast reservoir of nature requires an in-depth analysis and comparison of architecture and biology, both from a broader and deeper perspective and on a functional, structural and methodological level (Knippers and Speck 2012; Speck et al. 2015; Speck 2015, Imhof and Gruber 2015).

1.1 Biology as a Guide for Future Development in Architecture

Living beings have adapted to a constantly changing environment during evolution through mutation, recombination and selection. This also comprises the ability to react 'online' to environmental changes in a typically fast, efficient and highly

---

J. Knippers (✉)
Institute of Building Structures and Structural Design (ITKE), University of Stuttgart, Keplerstraße 11, 70174 Stuttgart, Germany
e-mail: j.knippers@itke.uni-stuttgart.de

T. Speck
Plant Biomechanics Group, Botanic Garden, University of Freiburg, Schänzlestraße 1, D-79104 Freiburg, Germany

Freiburg Centre for Interactive Materials and Bioinspired Technologies (FIT), University of Freiburg, Georges-Köhler-Allee 105, D-79110 Freiburg, Germany

K.G. Nickel
Applied Mineralogy (AMIN), University of Tübingen, Wilhelmstr. 56, 72074 Tübingen, Germany

J. Knippers et al. (eds.), *Biomimetic Research for Architecture and Building Construction*, Biologically-Inspired Systems 9, DOI 10.1007/978-3-319-46374-2_1

dynamic manner within their genetically encoded range of possible reactions. In the evolutionary process, the efficient usage of natural resources is a significant evolutionary advantage and therefore a guiding evolutionary principle. It leads to structures made of only a limited number of chemical elements (often light as to their atomic weight) and basic molecular building blocks, which are directly available as part of a natural material loop, mostly in the direct surroundings of the organism, and processed under ambient conditions, i.e. between −40 °C and +40 °C and at typical pressures of 1 bar (with the exception of the deep sea region).

Through genetically controlled self-organization, living organisms form hierarchically organized finely tuned and highly differentiated materials and structures that feature multiple networked functions and that sometimes fulfil contradictory functional requirements. Interestingly, the terms 'material' and 'structure', which are distinctive design categories in architecture and building construction, dissipate in nature as it is nearly impossible to discern between them in the hierarchically structured world of organisms, which ranges from the molecule to the largest living beings by ten orders of magnitude (Fratzl and Weinkammer 2007; Speck and Speck 2015; Wegst et al. 2015).

In this respect, biological structures differ fundamentally from most building constructions. The latter are made of a large variety of materials and form individual components. These are independently optimized for a few target functions at stationary boundary conditions. Technical materials are typically produced by extremely energy-consuming methods of manufacture, i.e. high temperature and pressure, and are frequently transported over huge distances, which often results in much worse specific sustainability values compared with natural materials. What's more, design principles such as adaptivity, multi-functionality or hierarchical structuring have so far been used in technology only to a very limited extent, even though they are omnipresent in nature (Knippers and Speck 2012) (Fig. 1.1).

During the last few decades, computational methods have been introduced into all fields of science and technology. In architecture, they enable the geometric differentiation of building components and allow the fabrication of porous, multi-layered or fibre-based materials with locally adjusted physical and chemical properties. Recent developments in simulation technologies focus on multi-scale models and the interplay of mechanical phenomena at various hierarchical levels. In the natural sciences, a multitude of quantitative methods covering diverse hierarchical levels have been introduced. The step towards digital technologies enables the direct

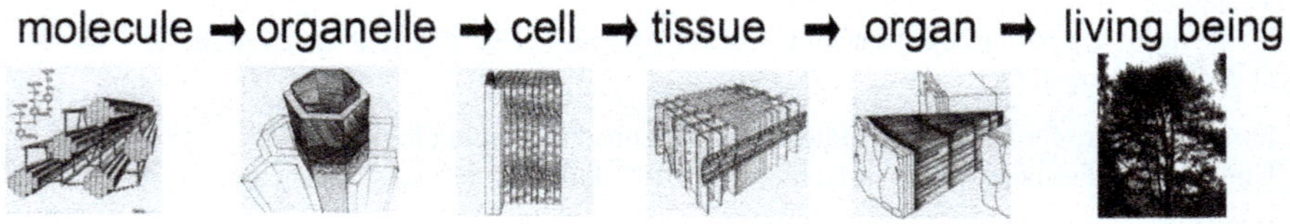

Fig. 1.1 Genetically controlled self-organisation and hierarchical structuring of a conifer tracheid and stem as role model for alternative design approaches in building construction (Adapted from Speck and Speck 2015)

exchange of information between fields of science that, until now, have been widely separated and is leading to a new transdisciplinary approach to biomimetic research in technology and architecture (e.g. Schleicher et al. 2015; Ahlquist et al. 2015; Oxman et al. 2015; Menges 2015).

Biology provides guidance not only for the development of individual bio-inspired technical features and solutions, but also for methodological strategies. Architectural design and biological evolution are both open-end processes in which evaluation criteria and developmental targets emerge and are subject to constant change. Biological organisms adapt via evolution through mutation, recombination and selection by developing multifunctional and (self-)adaptive solutions. The result is a compromise, satisfying partially conflicting requirements simultaneously, just as is required of a successful architectural design. This situation can often be well described by a pareto-optimum in which one character/function cannot be improved without causing a deterioration in another or even in divers others. As a consequence, biological solutions are very rarely optimal (as in, for example, the eye as a one-photon detector) but rather optimized for fulfilling multiple functions at the same time. A detailed analysis of biological solutions often reveals that they can been interpreted as 'good enough' to allow survival and propagation in a particular ecological niche and to outcompete rival species by using a minimum of resources with regard to materials and energy (evolutionary preference of 'cheap but good-enough' solutions).

These considerations show that evolution in biology follows guiding principles that have the potential to push architecture and building technologies towards a (more) sustainable direction: The efficient utilization of scarce natural resources and their efficient conversion into physiological performance is one of the main selection criteria of biological evolution. All processes in nature are essentially based on the usage of solar energy in various forms. Living beings are part of larger energy and material loops in the bio- and geosphere. All 'waste' is used in a cascade-like manner by other organisms. These are general principles that are also pursued in today's architecture and building technology.

1.2 Contemporary Biomimetic Research in the Realm of Architecture and Building Construction

This book comprises the first survey of the research topics addressed by the Collaborative Research Centre SFB-TRR 141 'Biological Design and Integrative Structures'. Initially, the SFB-TRR 141 will focus on the realm of 'materials and structures', because a long record of biomimetic research and extensive preliminary work exists in this field. In the subsequent periods, the perspective will be widened to a more holistic analysis of nature and architecture that will also include aspects of energy management and energy conversion (Fig 1.2).

The key questions of advanced biomimetic research as presented in the following chapters of this book are:

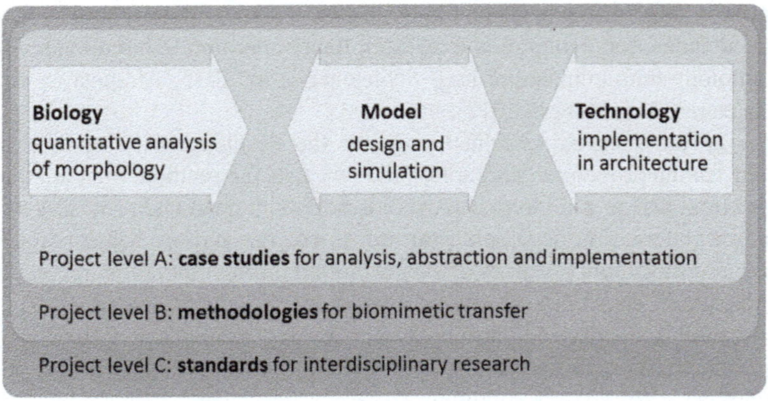

Fig. 1.2 Different levels of biomimetic research projects (Illustration by Jan Knippers, Thomas Speck and Klaus Nickel)

1. How can the morphology of hierarchically organized, multifunctional and adaptive biological structures be modelled and simulated? The computational simulation enables, on the one hand, a deeper understanding of biological structures and of their evolutionary development in the process of reverse biomimetics. On the other hand, it serves in parallel as a basis for successful transfer into biomimetic technical applications and for the development of novel biomimetic materials and structures in the field of building constructions (cf. Speck and Speck 2008).
2. How can computational methods be used to transfer the design and construction principles found in nature to the macro-scale of building construction? The aim here is not only to increase the efficiency of building structures, but also to develop new features such as multi-functionality, hierarchical structuring or adaptivity that until now have not been used in building construction (cf. Knippers and Speck 2012).
3. Under which conditions does the abstraction and transfer of biological structures lead to a more sustainable technology that inherits not only the functional, but also the ecologically beneficial properties of the natural role models? The development in building construction is more and more directed towards resource efficiency, the sustainable use of building materials and closed material loops, as are self-evident for most natural systems (cf. Antony et al. 2014).

The variety of highly different biological role models was chosen in the following chapters to enable us to study and compare the influence of diverse parameters on methodology and on the success of abstraction, technical implementation and reverse biomimetics. The selected role models differ in scale from nano- to macro-dimensions and in materiality from porous to solid and/or fibre-based. Moreover, the investigated functionalities differ widely from load bearing and energy dissipation to movement or media transport. Some role models follow classic lines of biomimetic research, such as the skeleton of the sea urchin, whereas others tackle new problems, such as the elastic deformation of plants.

Advanced biomimetic research is conceptualized as a dialogue between the disciplines: the functionally important features of the biological systems are abstracted in a model representing the properties of interest and the underlying design principles and structures. This model serves as a basis for simulating functional morphology and biomechanics and renders the basis for technical implementation in biomimetic solutions. In the process of 'reverse biomimetics' the results represent the basis for further investigation and a more detailed understanding of the biological systems. This in turn helps in an iterative process for improved transfer and implementation into biomimetic products (Speck and Speck 2008).

Whereas 'biomimetics' enjoys a positive image among the public and is often used as labelling in architecture it still does not represent an established and well-defined scientific discipline. One reason may be the lack of resilient and evaluated methodologies that can serve as validation of the systemic viability of biomimetic research. The overall goal of the research presented in this book is to contribute to the conceptualization and foundation of biomimetics as a scientific discipline first in the context of architecture and building technologies.

References

Ahlquist S, Kampowski T, Torghabehi OO, Menges A, Speck T (2015) Digital framework for the computation of complex material and morphological behavior of biological and technical systems. Comput Aided Des (Spec Issue Mater Ecol Des Comput Issues)60:84–104

Antony F, Grießhammer R, Speck T, Speck O (2014) Sustainability assessment of a lightweight biomimetic ceiling structure. Bioinspir Biomim 9. doi:10.1088/1748-3182/9/1/016013

Fratzl P, Weinkammer R (2007) Prog Mater Sci 52:1263–1334

Imhof B, Gruber P (eds) (2015) Built to grow – blending architecture and biology. Birkhäuser Verlag, Basel

Knippers J, Speck T (2012) Design and construction principles in nature and architecture. Bioinspir Biomim 7. doi:10.1088/1748-3182/7/1/015002

Menges A (ed) (2015) Special issue on material synthesis: fusing the physical and the computational. AD Archit Des 85/5:1–136

Oxman N, Ortiz C, Gramazio F (eds) (2015) Special issue on material ecology: design and computational issues. Comput Aid Des 60:1–118

Schleicher S, Lienhard J, Poppinga S, Speck T, Knippers J (2015) A methodology for transferring principles in plant movements to elastic systems in architecture. Comput Aid Des (Spec Issue Mater Ecol Des Computat Issues)60:105–117

Speck T (2015) Approaches to bio-inspiration in novel architecture. In: Imhof B, Gruber P (eds) Built to grow – blending architecture and biology. Birkhäuser Verlag, Basel, pp 145–149

Speck T, Speck O (2008) Process sequences in biomimetic research. In: Brebbia CA (ed) Design and Nature IV. WIT Press, Southampton, pp 3–11

Speck T, Speck O (2015) Von der Klette zum Klettverschluss – Bionik oder wie wir von der Natur die Technik der Zukunft abschauen. Naturwissenschaftliche Rundschau 68/10:510–523

Speck T, Knippers J, Speck O (2015) Self-x-materials and -structures in nature and technology: bio-inspiration as driving force for technical innovation. AD Architect Des 85/5:34–39

Wegst UGK, Bai H, Saiz E, Tomsia AP, Ritchie R (2015) Bioinspired structural materials. Nat Mater 14:23–36

Chapter 2
From Minimal Surfaces to Integrative Structures – The SFB-TRR 141 in the Light of the Legacy of Frei Otto and the SFB 230 'Natürliche Konstruktionen'

Jan Knippers

The research projects presented in this book have been conducted under the framework of Collaborative Research Centre SFB-TRR 141 'Biological Design and Integrative Structures – Analysis, Simulation and Implementation in Architecture' funded by the DFG ('Deutsche Forschungsgemeinschaft', i.e. the German National Science Foundation) since October 2014 at the Universities of Stuttgart, Tübingen and Freiburg. The motivation to write the proposal for SFB-TRR 141 arose from the inspiring experience of earlier interdisciplinary research that had led everyone involved towards new findings far beyond the reach of their individual disciplines. Our aim was to establish a collaborative framework for previously isolated activities. In this sense, the SFB-TRR 141 is purely driven by scientific enthusiasm.

Somewhat surprisingly from this perspective, the SFB-TRR 141, at least in the realm of architecture and structural engineering, is widely perceived as the successor of the famous 'SFB 230 – Natürliche Konstruktionen, Leichtbau in Architektur und Natur' funded by the DFG between 1984 and 1995 at the Universities of Stuttgart and Tübingen. This was not intended and the marked reaction, especially from the community of architects, was not expected. Thus, it makes sense to discuss the relevance of SFB 230 for SFB-TRR 141 in more detail.

The SFB 230 was based on the groundbreaking work of the research group 'Biologie und Bauen' that Frei Otto together with the biologist Johann Gerhard Helmcke started in the early 1960s at the Technical University of Berlin. After establishing the IL ('Institut für leichte Flächentragwerke', 'Institute for Lightweight Structures') at the University of Stuttgart in 1964 and completing large construction projects as the German Pavilion at EXPO 1967 in Montreal (in collaboration with Rolf Gutbrod), the roof structures for the Olympic Games in Munich 1972 (with

J. Knippers (✉)
Institute of Building Structures and Structural Design (ITKE), University of Stuttgart, Keplerstraße 11, 70174 Stuttgart, Germany
e-mail: j.knippers@itke.uni-stuttgart.de

J. Knippers et al. (eds.), *Biomimetic Research for Architecture and Building Construction*, Biologically-Inspired Systems 9, DOI 10.1007/978-3-319-46374-2_2

Günter Behnisch and Jörg Schlaich) and the Multihalle in Mannheim in 1975 (with Carlfried Mutschler, Joachim Langner and Ted Happold), Frei Otto turned back to his early interest in fundamental research into 'Natural Constructions' and initiated the collaborative research centre SFB 230 at the Universities of Stuttgart and Tübingen.

Frei Otto's initial concept was that of an open platform to strengthen and connect individual activities in the field. Following the all-encompassing lines of his previous research, the SFB 230 was topic-wise very broad: 'man and nature', 'evolution, optimization and self-organization', 'skeleton, truss structures, branchings', 'pneumatics and minimal surfaces', 'urban systems and infrastructure of man and animals' were some of the key words used in the application for the first funding period of SFB 230 from 1984 to 1986 (Otto et al. 1983). Frustrated by the burden to merge and consolidate the work of a larger group of colleagues, Frei Otto stepped back as a speaker for the SFB 230 as early as 1986 (Reiner 1996). Later, the SFB 230 was pushed by the DFG to establish a more rigid organizational framework and to focus on selected scientific topics, mainly computational form-finding and optimization strategies, and on their potential to study natural morphologies.

What relevance does the SFB 230 have for the SFB-TRR 141? Where does our current collaborative research centre follow the lines of the SFB 230 and where does it differ?

The initial proposal for the first funding period of SFB 230 was very much based on Frei Otto's conception that the methods for the design, analysis and construction of lightweight and minimal systems have first to be propelled forward within the boundaries of the respective disciplines of engineering and architecture with the utmost effort, before they can be used to study and analyse systems of nature. The proposal expressly declares that the intention of SFB 230 is not the transfer of knowledge from biology to technology but the other way around (Otto et al. 1983, p 26 ff). 'Der umgekehrte Weg' is a phrase that Frei Otto used in the application for SFB 230 and also later frequently. It can be translated as 'reverse biomimetics' today. This approach is also reflected by the organizational scheme of SFB 230: the research projects themselves were conducted on a disciplinary basis within the framework of the respective scientific methodology. The major task of the SFB 230 was to provide a common platform for interdisciplinary discussions, seminars and symposia or to extend the conception of Frei Otto's IL beyond the boundaries of his institute.

Since the time when the SFB 230 was conceptualized, fundamental steps towards digital technologies have been made in all fields of science and technology. In biology, new analytical methods such as Computer Tomography Imaging (CT, μCT), Magnetic Resonance Imaging (MRI) and many more have been introduced. In architecture and building construction, computational design, simulation and fabrication are omnipresent today. These advances mean more than merely obtaining deeper insights into biological systems or better tools for the design and construction of more complex building structures. The step towards a 'digital model' provides a common basis for the exchange of knowledge across the disciplines and enables direct communication between, so far, widely separated fields of science. In this sense, simulation technologies become increasingly relevant as they provide both a scientific approach to the quantitative analysis of the form-structure-function relationships of biological role models and a platform for the abstraction of their structural and functional principles and for their implementation in building construction and architecture. The transfer of data and the organization of the flow of information in both directions, from biology to technology but also vice-versa, has become the main aspect of research. One of the central aims of the SFB TRR 141 is to explore, use and improve these novel modes of communication across the disciplines, enabled through digital imaging, modelling and simulation. This approach requires, in contrast to SFB 230, multi-disciplinary teamwork at all levels of biomimetic research, including all stages of the individual partial projects of the programme.

For Frei Otto, 'lightness' was the overarching paradigm that connected the widely separated fields of research within the SFB 230 ('Gut konstruieren heißt .... leicht bauen.' Reiner 1996, p 22 ff). A 'lightweight structure' in the sense of a 'material-efficient' system was seen as being equivalent to an energy-saving, environmentally friendly and cost-effective design. In addition, lightweight structures were perceived as socially relevant, because they added value and followed timeless design principles. In general, 'lightness' was seen as the governing criterion guiding architecture and technology into a sustainable future. This perspective was shared by many architects and engineers at the end of the twentieth century, but no one postulated it as radically as Frei Otto.

Today, the awareness of the multiple networked and barely controllable mutual dependencies of our globalized world is becoming increasingly prevalent. This requires the permanent reconsideration of inherited paradigms, including those in architecture. A lightweight structure made of non-recyclable and primary-energy-intense materials, requiring non-local fabrication processes, does not necessarily contribute to sustainable development. Thus, the work of Frei Otto might be conceived as the end-point of a century-long search for efficiency. Even though his lightweight systems still offer the potential for further optimization from an engineering perspective, the challenges of our time are different: How can we mediate between the ever more cross-linked and often conflicting social and ecological requirements that our built environment has to fulfil in the twenty-first century.

This change of perspective might also explain the reception of the work of Frei Otto and the SFB 230. On a conceptual level, the processes of self-organization driven by physical laws offered a radical alternative to the 'form-first' design approach, prevalent in Western culture for centuries. Therefore, the impact of the ideas of Frei Otto on the discourse of architecture was immense. Even today, many avant-garde architects, including some from whom one would not expect it, refer, in their theoretical conceptualizations, to Frei Otto's integral and holistic view on design.

In a remarkable contrast, his own practical work as an architect was limited to only a few, yet often cited and iconographic buildings. The reason for this is maybe that Frei Otto's ideas for form-finding and self-organization are based on a single design criterion, namely the equilibrium of physical forces, whereas architectural design has to fulfil a multitude of criteria simultaneously in accordance with the specific requirements of site and programme. One can provokingly argue that soap bubbles and other minimal systems are not only 'non-architectural', but also 'non-natural'. Even a rather simple construction – from an architectural point of view – such as a spider's web is not a mechanically optimized system but has to fulfil a variety of conflicting requirements. A biological organism is a trade-off, constantly adapting to ever-changing environmental constraints and being just good enough to survive and propagate successfully in a particular ecological niche; it is not an optimized system in the sense of engineering.

What does this mean for the conceptual layout of the current SFB-TRR 141? The general aim of research has shifted from the analysis and development of physically driven strategies of self-organization and self-optimization under the guiding paradigm of 'lightness' to processes of communication and integration between a multitude of – sometimes conflicting – design criteria. Central questions comprise: How can current FEM methods focusing on individual phenomena be integrated into a multi-functional simulation covering the mutual dependency of various features simultaneously? How can the interplay between the scales of hierarchically structured materials be analysed and used for the development of new construction materials? How can buildings adapt and react to changing environmental conditions as natural structures do so successfully? How can computational technologies be used to transfer design and construction principles of nature to architecture and vice versa? These are some of the key question of the current biomimetic research addressed in the SFB TRR 141.

One can conclude that biological systems have the potential to compromise between partially conflicting requirements and to adapt to non-steady environmental conditions. Therefore, they might guide future development in architecture and technology. From this perspective, Frei Otto's concept of design processes driven by natural laws is even more valid than ever. However, the increasing challenges for our built environment require their extension towards the simultaneous integration of a multitude of design criteria beyond 'lightness' and the mediation between them. The technical means for this are provided by the fundamental progress made in computational technologies. The ground has been prepared by the SFB 230 and Frei Otto's work and will not only be extended, but also reinvented in the framework of the SFB-TRR 141.

References

Otto F et al (1983) Natürliche Konstruktionen – Leichtbau in Architektur und Natur. Finanzierungsantrag 1984–86 für Sonderforschungsbereich 230, Universität Stuttgart, 1983

Reiner R (1996) Komplexität als Programm. In: Teichmann K, Wilke J (eds) Prozess und Form 'Natürlicher Konstruktionen'. Ernst und Sohn, Berlin

Chapter 3
Bionics and Biodiversity – Bio-inspired Technical Innovation for a Sustainable Future

Wilhelm Barthlott, M. Daud Rafiqpoor, and Walter R. Erdelen

Abstract Rethinking the relationship between *Homo sapiens* and Planet Earth in the Anthropocene is fundamental for a sustainable future for humankind. The complex Earth system and planetary boundaries demand new approaches to addressing our current challenges. Bionics, namely learning from the diversity of life for nature-based technical solutions, is an increasingly important component.

In this paper, we address the interrelated aspects of the uneven geographic distribution of biodiversity, the issue of the continued erosion of biodiversity translating into a loss of the "living prototypes" for bionics, the relationship between bionics and biodiversity and the North-south gradient in institutional capacity related to biodiversity and bionics-related areas. World maps illustrating these points are included. In particular, we discuss historical aspects and complex terminological issues within bionics or rather bionics-related disciplines, the role of evolution and biodiversity as contributors to the *fabric* of bionics and the contribution of bionics to the attainment of sustainable development.

The history of bionic ideas and the confusing terminologies associated with them (the term bionic was coined in 1901) are discussed with regard to research, design and marketing. Bionics or Biomimetics, as we understand it today, dates back to the period between 1800 and 1925 and its proponents Alessandro Volta (electric battery), Otto Lilienthal (flying machine), and Raoul Francé (concepts). It was virtually reinvented under the strong influence of cybernetics in the 1960s by H. v. Foerster and W. McCulloch. The term biomimetics arose simultaneously with a slightly different connotation. "Bioinspiration" is a convenient modern overarching term that embraces everything from bionics and biotechnology to bioinspired fashion design. Today, marketing strategies play a crucial role in product placement within an increasingly competitive economy. The majority of so-called "biomimetic" products, however, only *pretend* to have a bionic origin or function; we have introduced the term "parabionic" for such products.

W. Barthlott (✉) • M.D. Rafiqpoor • W.R. Erdelen
Nees Institute for Biodiversity of Plants, Venusbergweg 22, 53115 Bonn, Germany
e-mail: barthlott@uni-bonn.de

J. Knippers et al. (eds.), *Biomimetic Research for Architecture and Building Construction*, Biologically-Inspired Systems 9, DOI 10.1007/978-3-319-46374-2_3

Life arose almost four billion years ago. Today's relevant living prototypes for bionics have a history of more than one billion years of evolution, in essence a process of "technical optimization" governed by mutation and selection. In one specific example, we provide evidence that superhydrophobicity, an important biomimetic feature, has been in existence since at least the Paleozoic period, the time when life conquered land.

Bionics might be a major contributor to future nature-based technological solutions and innovations, thus addressing some of humankind's most pressing issues. Bionics and related fields may become a major component of the current "great transformation" that humanity is experiencing on its trajectory towards sustainable development.

3.1 Introduction

The new millennium has seen enormous efforts of the international community, in particular from the United Nations, towards making sustainable development a reality at all levels: local, regional and global. The Brundtland Report *Our Common Future* has presented the definition most frequently cited and used for sustainable development: "*Sustainable Development is development that meets the needs of the present without compromising the ability of future generations to meet their own needs*" (WCED 1987).

Bionics can be seen as a relatively new field eventually contributing to the attainment of sustainable development as an "instrument provider" and a gate opener to an era of new technologies. However, "imitating living nature" without great success dates back to Daedalus and Icarus and Leonardo da Vinci. The first effective and fundamental result was the electric battery after the "living prototype" of the Electric Ray (*Torpedo torpedo*) by Alessandro Volta in 1800. It was followed by the first successful air planes built by Otto Lilienthal after 1889 based on birds as biological role models. *"Bionik"* was established as a science by Raoul Francé in 1920 and re-invented as "*Bionics*" (the term "bionic" was coined as early as 1901 with a different context by H.S. Williams) at the Dayton Symposium on "Living Prototypes" (Robinette 1961).

More recent milestones in the international arena have been (1) the implementation of the Kyoto Protocol in 2005 on reducing greenhouse gas emissions into the atmosphere, (2) the completion of the Millennium Ecosystem Assessment (MEA 2005a), and the Rio + 20 Summit in 2012. The later agreement was reached to develop universal sustainable development goals (SDGs) as a central element of a new 2030 Development Agenda (United Nations 2015). These were to replace the Millennium Development Goals and to continue a process towards a green economy, an issue closely related to the discussions in this paper (Barthlott et al. 2014). Until 2030 and certainly also beyond, the SDGs and their eventual successors will provide the framework for the contribution of bionics and neighbouring disciplines to the global development agenda.

Biodiversity, climate change and the global challenge of carbon neutrality set the scene for our discussion of bionics and biodiversity within the broader framework of sustainable development, in particular after the climate summit in Paris, in 2015. In other words, biodiversity provides the "evolved blueprints" for bionics. Biodiversity is the inspiration for innovation-based and nature-inspired technological research and could be an essential component of sustainable development itself. These interlinkages form the foundation of this paper, which aims at "putting bionics into context".

3.2 Biodiversity – Matter of Multiple Dimensions

3.2.1 The Dimensions of Biodiversity

The term biological diversity and its definition were first published by Elliott Norse and Roger McManus in 1980. The compressed term biodiversity became internationally known after the publication of E.O. Wilson's book "Biodiversity" (Wilson 1988), based on the National Forum on BioDiversity in September 1986. Further details of the history of the term biological diversity or biodiversity are given in Farnham (2007).

Biodiversity is an important feature of all ecosystems, irrespective of whether these are relatively undisturbed or are affected to various extents by human activity. Biodiversity is the foundation, moreover, of all ecosystem services (see Harrison et al. 2014). These services are vital to human well-being. In particular, "*biodiversity contributes to security, resiliency, social relations, health, and freedom of choices and actions*" (Millennium Ecosystem Assessment 2005b).

Life on Planet Earth, and hence biodiversity, is a product of over 4 billion years of evolution (e.g. Bell et al. 2015). Surprisingly little, however, is known about biodiversity. Some 1.8 million organisms have been described scientifically. Currently, efforts are underway to present a comprehensive tree of life (Hinchliff et al. 2015). Nevertheless, well over twenty years ago, Lawton (1993) made a provocative statement that is still relevant today: "... I have never seen anybody discuss what we actually know about these 1.7 million species that do have names. Overwhelmingly, the answer will be nothing, except where they were collected and what they look like".

Total species numbers of extant species had been previously estimated between 3.6 and 112 million. More recent estimates centre around 10 million (Mora et al. 2011; see Caley et al. 2013 for a critical discussion). Based on predictions from scaling laws, Locey and Lennon (2016), for instance, produced a staggering estimate of 1 trillion (10^{12}) for the number of microbial species on Earth.

As more than 80 % of all species – based on the conventional data- are unknown (Fig. 3.1) and as major groups of organisms are only poorly understood, this should

Fig. 3.1 Dragon trees (*Dracaena*) are particularly interesting living prototypes because of their biomechanically remarkable branching patterns. Many of these impressive trees are the last living remnants and highly threatened. But surprisingly, in the last ten years some six new species have been discovered. Above: *Dracaena cinnabari*, endemic to the Socotra Archipelago (Photograph: Bruno Mies)

be considered as a rather conservative estimate. In particular, this applies to larger segments of the so-called lower organisms, i.e. the fungi, bacteria and archaea.

Based on a 10 million total estimate, arthropods make up around 5 million (50 %) of all species (Fig. 3.2). However, total terrestrial arthropod species have been estimated at 6.8 million, with a range of 5.9-7.8 million (Stork et al. 2015).

This contrasts with the much better understood vertebrates, of which only around 66,500 species are known to science (IUCN 2015). This reflects over 80 % of the total estimate for the number of extant species.

Terrestrial (vascular) plants, comprising the flowering plants, gymnosperms and ferns, are comparatively poor in species numbers. These are estimated at only 370,000 species (compare Fig. 3.2). Plants have been well studied for centuries and, as a result, the percentage of total species numbers known (over 90 %) is extremely high compared with other taxonomic groups. More recently, Pimm and Joppa (2015) estimated flowering plants to comprise 450,000 species. Out of these, two thirds occur in the tropics and one third of all species are at risk of extinction.

Total species numbers are poor indicators of the status of global biodiversity at the ecosystem level. Of vital importance for ecosystem functioning are the major functional groups, viz. the producers, consumers and decomposers. Arthropods and other animals belong to just one of the functional groups within ecosystems, namely the consumers. This group strongly depends on the producer species, which

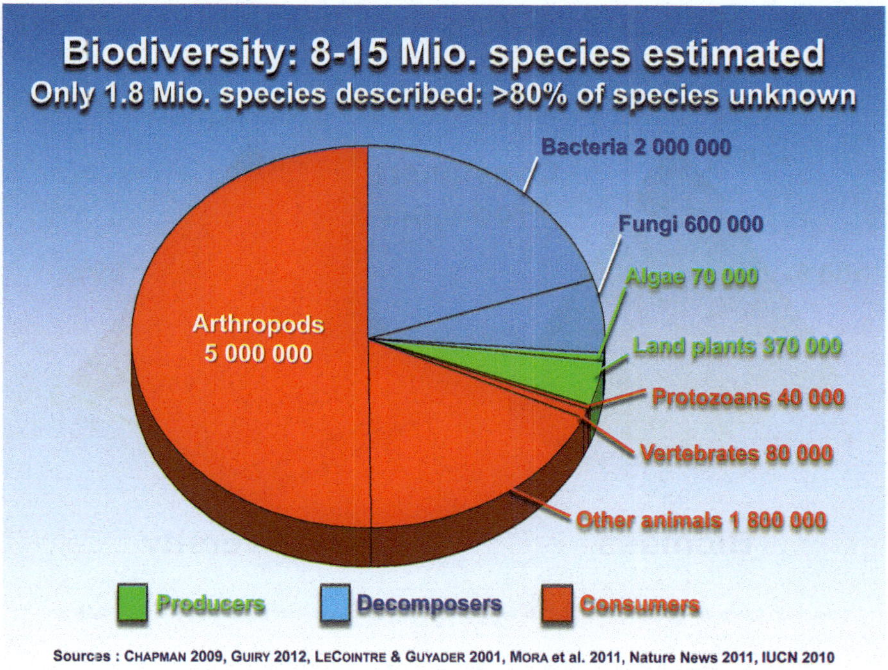

Fig. 3.2 Estimated numbers of species among major taxa and functional groups (Modified from Barthlott et al. 2014. See text for further explanation. For source citations see reference)

can be considered as "the giant powerhouse covering the planet as a global green solar collector: the higher plants" (Barthlott et al. 2014, Fig. 3.3). Producers and consumers also strongly interact with the third functional group within ecosystems, the decomposers.

A striking inverse relationship exists between species numbers and biomass amongst the three functional groups (see Fig. 3.3). Based on the total species estimate of 10 million, species diversity of consumers would account for some 69 % of all species, whereas the plants as producers make up only 5 % of all species. Producers, however, would comprise over 80 % of the total biomass. Green plants form the most important structural and functional elements of terrestrial ecosystems.

3.2.2 The Geographic Distribution of Biodiversity

Biodiversity is not homogeneously distributed over the globe. However, patterns are apparent, the best known being the increase in number of species and types of ecosystems from polar to temperate and tropical regions; this has been recognized for more than three centuries but is still not fully understood (Fig. 3.4, and

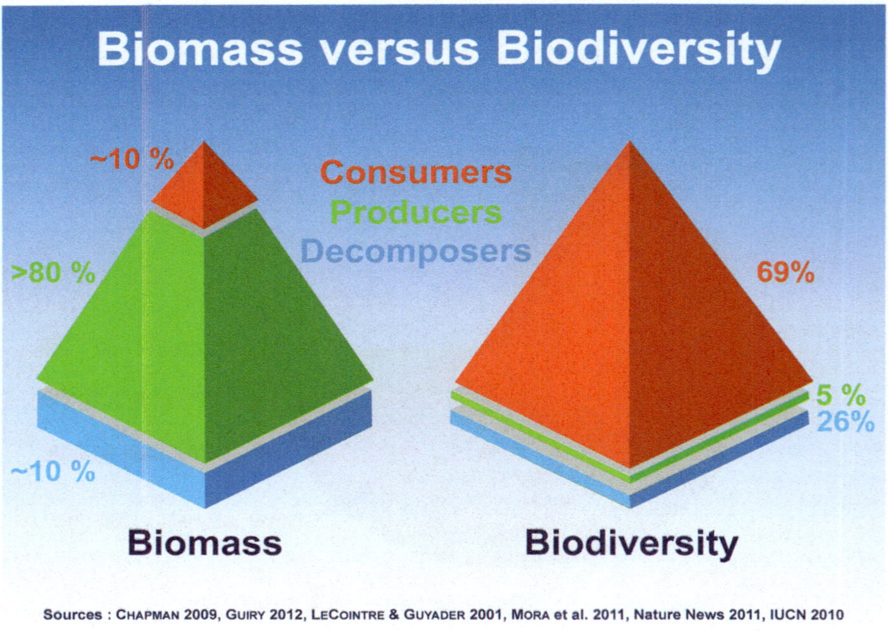

Fig. 3.3 Relationship between biomass and estimated species numbers (biodiversity) among functional groups (From Barthlott et al. 2014. See text for further explanation. For source citations see reference)

Brown 2014; for a more detailed discussion see Fine 2015). The term "geographic distribution of biodiversity" means, in the context of bionics, the distribution of "living prototypes".

This contribution focuses on vascular plants, which are well known enough to map geographically the global distribution of their species diversity (see Fig. 3.4). The map draws on information of several thousand flora inventories, checklists and databases (for details see references in Barthlott et al. 2014).

Figure 3.4 clearly shows discernible patterns. Polar and sub-polar regions are relatively poor in species numbers; the highest diversities are found in the subtropics and tropics. The overall picture, however, is much more complex: some regions in the tropics are comparatively poor in species, whereas areas in the temperate region might show high species diversities.

The pooling of areas with over 3000 species per 10,000 km^2 leads to a clustering of species diversity of terrestrial higher plants into 20 global centres of biodiversity (Fig. 3.4). These diversity centres largely represent lowland and mountain areas in the tropics and subtropics located in the so-called mega-diverse countries. Vital for global conservation efforts, these are, in most cases, developing countries or countries in transition. Furthermore, endemism and its link to island systems play a crucial role as a qualitative factor explaining these patterns. An important aspect related to institutional capacity is the inverse relationship between national research capacities in biodiversity and species diversities (discussed below).

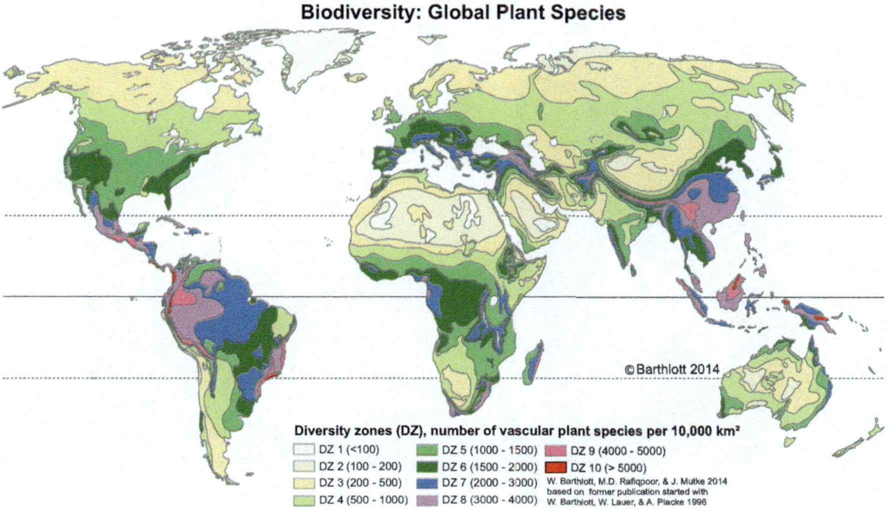

Fig. 3.4 Uneven distribution of global biodiversity: species numbers of plants per 10,000 km^2. Areas with more than 3000 species depict the centres of biodiversity. The red colours (DZ 8-DZ 10) delimit the global biodiversity centres (From Barthlott et al. 2014)

3.2.3 Status of Biodiversity – International Efforts

The most important normative international instrument for biodiversity is the Convention on Biological Diversity (CBD). The CBD defines biodiversity as "*the variability among living organisms from all sources including terrestrial, marine and other aquatic ecosystems and the ecological complexes of which they are part; this includes diversity within species, between species and of ecosystems*" (SCBD 2010). The CBD came into force in 1993. As stated in the *Global Biodiversity Outlook 3 (GBO 3)*, the objectives of the Convention are "*The conservation of biological diversity, the sustainable use of its components and the fair and equitable sharing of the benefits arising out of the utilization of genetic resources, including by appropriate access to genetic resources and by appropriate transfer of relevant technologies, taking into account all rights over those resources and to technologies, and by appropriate funding*" (SCBD 2010). Currently, there are 196 Parties (168 Signatories) to the CBD including the European Union (for further details, see www.cbd.int accessed 3 January 2016).

The Convention does not explicitly mention bionics. However, its links to bionics are its relationship to biotechnology and the Protocol on Access and Benefit Sharing (ABS Protocol, SCBD 2011). Biotechnology is defined under Article 2 of the Convention as "*any technological application that uses biological systems, living organisms, or derivatives thereof, to make or modify products or processes for specific use*" (UN 1992). Both of these entry points would certainly warrant further

Fig. 3.5 Speciation, migration, and extinction are shaping biodiversity. Many of the larger vertebrates became extinct during the history of mankind such as two races of elephants in North Africa and the Near East, both still mentioned in the bible. Shown here is an egg of the Elephant Bird (*Aepyornis maximus*), native to Madagascar until the seventeenth century. These birds were more than 3 m (10 ft) tall and weighed close to 400 kg (880 lb). *Aepyornis* became extinct after humans settled in Madagascar (seventeenth century). The birds "survived" as legendary bird "roc" or "rukh" in "One Thousand and One Nights". (Photograph: W. Barthlott)

in-depth discussion for a better understanding of the actual and potential future relationship between the Convention as *the* normative instrument with regard to biodiversity and bionics.

"Ideas" for bionics may be at risk as biodiversity continues to be lost at unprecedented rates (see Figs. 3.1 and 3.5 and below). Major international efforts were made to significantly reduce biodiversity loss by the "2010 Target". This target has not been met (see SCBD 2010). It was followed by a package of measures agreed upon in 2010, at the meeting of the Parties to the Convention on Biological Diversity (COP 10) in Nagoya, Japan (e.g. Erdelen 2014). More recently, the Global Biodiversity Outlook presented a mid-term review of the process towards reaching the Aichi Targets, one of the major outcomes of COP 10, with a timeline of 2020 (SCBD 2014).

3.3 Bioinspiration: Conflicting Terms and Views Among Research, Design, and Marketing

"Bionics", "Biomimetics", "Biologically Inspired Design" (Goel et al. 2014), "Bioinspiration" and "Biomimicry" are terms that are in common use and that vary in their meaning and usage. This might be an impediment to addressing multifaceted issues requiring transdisciplinary approaches. A common language is helpful if scientists, engineers, architects or fashion designers collaborate in product development schemes based on living organisms or their derivates.

Biologically Inspired Design (BID) sometimes shortened in "bio-inspired design", "bioinspiration" and "biodesign" is at least as old as the oldest biomorphic cave paintings, which date back to 40,000 YBP. During the last decade, the term bioinspiration has become a colloquial and seemingly easily understood term in areas such as science, marketing, and the fashion industry. The term *biological* relates to biology, life or living organisms. The term *design* has many different connotations. Design might be used in fields such as the arts, decoration and marketing, e.g. fashion design. In science and engineering, it might be related to the designing of materials, machines and tools or processes.

The ambiguous and thus very convenient term *bioinspired* principally applies to everything with a design related to life: from an ancient Greek acanthus capital, the story of Daedalus and Icarus to the floral design on a wedding dress; from an artificially flavoured strawberry yoghurt to the fascinating high-tech bionic Soma-Building at the 2012 Expo in South Korea (Fig. 3.6), all are inspired by nature. Accordingly, we suggest subdividing *bioinspiration* into four elements: biodecoration, biotechnology (including bioengineering), bionics (including biomimetics) and parabionics.

Biodecoration We suggest using the term *biodecoration* for elements of bioinspired design that are non-functional and, in many cases, an optically attractive decorative element in arts and design. Examples are the tail fins of a cadillac car (Fig. 3.11), the "Queen's Vagina" in the gardens of Versailles in 2015 or Palm Island off the coast of Dubai (Fig. 3.7 and 3.8). The most often used natural model for biodecoration are the spectacular tail feathers of the Indian Peacock (Fig. 3.7 and 3.8). Art Nouveau is a biodecoration style based on biomorphism. The eminent evolutionary biologist Ernst Haeckel named one of his publications *Kunstformen der Natur* (1898–1901), which inspired generations of designers (Figs. 3.9 and 3.10).

Biodecoration does not pretend to be functional and thus bionic but might have non-physical, economic or psychological functions. A classic example is the bioinspired *and* functional Trojan Horse: camouflage – but not biomimetic or bionic. The impressive 1960s Cadillac tail fins (Fig. 3.11) are biodecorational; meeting the demand of the customers, they improved sale rates. The world heritage icon of biodecoration might be the basilica Sagrada Familia in Barcelona by Antoni Gaudi.

Biotechnology and Bioengineering are closely related to bionics and the disciplines overlap. The term *Biotechnology* was coined by Karl Ereky in 1919 (overview in Goujon 2001). It refers to *the use of living organisms or their derivates to make or modify products* such as bread, wine or pharmaceuticals. "Yeast is a machine", Max Delbrück explained to beer brewers in 1884. Polymer science is biotechnology, originally derived from biopolymers produced by organisms; this is how Jöns Berzelius defined the term in 1833.

Fig. 3.6 "One Ocean", building of the Expo 2012 in South Korea, is a masterpiece of bionics. The moving lamellae are based on the sophisticated opening mechanism of the bird-of-paradise flower *(Strelitzia)*. The hinge-less flapping technology was developed by the groups around Jan Knippers and Thomas Speck and the kinetic facade was designed by the Austrian Soma architects and Knippers Helbig Engineering in Stuttgart. Even the hierarchical-structures composed of glass fiber reinforced polymer material for the kinetic façade are influenced by the hierarchical structured material from biological models (Photograph: Riccardo La Magna)

Figs. 3.7 and 3.8 (*left*) and (*right*): The "train" of the Indian peacock (*left*), made up of elongated upper tail coverts, is one of the most frequently used models for biodecoration. The "train" is related to the reproductive behaviour and fitness of male peacocks. (Photograph: BS Thurner Hof; http://is.gd/GdYHSz). Palm Island, off the shore of Dubai (*space view*), is a biologically inspired design (BID) derived from a date palm tree. The design is pure biodecoration. (Photograph: Nasa-Space Bild Palm City Dubai, ISS022-E-024940 (13 Jan. 2010), http://is.gd/GNL3jz)

Figs. 3.9 and 3.10 (*left*) and (*right*): Ernst Haeckel and his *Art Forms in Nature*. In Jena, the eminent biologist Ernst Haeckel studied microscopic organisms and published his four-volume treatise *Kunstformen in der Natur*. Haeckel was particularly fascinated by fragile but stable light-weight constructions. The entrance building of the World Exhibition in Paris 1900 was the first construction based on his studies. Many were to follow later (compare Figs. 3.19 and 3.20) (Adapted from plates of Haeckel 1899). (*right*): The "Wonder of Jena": the first light-weight geodesic dome was constructed by Walther Bauernfeld in 1927 for the Zeiss Company and resembled Haeckel's drawings. Geodesic domes were built in almost all major cities in Europe and the United States but were perfected by Buckminster Fuller only after World War II (Photograph from a commercial postcard http://is.gd/kIgPOT)

Fig. 3.11 Bioinspired tail *fins* of Cadillac cars are biodecoration. Design without function, driven by the demands of car drivers in the 1960s. (Photograph: G. Schlimm)

The term *Bioengineering* was coined in 1939 at the MIT Biological Engineering and Biotechnology Units and, in the US, its meaning often comprises biotechnology and even bionics (see Jack Steel in Robinette 1960). In other countries such as Germany, the meaning of bioengineering is restricted to the application of engineering principles for the processing of biological materials. However, an overlap is indeed notable between bioengineering and bionics. For instance, the wood-based production of modern paper, invented by Friedrich Gottlieb Keller in 1840, is a biotechnological process, although Keller's production process was not only based on material properties, but also on his observations of the way in which wasps build their papery nests.

Bionics and Biomimetics Bionics is commonly defined as the "study and application of construction principles and systems in living organisms for the design of (*sustainable*) **technological** solutions". Further details and discussions can be found, for example, in (Nachtigall 1998; Rechenberg 1971; VDI 2012; Vincent 2009a; Vincent 2009b; Gleich et al. 2010; Bertling 2014).

Bionics and biomimetics are considered to be synonyms (see Sect. 3.4.1). Bionics tends to be used in regard to mechanical constructions, whereas biomimetics, possibly used more often today, tends to be used for materials and systems. For example, a building may be referred to as bionic (see Fig. 3.6) but its surface as biomimetic (Fig. 3.12). "Inspired by nature", a phrase often used, is not clearly

Fig. 3.12 Biomimetic architecture and sustainability: the Technoseum (a technical museum in Mannheim, Germany) has a self-cleaning Lotus-Effect® surface. The biomimetic coating is sustainable and lasts several decades – facades have to be repainted after this time. The conventional building itself may last more than a century. Time-limited sustainability is a principle of living prototypes: Lotus leaves last only for a short time span and are permanently renewed. Their rhizomes, however, may grow for centuries, and Lotus seeds may even germinate after more than 1200 years. Source: "Technoseum, zooey braun"

defined: a nuclear reactor is inspired by nature, i. e. natural nuclear fission or the nuclear fusion in our sun, which provides the energetic base for the existence and diversity of life on our planet. But a nuclear reactor is not bionic.

Thus, it is even more difficult to define bionic products than bionics itself. Biological species, the *living prototypes* of bionics, are "sustainable" with regard to their "evolved" life spans (see the following chapters). Part of the definition of bionic products should be their sustainability and whether this reflects their original life spans attributable to the process of evolution or their "new" life spans after technical modification of the characteristics of the role model (prototype).

Parabionics The terms bionics and biomimetics are increasingly used for products and processes that are non-bionic (internet research and personal observations). This might be because "marketing considerations" have intruded into these domains. This watering down of established terminology has led to mixed feelings in the scientific communities working in these fields. For obvious reasons, no reliable statistics are available. By investigating internet sources (the term *bionic* has about 26 million hits on Google, additional data in Barthlott et al. 2016) and from personal experience gathered over three decades, the term bionic seems to be predominantly used as a marketing tool for non-bionic products, estimated to be the case for possibly over 80 % of products marketed.

Therefore, we suggest the use of the term *parabionics* for products or processes pretending to be bionic but having neither a bionic function nor a bionic origin. It should not be confused with the occasionally used term "Pseudo-Bionik" in the German grey literature (e.g. Ingo Rechenberg), which has a different meaning and connotation. Parabionics plays many important roles from product marketing (Fig. 3.13, see also Barthlott et al. 2016) to fundraising in research programs. As a biologist, one is tempted to classify this phenomenon under "mimicry", namely deceptive products that pretend to be something else, as in the Wasp Spider *Argiope* in Fig. 3.14.

Nature-Based Solutions – A Closing Remark An emerging term gaining momentum at international levels is "nature-based solutions" (NBS). This term is a major element of programs of international organizations such as the European Union and the World Conservation Union (IUCN). NBS is introduced here to complete this discussion of terms related to bionics. The term NBS was first used in contexts such as mitigation and adaptation to climate change, biodiversity conservation and sustainable livelihoods. NBS is closely related to concepts underpinning natural systems agriculture, natural solutions, ecosystem-based approaches, green infrastructure and ecological engineering (see (Eggermont et al. 2015) and references therein). As the authors point out, NBS might need to be re-evaluated based on scientific criteria and is probably in need of a thorough epistemological analysis, as the term is currently seen differently amongst stakeholders.

IUCN has identified nature-based solutions as a major program for the time period 2013–2016 (IUCN 2012). This approach is meant to address, within the

Fig. 3.13 Parabionics – mimicking bionics. Package of a fluorized polymer coated (e.g. "Teflon®") baking pan. The product suggests that it is bionic and refers to the Lotus Effect® in detail. It is named "Lotex-everclean". However, it does not differ from the billions of other pans in use since the 1960s. We expect the majority of "Bionic Products" currently on the market are in fact parabionic: bionics is purely a marketing tool. (Photograph taken from a package bought in a department store)

organization's ecosystems-related mandate, a wide range of global development challenges through the application of "IUCN's knowledge to sustainable development, empowerment and poverty reduction" (p. 5). These include for example: climate change, disaster risk reduction, food security, economic and social development, in addition to health and access to energy (details in IUCN 2012).

3.4 From Bionik to Bionics: History Revisited Between Science, Politics and Science Fiction

The idea of emulating the functions of biological models is as old as mankind (Bertling 2014). Early attempts were not very successful. Ovid's story of Daedalus and Icarus and Leonardo da Vinci's design of flying machines and other devices (Seireg 1969) did not translate into technical success stories. Most of the supposedly early bionic ideas were wishful thinking. Bionics in a modern sense seems to start with the invention of the electric battery by Alessandro Volta around 1800.

Fig. 3.14 Mimicry, but not biomimicry. The conspicuous Wasp Spider (*Argiope bruennichi*) imitates wasps and hornets, although the spider is completely harmless. Biologists have known for well over 100 years that such patterns of similarity of one species to another protect one or both species and have termed it "mimicry". Such stories of cheats and deceits are the stories of parabionics (see Fig. 3.13). *Argiope* builds very stable webs, like those of the *Nephila* spiders, which are used for the production of biomimetic technical spider silk, which is a good example of the multifunctionality of characteristics used for bionic applications and the multiplicity of prototypic information of an individual species. (Photograph: W. Barthlott)

3.4.1 Bionik- the Period from 1800–1960

The dream of flying and the use of electricity were the two fundamental forces for the foundation of what we call today *bionics* or *biomimetics.*

Electricity is a phenomenon which had been observed for thousands of years in amber – even its name is derived from this biological material: electron (ἤλεκτρον) is the Greek word for amber. The invention of static electricity storing devices in 1745/1746 the so-called "Leyden jars", opened a new field for experiments. Multiple jars could be connected in sequence, named with the military term "battery", to enhance their power. The Torpedo fish (today we call it Electric Ray) was known in the Mediterranean for millennia for its strange and dangerous properties of unknown nature. This was revealed by John Walsh as "electric property" in 1773, in a letter published to Benjamin Franklin.

Henry Cavendish was the first to see a similarity of the multiple repeating elements of the anatomy of the Torpedo as compared with multiple Leyden jars. The title of his letter published in the Philosophical Transactions of the Royal Society in London on January 1st in 1776 appears to be a biomimetic concept: "*An account of some attempts to imitate the effect of the Torpedo by electricity*". However, based on the Leyden jar static electricity the construction of a "battery" is not possible. The fundamental solution came from Alessandro Volta. Inspired by Cavendish and his own work on the torpedo in combination with the experiments of his rival Luigi Galvani in the 1780s on frog legs. He constructed an electrochemical battery: multiple copper and zinc plates separated by brine-soaked paper disks. A steady current could be produced: the "Volta Pile" was demonstrated to the Royal Society in London in 1800 (Volta 1800). The entire nineteenth century electrical industry was powered by batteries based on Volta's concept (Pancaldi 2003) until the invention of the dynamo in the 1870s. Volta (1800) and his living prototype *Torpedo* is the fundamental historical key event of biomimetics.

In architecture the Crystal Palace is considered to have been the first "bionic" building. The Palace opened in 1851 in London for the Great Exhibition. The cast-iron and plate-glass Palace was designed by Joseph Paxton, and is supposed to be inspired by the leaves of the Giant Waterlily (*Victoria amazonica*). However, modular wood and glass greenhouses had been in use as "Orangeries" since the seventeenth century. Paxton was the chief gardener and designer for the Duke of Devonshire. As early as 1836, he constructed the largest glass building in the world: the "Great Stove" at Chatsworth House – long before *Victoria* was first cultivated in Europe in 1849 (for historical details see Anisko 2013).

In contrast to Crystal Palace, the entrance building for the 1900 World Exhibition in Paris by René Binet was bionic based on the drawings of microorganisms (*radiolarians*) by Ernst Haeckel (e.g. Coineau et al. 2015; Kresling 1994). The eminent zoologist Ernst Haeckel, through his monumental volumes *Kunstformen der Natur* (1899–1904; *"Art Forms in Nature"*), has influenced generations of bioinspired designers. He was particularly fascinated by light-weight mostly microscopic structures (see Figs. 3.9 and 3.10). He held the Chair of Zoology at the University of Jena- and in Jena, the world's first geodesic dome (Figs. 3.9 and 3.10) was built by Walther Bauernfeld : "The Wonder of Jena" (Breidbach 2011). The construction comprised complex structures with modular polyhedral units derived from Platonic solids. The most successful Zeiss-Bauernfeld Geodesic Domes were built before World War II in cities such as Berlin, Düsseldorf, Paris, Rome, Chicago, Los Angeles and New York; some are still in use today. Bauernfeld's idea was perfected and popularized by R. Buckminster Fuller in the 1950s (e.g. Gruber 2011). Light weight constructions are an important element of bio-inspired architectural constructions (Fig. 3.15).

Another likely early bionic engineer was Michael Kelly, one of the proposed inventors of barbed wire (e.g. Nachtigall and Wisser 2014). His idea was to use the spiny branches of the North American Osage Orange (*Maclura pomifera*) as a model, he patented it in 1868. However, barbed wire had been in existence before and was already patented in 1845 – and the basic idea presented in the valid patent had nothing to do with bionics.

Fig. 3.15 Bio-inspired lightweight modular tensile structures became popular after the 1950s: Bionics was an important field in Russia: Yuri Lebedev (*second from right*) and Frei Otto (*right*) demonstrating light weight architectural models in Moscow c. 1975. (Photograph: Private Collection Berlin)

After Volta's fundamental work around 1800 Bionik or "bionics" as a science, in the sense we define it today, was established in the decades between 1880 and 1920 and expanded rapidly. For historical reasons we use *Bionik* as the German abbreviation for *Biotechnik* (Francé 1920) and refer to Bionics as in the Dayton Symposium 1960 (see next chapter).

Not surprisingly, Bionik was related to flying devices. Their complex and long history is closely linked to observations of bird flight. The focal point was the "simple" gliding of large birds – not the admirable, but over-complex hovering of hummingbird), and the wind dispersal of plant seeds.

It was Otto Lilienthal, who, based on his own research ("Der Vogelflug als Grundlage der Fliegekunst", 1889), built the first gliders after 1890, together with his brother Gustav. He was the first person to make well-documented, repeated and successful flights from 1894 onwards (Fig. 3.16). He built at least a dozen models of monoplanes, wing flapping aircraft and biplanes and obtained the first world-wide bionic patent for his machines in 1893 (Deutsches Reichspatent No. 77916). Otto Lilienthal's achievements summarized in his brother Gustav's book "Biotechnik des Vogelfluges" (1924), mark the beginning of "Biotechnik" (Bionik).

Plants also produce highly sophisticated gliders: wind-dispersed seeds can fly elegantly over very long distances, as in the tropical climber *Alsomitra (Zanonia) macrocarpa*. Ingo Ettrich constructed a monoplane aircraft in 1903 based on

Fig. 3.16 The beginning of modern Bionics: the first successful flying machines were built by Otto Lilienthal from 1890 onwards. They were based on his biophysical study of the bird flight from 1889 and patented in 1893. Airplane construction is based on bionic ideas up until today. The photograph shows Lilienthal on one of the first flights in August 1894 (Photograph: O. Anschütz, http://is.gd/Sc8HJ5)

the *Zanonia* seed; the aircraft went into commercial production ("Etrich-Taube"). However, it was the motor powered aircraft of the Wright brothers that subsequently led to the establishment of aviation. History has forgotten that the invention and development of airplanes is based on the analysis of bird flight. Airplane construction has since undergone a long and continuous technical evolution, although a modern Boeing 767 is still based on the original bionic ideas of Otto Lilenthal 1890 – and the batteries based on Volta 1800 are still on board.

In 1920 the Austrian botanist Raoul Francé coined the name for the new field "Biologie und Technik" *Biotechnik*, to which the German abbreviation *Bionik* refers. Francé (Francé 1924, Pichler 2005; Roth 1983) was a creative thinker and, apart from Biotechnik, he coined terms such as edaphon and edaphic, which are commonly used today. He was aware of the importance of bio-inspired technical design: one of the chapters in his highly popular book "Die Pflanze als Erfinder" (1920) was "der Sieg des biotechnischen Gedankens" ("the success of the bionic approach"). Francé was a prolific writer; his books were best-sellers in their time. The English translation of his bionic book ("Plants as inventors" 1926) *"brought the term Biotechnics into everyday usage*" (Goujon 2001). In Germany, the Nobel laureate Wilhelm Ostwald (1929) became interested in Bionik. At the University of Halle, the "Forschungsstelle für Biotechnik" ("Biotechnic Laboratory") was established by Alf Giessler who, under heavy National Socialist influence and ideology, wrote a popular textbook entitled "Biotechnik" (1939). This was a first survey of pre-war Bionik, including simple neuro-electricity related ideas. Giessler provides a comprehensive survey of living prototypes, including the adhesion mechanisms of seeds and burrs equipped with miniature hooks. However, it took the Swiss engineer and inventor George de Mestral who, when covered with burrs as experienced by generations of people in previous centuries, – observed in 1941 the way that the burrs (*Arctium*) clung to his trousers and his dog. De Mestral was the first to think of a technological transfer; in 1958, he developed the bionic *hook and loop* fastener under the trade mark Velcro®.

Biotechnik was also popular in philosophy: Lewis Mumford (Technics and Civilisation, 1934), influenced by Patrick Gedde, evokes a "Biotechnic Era" referring to Raoul Francé. The term "*Biotechnik*" was first used in a somewhat different context by the respected novelist Rudolf Goldscheid in Vienna as early as 1911 in his biotechnical philosophy (Bud 1993; Goujon 2001). More surprisingly still and obviously overlooked in the bionic literature, the term "bionic" was first introduced in science as early as 1901 by the important US geologist and paleontologist Henry Shaler Williams, but in a different context. The term is comprehensively defined by Williams in a chapter headed "*The Bionic Value of Fossils*" (1903) (compare also with Dicks 2015).

Bionik research continued to be conducted in Europe after World War II (Fig. 3.15). One of the first epicentres was the Technical University (Technische Hochschule) in Berlin. Heinrich Hertel, like Wernher von Braun, was a leading aeronautic engineer before World War II, he was involved in the development of the giant Ju 390 and the first turbojet He 178 aircraft. He was one of the masterminds of bionics in aeronautics. In 1955, he accepted the chair for Aeronautics at the Technische Hochschule Berlin. His superb book "Biologie und Technik" (Hertel 1963), including its English edition of the same year, is almost forgotten or ignored today. The young architect Frei Otto finished his dissertation in 1954 and founded the research group "Biologie und Bauen" in 1961, in Stuttgart. He was considered as one of the most influential structural engineers for bionic light-weight constructions and experimental architecture (Fig. 3.17, see also Figs. 3.9 and 3.10 and 3.15) (Nerdinger 2005). In the following decades, Stuttgart became the centre of experimental architecture. Light-weight constructions in microscopic algae (diatoms) were analysed in Berlin by his close colleague Johann G. Helmcke (e.g. Helmcke and Otto 1962, Helmcke 1984). An "evolution strategy" for was developed by Ingo (Rechenberg 1965; Rechenberg 1971; Rechenberg 1994); he accepted the first chair for "Bionik und Evolutionstechnik" in 1973. Towards the end of the 1960s, the physicist Herman Haken developed the field of synergetics, the theory of self-organization. In 1969 Werner Nachtigall established his Bionik working group in Saarbrücken and subsequently initiated his program "Technische Biologie und Bionik". His textbooks remain most influential (e.g. Nachtigall 1998; Nachtigall 2005). Materials science came surprisingly late into the field of bionics with the discovery of hierarchically structured superhydrophobic self-cleaning Lotus surfaces in Heidelberg (Barthlott and Ehler 1977, Barthlott 1990, 1992) and was applied technically (Fig. 3.18) only after 1997 (Barthlott and Neinhuis 1997, historical survey in Barthlott et al. 2016, 2017). Dietrich W. Bechert at the Technical University of Berlin contributed to the field with research in drag-reducing riblets derived from shark skins (e.g. Reif 1981; Bechert and Bartenwerfer 1989), in an unbroken tradition of Bionik up until today (Bertling 2014, Bhushan 2016). The Lotus research led to the unexpected geometry and physics of the microscopic egg-beater-shaped hairs in the drag-reducing air-retaining surfaces of the floating fern *Salvinia* (Figs. 3.19 and 3.20).

Fig. 3.17 The Giant Arum (*Amorphophallus titanum*) reaches 3 m in height and has the largest blossom in the plant kingdom. The giant pleated "petal" (spathe) weighs less than 4 kg: the largest light-weight construction amongst plants or even in organisms. In its habitat, the tropical rainforest, it attracts visiting insects like a torch: the central column heats up periodically at night and works as a convection device evaporating an unpleasant odour (for humans) to attract pollinators. Several new species of *Amorphophallus* have been discovered during the last decades. (Photograph: W. Barthlott)

Biotechnik or *Bionik*, based on the work of Francé and Lilienthal, was an emerging and popular field in Germany and Austria as early as the period between the two world wars. The young Heinz von Foerster (see e.g. Foerster and Glasersfeld 1999) studied technical physics at the Institute of Technology in Vienna. He was influenced by his uncle and mentor Ludwig Wittgenstein, an aeronautical engineer who graduated from the Technische Hochschule Berlin and Manchester, better known as the author of the *Tractatus logico-philosophicus*. Heinz von Foerster (1911–2002) must have known the works of the very popular Raoul Francé (1874–1943). Both lived in Austria in the crucial years between the wars. It seems inevitable that the young student was not aware of the Biotechnik development. Heinz v. Foerster came to be the key player in the re-invention of bionics in 1960 in the United States.

3.4.2 The Re-invention of Bionics and the Dayton Symposium 1960

Between 1890 and 1940, Biotechnik (Bionik) had become a well-established discipline: *"Bionics is the English translation of Francé's interest in the study of*

Fig. 3.18 The Lotus (*Nelumbo nucifera*) leaf has been the symbol of purity for millennia in Asian religions. We revealed the biophysical mechanism behind it in the 1970s and it was named the "Lotus-Effect®" in the 1990s. The first Lotus-Effect® product, the façade paint Lotusan®, came on the market in 1999 (Photograph: left W. Barthlott, right Sto AG)

Figs. 3.19 and 3.20 Drag-reducing surface of the Giant Floating Fern *Salvinia molesta*. Biology provides surprising technical solutions for engineers. Microscopic egg-beater-shaped superhydrophobic hairs with hydrophilic pins stably hold a thick air layer. Technical prototypes immersed under water can remain dry for years. This biophysical mechanism can be used to reduce friction. By using coating materials based on the Salvinia-effect, friction of large container ships can be reduced some 30 %, translating into fuel savings of some 3 % (Photographs: W. Barthlott (*left*) and Bertling/Fraunhofer/BBMBF (*right*))

the formation and principles of operation of living things to apply the knowledge gained to the development of physical systems. This vision interest was translated as "cybernetics" by visionaries such as Norbert Wiener whose own parallels between machines and organisms harked back to Francé's time" (Goujon 2001).

Thus began a new development of Bionik under the name “Bionics” in the United States during the second part of the twentieth century. Units and training courses on biological engineering and biotechnology had previously been set up at the MIT in Cambridge, Massachusetts, in 1936 and at the University of California in Los Angeles in 1947. In 1948, a new bioinspired concept and term was created by Norbert Wiener: *Cybernetics: Or Control and Communications in the Animal and the Machine* (see Kline 2015). In 1949, the polymath Heinz von Foerster, supported by the influential neurophysiologist Warren McCulloch, moved from post-war Vienna to the centre of one of the most important scientific movements of the time, the University of Illinois, Urbana. McCulloch, v. Foerster, Norbert Wiener, John v. Neumann and Margaret Mead were leading figures at the Macy Conference in 1949. In 1958, Foerster became the first director of the *Biological Computer Laboratory* (BCL) (Müller and Müller 2007) in Urbana, financially supported by military resources and part of the military-industrial complex (Müggenberg 2011, 2014). System theory, self-organization and what was later called “bionics” were focal areas of research at the BCL. “Bionics” involved the analysis of biological processes, their formalization and their implementation on computers following the ideas of McCulloch and the Macy Conference. Some of the technical prototypes of BCL, such as the celebrated “perceptual machine” *NumaRete* (Lettvin et al. 1959; Asaro in (Müller and Müller 2007), were based on biological models; the NumaRete was based on the Northern leopard frog (*Lithobates pipiens*). At the time, they were called “*Living prototypes*”, whereas today, we can refer to them as “biological role models” (e.g. Bar-Cohen 2011).

Only a few hours by car from the BCL Urbana campus, a remarkable meeting took place in Dayton on 13 and 14 September 1960: the *Bionics Symposium (Robinette* 1961*)*. It is difficult to evoke the atmosphere of the political and military tensions during this Cold War Era. Following the invention of the first rocket, namely the V2 launched before World War II, and the first object to cross the boundary layer into outer space, a Russian satellite was in orbit: *Sputnik*. Anyone in the US, who doubted its existence, could walk into the backyard and see the Russian satellite in the sky after dawn; a short-wave receiver could pick up its beep-beep: the sputnik crisis. The first man in space was Yuri Gagarin in April 1961 only a few weeks after the publication of the Bionics Symposium proceedings. Chapters such as “*Russian Attitudes*” (Robinette 1961, p. 21–24) or questions such as “*Is bionics good for improving the defence posture of the United States?*” (Robinette 1961, p, 479) played an obvious and important role in the Dayton Symposium.

Over 700 scientists, engineers and military personnel met at the Wright Air Development Division (WADD) in Dayton. The key players and speakers were Heinz von Foerster, Warren McCulloch, Hans Oestreicher and the WADD chief scientist John Keto. For the organization and, in particular, for public relations, the deputy chairman Major Jack E. Steele played a crucial role. The 506-page symposium proceedings were edited by the technical employee Joan C. Robinette and published in March 1961 under the title: *Bionics Symposium. Living*

Prototypes – the key to new technology. It seems not to have been reprinted and even electronic copies of the proceedings are not available. As a possible result, a historical-political and conceptual analysis of the meeting has never been carried out, apart from initial attempts (e.g. Müggenberg 2011, 2014). The contributions are dominated by electronic and cybernetic themes. From a scientific point of view, the report is of limited interest for bionics today. Apart from the biophysicist Otto H. Schmitt, no single renowned biologist attended the Dayton meeting. There was little interest in a dialogue with biologists, as reflected in McCulloch's contribution to the discussion (Robinette 1961, p. 492): "*Hostility you will encounter always among the echt* (sic!) *biologists. You cannot expect them to give you ordinary team play*." Jack Steele adds "*a man with a passion for synthesis or creative design does not become a biologist*" (Robinette 1961, p. 488).

The derivation of the Dayton name "Bionics" remains speculative and for obvious reasons is not explained in the Dayton proceedings. Several participants and witnesses of the time have subsequently made contradicting statements on this matter. Jack Steele had a rather limited knowledge of the existing literature. When he finally discovered Lilienthal's book *Biotechnik* (published in 1924) he expressed his upmost surprise in a letter to Heinz von Foerster dated October 17th 1963 (Yale Archives; Müggenberg 2011).

The term "bionics" might have been invented "*on an August evening in 1958*" by Major Jack Steele (Goujon 2001; Asaro in Müller and Müller 2007), but first publicly mentioned in May 1960 (12th Ann. Nat. Aeronautic Conference) and finally published in Robinette (1961). Today, it is usually assumed in dictionaries, internet sources and Wikipedia (Chatfield 2013) to be derived from "*Biology* and *Electronics*". Sometimes it is referred to *Biology* and *Technology* synonymous with Biotechnology and with Bionik derived from Francé (1920). Alterntatively it might have been derived from the Greek "*bios*" and the suffix "*-ic*" but, then, it is identical with H.S. Williams' term "bionic" from 1901. The most plausible derivation seems to be from *Biology* and *Cybernetics*. The titles of the two proceedings of the following Dayton Bionics Symposia support this: in 1963, *Information processing by living organisms and machines* (Oestreicher 1964) and, in 1966, *Cybernetic problems in Bionics* (Oestreicher and Moore 1968).

Only two decades after Norbert Wiener's fundamental publication of 1948, cybernetics was discredited in the United States and fragmented into separate fields (Kline 2015). Because Heinz von Foerster understood "'*bionics' to be a logical continuation and expansion of cybernetics*" (Müggenburg 2014), the new term could have been the appropriate alternative, as Dayton 1960 was a cybernetic dominated symposium. The only derivation of the term provided in the first Dayton report is the ironic statement by Otto H. Schmidt: "*having accepted sputniks, we now call our field bio-niks*" (Robinette 1961, p. 483–486). Schmidt also states, in the same contribution, that bionics is "*not an especially new field . . . . . . it has been . . . in several laboratories for over a decade under such names as biological engineering or applied biophysics*".

The Dayton bionics conferences were continued in 1963 und 1966. It should come as no surprise that Schmitt used a new term "Biomimetics" in his subsequent Dayton contributions. According to various sources he might have coined "Biomimetics" in the late 1950s (Harkness 2001; Vincent et al. 2006). The term "biomimesis" was first used by W.S. McCulloch in February 1961 in various talks, e.g. in a published lecture "*What is new in biomimesis*" at a Resident Staff and Research Division Seminar, Illinois State Psychiatric Institute, Chicago, Feb. 19, 1961, even before the publication of the term "bionics" in the Dayton symposium report. Many additional terms arose early, e.g. *Molecular Bionics* (Foerster 1963, Oestreicher 1964).

Biomimetics is the term increasingly used today. Derived from biomimetics is the superfluous and somewhat inappropriate term "Biomimicry" (mimicry is well defined in biology: see Fig. 3.14). It was first used and published in a PhD Thesis (Merill 1982) and has been subsequently used in a popular book (Benyus 1997) and programs (see also Dicks 2015; Wahl 2015).

Like cybernetics before it, the term bionics became somewhat discredited. An anonymous reviewer of the 3rd Dayton Symposium 1966 proceedings (Oestreicher and Moore 1968) states *"the word"bionics" has conjured up . . . distinctly unpleasant images. It has had connotations of dilettantism, of engineers doing bad biology and bad biologist doing worse engineering, and military men seeking to construct unholy engines . . . "* (Bull. Math. Biophysics 31, pp. 619–620). However, the buzzword *Bionics* gained public attention: Heinz von Foerster gave newspaper interviews and travelled with the frog-eye inspired and celebrated NumaRete, one of the first parallel-computing machines, packed in a most elegant case, to radio stations and NBC television shows. His public-relations protegé Jack Steele presented the WADD *Electronic Mouse* for newspapers and magazines (e.g. "Bionics – Nature's Arsenal" in "Popular Electronics", October 1962).

However, bionics still remained a somewhat obscure item until the fateful intervention of mass media (Chatfield 2013) in form of television series, comic books, and movies: Cyborgs and Bionics. The idea of cybernetic organisms or in short "Cyborgs" (Clynes and Kline 1960), with both organic and biomechatronic body parts (Halacy 1965), was discussed at the Dayton symposium already. The science fiction writer and aviation specialist Martin Caidin published his famous novel *Cyborg* (1972). Based on it television series on "Bionics" ("The six million Dollar Man", "Bionic Woman") and comics (Fig. 3.21) appeared in the 1970s. Subsequent movie productions (e.g. "Bionic Showdown"1989) became popular world-wide with the cyborg movement (Gray 1995). Bionics, as fusion of human and machine, had gained a new and fateful connotation: Science Fiction. Caidin (1972) already makes explicit reference to Major Jack Steele: he was the best army public-relation officer for bionics, but in this field of science he never published a line. For the broad audience however the medical doctor Jack E. Steele (1924–2009) was seen as the *inventor of a new science* – up to his obituary in the Dayton Daily News of January 29th in 2009: "*The Father of Bionics died*".

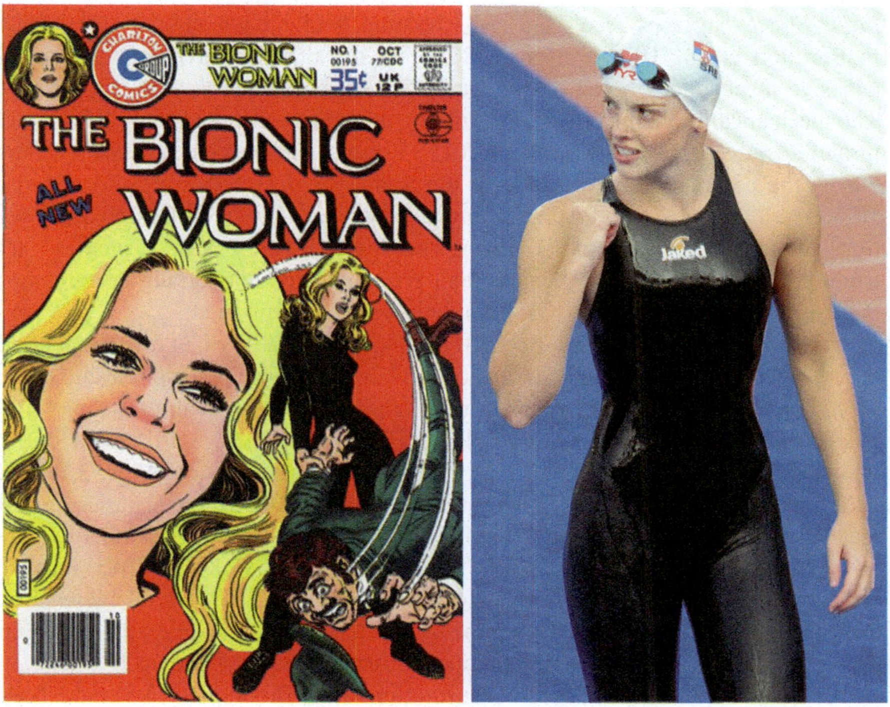

Fig. 3.21 Bionic women in fiction and reality. *Science fiction* popularized a different image of bionics beginning with M. Cardin's novel *Cyborg* (1972). *Left*: the title page of *The Bionic Woman* (October 1977). However, Olympic records in swimming today are a true triumph of bionics. The miniature swimming suits of Mark Spitz in the Munich Olympics 1972 are no longer in use – since 2000 swimmers have worn biomimetic drag reducing outfits. In 2009, at the World Swimming Championship in Rome, almost all records were broken using full-body biomimetic swimsuits. In the same year, the length of swimwear was strictly controlled by the International Swimming Federation (FINA). Today, almost full-body swimwear is back (*right*): somewhat smaller than in 2009, but still covering large parts of the body. At present, swimmers seem to use a high-tech combination of chemically heterogeneous surfaces based on the properties of shark riblets (drag reduction c. 3 %) and salvinia effect (drag reduction c. 30 %). For divers, drag reduction plays no role – divers in the 2016 Olympics in Rio de Janeiro were still dressed like Mark Spitz in 1972 (Sources: (*left*: Wikia, http://bit.ly/1TEp1t0, *right*: Wikipedia http://bit.ly/2bjM2qi))

3.4.3 From Bionics to Biomimetics and Parabionics

Bionics changed in the years after 1970 (e.g. Vincent 2009a; Vincent 2009b; Gleich et al. 2010; Bertling 2014). Classical robotics and aviation technology remained its essential components with cybernetics, evolutionary algorithms, computer modelling and mechanical engineering playing an increasing role, as did bioinspired applied communication concepts (e.g. particle swarm optimization PSO, ant colony optimization ACO). Surprisingly late, only after 1990, materials science became involved (Forbes 2005; Vincent 2012). Superhydrophobic and self-cleaning

Lotus-Effect® surfaces (see Fig. 3.18) became well known after 1997 (survey in Barthlott et al. 2016, Yan et al. 2011) as did the counter example, namely adhesive mechanisms based on the Gecko's foot (see the survey in Gorb 2009). The mechanical properties of bionic building materials are increasingly being used in architecture (Pohl and Nachtigall 2015; Vincent 2009a; Vincent 2009b; Addis 2015); the complex kinetic façade of the Soma Building (see Fig. 3.6) at Expo 2012 based on the sophisticated opening mechanism of the bird-of-paradise flower *Strelitzia* is a superb example (Knippers and Speck 2012). Processes such as *self*-healing, *self*-assembly, *self*-cleaning and *self*-organized networking are amongst the bioinspired "*selfies*" that are becoming increasingly important.

Based on the works of D.W. Bechert, friction reduction by riblet foils has caused public awareness: the America's Cups of 1987 and 2010 saw bioinspired ship hulls. From 1996 onwards, bionic swimming suits have been in use; the Olympic gold medals of Beijing 2008 are the result of using "bioinspired" *Fastskin*® technologies. Air-retaining Salvinia®-Effect (Barthlott et al. 2010) surfaces (see Figs. 3.19 and 3.20) might provide another solution for drag reduction. Gould (2015) has estimated that the global nano-coating market will reach a value of 14.2 billion US Dollars by 2019. Other estimations, such as the "Da Vinci Index" (San Diego 2013) provide estimations (425 billion US Dollars for the 2039 U.S. GDP) for all bioinspired technologies including biotechnological products. New institutions, research programs and networks such as the International Society of Bionic Engineering ISBE and BIOKON and many other institutions (e.g. Biomimicry 3.8) have been established. After a shift from Western Europe to the United States and Canada, capacities are growing enormously in East Asia, particularly in China (see map in Figs. 3.23 and 3.24).

The term "bionics" is increasingly being replaced by "biomimetics". Associated terminological issues have been discussed above. Regrettibly, "bionics" is increasingly used as a marketing tool ("parabionics", see Fig. 3.13). On the other hand, we keep forgetting that from the Velcro® tape to a modern electric battery or an aircraft, an astonishing number of technologies and products were originally based on bionic ideas.

3.5 Bionics and Evolution: One Billion Years of Free Research for Technical Innovations

Over 40 years ago, the famous geneticist and evolutionary biologist Theodosius Dobzhansky addressed the American Association of Biology Teachers in a talk entitled: "Nothing makes sense in biology except in the light of evolution". We are tempted to paraphrase his quotation and modify it into "Nothing makes sense in bionics except in the light of evolution".

Life on Planet Earth is over 4 billion years old. Unicellular prokaryotic organisms have existed for 3 billion years or longer. Prokaryotes in and of themselves are amazingly complex organisms with a number of physiological specializations. For instance, cyanobacteria used chlorophyll already about a billion years ago.

Prokaryotes are of major interest in *biotechnology,* specifically those found in hot springs or extremely cold water. Their enzymatic "machinery" is of particular importance for biotechnological applications and for materials science.

However, *bionics* largely builds on complex multicellular organisms that evolved less than one billion years ago. Palaeozoic trilobites might be considered as "extinct former living prototypes" but are potentially useful in the field of robotics. The first analysis of the riblets on the skin of *fossil* sharks (friction reduction water: Bechert and Bartenwerfer 1989; Reif 1981) and their bionic application millions of years after their extinction led to a success story.

Another new quality arose when the first terrestrial life forms appeared about half a billion years ago. This happened repeatedly and independently in major biological groups. Terrestrial organisms had to cope with the necessity of developing mechanical solutions to a new stronger "gravitational" environment.

Bionically interesting is the interrelationship with water. Life developed in water with wettable surfaces over several billions of year – the surfaces of all these primary marine or aquatic organisms are wettable (hydrophilic). We could show, using reconstruction methods based on molecular data (Barthlott et al. 2016) that extremely un-wettable "superhydrophobic" surfaces evolved very early amongst terrestrial organisms with the conquest of land about 430–500 Ma ago in the Ordovician or Silurian.

Gingko trees have been in existence almost unchanged since the Jurassic period. Their leaves are extremely water-repellent (superhydrophobic) because of specific chemical compounds and a hierarchical morphological structure. The leaves are covered by the complicated secondary alcohol nano-10-cosanol like the leaves of Lotus. Robust phylogenetic data suggest that nano-10-cosanol had evolved some 350 Ma ago and still is one of the most common chemical compounds in plants to generate unwettable surfaces. Superhydrophobicity seems to be one of the key innovations in the evolution of modern plants. Only since 1997 it gained the attention of material scientists and lead to a paradigm change in surface technologies (Barthlott et al. 2016).

Extant large-winged insects such as dragonflies have morphologically highly diverse superhydrophobic wing surfaces (Wagner et al. 1996). One of the largest insects known is *Meganeura monyi* (Fig. 3.22), a dragonfly that lived during the Carboniferous period, some 300 Ma ago. *Meganeura* had a wing span of around 70 cm. During the Carboniferous period, climate conditions were largely tropical. In such environments with frequent rainfall, a few raindrops adhering to the ultrathin wings of *Meganeura* would have made these structures non-functional. Therefore, *Meganeura*'s wings most likely had a hierarchically structured superhydrophobic surface. Interestingly, this argument reflects a process of back-casting from properties of an extant organism, understood through bionics, to those of fossil species that became extinct millions of years ago. How many potential living prototypes have been lost during the course of Earth's history? An estimated 99 % of all organisms that have ever lived on Earth have disappeared.

In conclusion, bionics has essentially used features that have evolved in multicellular organisms during a time span of about one billion years through a random process of mutation and various types of selection. Through this process,

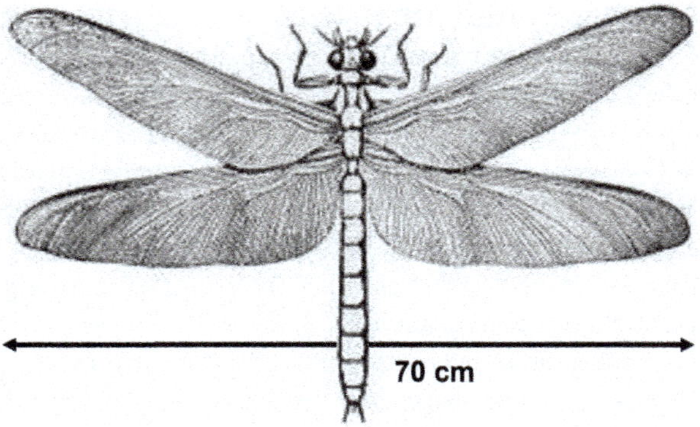

Fig. 3.22 Superhydrophic hierarchical structures had evolved 430–500 Ma ago when life conquered land. The largest insect that ever existed provides evidence for this idea: *Meganeura monyi* with its light-weight wings of some 70 cm in diameter. This ancient giant dragonfly lived during the wet and moist Upper Carboniferous period and would not have survived without extremely water repellent (physically only possible by chemical hydrophobicity combined with hierarchical structuring) wings, like modern dragonfly species (Illustration adapted from Brongniart 1884)

highly complex characters or organs such as the vertebrate eye have evolved. However, evolutionary processes have led to optimization with regard to selective pressures in specific environments. Contrary to technological processes, these macro-evolutionary processes have affected many characteristics and organisms at the same time and do not reflect a one-purpose and boundary-condition-independent "solution".

Extant biodiversity is the product of a long evolutionary process that largely took place in the Earth's past. As discussed in the following section, the diversity of life faces a major crisis caused by one species currently dominating Planet Earth: *Homo sapiens*. Man is even modifying natural processes. Thus, man has (in evolutionary terms, very recently) created its own geological epoch, the Anthropocene. Has Earth's history as we used to know it come to an end?

3.6 Biodiversity in Transition: The loss of Living Prototypes

3.6.1 Biodiversity Loss and International Measures Taken

Continued biodiversity loss is one of humankind's major current challenges. Despite tremendous efforts of the international community, human pressure on virtually all ecosystems of Planet Earth continues to erode biodiversity (Erdelen 2014). Current species extinction rates, for instance, have most likely increased between 1000 and 10,000 times the background rates during Earth's history (Pimm et al. 2014; Pimm and Joppa 2015). Scientists speak of the sixth mass extinction (Barnosky et al. 2011; Kolbert 2014).

The assessment of extinction risks carried out by the International Union for Conservation of Nature (IUCN) draws a bleak picture of the status of species assessed to date. The 2015 IUCN Red List of Threatened Species comprises a total of 77,340 species currently assessed (IUCN 2015). Of these, almost 30 % are considered threatened with extinction. However, two major uncertainties should be taken into account. First, despite enormous efforts, the number of species assessed is less than 5 % of the species known to science. Secondly, these may only represent a small fragment of the total number of species (see below). The IUCN Red List insufficiently covers major groups of organisms. Sobering information is also provided by WWF's latest Living Planet Report (WWF 2014) stating that "The state of the world's biodiversity appears worse than ever".

In 2002, the Parties to the CBD committed themselves to the "2010 Target", i.e. "*to achieve by 2010 a significant reduction of the current rate of biodiversity loss at global, regional and national level as a contribution to poverty alleviation and to the benefit of all life on Earth*" (SCBD 2010). The 2010 target was not met, as clearly illustrated in the *Global Biodiversity Outlook 3*, a report released in 2010 (SCBD 2010). The missing of the target triggered a massive scaling up of global biodiversity-related efforts. By the end of 2010, a new package of aims was decided upon at COP 10 of the CBD (for further details, see SCBD 2014; Erdelen 2014).

To address the science-policy issue in the domain of biodiversity, an Intergovernmental Science-Policy Platform on Biodiversity and Ecosystem Services (IPBES) was established in 2012 (see www.ipbes.net). An overview of its conceptual framework has recently been presented by (Díaz et al. 2015).

3.6.2 The Relationship Between Biodiversity and Bionics

As outlined above, despite all international efforts, serious concerns have been raised that biodiversity loss might not be brought to a halt within internationally agreed timelines (Leadley et al. 2014; SCBD 2014; Tittensor et al. 2014). As a reminder, the timeline for the Strategic Plan for Biodiversity and its associated Aichi Targets is 2020, the new post-2015 development agenda of the United Nations has a time horizon of 2030 and, lastly, according to the 2050 vision for biodiversity, biodiversity loss should ultimately be stopped by 2050 (SCBD 2014).

Interestingly, a position paper on the future of bionics, published by the Association of German Engineers (VDI) and the German BIOKON network, points out the interrelationship between the (German) National Biodiversity Strategy and the role of bionics. Bionics is seen as a facilitator for the development of sustainable products and for fostering environmental awareness and as a guiding principle for the transformative process towards a green economy (VDI and BIOKON 2012).

Without doubt, the strongest human impacts and loss of species and ecosystems are recorded for the more species-rich regions and areas with high species endemism. This is reflected *inter alia* in the concept of "hotspots of biodiversity" (Myers et al. 2000, see also Barthlott et al. 2014 and Fig. 3.4). In other words, (1) most species have small geographic ranges, (2) most species tend to be scarce within those ranges, (3) most of these species co-occur, and (4) most of these areas of co-occurrence of species (commonly referred to as centres of endemism) are the areas most threatened by human activities (see discussions in Brooks et al. 2002; Pimm et al. 2014; Pimm and Joppa 2015).

For an understanding of the current and future relationship between bionics and biodiversity, we first urgently need to compare the geographic patterns of species and ecosystem loss with the occurrence of species "used" for or inspiring bionic applications. Most species are still unknown to science and bionic applications to date certainly reflect only the tip of the iceberg of the immensity that biodiversity has to offer (see Fig. 3.17), an iceberg that is melting away with increasing speed.

Bionics utilizes evolutionary products related to various kinds of characteristics amongst and between species and does not use species or individuals "as a whole". With regard to sharks, for instance, skin morphology, sensory system, and tooth replacement mechanisms might offer models for entirely different bionic applications. Indeed, bionic applications might be based on virtually all levels of hierarchical organizations within living systems, including the molecular level, internal and external morphology, physiology and behaviour and all levels of organization of ecological systems. In other words, in the face of the continued erosion and loss of biodiversity and ecosystems, we might lose "opportunities" from all organizational levels of biodiversity, i.e. the genetic, species, (see Figs. 3.1 and 3.5) ecosystem and even landscape level. On the other hand, bionic templates can be found at generic or even higher taxonomic levels (e.g. superhydrophobic surfaces) and might thus not have a 1:1 relationship to the loss of individual species. Therefore, the distribution of characteristics or systemic properties within the diverse organization of life on Earth is of major relevance for bionics-related research, rather than the focus on a particular species *per se*. At the same time, however, species are the central units used for "learning from nature".

Thirdly, if most of our bionic applications stem from highly specialized organisms and thus eventually from more complex ecosystems, and if that specialization is highest where biodiversity is highest, we can expect a pattern that is highly congruent between species, the diversity of ecosystems and the occurrence of "carriers" of bionics-relevant information (see Figs. 3.23 and 3.24). This is complemented by the intrinsic extinction vulnerabilities of species being expected to be higher in tropical species, a result of increased sensitivity of specialized species and greater niche-packing (see discussions in Gallagher et al. 2015; Stroud and Feeley 2015). As a result of human-induced environmental disturbances, highly diverse communities might therefore be particularly at risk of losing species. As Stroud and Feeley (2015) put it: "Given this potential downside to diversity, we argue that there is additional motivation to prioritize conservation of high-diversity communities in the tropics".

Lastly, as a result of the complexity of species as evolutionary products, on the one hand, and technology being a rather simplified means to address humankind's modern needs, on the other, fundamental differences exist between "products of nature" and "nature-based bionic products", respectively. One key difference is the high phylogenetic age of organisms and their evolutionary history, as outlined above. Others are the principal differences between structural biomaterials used by plants and animals, the multifunctionality of biogenic materials in organisms, the different utilization of materials in organisms and in modern technology (bottom-up vs. top-down) and the fact that nature, for instance, does not produce materials under high temperature or pressure regimes.

Moreover, the life cycles of materials in organisms depend on functional contexts; the teeth and skin of vertebrates are extreme examples of structures, functions and associated materials with extremely long and extremely short life expectancies, respectively. Finally, the growth patterns of biomaterials can vary enormously and last for the whole life span of an organism or need to be replaced several times during life. In summary, only careful comparisons, planning and designing processes based on the sound knowledge of the "natural template" can ensure that the "nature-based technological product" meets our expectations. This includes patterns and processes related to energy efficiency, life expectancy, recycling, resource utilization and options for eco-design, to name just a few (for further information, see, for example, the discussion in Drack 2002). Thus, the characteristics of organisms used in bionic applications have had a complex evolutionary history and probably need to be seen in their multifunctional and complex interrelationship within an organismic entity and/or its external environment before the process towards developing bionic products can be started in an appropriate manner.

3.6.3 *Patterns and Origins of Bionic Applications*

As discussed above, bionic applications "use" characteristics from virtually all major groups of organisms. Characters that can be "exploited" for their usefulness might include the whole range of functional organismic levels including genetics, physiology, internal and external morphology, behaviour and ecology.

Bionic applications might also have clearly identified relationships to the geographic distribution of the "information carriers", i.e. biodiversity at all its levels as the fabric for bionic applications. Accordingly, the information, materials and properties that are sought after are closely linked to the location at which the relevant organisms or their respective characteristics are found, i.e. their geographic distribution. This in itself is more complex than it may seem at first. Aquatic organisms such as sharks and the character "shark skin" might be widely distributed. Others such as the *Strelitzia species* or the Giant Arum (see Fig. 3.17) have a highly localized geographic distribution.

We suspect that target organisms for future bionic applications will mostly be found where biodiversity is high, i.e. in subtropical and tropical regions and/or the so-called centres of biodiversity or hotspots of biodiversity (see Figs. 3.4 and 3.16). This is reflected in the fact that bionic researchers tend to work in highly diverse tropical areas (e.g. Gebeshuber and Macqueen 2014; Gruber 2013).

As has been recognized for over 15 years, these centres not only contain exceptionally high numbers of endemic (= unique) species, but also experience exceptionally high rates of biodiversity loss (see e.g. Myers et al. 2000). The island of Madagascar is an excellent example (see Fig. 3.5). Furthermore, if our ignorance of biodiversity relates in particular to the mega-diverse regions of our planet in which biodiversity loss is also very high, we may experience a tremendous loss of "nature's treasure trove of ideas" before we can even think about using them for bionics-based technological research and eventual product development. Additionally, highly diverse ecosystems such as tropical rainforests contain communities with highly specialized and unique organisms (see discussion above). These specializations are often the trigger for innovative thinking through bionic applications and for the development of the relevant technological processes and products. Complicating things further is the issue of the "diversity of biodiversity": Is research into potential bionic applications carried out in a balanced manner, focusing on both the relevance of the richness in species and that of evolutionary richness (see Mooers 2007)?

A key question for bionics is: which functional characteristic of which taxonomic group best serves for the development of bionic products? For instance, we may tap into knowledge about active flight in organisms as diverse as bats, birds and insects or about passive flight in a variety of "gliders" of the plant (seed dispersal) and animal (gliding snakes, lizards and squirrels – but also birds) kingdoms. Lilienthal was more interested in the soaring of an albatross than in the hovering of a hummingbird. What counts is the selection of the *right* role model and its dimensions: the albatross, humming-bird and hover fly are quite different organisms that reflect different solutions to "evolutionary demands" for flight in different environments and functional contexts.

The role model function of the Northern Leopard Frog, which inspired the NumaRete happened to be "detected" by the researchers involved in developing the machine: It was the most common species near the Biological Computing Laboratory. The ideal biological role model could have been the Giant Squid *Architheutis* with eyes of the size of a soccer ball but this is an organism of the deep sea and was not available for studies in a laboratory. In 1952, Alan L. Hodgkin and Andrew F. Huxley developed the famous conductance-based and Nobel-Prize-winning model of the way that action potentials in neurons are initiated and propagated, a basis for understanding our own body and brain. Huxley was also an experienced biologist and chooses to analyse the *squid giant axon* described already in 1909. He did not choose axons of a laboratory mouse. In other words, he used the most appropriate biological role model.

The study of extreme adaptations is most awarding (see Fig. 3.17). To give an example, an engineer can learn more about the limits of automobile constructions by analysing an exotic Formula I racing car (a *rare species*) than a family-car in daily use. The self-cleaning effect of certain superhydrophobic surfaces is another example. The plant studied was the exotic Lotus plant (*Nelumbo nucifera,* see Fig. 3.18), the best role model for studying superhydrophobicity for bionic purposes. However, this was discovered only after some 16,000 different plant species had been studied and analysed over a period of more than twenty years, namely 1973 to 1994 (Barthlott et al. 1998; Neinhuis and Barthlott 1997). Furthermore, it took an additional three years until the effect was understood in its structural and functional dimensionality (Barthlott and Neinhuis 1997). *A posteriori*, it became evident that any leaf of a cabbage plant or of a tulip could have served as the model for studying the *Lotus* effect. However, the remarkable self-cleaning properties of the Sacred Lotus Plant had already been described in the Bhagavad-Gita and other ancient scripts.

Geckos, found throughout the tropical and warm temperate regions, are one of the most species-rich and widespread groups of vertebrates. Their phylogenetic age is estimated at around 200 Ma. They comprise about 1500 extant species. Around 60 % of them have adhesive toe pads. Gekkotan adhesive pad systems have been in existence since at least the mid-Cretaceous, i.e. during the last 100 Ma (Arnold and Poinar 2008). They have been gained and lost repeatedly over the course of gecko evolution (Gamble et al. 2012). Today, we understand the complex physical properties and nanostructures of the gecko foot and even biomimetic products have been developed. Again, it was a large conspicuous exotic organism that drew our attention to a phenomenon (adhesion) that can also be observed in an organism living next to our laboratories (houseflies) in many different "technical" modifications.

Engineers usually have limited knowledge of biological systematics. Likewise biologists normally are not very familiar with the engineering sciences. This has led to a demand of lists of mechanisms and possible biological role models (species) for bionic research. Ideally this information should translate into "shopping lists" for data-mining (see e.g. Nachtigall 2005; Nagel 2014; Vincent et al. 2006). Anticipating a future bionic product is based either on sound biological knowledge or a result of transdisciplinary collaboration between researchers from different disciplines such as engineers and plant or animal systematists and morphologists. However, organismic biology is and will always continue to be the foundation of bionics. This poses new challenges to institutions housing biological collections such as natural history museums, herbaria or botanical gardens. As information sources for bionic-related research new "service functions" might even be envisaged which may translate into what these institutions may be in need of: Improved awareness of their role and importance for civil society at large.

In sum, living prototypes are the "evolved masterminds" for bionics-related applications. Ultimately, it is a complex, creative and successful research process which translates this product of evolution into products needed by human society.

3.7 Bionics and Biomimetics – Change and Challenge for the Anthropocene

3.7.1 Biodiversity and Bionics – The Issue of Institutional Capacity

A comparison between the global distribution of biodiversity and the distribution of biodiversity-related research infrastructure reveals an interesting pattern (Figs. 3.23 and 3.24). Whereas species diversities are highest in the mega-diverse countries of the subtropics and tropics, these countries are comparatively poor in relevant institutional capacity. This becomes clearly evident when the "geography" of major research institutions is projected onto our biodiversity map (see Fig. 3.4). Research capacities are highest in the "North", i.e. in industrialized countries of the temperate

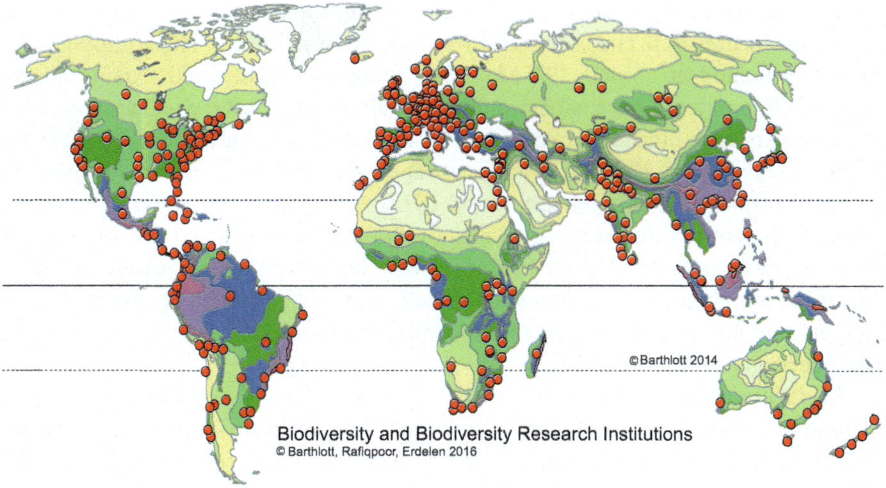

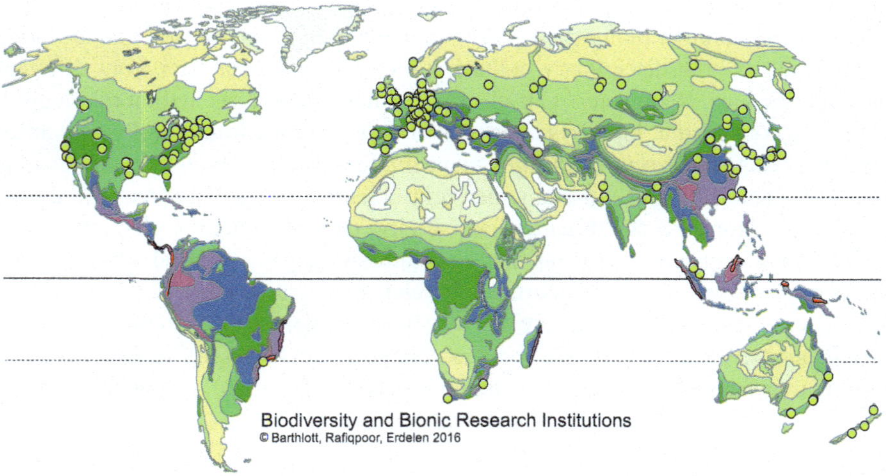

Figs. 3.23 and 3.24 (*Top*) Biodiversity and Biodiversity Research Institutions locations (*red dots*) overlaid onto the global biodiversity map (see Fig. 3.4). (*Bottom*) The locations of the Biodiversity and Bionic Research institutions (*yellow dots*) are also overlaid onto the global biodiversity map. The pronounced north/south gradient reflects the capacity differences between industrialized and developing countries. The gradient appears to be steeper in the case of bionics. This may be linked to the strong dependence of bionics-related research on the availability of high-tech laboratory environments. Another factor could be the long tradition of bionics-related research in the North, as well as the recent growth of capacities in South and Southeast Asia. See text for further discussion. Sources: (Barthlott et al. 1996, 2007, 2014). The bionic research institution location data is based on personal contributions, institutional member lists of ISBE and BIOKON, as well as data collected from the following publications: Cross, J.L (1995); San Diego Zoo Global and Fermanian Business & Economic Institute Point Loma (2013)

regions, both north and south of the equator. It is in the developing countries and the countries in transition, however, that ecosystems are most complex and richest in species numbers. To close this North/South gap and in view of the global importance and pressing nature of the biodiversity issue, we need a process of massive scale-up in research and education capacities and to raise awareness in the South. Even terms such as “responsibility” and “moral obligations” have entered the debates of this North/South issue of a widening gap between the highly developed and less developed world. Indeed, there are striking similarities between international discussions of biodiversity and climate change. Innovative schemes such as South-South or triangular schemes of collaboration including new regimes of cash-flow are of the essence to meet the capacity needs of the South, particularly in view of the continued loss of biodiversity.

Traditionally, biomimicry, biomimetics and biologically inspired design, commonly referred to as B3D, have evolved from extensive research carried out in Britain, China, France, Germany, Japan and the USA (Hoeller et al. 2013).

3.7.2 Bionics, Sustainability and Sustainable Development

As discussed above, closer examination of the field of bionics not only reveals interesting new insights into its terminological, epistemological and normative aspects, but also highlights core concepts related to sustainability and sustainable development.

3.7.2.1 Bionics and Sustainability

A closer look at the interrelationships between bionics and sustainability sheds new light on the term “sustainability”. Evolution has framed the templates for bionics and, as a result, also the life spans of the respective characters or characteristics that are found in nature and used for technological purposes. For instance, Lotus leaves are designed for a life span of several years only, whereas the seeds of the Lotus plant can still germinate after having been dormant for over 1400 years. Nature’s prototypes are sustainable but only for the time spans and environments for which they have been designed through the process of evolution.

Buildings painted with Lotus coatings (see Fig. 3.12) need re-painting about every twenty years. In other words, the desired function has an in-built limited duration. Bionics-based applications might therefore be sustainable only for given periods of time. Sustainability is time-bound and not forever; it is in fact a product of evolutionary forces that have selected exactly in favour of such characteristics. The introduction of the “limited duration” or “time bound” argument into the discussion of sustainability or sustainable development represents a new dimension to international debates, not only with regard to the results and applications of bionics-related research.

As a matter of fact, the temporal or time-limited character of bionic applications might even be stacked to reflect the variability in the life spans of the various compartments within human-devised systems. An example is the time that elapses until a building needs re-painting and the material-derived life span of the building itself. At a higher level of organization, this can apply to the question of sustaining the process of constructing buildings the traditional way and, at the system level, of sustaining urban systems in their entirety. Sand and pebbles, the major non-renewable resources used in the building sector, have become one of the most heavily exploited resources on Earth. Their exploitation even exceeds that of another important natural resource, namely water. This non-sustainable resource utilization cannot continue unabated, particularly in view of the important ecological and other service functions that sand has for ecosystems, and ultimately, for humans (for an overview, see Perera 2015).

Again, in order to sustain human populations in an increasingly urbanized world, we are far from having used "template nature" to the full possible extent. The famous old-growth temperate redwood rainforests (Fig. 3.25, left) of northern California (USA) can serve as a model with an almost endless life span and the extremely low turnover times of its major elements, the redwood trees (*Sequoia sempervirens* and *Sequoiadendron giganteum*). These are the tallest and one of the most massive tree species on Earth, and they "make the system". We might even imagine a bold vision for the future, derived from the Redwood Forest Ecosystems: re-growing urban conglomerations.

On the contrary, immensely growing urban systems (Fig. 3.25, right) house an increasing segment of the human population. The buildings, however, have a life expectancy of less than 100 years. Modern urban systems and, in particular, the growing number of megacities thus have a much higher turnover rate when compared with redwood ecosystems, requiring a permanent supply with essential resources, maintenance and eventually even re-building to sustain them over longer periods of time.

Fig. 3.25 Sustainability; a two-thousand-year-old Giant Redwood (*Sequoiadendron giganteum*) forest in Sequoia National Park in California (*left*). In contrast to the skyline of a modern city with buildings of an age less than one decade (*right*). The redwood forests could continue to grow another thousand years or more; the skyscrapers will probably last less than one century (Photographs: left M. Neumann; right Wiki-Commons: http://is.gd/l9TE1s)

In conclusion, the relationship between bionics and sustainability is not simple or straight-forward but highly complex. This may be of major importance for the discussion of sustainability at higher levels within our hierarchically structured world. Bionics and bioinspired design should not be seen as good or sustainable per se but their templates and ultimately the products created should be examined carefully before reaching any value-related or sustainability-related general conclusions.

3.7.2.2 Bionics and Sustainable Development

We conclude with the following questions. Can bionics contribute to realizing sustainable development, faster than and independently of fossil-fuel-based energies? Can this inform and benefit the latest efforts of the international community towards sustainable development, including in particular the move towards a green economy (see Barthlott et al. 2014 and references therein, Daly 2015)?

These are not easy questions; here we only can present our initial thoughts. Several entry points can be proposed for bionics into the sustainable development debate. Examples are (1) the obvious link between bionics and biodiversity (see above), (2) the link between climate change, low-carbon technologies and bionics (not further discussed here) and, in more general terms, (3) the potential contributions of bionics to the 2030 Development agenda and, in particular, the newly crafted Sustainable Development Goals or SDGs (UN 2015).

With regard to the latter, SDG 9 "Build resilient infrastructure, promote inclusive and sustainable industrialization and foster innovation" and the entry point "technology" under SDG 17 "Strengthen the means of implementation and revitalize the Global Partnership for Sustainable Development" are of particular relevance for our discussion (see UN 2015).

At the European level, bionics is playing an increasingly important role within the EU Horizon 2020 Vision, the EU's research and innovation program and the concomitant investment into nature-based solutions for Europe's Sustainable Development. According to the European Commission, nature-based solutions are "*living solutions inspired by, continuously supported by and using nature, which are designed to address various societal challenges in a resource-efficient and adaptable manner and to provide simultaneously economic, social, and environmental benefits*" (see Maes and Jacobs 2015). However, here again, terminological and epistemological aspects should be carefully examined to identify clearly the relationship between the broader concept of nature-based solutions and its association with bionics or bionics-based products.

Under the heading "Systems engineering and engineering for the Earth system" UNESCO's Engineering Report says that "*the rediscovery of holistic thinking is perhaps not surprising and indeed overdue, prompted, for example, by the renewed interest in biomimetics that links engineering and technology with natural life structures and systems. This marks a belated return to the biomimetics of Leonardo da Vinci ... although this rediscovery has been facilitated by the development of computer science and technology and new materials*" (UNESCO 2010, p. 260).

In fact, we now have enormous opportunities to accelerate the pace of innovation, science, technology and engineering through bionics, i.e. through a change towards research that uses inspiration from nature. A deeper understanding of the way that biological systems work can create new insights and approaches for a number of key aspects of sustainable development, such as water, energy, health, food security and materials.

To be ready for this role in the future, however, some fundamental changes, if not paradigm shifts, are required. These include (1) a re-orientation of research in the sciences, technology and engineering towards a more in-depth examination of what nature has to offer and how we can turn this into bionic products. We need (2) to link this effort to increased activities in the basic sciences, to education and to increase capacities as the *sine qua non* underpinning the relevant research efforts and to form international transdisciplinary collaborations to reinforce the contribution of bionics to the Global Sustainable Development Agenda. Another decisive point is (3) a significant change along the science-policy axis. This requires several significant changes in science policy and associated measures in order globally to coordinate efforts as suggested several years ago by Gyllenberg et al. (2012).

In their background paper for the Rio + 20 Summit in 2012, Drexhage and Murphy (2010) have reached the following conclusions about the complex process of making sustainable development a reality: What is needed is "*systemic change … in the way the world does business* and "*the opportunity is ripe to move beyond incrementalism to real systemic change*". Our hope is that bionics will bring its capabilities to the fore and shape the continued process towards an aspect that is needed more urgently than ever, i.e. "an integrated approach that reconnects human development with the biosphere" as recently put by Johan (Rockström 2015).

3.8 Conclusions

Biodiversity is a seemingly "endless source of information" and provides the fabric for innovation and ultimately for nature-based solutions. At the same time, this source of information is being eroded at rates unprecedented throughout the history of Planet Earth. This loss translates into the loss of principally accessible information from all that has evolved on our planet and might be crucial for sustaining our future. Bionics and the associated market growth for biomaterials in their broadest sense are steadily growing and might be indicative of future demands for "evolved products and processes". Bionics-based innovation is essentially a learning process from nature.

This paper has critically discussed the notions that underscore bionics-based innovation and the meaning of sustainability for products "derived from nature". This is a requirement that has become ever more pressing, in particular with regard to the human-induced impact on the Earth's systems in our era of the Anthropocene.

Moreover, with the increasing recognition of bionics and biomimetics as key elements of future technologies, a mushrooming of terms related to this discipline or rather to a set of disciplines is notable. This is *inter alia* reflected in a growing interest in the field and investments from a number of global companies. Market sizes for products and construction projects applying biomimetics are estimated to have grown from USD 1.5 billion between 2005 and 2008 to about USD 1 trillion in 2015 (see (Hwang et al. 2015) for a more detailed discussion).

We have discussed the confusing terminology currently in use and the overarching term "biologically inspired design" (BID) is defined to include all processes inspired by biology: biotechnology, bionics and biomimetics. However, BID also includes non-functional design, e.g. in bioinspired fashion or Art Nouveau Architecture, a category that we name "biodecoration". Bionics and Biomimetics are considered to be synonymous. The history of these terms with emphasis on the decisive period 1960–1970 is analysed. For all the many products on the market that are called "bionspired", "bionic", or "biomimetic" for marketing reasons but without having a bionic background, we suggest the term "parabionic". Economically, parabionic products play an important role.

With regard to the normative aspects of the "bionics discussion", we have frequently encountered an overly simplified view of bionics and bio-materials, with them being perceived as "good *per se*". On a broader scale, this translates into a broad classification of nature as good (and sustainable) and technology as bad. Indeed, living organisms are sustainable only as long as their "characteristics" or "Merkmale" (characters) meet ecological and evolutionary demands for which "evolution has designed them". Major extinction events during the history of our planet have shown the speed at which these demands can change.

Since the commencement of the discussion about planetary boundaries, less than a decade ago, serious concerns have been raised that the boundary for biodiversity might already have been transgressed, leading to the irreversible destruction of critical ecosystem functions. Despite all efforts by the international community to bring biodiversity loss to a halt, the evidence indicates that the goals and associated timelines that we have set ourselves may most likely not arrest what has been estimated to be 1000 to 10,000 times the natural extinction rate. In more general terms, the world "urgently needs a great transition that rapidly bends the curve of negative global environmental change" (Rockström 2015).

With continued population growth, the human impact on biodiversity and ecosystem services might soon transcend critical thresholds. We risk losing forever the abundance that nature has to offer humankind in its quest for its own sustainable future. We should not forget that we have only one Earth as our resource base, and not several such planets. Will our species in the long run turn into Earth's major parasite, ultimately destroying its own natural resource base? Will visitors of the future Earth ask why we named ourselves "sapiens"?

Acknowledgements We acknowledge the help and support of many colleagues and friends in preparing the complex text. The first author (WB) had the chance to meet some of the pioneer workers mentioned in the text (e.g. Ingo Rechenberg, Johann Helmcke and Frei Otto). He met the inspiring Heinz von Foerster 1993 on the occasion of the NeuroWorld symposium in Düsseldorf but missed the singular chance to ask him about the origin of the term "bionic" because, at that time, he was unaware of von Foerster's crucial role in the Dayton Bionic Symposium of 1960.

WB acknowledges information from multiple discussions with his colleagues, friends and students such as Christoph Neinhuis, Thomas and Olga Speck, Armin B. Cremers, Stanislav Gorb, Claus Mattheck, Fredmund Malik, Rainer Erb, Bharat Bhushan and the late Günther Osche.

Many of the ideas presented in this paper benefitted from Walter Erdelen's extensive international experience, in particular his work as Assistant Director-General for Natural Sciences (2001–2010) and subsequently as strategic adviser at UNESCO. He expresses his sincere thanks to UNESCO Member States and their representatives, up to highest political levels, former staff of the Natural Sciences Sector and colleagues in the Organization with whom he specifically collaborated as Head of Delegation to the World Summit on Sustainable Development (2002).

We are grateful to the Hon. Margaret Austin and Patrick Lim, Callaghan Institute New Zealand, for information on the situation of bionics in New Zealand and to Jacques G. Richardson for constructive comments on earlier drafts of the manuscript.

We acknowledge the reviewers for most valuable comments and the help of Danica Christensen in rereading the English version. Last but not least, the text is shaped by our own experience and work in Bionics and Biodiversity, which was supported by the Deutsche Bundesstiftung Umwelt DBU, the German Research Council DFG, the Federal Ministry for Science and Education BMBF, and the Academy of Science and Literature in Mainz.

References

Addis B (2015) Buildings – 300 years of design, engineering and construction. Phaidon Press, London

Anisko T (2013) Victoria, the seductress. Longwood Garden Press, Philadelphia

Arnold EN, Poinar G (2008) A 100 million year old gecko with sophisticated adhesive toe pads, preserved in amber from Myanmar. Zootaxa 1847:62–68

Bar-Cohen Y (2011) Biomimetics: nature-based innovations. CRC Press. Biomimetic series 778 pages. ISBN 9781439834763

Barnosky AD, Matzke N, Tomiya S et al (2011) Has the Earth's sixth mass extinction already arrived? Nature 471:51–57

Barthlott W (1990) Scanning electron microscopy of the epidermal surface in plants. In: Claugher D (ed) Application of the scanning EM in taxonomy and functional morphology, Systematics Associations' special volume. Clarendon Press, Oxford, pp 69–94

Barthlott W (1992) Die Selbstreinigungsfähigkeit pflanzlicher Oberflächen durch Epicuticularwachse. In: Rheinische Friedrich-Wilhelms-Universität Bonn (Hrsg) Klima- und Umweltforschung an der Universität Bonn, pp 117–120

Barthlott W, Ehler N (1977) Raster-Elektronenmikroskopie der Epidermis-Oberflächen von Spermatophyten. Tropisch-subtropische Pflanzenwelt 19. Akad. Wiss. Lit. Mainz. Franz Steiner Verlag Stuttgart, 105 S

Barthlott W, Neinhuis C (1997) Purity of the sacred lotus, or escape from contamination in biological surfaces. Planta 202:1–8

Barthlott W, Lauer W, Placke A (1996) Global distribution of species diversity in vascular plants: towards a world map of phytodiversity. Erdkunde 50:317–328

Barthlott W, Neinhuis C, Cutler D, Ditsch F, Meusel I, Theisen I, Wilhelmi H (1998) Classification and terminology of plant epicuticular waxes. Bot J Linn Soc 126:237–260

Barthlott W, Hostert A, Kier G et al (2007) Geographic patterns of vascular plant diversity at continental to global scales. Erdkunde 61(4):305–315

Barthlott W, Schimmel T, Wiersch S, Koch K, Brede M, Barczewski M, Walheim S, Weis A, Kaltenmaier A, Leder A, Bohn HF (2010) The Salvinia paradox: Superhydrophobic surfaces with hydrophilic pins for air-retention under water. Adv Mater 22:1–4. doi:10.1002/adma.200904411

Barthlott W, Erdelen WR, Rafiqpoor DM (2014) Biodiversity and technical innovations: bionics. In: Lanzerath D, Friele M (eds) Concepts and Values in Biodiversity. Routledge, London/New York, pp 300–315

Barthlott W, Mail M, Neinhuis C (2016) Superhydrophobic hierachically structured surfaces in biology: evolution, structural principles and biomimetic applications. Phil Trans R Soc A 374:20160191. doi:http://dx.doi.org/10.1098/rsta.2016.0191

Barthlott W, Mail M, Bhushan B, Koch K (2017) Plant surfaces: structures and functions for biomimetic applications. In: Bhushan B (ed) Springer handbook of nanotechnology, Chapter 36, 4th edn. Springer Publishers (in print)

Bechert DW, Bartenwerfer M (1989) The viscous flow on surfaces with longitudinal ribs. J Fluid Mech 206:105–129. Cambridge University Press, doi: http://dx.doi.org/10.1017/S0022112089002247

Bell EA, Boehnke P, Harrison TM, Mao WL (2015) Potentially biogenic carbon preserved in a 4.1 billion-year-old zircon. Proc Natl Acad Sci U S A 112(47):14518–14521. doi:10.1073/pnas.1517557112

Benyus J (1997) Biomimicry. William Morrow, New York

Bertling J (2014) Bionik als Innovations-Strategie. In: Herstatt C, Kalogerakis K, Schulthess M (eds) Innovationen durch Wissenstransfer. Springer, Heidelberg/New York, pp 140–184

Bhushan B (2016) Biomimetics – bioinspired hierarchical-structured surfaces for green science and technology. Springer, Heidelberg/New York

Breidbach O (2011) Ernst Haeckel, Walther Bauernfeld und die Konstruktionsidee des Jenaer Planetariums. In: Meinl H et al (eds) Die Weltenmaschine – Beiträge zur frühen Geschichte des Zeiss-Planetariums. Ernst-Abbe-Stiftung, Jena, pp 45–62

Brongniart C (1884) Sur un gigantesque Neurorthoptère, provenant des terrains houillers de Commentry (Allier). C R Hebd Seances Acad Sci 98:832–833

Brooks TM, Mittermeier RA, Mittermeier CG et al (2002) Habitat loss and extinction in the hotspots of biodiversity. Conserv Biol 16:909–923

Brown JH (2014) Why are there so many species in the tropics? J Biogeogr 41:8–22

Bud R (1993) The uses of life – a history of biotechnology. Cambridge University Press, New York

Caidin M (1972) Cyborg. Warner Paperback Library – Warner Books, New York

Caley MJ, Fisher R, Mengersen K (2013) Global species richness estimates have not converged. TREE 29:187–188

Cavendish H (1776) An account of some attempts to imitate the effects of the Torpedo by electricity. Phil Trans R Soc Lond 1776:196–225

Chapman AD (2009) Numbers of Living Species in Australia and the World, 2nd edn. Australian Government, Department of the Environment, Water, Heritage and the Arts. Canberra. http://is.gd/k8ljSQ

Chatfield T (2013) Netymology: from apps to zombies – a linguistic celebration of the digital world. Quercus Publishing, London

Clynes ME, Kline NS (1960) Cyborgs and space. Astronautics, pp 24–27, 74–76

Coineau Y, Darmanin C, Guittard F (2015) Superhydrophobic and superoleophobic properties in nature. Materials Today 18(5):273–285

Daly H (2015) Economics for a Full World. Great Transformation Initiative (June 2015).

Díaz S et al (2015) The IPBES Conceptual Framework – connecting nature and people. Curr Opin Environ Sustain 14:1–16

Dicks H (2015) The philosopy of biomimicry. Phil Technol 29:223–243, Springer. doi:10.1007/s13347-015-0210-2. http://bit.ly/2ad6REM

Drack M (2002) Bionik und Ecodesign – Untersuchung biogener Materialien im Hinblick auf Prinzipien, die für eine umweltgerechte Produktgestaltung nutzbar sind. Dissertation an der TU Wien, Austria

Drexhage J, Murphy D (2010) Sustainable Development: From Brundtland to Rio 2012. Background paper prepared for consideration by the High Level Panel on Global Sustainability at its first meeting, 19 September 2010. United Nations, New York.

Eggermont H et al (2015) Nature-based solutions: new influence for environmental management and research in Europe. Gaia 24:243–248

Erdelen WR (2014) The future of biodiversity and sustainable development: challenges and opportunities. In: Lanzerath D, Friele M (eds) Concepts and values in biodiversity. Routledge, London/New York, pp 149–161
Farnham TJ (2007) Saving nature's legacy: origins of the idea of biological diversity. Yale University Press, New Haven/London
Fine PVA (2015) Ecological and evolutionary drivers of geographic variation in species diversity. Annu Rev Ecol Evol Syst 46:369–392
Foerster Hv, Glasersfeld Ev (1999) Wie wir uns erfinden. Carl Auer Verlag, HeidelbergFoerster, H v (1963) Bionics. In: McGraw-Hill yearbook science and technology. McGraw-Hill, New York, pp 148–151
Forbes P (2005) The Gecko's foot. Bio-inspiration: engineered from nature. Fourth Estate, London
Francé RH (1920) Die Pflanze als Erfinder – Franckh'sche Verlagshandlung, Stuttgart (engl. Edition: Plants as inventors. Simpkin and Marshall, London 1926)
Francé RH (1924) Der Begründer der Lebenslehre, Raoul H Francé. Eine Festschrift zu seinem 50. Geburtstag, Heilbronn
Gallagher AJ, Hammerschlag N, Cooke SJ et al (2015) Evolutionary theory as a tool for predicting extinction risk. TREE 30:61–65
Gamble T, Greenbaum E, Jackman TR et al (2012) Repeated origin and loss of adhesive toepads in geckos. PLoS One 7(6):e39429. doi:10.1371/journal.pone.0039429
Gebeshuber IC, Macqueen MO (2014) What is a physicist doing in the jungle? Biomimetics of the rainforest. Appl Mech Mat 461:152–162
Giessler A (1939) Biotechnik. Quelle und Meyer, Leipzig
Gleich A, Pade C, Petschow U, Pissarskoie E (2010) Potentials and trends in biomimetics. Springer, Heidelberg/New York
Goel AK, McAdams DA, Stone RB (eds) (2014) Biologically inspired design. Springer, Heidelberg
Gorb S (2009) Functional surfaces in biology, 2 vols. Springer, Heidelberg
Goujon P (2001) From biotechnology to genomes. World Scientific Publishing Co Pte Ltd. ISBN 978-981-02-4328-9
Gould J (2015) Learning from nature's best. Nature 519:S2–S3. doi:10.1038/519S2a
Gray CH (1995) An interview with Jack Steele. In: Gray (ed) The Cyborg handbook. Routledge, New York, pp 453–467
Gruber P (2011) Biomimetics in Architecture. Springer, Heidelberg
Gruber P (2013) Was macht die Architektin im Dschungel? Bautechnik 90:1–9
Guiry MD (2012) How many species of Algae are there? J Phycol (48)5:1057–1063, doi:10.1111/j.1529-8817.2012.01222.x
Gyllenberg M, Akay A, Hynes M (2012) Nature-inspired science and engineering for a sustainable future. Science policy briefing 44. European Science Foundation, Strasbourg
Haeckel E (1899–1904) Kunstformen der Natur. Leipzig, Wien
Halacy DS (1965) Bionics – the science of living machines. Holiday House, New York
Harkness JM (2001) A lifetime connections – Otto Herbert Schmitt 1913–1998. Phys Perspect 4(4):456–490
Harrison PA, Berry PM, Simpson G et al (2014) Linkages between biodiversity attributes and ecosystem services: a systematic review. Ecosyst Serv 9:191–203
Helmcke JG (1984) Diatomeen, morphogenetische Analyse und Merkmalssynthese an Diatomeenschalen (ein Versuch). In: Bach K, Burkhard B (eds) Diatomeen 1, Schalen in Natur und Technik. Cramer, Stuttgart, pp 10–207
Helmcke JG, Otto F (1962) Lebende und Technische Konstruktionen in Natur und Technik. Deutsche Bauzeitung 67(11):855–861
Hertel H (1963) Biologie und Bauen. Krauskopf, Mainz. (Englisch: Structure-Form-Movement. Reinhold, New York
Hinchliff CE et al (2015) Synthesis of phylogeny and taxonomy into a comprehensive tree of life. Proc Natl Acad Sci U S A 112(41):12764–12769
Hoeller N, Goel A, Freixas C et al (2013) Developing a common ground for learning from nature. Zygote Q 7:134–145

Hwang J, Jeong Y, Park JM et al (2015) Biomimetics: forecasting the future of science, engineering, and medicine. Int J Nanomed 10:5701–5713
IUCN (2010) The World Conservation Union 2010. IUCN Red list of threatened species. Summary statistics for globally threatened species. Table 1: numbers of threatened species by major groups of organisms (1996–2010). http://is.gd/KKIe5l
IUCN (2012) The IUCN programme 2013–2016. Gland, Switzerland
IUCN (2015) Red List version 2015.2, update of 23 June 2015. Retrieved, Gland, Switzerland
Kline RR (2015) The cybernetic moment. Johns Hopkins University Press, Baltimore
Knippers J, Speck T (2012) Design and construction principles in nature and architecture. Bioinspiration and Biomimetics 7. doi:10.1088/1748-3182/7/1/01500
Kolbert E (2014) The sixth extinction: an unnatural history. Henry Holt and Company, New York
Kresling B (1994) Bionics and design: witnesses to the evolution of this approach. ELISAVA (Escola Superior de Disseny) TdB, Barcelona
Lawton J (1993) On the behaviour of autecologists and the crisis of extinction. Oikos 67:3–5
Leadley PW, Krug CB, Alkemade R et al (2014) Progress towards the Aichi biodiversity targets: an assessment of biodiversity trends, policy scenarios and key actions. Secretariat of the Convention on Biological Diversity. Technical series 78. Montreal
LeCointre G, Le Guyader H (2001) Classification phylogenetique du vivant. Belin, Paris
Lettvin JW, Maturana HR, McCulloch WC, Pitts WH (1959) What the frog's eye tells the frog's brain. Proc IRE 47:1940–1951
Lilienthal O (1889) Der Vogelflug als Grundlage der Fliegekunst. Gaertners, Berlin/London/Heidelberg/New York/Dordrecht
Lilienthal G (1924) Biotechnik des Vogelfluges. Voigtländer, Leipzig
Locey KJ, Lennon JT (2016) Scaling laws predict global microbial diversity. www.pnas.org/cgi/doi/10.1073/pnas.1521291113
Maes J, Jacobs S (2015) Nature-based solutions for Europe's sustainable development. Conserv Lett 2015:1–4. doi:10.1111/conl.12216
Merill CL (1982) Biomicry of the dioxygen active site in the copper proteins hemocyanin and cytochrome oxidase. PhD thesis, Rice University, Huston
Millennium Ecosystem Assessment (MEA) (2005a) Ecosystems and human well-being: synthesis, Island Press, Washington, DC
Millennium Ecosystem Assessment (MEA) (2005b) Ecosystems and human well-being: biodiversity synthesis. World Resources Institute, Washington, DC
Mooers AØ (2007) The diversity of biodiversity. Nature 445:717–718
Mora C, Tittensor DP, Adl S et al (2011) How many species are there on earth and in the ocean?. PLoS Biol 9(8). Online. Available:http://is.gd/tlUZmB
Müggenberg J (2011) Lebende Prototypen und lebhafte Artefakte. Die (Un-) Gewissheiten in der Bionik. ilinx 2. http://is.gd/aIAXtU
Müggenberg J (2014) Clean by nature. Lively Surfaces and the Holistic-Systemic heritage of Contemporary Bionik – communication +1 Vol 3, Article 9. doi:10.7275/R5MK69TR
Müller A, Müller K (eds) (2007) An unfinished revolution? Heinz von Foerster and the Biological Computing Laboratory (BCL) 1958–1976. Edition Echoraum, Vienna, pp 277–302
Mumford L (1934) Technics and civilization. Harcourt, New York
Myers N, Mittermeier RA, Mittermeier CG et al (2000) Biodiversity hotspots for conservation priorities. Nature 403:853–858
Nachtigall W (1998) Bionik. Springer, Heidelberg
Nachtigall W (2005) Biologisches design. Springer, Heidelberg
Nachtigall W, Wisser A (2014) Bionics by examples. Springer, Heidelberg
Nagel JKS (2014) A Thesaurus for bioinspired engineering design. In: Goel AK et al (eds) Biologically inspired design. Springer, London. doi:10.1007/978-1-4471-5248-4_4
Nature News (2011) Number of species on Earth tagged at 8.7 million. Published online 23 August 2011. doi:10.1038/news.2011.498

Neinhuis C, Barthlott W (1997) Characterization and distribution of water-repellent, self-cleaning plant surfaces. Ann Bot 79:667–677

Nerdinger W (ed) (2005) Frei Otto, complete works. Birkhäuser, München

Oestreicher H.L (ed.) 1964 Information processing by living organisms and machines. In: Proceedings of the 2nd Dayton bionics symposium, 1963

Oestreicher H, Moore DR (eds) (1968) Cybernetic problems in Bionics. In: Proceedings of the 3rd Dayton bionics symposium, 1966, Gordon and Breach, New York

Pancaldi G (2003) Volta - Science and culture in the age of enlightenment, xx + 381 pp. Princeton University Press

Perera K (2015) Sand, ein knappes Gut. In: Atlas der Globalisierung – Weniger wird mehr. Berlin: Le Monde diplomatique / taz Verlags- und Vertriebs GmbH, pp 72–75

Pichler F (2005) The contribution of Raul Francé: Biocentric modelling. In: Weibel P (ed) Beyond art: a third culture. Springer, Heidelberg, pp 371–375

Pimm SL, Joppa LN (2015) How many plant species are there, where are they, and at what rate are they going extinct? Ann Missouri Bot Gard 100:70–176

Pimm SL, Jenkins CN, Abell R et al (2014) The biodiversity of species and their rates of extinction, distribution, and protection. Science 344:1246752. doi:10.1126/science.1246752

Pohl G, Nachtigall W (2015) Biomimetics for architecture & design. Springer, Heidelberg/New York

Rechenberg I (1965) Cybernetic solution path of an experimental problem. Royal Aircraft Establishment, Library Translation No. 122, Farnborough

Rechenberg I (1971) Evolutionsstrategie: Optimierung technischer Systeme nach Prinzipien der Biologischen Evolution. Dissertation, TU Berlin

Rechenberg I (1994) Evolutionsstrategie'94. Frommann-Holzboog, Stuttgart

Reif WE (1981) Oberflächenstrukturen und Skulpturen bei schnell schwimmenden Wirbeltieren. Paläontologische Kursbücher 1:141–157

Robinette JC (ed) (1961) Living Prototypes – the key to new Technology. – Proceedings of the Bionic Symposium 13.-15. Sept. 1960, Wright Air Development Division (WADD), Dayton, WADD Technical Report No. 60–600, 506 (499 + vii), Dayton, Ohio

Rockström J (2015) Bounding the Planetary Future: Why We Need a Great Transformation. Great Transformation Initiative. Tellus Institute, Boston. http://is.gd/k6zL9c

Roth RR (1983) The foundations of Bionics. Persp Biol Med 26(2):229–242

San Diego Zoo Global and Fermanian Business & Economic Institute Point Loma (Eds) (2013) Bioinspiration: An Economic Progress Report. San Diego: 1–45

Secretariat of the Convention on Biological Diversity (SCBD) (2010) *Global Biodiversity Outlook 3*, Montreal

Secretariat of the Convention on Biological Diversity (SCBD) (2011) Nagoya Protocol on Access to Genetic Resources and the Fair And Equitable Sharing of Benefits Arising from their Utilization To The Convention On Biological Diversity, Montreal

Secretariat of the Convention on Biological Diversity (SCBD) (2014) *Global Biodiversity Outlook 4*, Montreal

Seireg A (1969) Leonardo da Vinci – the bio-mechanician. In: Bootzin D, Muffley HC (eds) Biomechanics. Plenum, New York

Stork NE, McBroom J, Gely C, Hamilton AJ (2015) New approaches narrow global species estimates for beetles, insects, and terrestrial arthropods. Proc Natl Acad Sci U S A 112:7519–7523

Stroud JT, Feeley KJ (2015) A downside to diversity? A response to Gallagher et al. TREE 30:296–297

Tittensor DP, Walpole M, Hill SLL et al (2014) A mid-term analysis of progress toward international biodiversity targets. Science 346:241–244

UN (United Nations) (2015) Transforming our world: The 2030 agenda for sustainable development. United Nations, New York

UNESCO (United Nations Educational, Scientific and Cultural Organization) (2010) Engineering: issues, challenges and opportunities for development. UNESCO, Paris

United Nations (1992) Convention on biological diversity. The United Nations, New York
United Nations (2015) Transforming our world: the 2030 agenda for sustainable development. The United Nations, New York
VDI (2012) Biomimetics – conceptions and strategy. VDI (Verein Deutscher Ingenieure), Richtlinie 6220, Düsseldorf
VDI (Verband Deutscher Ingenieure) and BIOKON (Bionik-Kompetenznetzwerk) (2012) Die Zukunft der Bionik: Interdisziplinäre Forschung stärken und Innovationspotentiale nutzen. VDI und BIOKON: Positionspapier
Vincent J (2009a) Biomimetics in architectural design. In: AD Architectural design, special issue patterns of architecture. Wiley, New York, pp 74–78
Vincent JFV (2009b) Biomimetics – a review. J Eng Med Proc Inst Mech Eng 223(8):919–939
Vincent J (2012) Structural biomaterials. Princeton University Press, Princeton
Vincent J, Bogatyreva OA et al (2006) Biomimetics: its practice and theory. J Roy Soc Interf 3(9):471–482
Volta A (1800) On the electricity excited by the mere contact of conducting substances of different kinds. Phil Trans Roy Soc London 403–431
Wagner T, Neinhuis C, Barthlott W (1996) Wettability and contaminability of insect wings as a function of their surface sculptures. Acta Zool 77(3):213–225
Wahl CD (2015) Bionics vs. biomimcry: from control of nature to sustainable participation in nature; wit design & nature paper. http://bit.ly/29Scjbi
WCED (World Commission on Environment and Development) (1987) Our common future. Oxford University Press, Oxford
Williams HS (1903) Correlation of geological faunas. GPO, Washington, DC
Wilson EO (ed) (1988) Biodiversity. National Academy Press, Washington, DC
World Wide Fund for Nature (WWF) (2014) Living planet report. WWF, Gland
Yan YY, Gao N, Barthlott W (2011) Mimicking natural superhydrophobic surfaces and grasping the wetting process: A review on recent progress in preparing superhydrophobic surfaces. Adv Colloid Interf Sci 80–105. doi:10.1016/j.cis.2011.08

Chapter 4
Insect-Inspired Architecture: Insects and Other Arthropods as a Source for Creative Design in Architecture

Stanislav N. Gorb and Elena V. Gorb

Abstract Materials, structures, surfaces and buildings of insects and other arthropods are of great scientific interest. Moreover, basic knowledge about the functional principles of these structures is also highly relevant for technical applications, especially in architecture. Some of the greatest challenges for today's architecture are multifunctionality and sustainability. Insects have solved these problems during their evolution. Zoologists, entomologists and animal morphologists have collected a huge amount of information about the structure and function of such living constructions and surfaces. This information can be utilized in order to mimic them for applications in architecture. The main technology areas, in which insect solutions to problems can be applied, are the following: (1) new materials, (2) constructions, (3) surfaces, (4) adhesives and bonding technology, (5) optics and photonics. A few selected examples are discussed in this chapter, but having more than one million described species as a source for inspiration, one may expect many more ideas from entomology for insect-based biomimetics in architecture. The incorporation of additional biological knowledge into the design of artificial systems will improve their performance. However, biologists still do not have a complete understanding of how insect materials are constructed, what their performance is, how insect surfaces function, etc. Hence, many technological areas will benefit from additional entomological research. Additionally, most of the huge variety of insects and their systems have been not previously studied at all. This is the reason that the screening for new systems with interesting properties in biology seems to remain an extremely important research field in the near future.

S.N. Gorb (✉) • E.V. Gorb
Department of Functional Morphology and Biomechanics, Zoological Institute of the Kiel University, Am Botanischen Garten 9, D-24118 Kiel, Germany
e-mail: sgorb@zoologie.uni-kiel.de

J. Knippers et al. (eds.), *Biomimetic Research for Architecture and Building Construction*, Biologically-Inspired Systems 9, DOI 10.1007/978-3-319-46374-2_4

4.1 Introduction

Insects are among the most diverse groups of animals on Earth, including more than a million described species and representing more than half of all known living organisms (Chapman 2006). The number of extant species is estimated to be between six and ten million (Erwin 1982, 1997; Novotny et al. 2002; Chapman 2006) and approximately represents over 90 % of animal life forms (Erwin 1982). Insects can be found in nearly all environments. During their evolution, insects and related arthropods have evolved a huge variety of shapes and structures. Despite often looking miniature and fragile, they can nonetheless deal with extreme mechanical loads. Many functional systems responsible for their evolutionary success are based on a variety of ingenious materials and structural solutions. The rich sensory equipment of insects, including their compound eyes, chemoreceptors, mechanoreceptors and infra-red (IR) receptors, taken together with their rather compact brain reveals self-adaptive control patterns and often supports remarkable behavioral features.

Studies revealing the functional principles of insect structures, materials, sensors, actuators, locomotion and control systems are, on the one hand, of major scientific interest, since we can learn about the mechanisms behind the structures and their biological roles. On the other hand, this knowledge is also highly relevant for various engineering applications including those in the field of architecture. Two of the greatest challenges for today's architectural designs are energy saving and sustainability. During their evolution, insects and other arthropods have solved many problems dealing with their external lightweight skeleton. Additionally, these animals build a number of remarkable constructions based on silk, glue and surrounding non-living materials (Hansell 2007; Gruber 2011). Zoologists, entomologists and morphologists have collected a vast amount of information about the structure and function of such animal-made buildings and about the materials and structures utilized in their bodies. This information can be used to mimic them for diverse applications in modern architecture. The main fields, in which insect-inspired solutions can be applied to architecture, are as follows: (1) new materials, (2) constructions, (3) surfaces, (4) adhesives and bonding technology, (5) optics and photonics (Fig. 4.1). Possible innovations might also appear on the boundary between insect science and the areas listed above (Gorb 2011). Some selected examples are discussed in this chapter, but with more than one million described species (about one half of the eukaryotic organisms) as a source for inspiration, many more ideas might arise from the study of insects to be used for biomimetics in architecture.

4.2 Buildings and Constructions of Insects

Among the animals capable of building, social insects have the largest workforces to generate complex constructions. Large colonies of social insects with their great overall behavioural complexity build enormous architectures with channels and spaces permeating them and bringing oxygen to their core and carrying carbon

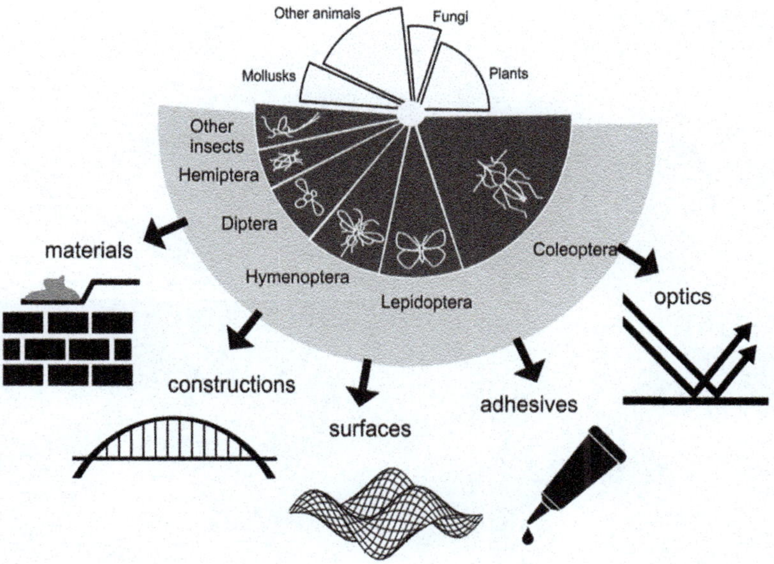

Fig. 4.1 Diagram demonstrating insect diversity (*dark grey* segment of the pie) as a source of biomimetic ideas for application in architecture (Adapted from Gorb 2011)

dioxide away (Hansell 2007). In termite mounds, the multitude of chambers of the living area is linked by apertures and short corridors that can be regarded as the capillaries of the circulation system. These structures produced by termites, for example, have provided inspiration for solar-driven thermoregulating ventilation systems in Europe and Africa. One recent well known example is the ventilation system designed by the company Arup for the East-Gate Hall in Harare, Zimbabwe (Pohl and Nachtigall 2015).

Honeycomb structures occur often in nature and can be actively constructed by mostly social insects such as bees and wasps from completely different materials, e.g. wax and paper. This geometrical pattern is well known to represent the most densely packed units in two-dimensional space, which is an interesting principle for its implementation in architecture. Technical honeycomb structures can be made of a stronger variety of materials, such as plastics, ceramics and metals, and by using a diversity of processing techniques, such as the cutting of hexagonal sheets and subsequent gluing, the insertion of strips of glue between the sheets and subsequent stretching or the application of moulding techniques, especially when polymers are used (Gruber 2011; Pohl and Nachtigall 2015). Honeycomb-patterned lightweight materials can be used in architecture for the core of sandwich panels and composite designs. Because of their large surface area, they are interesting for applications as cooling structures.

For example, Dr. Mirtsch GmbH has developed a broad variety of three-dimensional (3D) metal tins having various thicknesses of sheaths. These structures often originate from a highly specific method of production by using self-

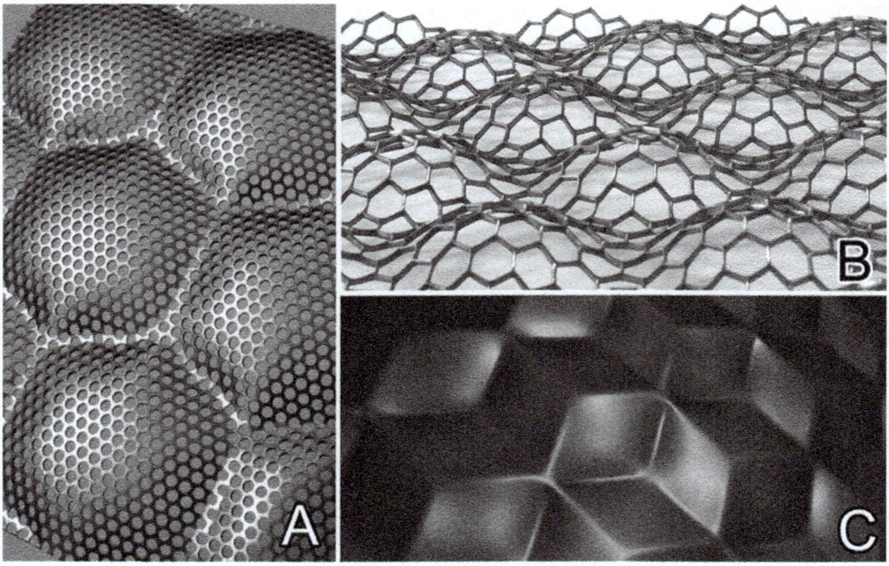

Fig. 4.2 Technical honeycomb structures for use in architecture. Dr. Mirtsch GmbH, Courtesy of Frank Mirtsch (www.woelbstruktur.de)

organization. The structures are rather impressive not just because of their beautiful optical appearance, but mainly due to their convincing technical properties (Fig. 4.2). Among their functional features, the most striking are (1) much stronger bending stability at much lower material use (ultra-lightweight), (2) enhanced crash-resistance properties attributable to stronger energy dissipation, (3) further enhancement of bending stiffness because of their combination into sandwich-like structures, and (4) their ability for combination with other, also dissimilar, materials.

Numerous honeycomb constructions can be found in external facade architecture (Gruber 2011; Pohl and Nachtigall 2015). However, honeycombs are also widely used in interior architecture, such as The Hive (Ben Huckerby Design by Kyle Minnock) (Minnock 2016). The sources of inspiration for this design were the structures, forms and characteristics associated with insects. The project involved not only in-depth research into the structures of honeybee combs, but also a large amount of practical experimentation with card models and other craft materials. Experimentation with light, form, colour and texture was also carried out to achieve the final concept (Minnock 2016) (Fig. 4.3).

Many insects (Lepidoptera, Trichoptera, Hymenoptera, Neuroptera) and other arthropods construct silky shells called cocoons (Fig. 4.4a, b). Usually, larval insects do this in order to additionally protect their pupal stage. In insects, silk is usually produced by labial glands located in the region of the mouthparts. Cocoons can be soft or rigid, dense or loose, and have diverse colours and numbers of layers. Cocoons are also used by spiders, but mainly for egg clutch protection. Spider cocoons are typically two-layered: the inner layer is softer, whereas the external one is more rigid. Additionally, orb-web spiders wrap their prey items in cocoon-like

Fig. 4.3 The Hive, insect-inspired interior architecture by Kyle Minnock, Leeds, United Kingdom (Minnock 2016)

shells to prevent prey mobility, especially, when the prey is larger than the predator itself.

The water spider *Argyroneta aquatica* spends most of its life under water. It constructs a special reinforced kind of web construction that holds an air bubble underwater and provides a safe and stable habitat for the spider. First, the spider builds a horizontal sheet web, under which the air bubble is placed. In a further step, the air bubble is sequentially reinforced by the spider laying a hierarchical arrangement of fibres from within (Fig. 4.4c). Such a two-layered construction is stable and can withstand mechanical stresses caused by water currents.

The underwater construction of the water spider and its natural production process show the way that adaptive fabrication strategies can be utilized to create efficient fibre-reinforced structures. The ICD/ITKE Research Pavilion 2014–2015 was inspired by this biological prototype. The pavilion demonstrates the architectural potential of a novel building method. A new robotic fabrication process was developed by using an initially flexible pneumatic formwork that was gradually stiffened by reinforcement with carbon fibres from the inside (Fig. 4.4d–g). The resulting lightweight fibre composite shell forms a pavilion with unique architectural qualities.

This design concept is based on the study of biological construction processes for fibre-based structures. In biology, fibre-reinforced structures are made in a highly material-effective and functionally integrated way. The prototypical project at the ICD/ITKE was the result of 1.5 years extensive development by researchers and students of architecture, engineering and natural sciences. The process developed in this project is of high relevance for applications in architecture, as it does not require complex formwork and is capable of being adapted to the various demands of individual constructions.

In order to transfer the biological way of building to the application in the architecture, a novel process was developed: an industrial robot was placed within an air-supported membrane envelope made of the polymer ethylene tetrafluoroethylene (ETFE). This inflated soft shell was initially supported by air pressure and gradually reinforced from inside with carbon fibre by using the robot (Fig. 4.4d, e). The

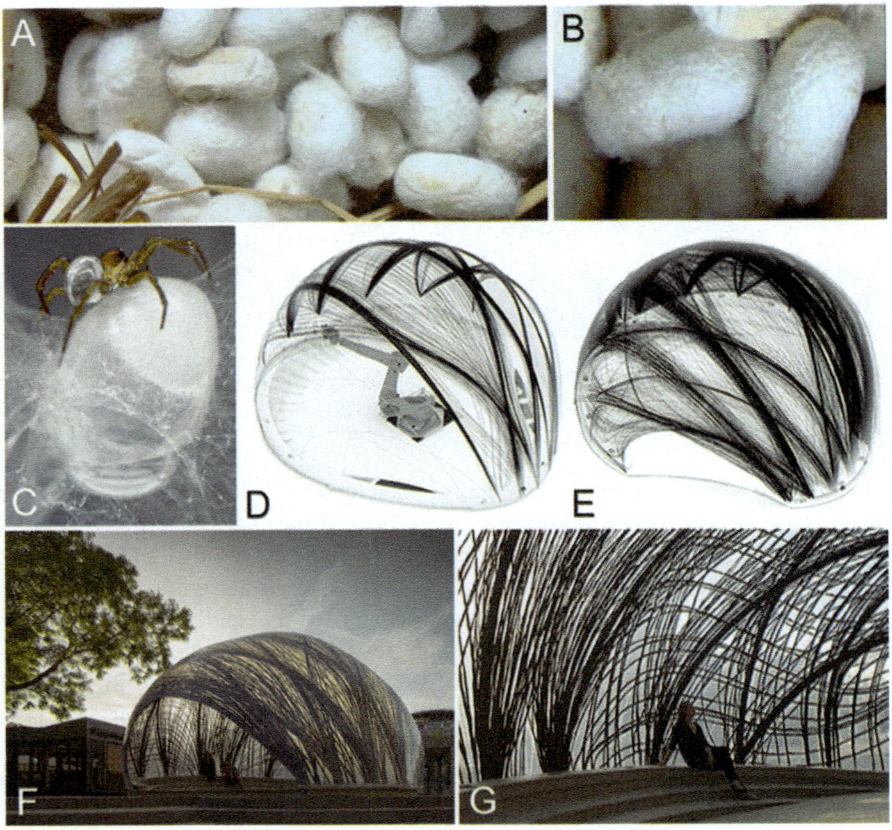

Fig. 4.4 Silk-based arthropod constructions and the use of similar concepts in architecture. (**a**, **b**) Cocoons of the silkworm *Bombyx mori*. (**c**) Air-filled bell-like underwater construction of the water spider *Argyroneta aquatica*. (**d**–**g**) The ICD/ITKE Research Pavilion 2014–2015 designed by the Institute for Computational Design (ICD) and the Institute for Building Structures and Structural Design (ITKE) at the University of Stuttgart was inspired by the silky construction of the spider. (**c**–**g**) Courtesy of ICD and ITKE, Stuttgart, Germany

construction gradually stiffened into a self-supporting monocoque structure. The carbon fibres were only selectively applied, wherever they were required for structural reinforcement. The pneumatic formwork was used at the same time as a functionally integrated building skin.

4.3 Arthropod Skeleton Is Fibre-Reinforced Composite

The continuous external skeleton (exoskeleton) of arthropods is made of chitin fibres embedded in a protein matrix (Hepburn 1985). The chemical, structural and mechanical properties of such a composite material can vary to a large extent and, thereby, allow local functional adaptations in the different areas of the insect's body (Hepburn and Chandler 1976, 1978; Vincent and Wood 1972; Vincent 1981).

The arthropod exoskeleton is an interface between a living animal and the environment and, therefore, serves many functions. (1) It limits the dimensions of the exoskeleton and is a basis for muscle insertion (mechanical function and function of locomotion). (2) It is an important element in organism defence against a variety of external factors, such as mechanical stress and dry, wet, cold or hot environments. (3) It takes part in the transport of diverse epidermal secretions and serves as a chemical reservoir for the storage of metabolic waste products. (4) A variety of cuticular structures are parts of mechano- and chemoreceptors. (5) The cuticle, its coloration pattern and chemical components are important for thermoregulation and are often involved in diverse communication systems. (6) Specialized cuticular protuberances might serve a variety of functions, such as oxygen retention, food grinding, body cleaning (grooming), etc. (see below).

Cuticle is a layered composite material (Fig. 4.5) that consists of two principal components: chitin and protein. The arrangement of chitin molecules usually varies within its different morphs. Chitin microfibrils are always associated with protein in a chitin-protein complex. Chitin-bearing cuticular microfibrils have a complex pattern of orientation in the 3D space of the cuticle. Some types of cuticle are rather stiff and others, such as arhtrodial membranes, are flexible.

Arthrodial and caterpillar-like cuticles are materials of great flexibility, great extensibility and reasonable strength. The interaction between microfibrils and protein matrix is very loose here. Under tension, these cuticles are visco-plastic and show so-called «necking», as do steel specimens under the same conditions

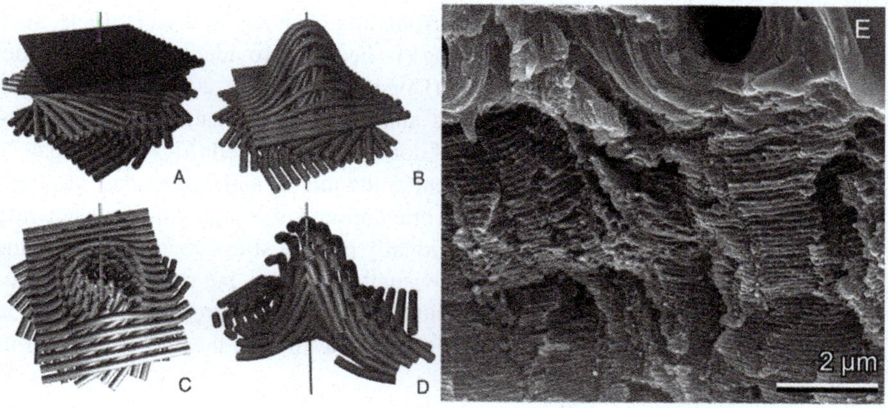

Fig. 4.5 Insect cuticle structure. (**a–d**) Model of the orientation of the microfibrils within one lamella in the helicoidal flat cuticle (**a**) and in the helicoidal cuticle of a surface outgrowth (**b–d**) (**b**) The whole model of fibre orientation in such a cone helicoid. (**c**) Transverse section of the cone helicoid. (**d**) Longitudinal section of the cone helicoid (Gorb 1997a, b). (**e**) Fracture through the cuticle containing surface microstructures at the rear side of the head in the dragonfly *Aeshna mixta*. Note the layered structure of the exocuticle, which forms such a cone helicoid as depicted in **b–d** (Gorb 2011)

(Hepburn and Chandler 1976). However, remarkable variation exists with regard to the extent, to which such cuticles can be extended (Vincent and Wood 1972). Some of these membranous cuticles are strongly folded, such as those found in the abdominal membranes of the tsetse fly *Glossina morsitans* (Hackman and Goldberg 1987) or in the neck membrane of adult dragonflies (Gorb 2000).

Insect cuticle demonstrates, in various functional systems, a gradient of material properties that can range from very stiff areas of the condyli of joints to membranous areas between leg segments. These gradients depend on the fibre density, fibre orientation, polymerization degree of the matrix and thickness of single layers.

Currently, composite materials are widespread in various areas of technology. However, the fibre diameter remains at the range of few micrometers. They also normally have random or preferable orientation within the matrix. Insect cuticle may inspire engineers, for example, with its helicoidal arrangement of fibrils in successive layers and with its gradient-like materials properties. Moreover, fibre orientation in insect cuticle often depends on the local structure geometry (Figs. 4.5 and 4.6a, b). Natural materials, therefore, consist of only a few basic components that are geometrically, physically and chemically differentiated and, in this respect, they are fundamentally different from most architectural constructions (Knippers and Speck 2012). This is the reason that the specific fibre orientations found within insect structures might inspire novel constructions in architecture.

A good example of local geometry-dependent fibre orientation is the architecture of the pore channel (Fig. 4.6a) used by architects and engineers as an inspiration for novel types of building construction. The ICD/ITKE Research Pavilion 2013–2014 (Dörstelmann et al. 2014; Parascho et al. 2014) is the result of multidisciplinary research. A strategic pre-selection of biological models based on natural lightweight fibre structures with specific anisotropic features preceded the project (Fig. 4.6a, b). Natural fibre composite structures, surrounding channels in insect cuticle, such as those of the elytra (protective forewing) in beetles, provided a versatile model for performative lightweight structures (Dörstelmann et al. 2014). The material efficiency of the system is based on the anisotropic organization of chitin fibre composite material, which forms a kind of double-layered shell (Figs. 4.6c and 4.7).

Biological architecture, which is present at the micro- and nanoscales, has been translated to the macroscale. Various fibre orientations surrounding the hole not only stabilize the entire construction, but additionally prevent stress concentration in the hole. This kind of architecture not only is visually beautiful, but provides enormous stability at the minimum material expenditure in spite of the presence of holes. On the material scale, structural integration is achieved through the use of continuous fibre.

4.4 Folding Mechanisms

Insect wings are usually thin and delicate outgrowths of the body wall. In order to prevent their damage, some groups, such as beetles, earwigs or bugs, have fore wings that serve a protective function and are therefore thickened and stiff. The hind

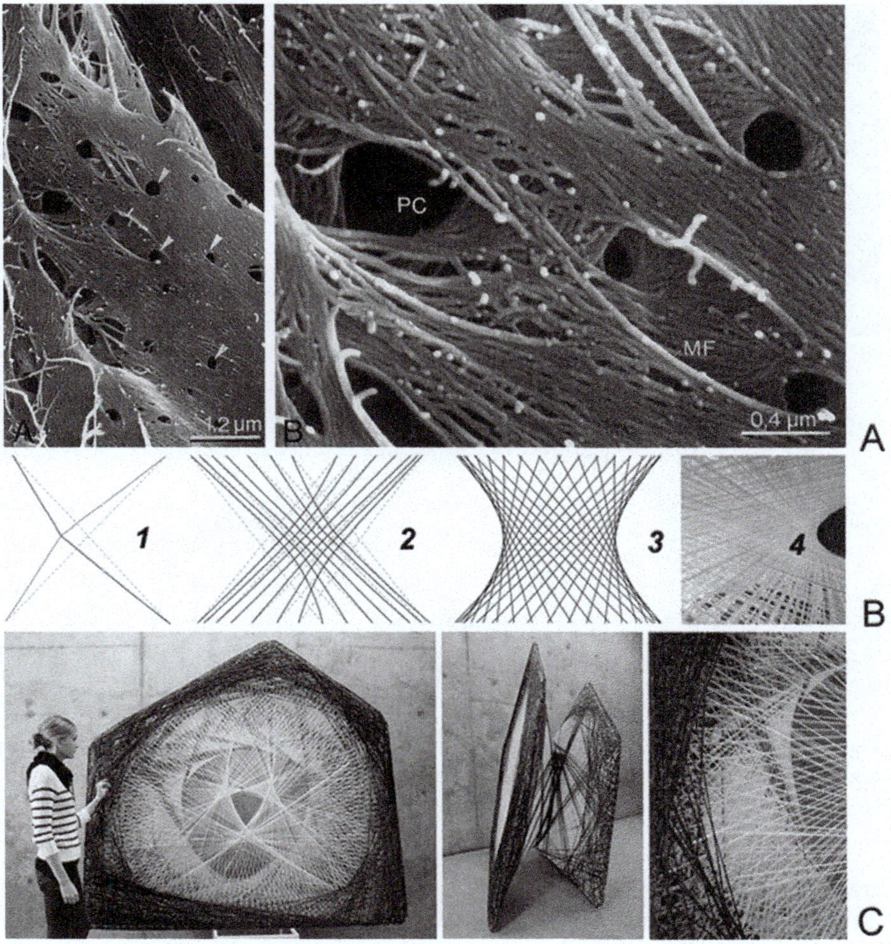

Fig. 4.6 (**a**) Microfibril orientation around the porous channels in the insect cuticle (scanning electron micrograph): *MF* microfibrils, *PC* and *white arrowheads*, porous channels (Gorb 1997a, b). (**b**) Coreless winding: (*1*) two fibres connecting four non-planar points; (*2*) fibres deforming under the tension of subsequently wound fibres; (*3, 4*) anticlastic curvature induced by helicoidal winding. (**c**) Single component and fibre detail (Dörstelmann et al. 2014; Parascho et al. 2014). (**b, c**) Courtesy of ICD and ITKE, Stuttgart, Germany

wings must exhibit a certain area in order to be aerodynamically functional and they are indeed larger than the thickened forewings. The only possibility for the hind wings to be completely covered by the fore wings is to be folded (Haas et al. 2000a) (Fig. 4.8a–c).

The folding pattern depends on the wing venation pattern and the material properties of the structures involved. Consequently, the morphology of wings in insects with an additional folding function differs from the wings without the folding capability. The design of foldable wings is a compromise between flight and folding (Haas et al. 2000b). For example, the hind wings of earwigs (Dermaptera) are strongly folded and covered by the small, thickened and sclerotized fore wings. The

Fig. 4.7 ICD/ITKE Research Pavilion 2013–2014 (Dörstelmann et al. 2014; Parascho et al. 2014) inspired by fibre orientation around pores in insect cuticle. Courtesy of ICD and ITKE, Stuttgart, Germany

area of the unfolded wing is ten times larger than that of the folded wing (Haas et al. 2000a). The folding pattern in earwigs is rather complicated and is made possible by the combination of muscle activity, the pleating pattern, and the specific distribution of resilin, a rubber-like protein, within the folds (Fig. 4.8d).

A specific pleating pattern is responsible for high elasticity of the wing. Three-dimensionally pleated wings have an asymmetric torsional rigidity (Wootton 1991). Another mechanism is based on a gradient-like distribution of the rigid and soft resilient materials of the thin membranous areas of wing cells (Haas et al. 2000a, b).

Pleated structures are widely used in architecture (Gruber 2011) because of their flexural rigidity and lightweight properties. However, the number of real functional folding constructions in architecture is not that high. One great example of such a kinematic structure is the three-field Bascule Bridge in Kiel Horn, Germany (Fig. 4.9a–c). The bridge is moved through a complex cable system of numerous ropes, winches and rollers and, simultaneously, each change to the construction's position has to be stabilized to counteract wind loads potentially coming from all directions (Knippers and Schlaich 2000; Knippers and Speck 2012). The architectural intent was to incorporate real folding mechanics and its functionality. However, this bridge is a unique item: it was planned and built

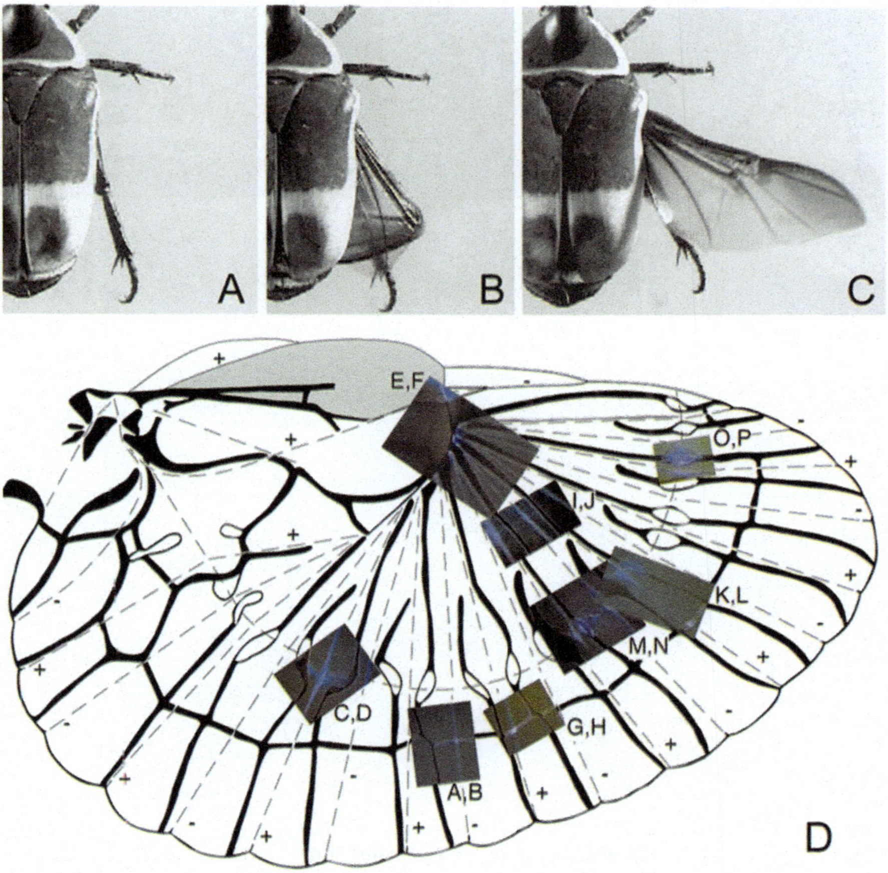

Fig. 4.8 Insect wings folding structures. (**a–c**) Beetle *Pachnoda marginata*, single frames from a video sequence of the wing unfolding (from Haas et al. 2000b). (**d**) Distribution of resilin, a rubber-like protein, in a hind wing of the earwig *Forficula auricularia*. Blue autofluorescence shows the presence of resilin (From Haas et al. 2000a)

without any reference projects or prototypes (Knippers and Speck 2012). This means that the design of the bridge was not inspired by biology, although an increasing amount of data from the mechanics of biological folding structures might help to improve the robustness, reliability and appearance of such structures in architecture.

Another interesting example of folding structures is the Zoomlion Headquarters Exhibition Center located in the city of Changsha, Hunan Province, China (Fig. 4.9d–f) (AmphibianArc 2012). The total building height is 26 m. Since Zoomlion is one of China's leading manufacturers of heavy machinery equipment, one of the criteria for the design of its exhibition centre was the incorporation of nature-inspired mobility into its architecture.

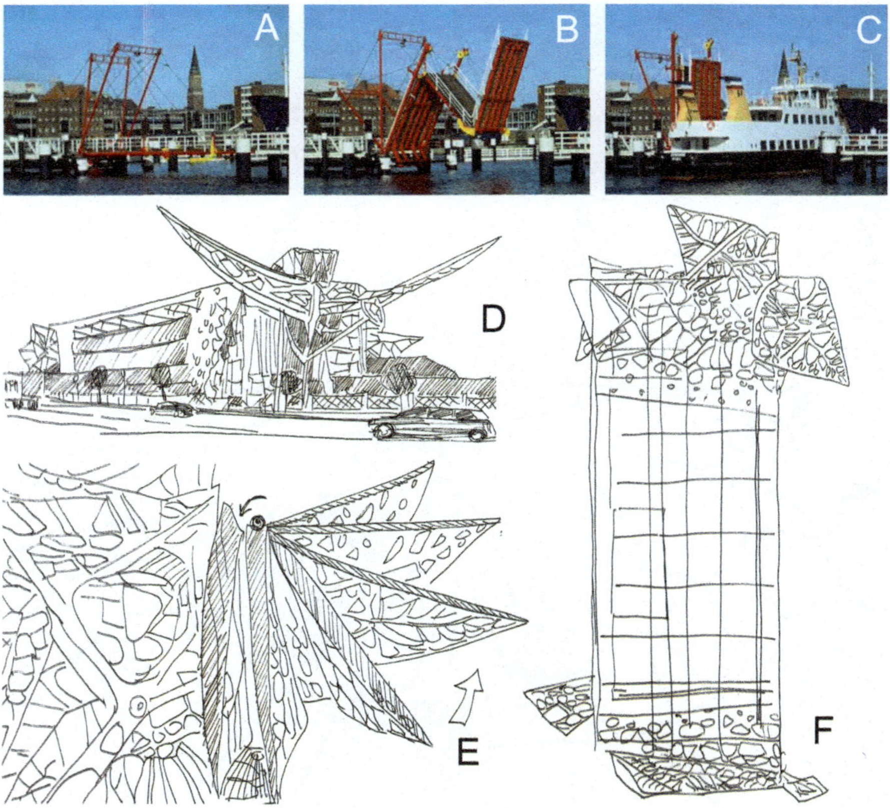

Fig. 4.9 Folding structures in architecture. (**a–c**) Folding bridge in Kiel (1998). Architect: von Gerkan Marg and Partner; Engineer: Schlaich Bergermann and Partner (photo: Klaus Frahm, from Knippers and Speck 2012). (**d–f**) The project Zoomlion Headquarters Exhibition Center for the city of Changsha (Hunan Province, China) by amphibianArc (AmphibianArc 2012)

The skin of the building is made of steel and glass. The most unique aspect of the project is the building's ability to change shape. The double skin system throughout the building is responsible for this transformability. The inner skin takes care of the enclosure and building systems, whereas the outer one contains controllable portions that can be opened or closed to mimic various animal forms, including that of a butterfly (AmphibianArc 2012). This reflects the company's philosophy of maintaining a balance between nature and technology. The façade is inspired by the wing constructions of insects such as butterflies and dragonflies. To achieve the systematic and organic nature of the patterns found on the wings of these insects, the architects used parametric modelling tools to generate and design the façade (AmphibianArc 2012).

4.5 Surfaces and Textures

Insect surface structures can serve many different functions, such as air retention, food grinding, body cleaning, etc. (Fig. 4.10). Some examples of these surfaces seen in a scanning electron microscope are given in Fig. 4.11.

Because of the structural and chemical complexity of insect surfaces, exact working mechanisms have been studied in only a few systems. Due to a broad diversity of surface functions, inspirations from entomology are currently in focus in a broad range of research topics in the engineering sciences including adhesion, friction, wear, lubrication, filtering, sensorics, wetting phenomena, self-cleaning, anti-fouling, thermoregulation, optics, etc. Since insect surfaces are multifunctional, it makes them even more interesting from the perspective of potential applications in architecture. In Fig. 4.11, we see a rather unspecialized polygonal surface on a beetle tarsus (a), anti-reflectors in the fly (b), self-cleaning scales on the dorsal surface of the beetle (c–d), drag-reducing wing surface of the fly (e), food filter of the fly (f), respiratory filter of the beetle (g), grinding teeth in the fly (h) and air retaining coverage in water bugs (i, j) (Gorb 2011).

One of the challenges in the design of moving parts is the fabrication of joints allowing the precise motion of parts about rotational axes. An important problem in any type of mobile joint is the high friction and wear rate. Wear of the interacting surfaces is a consequence of friction and affects the material's contact points by their becoming deformed or being torn away. Friction and wear are strongly correlated processes, by which the points of the surfaces that are in contact change their topography continuously. Conventional methods of lubrication cannot always be used. Friction reduction in some man-made mechanical systems is based on the different hardness of the elements in contact (Miyoshi 2001, Li et al. 2004), on

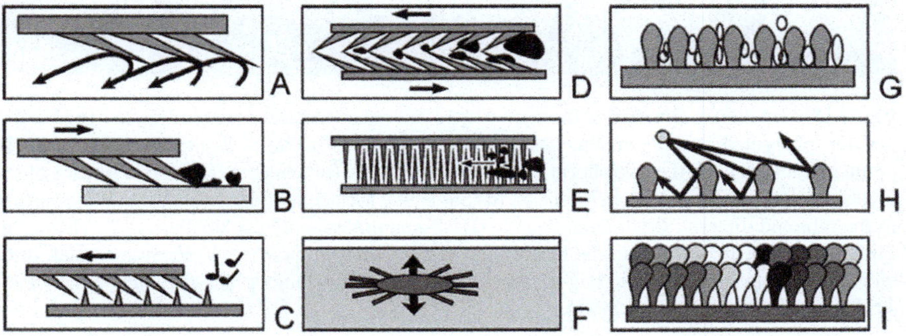

Fig. 4.10 Diagram of functions of cuticular microstructures in insects. (**a**) Aerodynamically active surfaces. (**b**) Grooming. (**c**) Sound generation. (**d**) Food grinding. (**e**) Filtration. (**f**) Hydrodynamically active surfaces. (**g**) Air retention. (**h**) Thermoregulation. (**i**) Body coloration pattern (Gorb 2001)

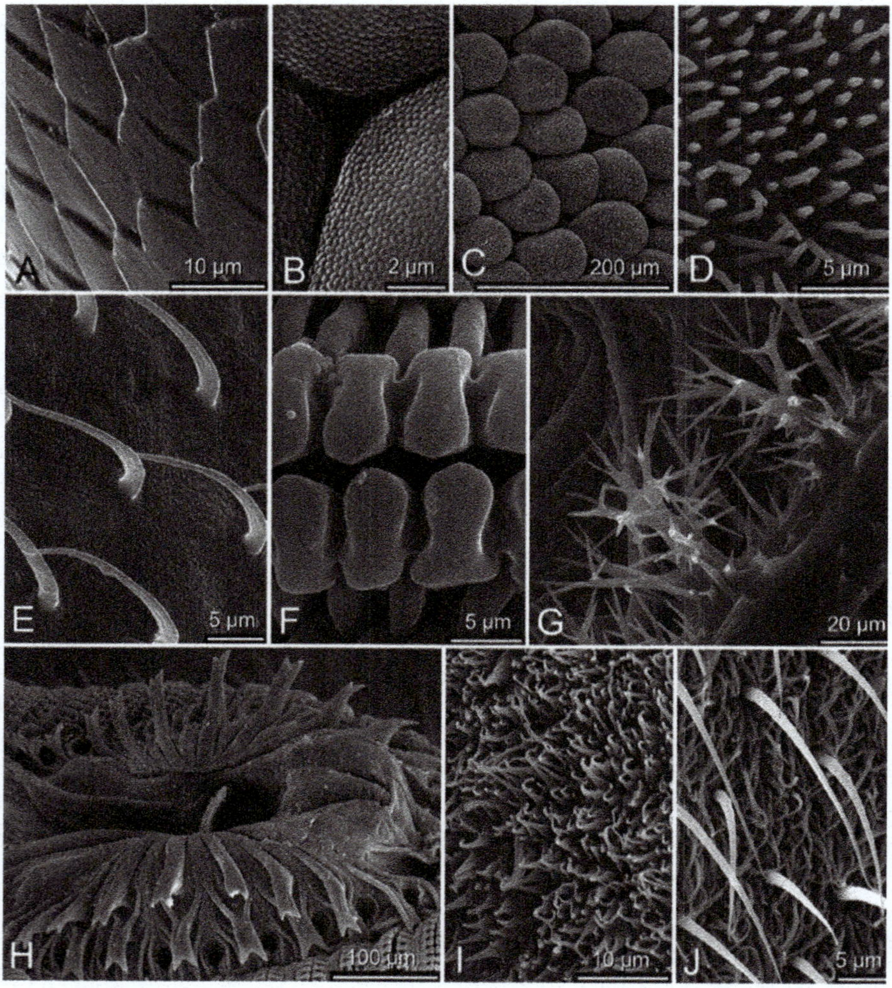

Fig. 4.11 Functional diversity of non-innervated cuticular protuberances in insects. (**a**) Unspecialized polygonal surface on the tarsus of the scarabaeid beetle *Melolontha melolontha*. (**b**) Ommatidial surface in the calliphorid fly *Calliphora vicina*. (**c**) Scales on the dorsal surface of the elytron in the scarabaeid beetle *Hoplia sp*. (**d**) Same; the surface of a single scale. (**e**) Wing surface in the bibionid dipteran *Bibio ferruginatus*. (**f**) «Pseudotrachea» of the labellum in *C. vicina*. (**g**) Filter system of the spiracle in the tenebrionid beetle *Tenebrio molitor*. (**h**) Prestomal teeth in *C. vicina*. (**i**) Plastron in the nepid bug *Ranatra linearis*. (**j**) Air-retaining hair coverage in the water-strider *Gerris lacustris* (Gorb 2011)

the use of hydrophobic surfaces, and on the application of surface texture, which minimizes the real contact area between two solid surfaces. Ideas from studying the surface properties of insect joints, which have stiffer outer layers located onto softer ones (Fig. 4.12b, c, e), the specialized microstructure covering the contact

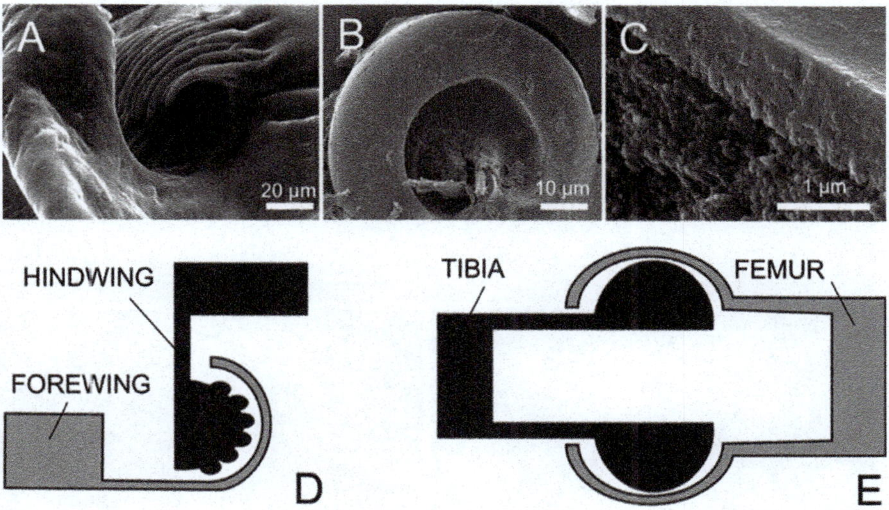

Fig. 4.12 Examples of micro-joints in insects. (**a**) Lateral view of the wing double-wave locking mechanism in the bug *Coreus marginatus* (forewing part). This is a monoaxial sliding joint that provides interlocking between both wings on the same side of the body in the anterior direction allowing them to slide in the medial and lateral directions (Perez Goodwyn and Gorb 2004). Sliding is possible along the axis perpendicular to the plane of the image **a**, whereas the motion is restricted in all directions in the plane of the image. (**b**) Medial aspect of the monoaxial rotating femoro-tibial joint (femoral part) of the leg in the beetle *Melolontha melolontha*. Rotational movement is possible about the axis perpendicular to the plane of the image **b**, whereas the motion is restricted in all directions in the plane of the image. (**c**) Fracture of the material of the joint in the beetle *M. melolontha*. (**d**) Diagram of the sliding joint shown in **a**. (**e**) Diagram of the femoro-tibial joint shown in **b** and **c** (Gorb 2011)

pair (Fig. 4.12a, d), and the particular fibre orientation in exocuticle might provide an interesting set of principles leading to a solution of this problem (Perez Goodwyn and Gorb 2004; Barbakadse et al. 2006).

In insect joints working under lower loading forces, but much higher frequencies than vertebrate joints (Wootton and Newman 1979; Gronenberg 1996), the joint surfaces usually present a combination of wavy and smooth counterparts (Fig. 4.12a, d).

Insect surfaces are often covered by a superhydrophobic (non-wettable by water) cuticle that has an external layer consisting of both cuticle microstructure and/or epicuticular waxes. The layer may contain wax projections with dimensions ranging from hundreds of nanometers to micrometers (Fig. 4.13). The roughness of such surfaces together with their hydrophobic properties decreases wettability, which is reflected in a greater contact angle of water droplets on such surfaces compared to smooth surfaces of the same chemical composition. In some plant surfaces, this property results in their ability to be cleaned by rolling drops of water (Barthlott and Neinhuis 1997, 1998). Similar insect structures, such as those on the wings

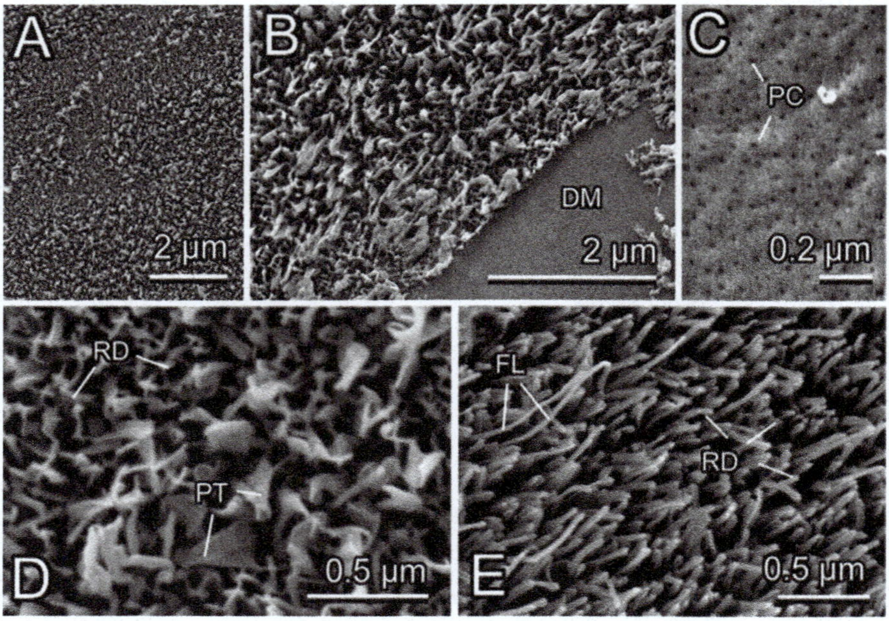

Fig. 4.13 Dorsal wing surface of the dragonfly *Calopteryx splendens*: young mature male surface of wing membrane covered with crystalline wax (**a**, **d**, **e**); wing membrane surface with damage (scratches) in the wax crystalline layer (**b**, **c**). *DM* damage/scratch, *FL* wax filaments, *PC* porous channels of cuticle, *PT* wax platelets, *RD* wax rods (Gorb et al. 2009)

of representatives of Odonata, Ephemeroptera and Neuroptera, are extremely non-wettable and self-cleaning (Wagner et al. 1996). Superhydrophobic coatings are widespread nowadays in modern architecture. One of the most recent examples is the so-called Lotus House in Daegu, South Korea, designed by smart architecture (Daegu) (Baunetz 2015). The name actually originates from the perforated walls resembling the lotus flower, but the self-cleaning superhydrophobic surface completes the impression of this famous plant.

Many aquatic and semiaquatic arthropods have sculptured surfaces involved in holding air under water for respiration. Such surfaces called plastrons usually contain fields of microtrichia, i.e. very small cuticle protuberances (Heckmann 1983). These structures appear convergently as an adaptation to aquatic environments in various arthropod taxa: Collembola, Lepidoptera, Coleoptera, Heteroptera, Diptera, Araneae and Diplopoda (Thorpe and Crisp 1947; Hinton 1976; Messner 1988). Some terrestrial insects, such as Aphididae (Auchenorrhyncha), also bear similar features in the form of bristles, mushroom-like spines or stigmal plates, which can protect their surfaces from moisture (Heie 1987). In water striders and some spiders, the anti-wetting surfaces of their legs and the ventral body side are involved in the locomotion mechanism of walking on water.

Dragonflies, which can spend quite a long time underwater and take off directly from the water surface, have elaborate stable superhydrophobic coatings on their entire body (Gorb et al. 2009) (Fig. 4.13). However, the most stable air-holding surfaces are known from strongly specialized aquatic bugs from the genera *Halobates* (Gerridae) and *Haloveloides* (Veliidae), the only open sea water dwellers (Perez Goodwyn 2009). Representatives from the genus *Halobates* have outstanding water protection structures. The microtrichia pile is composed of specialized prolongations. Each microtrichium has a thickened head, which is several times wider than the shaft and usually tilted to one side like a golf club. On the shaft itself, up to four perpendicular branches interlock the microtrichia, the shafts of which are 0.8–1.1 μm apart. Such surfaces, which can prevent wetting for a long time, are of interest for underwater buildings (Mazzoleni 2013) and in naval architecture (Tupper 2013).

Surface outgrowths can provide the multi-level reflection of sunlight. Such an ability of wing scales has been suggested to be an adaptation for cooling in butterflies (Grodnicky 1988). Body coverage by bristles, scales and hairs in the honey-bee *Apis mellifera* might be used for warming up (Southwick 1985). In species of curculionid beetles of the genus *Tychius*, which inhabit arid areas, cuticular scales have been suggested to be responsible for maintaining thermal balance (Karasev 1989). A water-loss-preventing function has been proposed for the leaf-like bristles at the body margins in Aphididae (Auchenorrhyncha) (Heie 1987).

The surfaces of some desert insects are covered with hydrophobic wax projections that presumably decrease water evaporation through the cuticle and aid in water collection by condensation (Parker and Lawrence 2001). This biomimetic idea is realised in the WarkaWater tower, which has been designed for the Ethiopian landscape (Bamboo Tower 2015). The bamboo-shaped building is designed to harvesting water out of the air, thereby providing a sustainable source of H_2O for developing countries. Created by Arturo Vittori and his team at Architecture and Vision, the tower can harvest water from rain, fog and dew. The WarkaWater functions by using mesh netting to capture moisture and to direct it into a hygienic holding tank accessed via a spout.

Insects provide an enormous amount of interesting textures for building design. Many 3D patterns, such as those observed on the covering wings of the carabid beetles, are almost an unexplored source of inspiration for design in architecture (Fig. 4.14).

Functional surfaces in architecture have many functional requirements that can be fulfilled by using ideas from biology. Since insects bear a huge variety of such microstructures, many of which have not even been previously described, a systematic approach to insect surface science would be highly desirable. An important step in this direction is the establishment of a database of insect functional surfaces. Furthermore, we need more experimental studies targeted to understanding the relationship between structure at various levels of organization and function. We believe that the enhancement of the pool of new ideas from biology will provide a great leap forward with regard to the surface technology of tomorrow.

Fig. 4.14 Surface textures of beetles from the family Carabidae for potential implications in architecture and design

4.6 Photonics

Ommatidial gratings are anti-reflective structures on the eyes of insects, especially those that are nocturnally active (Fig. 4.15). These protuberances are very small microtrichia (200 nm in diameter) that increase visual efficiency through decreased surface reflection in their density and increased photon capture for a given stimulus condition (Parker et al. 1998; Vukusic and Sambles 2003). Such a grating is particularly useful on a curved corneal surface, since it increases the transmission of incident light through the cornea compared to a smooth surface. For an increase in transmission and reduced reflection, a continuous matching of the refractive indexes n_1 and n_2 at the boundary of both adjacent materials is highly critical (Bernhard et al. 1965).

Grooming is an extremely important function for insects, some of which live in extremely dirty or dusty environments. Their rich sensory equipment of eyes and antennae has to be kept clean in order to be able to respond adequately to external signals. Many insects bear specialized cleaning structures (Schönitzer and Lawitzky 1987; Francouer and Loiselle 1988). Others rely on micro- and nanostructured surfaces with an anti-adhesive function. Ommatidial gratings are a multifunctional surface that employs self-cleaning by means of a real contact surface reduction mechanism (Peisker and Gorb 2010).

Structural coloration attributable to the presence of scales and bristles is well known in insects, such as butterflies (Ghiradella 1989) and beetles (Schultz and

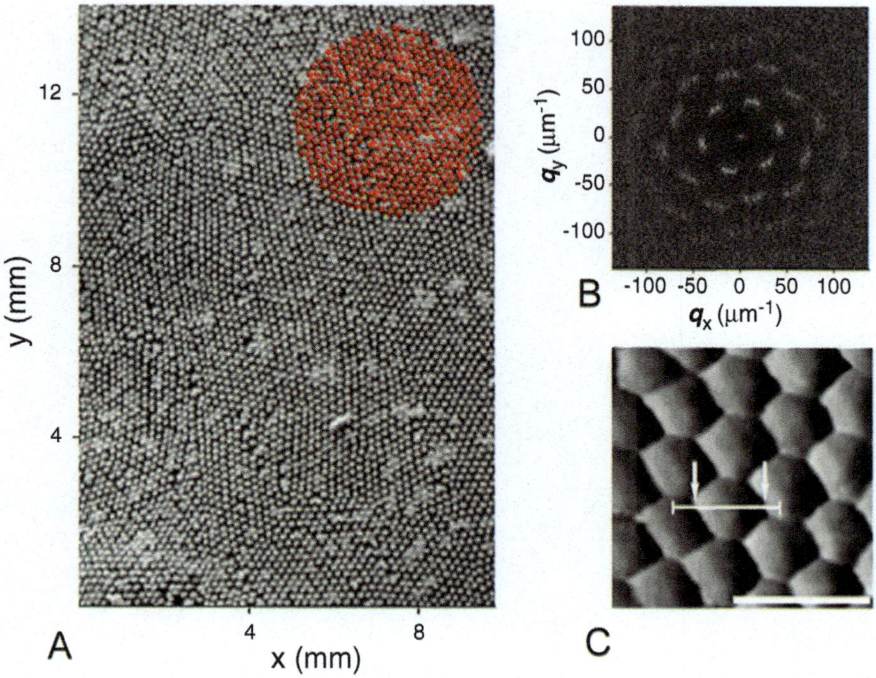

Fig. 4.15 Anti-reflector of the moth eye. (**a**) Scanning electron microscopy image of a single ommatidium surface of an eye in the moth *Manduca sexta*. (**b**) Fourier transformation of the encircled area highlighted in red in (**a**) (Kovalev et al. 2016). (**c**) Atomic force microscopic image of ommatidial nipples of the moth *Laothoe populi*. *White bar* marks the region used to indicate a single nanonipple. *Scale bar* = 500 nm (Peisker and Gorb 2010)

Hadley 1987). For example, scales of some curculionid beetles bear photonic crystals inside scale-like setae on their surfaces responsible for the lusterless appearance of the elytra (Fig. 4.16a–c). The coloration pattern serves for species communication and sex recognition and also for camouflage and mimicry. The most interesting type of structural coloration is called iridescence, which is well known in beetles and butterflies (Ghiradella et al. 1972; Huxley 1975) and has also been recently characterized for some dragonfly species (Gorb et al. 2015; Guillermo-Ferreira et al. 2015a, b) (Fig. 4.16d–f).

The iridescence is a result of optical interference within multilayer structures (Ghiradella 1991) that are rather complex in their architecture and can be incorporated into systems that can produce several different optical effects. Such effects include diffraction-assisted reflection angle broadening (Vukusic et al. 1999, 2000a), all-structural colour mixing and strong polarization effects (Vukusic et al. 2000b). Vinothan Manoharan and his collaborators from the Harvard School of Engineering and Applied Sciences have developed man-made colour technology that never fades (Park et al. 2014). Their innovative method recreates structural

Fig. 4.16 Photonic crystals of the dorsal green scale-like hairs in the male curculionid beetle *Phyllobius argentatus*. (**a**) Scanning electron microscopy micrograph of a cross fracture through the scale. (**b**) Transmission electron microscopy micrograph of an oblique ultrathin section of the scale (Gorb 2011). (**c**) Copulating female and male of *Phylobius* beetles (Image by Marcel Zurreck). (**d–f**) Three different examples of photonic systems in dragonflies: *Zenithoptera lanei* (Guillermo-Ferreira et al. 2015a), *Mnesarete pudica* (Guillermo-Ferreira et al. 2015b), and *Platycnemis phyllopoda* (Gorb et al. 2015)

colours in artificial materials. They fill microcapsules with an aqueous solution of small disordered particles. When the capsules dry out, they bring the particles closer together, producing a colour. Depending on the degree of solvent evaporation, the researchers can produce diverse structural colours for potential use in architecture.

4.7 Adhesives and Bonding Technology

Adobe is a building material made from earth and often organic material, the adhesive properties and strength of which are crucial for the stability of the adobe-based buildings. Adobe was among the earliest building materials and is used throughout the world. Adhesives in general and in architecture in particular currently have three main goals: (1) an increase in the reliability of glued contact; (2) the mimicking of natural environment-friendly glues and (3) the development of mechanisms for the application of a minute amount of glue to the surface (Hennemann 2000). An additional challenge is the use of substances and/or mechanisms that allow multiple attachments and detachments and enable attachment to a variety of surfaces.

The advantages of bonding in architecture in comparison to conventional joining technologies are numerous. Adhesive bonding enables the architect to go beyond the borders of the conventional. In particular, the joining of different materials can only be achieved by means of adhesives. By using bonding, damage to the parts that have to be joined can be greatly reduced. Even the smallest parts and very thin components (e.g. films) can be attached. Reduction of weight can be easily achieved: lightweight construction is possible. However, for two serious reasons, bonding is only used to an extremely limited extent for load-bearing connections. The main reasons are that long-term behaviour (a life time of 50 years is expected for any building construction) is usually not proven and, in the case of fire, capacity is immediately lost. The present section demonstrates some potentials of insect adhesive systems for inspiring the development of technical adhesives for use in the construction of buildings.

A variety of biological systems prevents the separation of two surfaces. These systems are often called *attachment devices* (Gorb 2001). Some of them are mainly based on mechanical principles, whereas others additionally rely on the chemistry of polymers and colloids (Scherge and Gorb 2001; Habenicht 2002). The inspirations from the first type are potentially more suitable for use in architecture. There are at least three reasons for using attachment devices in architecture: (1) they join dissimilar materials, (2) they improve stress distribution in the joint and (3) they increase design flexibility (Waite 1983). These reasons are also relevant to the evolution of natural attachment systems.

Many species of insects are supplied with diverse attachment devices, with the morphology depending on the species' biology and the particular function, in which the attachment device is involved. The evolutionary background and animal behaviour influence the specific composition of attachment systems in each particular species. Eight fundamental classes of attachment principles have been described in insects: (1) hook, (2) lock or snap, (3) clamp, (4) spacer, (5) suction, (6) expansion anchor, (7) adhesive secretion (glue) and (8) friction (Gorb 2001). However, various combinations of these principles also occur in existing attachment structures. Three types of adhesion at the organism level are known: (1) temporary adhesion allowing the organism to attach strongly to the substrate and detach quickly when necessary, (2) transitory adhesion permitting simultaneous attachment and movement along the substrate and (3) permanent adhesion involving the secretion of cement. These three types of adhesion do not have the same purpose and use different adhesive systems.

Insects are capable of walking on smooth and structured substrata, on inclines and vertical surfaces and some of them even walk on the ceiling. For example, flies and beetles rely on the hairy (setose) surfaces on their legs. This system additionally uses a secretion enabling hairs called tenent setae to attach and detach very quickly to diverse substrata. The hair design includes a mechanism that delivers the secretion in extremely small amounts directly to the contact area (Ishii 1987) and only then, when contact to the substrate is achieved. Tenent setae are relatively soft structures

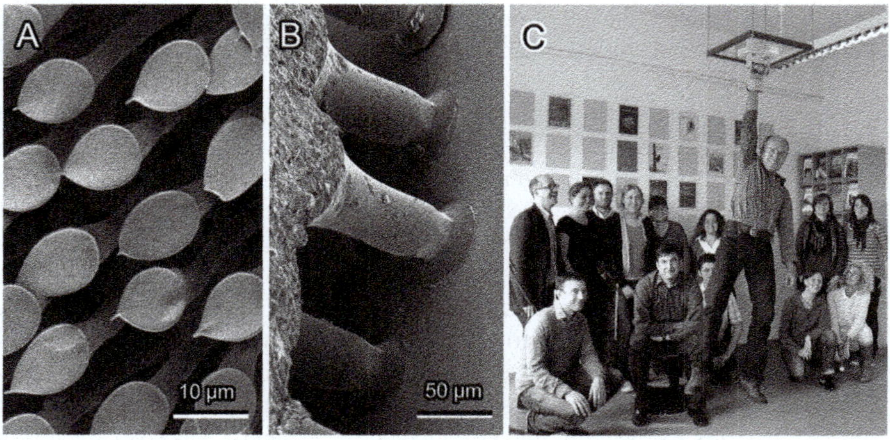

Fig. 4.17 Insect-inspired fibrillar adhesive. (**a**) Ventral surface of the tarsus in the male of the chrysomelid beetle *Gastrophysa viridula*. (**b**) Microstructured polymer foil adhering to a glass surface. Note specific mushroom-shaped geometry of terminal contact elements in both systems (Gorb et al. 2007). (**c**) Demonstration of the performance of the dry adhesive material shown in **b**: A piece of foil of 20 cm × 20 cm can hold about 65 kg weight on the ceiling

(Niederegger et al. 2002). In *Calliphora* flies, their tips are usually compressed, widened and bent at an angle of about 60° to the hair shaft (Bauchhenss and Renner 1977). Some male beetles possess mushroom-like terminal tips on their setae (Fig. 4.17a).

Various forces might contribute to the resulting attachment force: capillary adhesion and intermolecular van der Waals forces. A contribution of intermolecular interaction to overall adhesion has been shown in experiments examining the adherence of beetles (Stork 1980) and beetle setae (Stork 1983) to a glass surface. In the beetle *Chrysolina polita* (Chrysomelidae), the resulting attachment force directly depends on the number of single hairs contacting the surface. The contribution of intermolecular interaction and capillary force has been demonstrated for the fly *Calliphora vicina* in a nanoscale experiment by using an atomic force microscope (Langer et al. 2004). Attachment forces increase, when the contacting surfaces slide against each other. This might be the reason that flies placed on a smooth undersurface always move their legs in a lateral-medial direction (Wigglesworth 1987; Niederegger and Gorb 2003). During these movements, setae slide over the surface obtaining optimal contact.

The size of single points in hairy attachment devices gets smaller and their density higher as the body mass increases (Scherge and Gorb 2001; Arzt et al. 2003). The fundamental importance of multiple micro- and nanoscopical contacts for adhesion on smooth and rough substrata has been demonstrated experimentally (Peressadko and Gorb 2004; Gorb et al. 2007). A patterned surface, made out of polyvinylsiloxane, has significantly higher adhesion on a smooth substrate than a smooth sample made out of the same material. An additional advantage of patterned surfaces is the reliability of contact on various surface profiles and the increased defect tolerance of individual contacts. The forces generated by such a microstructured insect-inspired tape are sufficient to withstand the large forces (Fig. 4.17b, c) required for their use in architecture.

Specific advantages of using insect-inspired glue-free bonding technology in architecture is its good acoustic insulation, protection of bonded surfaces against corrosion, freedom of design, absorption of vibration, compensation of joint tolerances and thermal isolation. This type of adhesive provides a broad variety of customer-specific solutions; this is ideal when a previously existing profile design has to be retained. It guarantees easy installation and removal, while providing maximum adhesion to glass and other smooth surfaces. Easy removal of the tape allows an optimal recycling process.

4.8 Future Perspectives

Statistics shows that American companies generate, on average, 0.5 ideas per employee per year, whereas typical Japanese companies generate 9 ideas per employee per year; both numbers are relatively low. Our creativity is obviously limited. However, we can extend this by employing the great bank of ideas from living nature. Every organism on the Earth has evolved through adaptation and the survival of the fittest and, hence, organisms have retained only those evolutionary adaptations that make them strong.

What can be done to advance architecture by using insect-inspired biomimetics? First and obviously, additional research into insect materials, constructions and surfaces will help in the application of biological knowledge to recent challenges in architecture. The incorporation of additional biological knowledge into the design of artificial systems will improve their performance. Unfortunately, biologists still do not have a complete understanding of how insect materials are constructed, what their performance is, how insect surfaces function, etc. Hence, many technological areas will benefit from additional entomological research. Additionally, a huge variety of insects and their systems have never previously been studied. Screening for new biological systems with interesting properties therefore remains an extremely important research field in the near future.

Some 1.7 million different organisms are known to science. However, all estimations suggest that the actual number of organism on Earth is between 8.6 to over 20 million species: about 90 % of the residents of our planet are unknown. Despite the Convention on Biological Diversity (CBD) and other agreements, loss of habitats and dramatic extinction rates are unchanged or even increasing (see the chapter by Barthlott et al. in this book). We have a responsibility to maintain the idea bank of biodiversity in order to increase the quality of human life, while overcoming the degenerative forces that might destroy organisms and their environment (Hwang et al. 2015). Advances in biomimetics provide an additional reason for carrying this task out because, in doing so, we might harmonize relationships between biological evolution and technological development.

Acknowledgements The authors are very grateful to Jan Knippers (Stuttgart) and Oliver Betz (Tübingen) for their valuable comments on an early version of the manuscript.

References

AmphibianArc (2012) http://www.dezeen.com/2012/07/13/zoomlion-headquarters-exhibition-center-by-amphibianarc/

Arzt E, Gorb S, Spolenak R (2003) From micro to nano contacts in biological attachment devices. Proc Natl Acad Sci U S A 100:10603–10606

Bamboo Tower (2015) http://www.wired.com/2015/01/architecture-and-vision-warkawater/

Barbakadze N, Enders S, Gorb SN, Arzt E (2006) Local mechanical properties of the head articulation cuticle in the beetle *Pachnoda marginata* (Coleoptera, Scarabaeidae). J Exp Biol 209:722–730

Barthlott W, Neinhuis C (1997) Purity of the sacred lotus or escape from contamination in biological surfaces. Planta 202:1–8

Barthlott W, Neinhuis C (1998) Lotusblumen und Autolacke: Ultrastruktur pflanzlicher Grenzflächen und biomimetische unverschmutzbare Werkstoffe. In: Nachtigall W, Wisser A (eds) 4. Bionik – Kongress, München 1998. Gustav Fischer Verlag, Stuttgart/Jena/Lübeck/Ulm, pp 281–293

Bauchhenss E, Renner M (1977) Pulvillus of *Calliphora erythrocephala* Meig. (Diptera; Calliphoridae). Int J Insect Morphol 6(3/4):225–227

Baunetz (2015) http://www.baunetz.de/meldungen/Meldungen-Koreanisches_Apartmenthaus_4508247.html

Bernhard CG, Miller WH, Moller AR (1965) Insect corneal nipple array – a biological broad-band impedance transformer that acts as an antireflection coating. Acta Physiol Scand S 63(Suppl, 243):5

Chapman AD (2006) Numbers of living species in Australia and the world. Australian Biological Resources Study, Canberra. ISBN 978-0-642-56850-2

Dörstelmann M, Parascho P, Prado M, Menges A, Knippers J (2014) Integrative computational design methodologies for modular architectural fiber composite morphologies. In: Proceedings of ACADIA 2014, Los Angeles, pp 219–228

Erwin TL (1982) Tropical forests: their richness in Coleoptera and other arthropod species. Coleopt Bull 36:74–75

Erwin TL (1997) Biodiversity at its utmost: tropical forest beetles. In: Reaka-Kudla ML, Wilson DE, Wilson EO (eds) Biodiversity II. Joseph Henry Press, Washington, DC, pp 27–40

Francouer A, Loiselle R (1988) Evolution du strigile chez les formicides (Hymenopteres). Natur Can 115:333–335

Ghiradella H (1989) Structure and development of iridescent butterfly scales: lattices and laminae. J Morphol 202:69–88

Ghiradella H (1991) Light and colour on the wing: structural colours in butterflies and moths. Appl Optics 30:3492–3500

Ghiradella H, Aneshansley D, Eisner T, Silbergleid RE, Hinton HE (1972) Ultra-violet reflection of a male butterfly: interference colour caused by thin layer elaboration of wing scales. Science 178:1214–1217

Gorb SN (1997a) Porous channels in the cuticle of the head-arrester system in dragon/damselflies (Insecta: Odonata). Microsc Res Techn 37(5/6):583–591

Gorb SN (1997b) Ultrastructural architecture of the microtrichia of the insect cuticle. J Morphol 234:1–10

Gorb SN (2000) Ultrastructure of the neck membrane in dragonflies (Odonata). J Zool (Lond) 250:479–494

Gorb SN (2001) Attachment devices of insect cuticle. Kluwer Academic Publishers, Dordrecht/Boston/London

Gorb SN (2011) Insect-inspired technologies: Insects as a source for biomimetics. In: Vilcinskas A (ed) Insect biotechnology. Springer, Dordrecht, pp 241–264

Gorb SN, Sinha M, Peressadko A, Daltorio KA, Quinn RD (2007) Insects did it first: a micropatterned adhesive tape for robotic applications. Bioinsp Biomim 2:S117–S125

Gorb SN, Tynkkynen K, Kotiaho JS (2009) Crystalline wax coverage of the imaginal cuticle in *Calopteryx splendens* (Odonata: Calopterygidae). Int J Odonatol 12:205–221

Gorb SN, Appel E, Kovalev A (2015) Structural background of highly-visible white tibia in male *Platycnemis phyllopoda*. In: Abstracts of "34. Jahrestagung der GdO, 20.-22. März 2015 Braunschweig", Germany, p 14

Grodnicky DL (1988) Structure and function of the scale coverage of the wings in butterflies (Lepidoptera /Hesperioidea, Papilionoidea). Entomol Rev 67:251–256

Gronenberg W (1996) Fast actions in small animals: springs and click mechanisms. J Comp Physiol A 178:727–734

Gruber P (2011) Biomimetics in architecture: architecture of life and buildings. Springer, Vienna/New York

Guillermo-Ferreira R, Bispo PC, Appel E, Kovalev A, Gorb SN (2015a) Mechanism of the wing colouration in the dragonfly *Zenithoptera lanei* (Odonata: Libellulidae) and its role in intraspecific communication. J Insect Physiol 81:129–136. doi:10.1016/j.jinsphys.2015.07.010

Guillermo-Ferreira R, Gorb SN, Appel E, Kovalev A, Bispo PC (2015b) Variable assessment of wing colouration in aerial contests of the red-winged damselfly *Mnesarete pudica* (Zygoptera, Calopterygidae). Naturwissenschaften 102(3–4):13 pp. doi:10.1007/s00114-015-1261-z

Haas F, Gorb S, Wootton RJ (2000a) Elastic joints in dermapteran hind wings: materials and wing folding. Arthr Struct Dev 29:137–146

Haas F, Gorb SN, Blickhan R (2000b) The function of resilin in beetle wings. Proc Roy Soc Lond B 267:1375–1381

Habenicht G (2002) Kleben: Grundlagen, Technologien, Anwendung. Springer, Berlin

Hackman RH, Goldberg M (1987) Comparative study of some expanding arthropod cuticles: the relation between composition, structure and function. J Insect Physiol 33:39–50

Hansell M (2007) Built by animals: the natural history of animal architecture. Oxford University Press, New York

Heckmann CW (1983) Comparative morphology of arthropod exterior surfaces with capability of binding a film of air underwater. Int Rev Ges Hydrobiol 68:715–736

Heie OE (1987) Morphological structure and adaptations. In: Aphids: biology, natural enemies and control. Elsevier, Amsterdam, pp 393–400

Hennemann O-D (2000) Kleben von Kunststoffen. Anwendung, Ausbildung, Trend. Kunststoffe 90:184–188

Hepburn HR (1985) Structure of the integument. In: Kerkut GA, Gilbert LI (eds) Comprehensive insect physiology, biochemistry and pharmacology. Pergamon Press, Oxford et al., pp 1–58

Hepburn HR, Chandler HD (1976) Material properties of arthropod cuticles: the arthrodial membranes. J Comp Physiol A 109:177–198

Hepburn HR, Chandler HD (1978) Tensile mechanical properties and transconformational changes of chitins. In: Muzzarelli RA, Parisier ER (eds) Proceedings of the first international conference on chitin/chitosan. Massachusetts Institute of Technology, Cambridge, MA, pp 124–143

Hinton HE (1976) The fine structure of the pupal plastron of simulid flies. J Insect Physiol 22:1061–1070

Huxley J (1975) The basis of structural colour variation in two species of *Papilio*. J Entomol A 50:9–22

Hwang J, Jeong Y, Park JM, Lee KH, Hong JW, Choi J (2015) Biomimetics: forecasting the future of science, engineering, and medicine. Int J Nanomed 10:5701–5713

Ishii S (1987) Adhesion of a leaf feeding ladybird *Epilachna vigintioctomaculata* (Coleoptera: Coccinellidae) on a vertically smooth surface. Appl Ent Zool 22:222–228

Karasev VP (1989) Scale coverage of the curculionid beetles of the genus *Tichius* Germar (Coleoptera, Curculionidae). In: Dynamics of zoocoenozes and animal conservation in Belorussia, Minsk, p 85 (in Russian)

Knippers J, Schlaich J (2000) Folding mechanism of the Kiel Hörn Footbridge. Struct Eng Int 10:50–53

Knippers J, Speck T (2012) Design and construction principles in nature and architecture. Bioinspir. Biomim 7015002 (10 pp). doi:10.1088/1748-3182/7/1/015002

Kovalev A, Filippov A, Gorb SN (2016) Correlation analysis of symmetry breaking in the surface nanostructure ordering: case study of the ventral scale of the snake *Morelia viridis*. Appl Phys A 122:253. doi:10.1007/s00339-016-9795-2

Langer M, Ruppersberg P, Gorb SN (2004) Adhesion forces measured at the level of a terminal plate of the fly's seta. Proc R Soc Lond B 271:2209–2215

Li X, Madan I, Bozkurt H, Birkhofer H (2004) Lifetime of solid lubricated roller bearings. In: 14th international colloquium tribology, January 2004, Esslingen, Germany, vol 3, pp 1361–1364

Mazzoleni I (2013) Architecture follows nature – biomimetic principles for innovative design. CRC Press, Boca Raton

Messner B (1988) Funktionelle Morphologie der Insektenkutikula am Beispiel der Plastronatmer. Wiss Z E M Arndt-Univ Greifswald Math Naturwiss R 37:27–30

Minnock, K. (2016) http://www.benhuckerbydesign.co.uk/team/kyle-minnock

Miyoshi K (2001) Solid lubrication: fundamentals and applications. Marcel Decker Inc., New York/Basel

Niederegger S, Gorb S (2003) Tarsal movements in flies during leg attachment and detachment on a smoothe substrate. J Insect Physiol 49:611–620

Niederegger S, Gorb SN, Jiao Y (2002) Contact behaviour of tenent setae in attachment pads of the blowfly *Calliphora vicina* (Diptera, Calliphoridae). J Comp Physiol A 187:961–970

Novotny V, Basset Y, Miller SE, Weiblen GD, Bremer B, Cizek L, Drozd P (2002) Low host specificity of herbivorous insects in a tropical forest. Nature 416(6883):841–844

Parascho S, Knippers J, Dörstelmann M, Prado M, Menges A (2014) Modular fibrous morphologies: computational design, simulation and fabrication of differentiated fibre composite building components. In: Block P et al (eds) Advances in architectural geometry. Springer, Vienna, pp 29–45. doi:10.1007/978-3-319-11418-7_3

Park J-G, Kim S-H, Magkiriadou S, Choi TM, Kim Y-S, Manoharan VN (2014) Full-spectrum photonic pigments with non-iridescent structural colors through colloidal assembly. Angewandte Chemie Int Ed 53(11):2899–2903. doi:10.1002/anie.201309306

Parker AR, Lawrence CR (2001) Water capture by a desert beetle. Nature 414:33–34

Parker AR, Hegedus Z, Watts RA (1998) Solar-absorber antireflector on the eye of an eocene fly. Proc Roy Soc London B 265:811–815

Peisker H, Gorb SN (2010) Always on the bright side of life: anti-adhesive properties of insect ommatidia grating. J Exp Biol 213:3457–3462

Peressadko A, Gorb SN (2004) When less is more: experimental evidence for tenacity enhancement by division of contact area. J Adhes 80:247–261

Perez Goodwyn P (2009) Anti-wetting surfaces in Heteroptera (Insecta): Hairy solutions to any problem. In: Gorb SN (ed) Functional surfaces in biology, vol 1. Springer, Dordrecht, pp 55–76

Perez Goodwyn PJ, Gorb SN (2004) Frictional properties of contacting surfaces in the hemelytra-hindwing locking mechanism in the bug *Coreus marginatus* (Heteroptera, Coreidae). J Comp Physiol A 190:575–580

Pohl G, Nachtigall W (2015) Biomimetics for architecture & design. Springer, Cham, doi:10.1007/978-3-319-19120-1_7

Scherge M, Gorb SN (2001) Biological micro- and nanotribology. Springer, Berlin

Schönitzer K, Lawitzky G (1987) A phylogenetic study of the antenna cleaner in Formicidae, Mutillidae and Tiphiidae (Insecta, Hymenoptera). Zoomorphology 107:273–285

Schultz TD, Hadley NF (1987) Structural colors of tiger beetles and their role in heat transfer through the integument. Physiol Zool 60:737–745

Southwick EE (1985) Bee hair structure and the effect of hair on metabolism at low temperature. J Apicultural Res 24:144–149
Stork NE (1980) Experimental analysis of adhesion of *Chrysolina polita* (Chrysomelidae, Coleoptera) on a variety of surfaces. J Exp Biol 88:91–107
Stork NE (1983) The adherence of beetle tarsal setae to glass. J Nat Hist 17:583–597
Thorpe WH, Crisp DJ (1947) Studies on plastron respiration. I. The biology of *Aphelocheirus* and the mechanism of plastron retention. J Exp Biol 24:227–269
Tupper EC (2013) Introduction to naval architecture, 5th edn. Elsevier, Amsterdam
Vincent JFV (1981) Morphology and design of the extensible intersegmental membrane of the female migratory locust. Tiss Cell 13:18–31
Vincent JFV, Wood SDE (1972) Mechanism of abdominal extension during oviposition in *Locusta*. Nature 235:167–168
Vukusic P, Sambles JR (2003) Photonic structures in biology. Nature 424:852–855
Vukusic P, Sambles JR, Lawrence CR, Wootton RJ (1999) Quantified interference and diffraction in single *Morpho* butterfly scales. Proc R Soc Lond B 266:1403–1411
Vukusic P, Sambles JR, Ghiradella H (2000a) Optical classification of microstructure in butterfly wing-scales. Photonics Sci News 6:61–66
Vukusic P, Sambles JR, Lawrence CR (2000b) Structural colour: colour mixing in wing scales of a butterfly. Nature 404:457
Wagner T, Neinhuis C, Barthlott W (1996) Wettability and contaminability of insect wings as a function of their surface sculpture. Acta Zool 77:213–225
Waite JH (1983) Adhesion in byssally attached bivalves. Biol Rev 58:209–231
Wigglesworth VB (1987) How does a fly cling to the under surface of a glass sheet? J Exp Biol 129:363–367
Wootton RJ (1991) The functional morphology of the wings of Odonata. Adv Odonatol 5:153–169
Wootton RJ, Newman DJS (1979) Whitefly have the highest contraction frequencies yet recorded in non-fibrillar flight muscles. Nature 280:402–403

Chapter 5
In Search of Some Principles of Bio-mimetics in Structural Engineering

Bill Addis

Abstract There is a strong temptation to assume that biomimetics is merely a matter of copying a structural idea that is found in nature. However, this overlooks the process of structural design which is much more than creating a certain structural form. It also involves using appropriately approximate mathematical models of material properties and the loads that a structure must carry without suffering damage. Fundamental to the design process is the engineer's need to gain sufficient confidence in a design that it can be built, and will perform safely. There are many differences and a few similarities between natural and man-made structural systems; the relationship is complex. In order to understand and define the process of biomimicry, it will be important to establish some principles underlying this process – for example, How does the process of biomimicry work? What do effective examples of biomimicry have in common? To identify the principles, it would be rewarding to study successful examples of biomimetics, not only in structural engineering, but also in other fields such as aerospace and medical engineering.

5.1 What Underlies the Structural Design Process?

Everyone uses the natural and physical world around them to learn about structures and materials and their behaviour, especially as children: after all, how else could we learn? Man naturally makes analogies between one situation and another; sometimes these steps of inductive logic are valid and useful – what we often call 'correct'; sometimes they break down and are not useful – what we often call 'false'.

When building designers are designing a structure, they are influenced by two particular features:

- It is they who are creating the new structure (not God or Nature)
- Since the structure is at a larger scale than they can experience with their own muscles, direct 'first-hand' experience is not available to them. They must rely on induction, inference, abstraction and intellectualisation of the challenge at hand.

B. Addis (✉)
44, Fairfax Road, Teddington TW11 9BZ, UK

Architectural History and Heritage Preservation, University of Innsbruck, Innsbruck, Austria
e-mail: bill.addis@cantab.net

J. Knippers et al. (eds.), *Biomimetic Research for Architecture and Building Construction*, Biologically-Inspired Systems 9, DOI 10.1007/978-3-319-46374-2_5

In order to undertake this process satisfactorily, designers must raise their level of confidence sufficiently to the point at which they are happy that the completed structure will be safe and will perform as intended and, thus, that construction can begin. They may use a variety of processes or techniques to raise their confidence:

- They can seek inspiration from existing natural structures
- They can seek inspiration from existing man-made structures – precedent
- They can conduct tests or experiments
- They can use logical (deductive) arguments based on mathematical models of loads, materials and structures.

In each of these cases, the process is underpinned by a body of knowledge and experience, which provides a certain level of confidence that the resulting structure will be satisfactory, and the designer can use this knowledge and experience to explain or to justify that confidence. Hence, a designer might believe that "my structure will work because it is like a tree or a sea shell", that "my structure will work because it is essentially the same as other structures I know" or that "I have conducted satisfactory tests on a scale model of the structure and, hence, the full-size structure will perform satisfactorily".

Of course, such a process or technique is not guaranteed to work. A design inspired or justified by a tree may collapse for many reasons: the materials may be of inadequate quality; the loads on the structure might be different from those that a tree was 'designed' to withstand; whereas a structure may appear to be similar to a tree, it may be based on an erroneous understanding of how trees work as structures; other, hidden, non-structural factors may be significant, for example, the structure might be required to perform at temperatures outside the range in which trees perform well. Likewise, in the case of using small-scale physical models, the designer may be unaware that some aspects of structural behaviour can be scaled up in geometric proportion (e.g. traditional masonry structures), whereas others cannot (e.g. the behaviour of a beam or a thin-shell reinforced-concrete structure) (Addis 2013).

5.2 Some Developments in the History of Designing Structures

5.2.1 Designing Structures Before the Era of Structural Science

In ancient Greece, we find the European origins of mathematics and science. Through observation, systematic analysis and the application of both inductive and deductive logic (but generally not by conducting practical experiments), Greek scholars discovered and wrote down the ideas and the rules that lay behind geometry, numbers and music, for example. These rules were expressed in terms of the characteristics and proportions found in basic geometrical shapes, numbers and the harmonics of the vibrating strings of musical instruments. The world was described and explained in terms of simple forms, such as circles, squares, triangles, etc., and simple ratios, such as 1:2, 1:3, etc., $1:\surd 2$, $2:\surd 3$, $(1+\surd 5)/2$ (the Golden Ratio), and so on.

They argued that, because these ratios and patterns had been used to create the universe, then these same 'rules' could be used by man to create new artefacts – notably, in the current context, buildings. Evidence for this can be found not only in the many buildings surviving from this time, often including their decorations and sculptures, but also in the one surviving book from this era, namely the ten chapters written on designing cities, buildings, scientific instruments and war machines by Vitruvius.

Geometry, number and harmonics were thus used *both* to create the design *and* to help raise confidence in the design by providing a justification that it would be successful. Of course, such a theoretical justification was augmented by ensuring that proposed designs were also largely based on well-established building precedent.

Similar approaches were used in Gothic and Renaissance buildings. [Addis 1990] Indeed, the ancient ideas of proportion, geometry and harmony were highly respected and used by scientists well into the 17th century. Even Kepler (1571–1630) published 'corrected' Greek values for harmonic proportions when he was developing his model of the helio-centric universe.

5.2.2 Designing Structures in the Era of Structural Science

Although Galileo was not an engineer or building designer, he gave us some insights into the idea of 'structure' that have a remarkably modern feel to them. For example, he commented on the unreliability of scaling designs up in a simple geometric manner:

> From what has already been demonstrated, you can plainly see the impossibility of increasing the size of structures to vast dimensions either in art or in nature; likewise the impossibility of building ships, palaces or temples of enormous size in such a way that their oars, yards, beams, iron bolts and, in short, all their other parts will hold together; nor can nature produce trees of extraordinary size because the branches would break down under their own weight; so also it would be impossible to build up bony structures of men, horses or other animals so as to hold together and perform their normal functions if these animals were to be increased enormously in height; for this increase in height can be accomplished only by employing a material which is harder and stronger than usual, or by enlarging the size of the bones, thus changing their shape until the form and appearance of the animals suggest a monstrosity. ...
>
> To illustrate briefly, I have sketched a bone whose natural length has been increased three times and whose thickness has been multiplied until, for a correspondingly large animal, it would perform the same function which the small bone performs for its small animal. From the figures here shown you can see how out of proportion the enlarged bone appears. Clearly

> then if one wishes to maintain in a great giant the same proportion of limb as that found in an ordinary man he must either find a harder and stronger material for making the bones, or he must admit a diminution of strength in comparison with men of medium stature; for if his height be increased inordinately he will fall and be crushed under his own weight. Whereas, if the size of a body be diminished, the strength of that body is not diminished in the same proportion; indeed the smaller the body the greater its relative strength. Thus a small dog could probably carry on his back two or three dogs of his own size; but I believe that a horse could not carry even one of his own size (Galileo 1638, pp. 130–31).

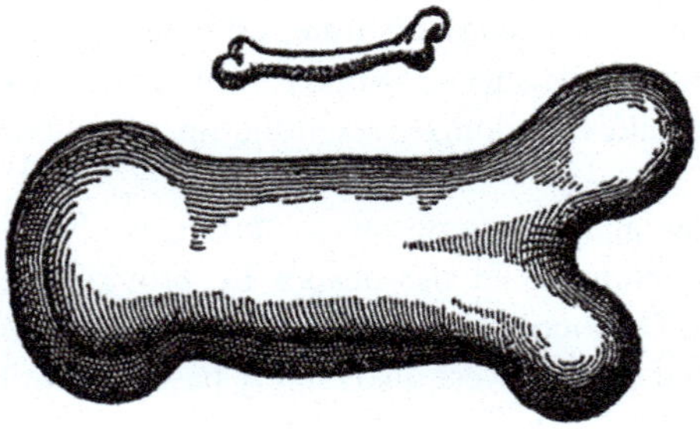

When discussing hollow structures, Galileo made a clear logical step from observations about animals to designing man-made artefacts and also makes the point that man discovered that for himself. This thought process expresses precisely the way that engineers today can abstract ideas and understanding from observations of the natural world by using nature not as a source of ideas to copy, but as an allegory and a stimulus for thinking about their engineering-design challenges in different ways:

> But, in order to bring our daily conference to an end, I wish to discuss the strength of hollow solids, which are employed in art—and still oftener in nature—in a thousand operations for the purpose of greatly increasing strength without adding to weight; examples of these are seen in the bones of birds and in many kinds of reeds which are light and highly resistant both to bending and breaking. For if a stem of straw which carries a head of wheat heavier than the entire stalk were made up of the same amount of material in solid form it would offer less resistance to bending and breaking. This is an experience which has been verified and confirmed in practice where it is found that a hollow lance or a tube of wood or metal is much stronger than would be a solid one of the same length and weight, one which would necessarily be thinner; men have discovered, therefore, that in order to make lances strong as well as light they must make them hollow (Galileo 1638, pp. 150).

From around 1800, engineers have been able to help to determine the sizes and shapes of structural elements – arches and abutments, beams and columns, roof and bridge trusses – based on structural mechanics. This application of science is effectively the modern equivalent of using geometry, number and harmonics. The result is that we now have a variety of 'ideal' or rational shapes for such structural elements, which reflect the structural mechanics used to determine them. Thus, we have fish-belly and hog-backed beams and girders that reflect the bending moment diagram for a uniformly distributed load; we have columns and struts that are thicker at mid span providing the most efficient form to resist the bending associated with buckling; we have arches in the form of inverted catenaries (for linear arch loads) or parabolas (for uniformly distributed loads); and we have pin-jointed trusses that are statically determinate, whose component members carry only tension or compression forces and whose cross sections correspond precisely to the magnitude of the forces that they carry. Finally, in the world of tensile membrane and cable-net structures, and their inverse, thin concrete shells, we also have forms that are very precisely determined and optimised to minimise internal stresses and so use minimum amounts of material.

However, these ideal shapes and sizes are driven by tenets of structural mechanics. Although, during certain periods of history, some architects have been keen to use them as expressions of various structural truths, most architects do not want to have their ideas and forms subjugated to laws of mechanics – this is felt to be undermining their freedom and creativity (Addis 1997). The Gateway Arch in St Louis, USA provides a case in point. Eero Saarinen asked the engineer Fred Severud to create the most efficient structural form for this 192 m high, steel, monumental arch. He responded with an inverted catenary shape that reflected the cross section, which varied according the wind loads that the arch had to resist. However, Saarinen found this proposal "too pointed, too steep". Severud then proceeded in the reverse direction, finding a shape that pleased Saarinen and then providing adequate strength and stiffness within the cross section of the arch. This anecdote is representative of much of the dialogue between engineers and architects. The point is that designers in general and architects in particular often seek only a loose connection between a source of inspiration and the finally chosen design. Architects often use their sources of inspiration as objective origins to underpin highly subjective designs, in order to give them greater legitimacy and to avoid their being seen as the result of mere whim and fancy. Sometimes, indeed, such 'origins' can appear quite late in the day, in the form of post-rationalisation.

5.3 Natural Structures and Man-Made Structures

Natural structures and man-made structures inevitably have many similarities as they both 'make use of' the same laws of physics, statics, elasticity and dynamics. In this respect, a similarity exists with the thinking of the ancient scholars. Once the 'laws' underlying geometry and music had been discovered, the same laws could reliably be used by man to design buildings. However, this did not mean that buildings were designed by copying geometry and music; they were designed by using the same underlying laws. When an engineer designs a concrete shell structure that looks like a sea shell, he is not copying or mimicking ('bio-mimicking') nature; rather he is making use of the same laws of equilibrium and structural mechanics that underlie both sea shells and concrete shells.

Table 5.1 Some differences between natural and man-made structures and materials

Natural structures and materials	Man-made structures and materials
Evolved over many thousands of generations by means of large numbers of random changes	Designed with relatively few, deliberate, entirely rational changes
Grew by following a predetermined set of genetic instructions	Designed intentionally to meet a unique, probably unprecedented set of requirements
Each organism has a single set of actions and boundary conditions to which it is programmed (so to speak) to respond	Many structures need to be adaptable and allow the possibility of having to respond to actions and boundary conditions (e.g. attributable to change of function or use) not considered at their conception
Most natural structures are capable of some self-repair and regeneration	Not capable of self-repair and regeneration
The overriding purpose is that a species must survive; the loss of individuals (even a great many of them) is acceptable as long as the population can survive	Total failure or collapse of even a single structure is unacceptable, although certain types of partial failure may be tolerated, especially when repair is feasible
Natural structures include no 'safety factor'	Always include a 'safety factor' applied to mathematical models of loads, structures and materials
Fire cannot usually be resisted	Their design is greatly influenced by the need to resist damage by fire.
An organic 'structure' may be influenced by the need to resist non-structural loads or actions, such as UV light or disease, which may not be apparent to the human observer.	Actions and loads are addressed deliberately, when known, and can be added or refined
Materials are generally low-strength, low-stiffness	Materials are generally high-strength, high-stiffness
Stresses in materials are generally of low magnitude; large strains are often tolerable	Stresses are generally of high magnitude; strains generally need to be kept very low
Unique 'designs' have developed in response to unique sets of actions and boundary conditions	The purpose of a structure (e.g. bridge, theatre) is determined before a precise performance specification is developed; many different designs can be conceived to meet a single set of performance specifications
Within timespans of a few generations, no new materials are developed	'New' materials may be used to perform 'old' functions, consequent upon progress in technology, discoveries and inventions and in designing materials to predetermined performance specifications

Nevertheless, there seem to be considerable differences between the two worlds (Table 5.1).

Apart from similarities consequent upon the laws of physics and mechanics, other similarities can be found between natural and man-made systems (Table 5.2).

Table 5.2 Some similarities between natural and man-made structures and materials

Natural and man-made structures and materials	Comment
Both are usually very resource-efficient	Natural systems are usually most efficient since lower 'safety factors' can be tolerated.
Over time, 'designs' for the same purpose develop towards greater resource-efficiency	Organic structures and materials develop by natural selection; man-made ones by learning from experience and using new unprecedented applications of mechanics and materials.
Both can be understood better by means of a 'systems approach'	For example, load-bearing system, fluid-flow system, heat-flow system; natural structures usually display a high degree of integration of these systems; many man-made structures have developed in this direction.
Some elements of both types of structure and material may belong to different systems	The bundled tubes of a plant stem serve a structural function and convey fluids; the concrete structure of a building can also serve as thermal mass.
Both are capable of a high degree of systems integration leading to greater resource efficiency	However, a high degree of systems integration in buildings leads to less flexibility and adaptability

5.3.1 Natural Structures as Inspiration or Determinant for Man-Made Structures

As engineers and designers, we can have great fun in finding similarities between natural structures and man-made structures and in discovering and even using extraordinary shapes in nature; we can have fun, but this has little to do with engineering design. Nevertheless, it is easy to see how an engineer who has designed a structure and then finds a similar structure in nature might be seduced into concluding that it may be fruitful to work the other way round.

There are some famous examples of natural structures being reported as inspiring the design of man-made objects. The engineer George de Mestral was inspired by the burrs on burdock seeds to invent Velcro, although it took 14 years between his initial idea and being granted the patent and several more before successful commercial exploitation. In the 1910s, Walther Bauersfeld was inspired by Ernst Haeckel's illustrations of *radiolaria*, in his 1862 book, to solve the geometric problem of creating a hemisphere by using a triangulated structure. However, the *radiolaria* was not a triangulated structure and, in fact, the geometric problem had already been solved by Leonard Euler in 1751 – if only Bauersfeld had known that!

We can also be amazed that nature seems to have pre-empted man in 'discovering' a wide range of engineering ideas, the subject of many popular books and a growing number of conferences (Hernández and Brebbia 2012). A few examples are:

- Bundled tubes in plant stems and high-rise buildings
- The Warren truss in the vulture's wing and statically determinate bridge girders
- Space frames in crystal structure and roof structures
- Toothed gears in the leg joint of *Issus coleoptratus* (an insect) and in machines
- The freely rotating joint in a bacterial flagellum and the bearing of a cart wheel
- Pneumatic structures in spiders' legs and air-inflated roof structures
- Fibre-reinforcement in muscles (with collagen) and fibre reinforced plastics
- Distributed material (non-uniform density) in bones and various sandwich boards.

The list could go on. An important question, though, is: how many of the examples in this list did man develop by emulating nature? The answer is – none of these. In every case, the natural example was discovered after the invention by man. One reason may, of course, be that there had been no determined research programme looking for materials and structures in the natural world that might have applications in engineering. Certainly the world that microscopes have revealed was not available to many of our ancestors.

In the busy world of the design engineer, it is difficult to imagine when it might be worthwhile, from a project point of view, for an engineer to spend time studying natural structures in great detail, rather than spending the same amount of time working directly on the real structure at hand.

Finally, recalling the attitude of many architects to the perceived constraints of structural mechanics, some of them will probably also reject ideas taken from nature as "constraining their creativity". And conversely, some designers will seek comfort in a justification for a shape or form that arises from biological origins, irrespective of whether it reflects, in any way, the ingenious structural mechanics and other bio-engineering achievements that emerge from a detailed study of living organisms.

5.3.2 Man-Made Structures as a Stimulus for Understanding Natural Structures

Perhaps the way that engineers think and work, together with their experience gathered while working with man-made structures, can help biologists look at their world through different eyes. Suppose, for example, we approach natural structures by trying to imagine what the design brief and design process might have been to produce the natural structure – as Rik Huiskes asked in his well-known paper "If bone is the answer, what is the question" (Huiskes 2000).

Whereas simply transporting the engineer's way of approaching design problems is probably not helpful, the engineer's vocabulary of structural forms can be immensely useful for understanding. A structural engineer looking at the structure of organisms can quickly recognise various structural functions or techniques. Curved surfaces or folder plates, for example, are often used to stiffen and strengthen thin-walled structures in compression or bending – as found in bundled tubes in plant stems, the petals of orchids, palm fronds, clam shells and so on. The engineer can thus provide theoretical models for estimating the load-bearing capacity of such organisms and hence, perhaps, better understand their evolution.

5.3.3 *Bio-mimetics as a Pedagogic Device*

Nature can, of course, be a good teaching tool – as Galileo mentioned in his book (cited above).

It may be helpful when teaching about man-made structures to draw parallels with nature – natural stone arches, spiders' webs, trees, and so on. However, this does not provide a significant insight into the structural design process. Design requires a function for a structure, defined loads, appropriate materials and the skill to make an idea work at the scale of the full-size building.

In *On Growth and Form*, D'Arcy Thompson used geometry to help understand the relationship between the natural world and the analytical world of mathematics (Thompson 1942). He looked at geometrical transformations during growth (but, curiously, not through evolution) to help understand change. He did not, however, suggest that organic change follows a mathematical process; his approach was largely pedagogical and provided biologists with a new way of looking at the process of change and of categorising form and change.

Likewise, the concepts of structural and materials engineering can give biologists an alternative way of cataloguing what they find in nature and, the other way round, the structures that we find in nature can give engineers an alternative epistemological model for structures. This is especially helpful with regard to the relationship between 'material' and 'structure' – a hierarchy of scale. At the atomic scale, we can find triangulated space frames looking like large roof structures. At an intermediate scale, we have an amorphous material (steel) and, at a higher level, the steel is shaped into a structural form (a hollow tube).

5.4 Final Thoughts

5.4.1 *Some Concerns*

My concern is that talking about 'bio-mimetics' in the context of the structural engineering of buildings may be little more than seeking new sources of structural and architectural forms that have a 'pseudo justification' by being based on nature. This post-rationalisation of certain forms can satisfy some architects but is not good structural engineering.

'Bio-mimetics' may also be little more than a modern version of the way in which the ancients used classical science – transforming 'If the Gods and Nature used geometry, number and harmony to create the world, then humans should follow this example' into 'If nature used materials science and structural science to create flowers / skeletons / trees, then we should use these forms too'. Such an argument is not good logic and is of little use to engineering design.

Finally, if a little cynically, there is the danger that 'bio-mimetics', like 'sustainability', is just a new 'band wagon' upon which to load various design processes and academic research; whereas it may make a proposal sound better, it may in fact only be the latest case of 'badge engineering'.

I genuinely hope I am wrong in all these concerns.

5.4.2 Moving Forwards

So what might a structural engineer's 'bio-mimetic' wish list contain? Self-repairing materials, perhaps? But human engineers would still need to be in control, if only for insurance purposes! Natural decay and recycling at the end-of-life, perhaps? But who decides when the end of life is, and how and by whom will structures and materials be recycled? Again, engineers will need to be in control.

If structural engineers are seeking sources of new ideas, there are two fields that might be more fruitful than seeking inspiration in natural materials and structures. To those who study the subject, the history of structures, from ancient Greece to modern times, never ceases to amaze in the variety, ingenuity and sophistication of the ideas revealed as being used by our ancestors. [Addis 2007] Similarly, there are several industries that are 'higher-tech' than building construction and on which extraordinary amounts of time, money and high-quality research have been committed. Aerospace is perhaps the most obvious; the materials and structures that they have developed are often truly astonishing to building engineers. But then, it all comes down to money.

There are many good examples of engineers using nature in developing new designs. An excellent current example is the study of owls' wings and the way that they reduce noise in flight. This is opening up great advances for the design of aero-engines, wind turbines and fans.[1] However, this is not 'bio-MIMETICS'; it is high-quality cross-disciplinary research.

I conclude by suggesting the compilation of three inventories in order to help develop a better understanding of how nature has been and can be effectively used to inspire structural materials and structural engineering designs:

- A comprehensive historical review of work on bio-structures and materials (at least as far back as Galileo)
- A comprehensive review of previous work on 'bio-mimetics' in other engineering fields, especially aerospace
- An inventory of successful (and failed?) examples of using 'bio-mimetics' as a process for transferring ideas from nature into engineering design and of using

engineers' ways of thinking to illuminate the way that nature uses materials and structures to resist and carry loads.

Taken together, this research would help to reveal some *principles* that underlie the idea and practice of 'bio-mimetics' and, by using nature, to inspire structural design, whereas structural engineering knowledge could be used to understand and explain natural structures. How does the process work? What can be done to make it work better? What does a 'success' in such cross-disciplinary work look like?

Note

1. See http://www.cam.ac.uk/research/news/silent-flights-how-owls-could-help-make-wind-turbines-and-planes-quieter Consulted 29 April 2016.

References

Addis W (1990) Structural engineering – the nature of theory and design. Ellis Horwood, Chichester

Addis W (1997) Free will and determinism in the conception of structures'. J Int Assoc Shell Spat Struct 38(2):83–89

Addis B (2007) Building: 3000 years of design, engineering and construction. Phaidon, London/New York

Addis B (2013) "Toys that save millions": a history of using physical models in structural design'. Struct Engineer 91(4):11–27

Galileo G (1638) Dialogues concerning two new sciences. English translation, Macmillan, New York, 1914. Available at http://oll.libertyfund.org/titles/galilei-dialogues-concerning-two-new-sciences. Consulted 29 April 2016

Hernández S, Brebbia CA (eds) (2012) Design and nature VI: comparing design in nature with science and engineering, transactions on ecology and the environment, Wessex Institute of Technology, England. http://www.witpress.com/elibrary/wit-transactions-on-ecology-and-the-environment/design-nature

Huiskes R (2000) If bone is the answer, then what is the question? J Anat 197:145–156

Thompson DW (1917) On growth and form (2nd edn., 1942). Cambridge University Press, Cambridge

Chapter 6
Fundamentals of Heat and Mass Transport in Frost-Resistant Plant Tissues

Lukas Eurich, Rena Schott, Arndt Wagner, Anita Roth-Nebelsick, and Wolfgang Ehlers

Abstract During their evolutionary history, plants have developed an amazing resistance to various weather conditions, in particular with regard to temperatures below freezing point. In contrast, a phase change of the pore-space content from (liquid) water to (solid) ice within standard construction materials frequently leads to damage. Therefore, our vision is to transfer, at least in part, some of the strategies and structural properties from frost-resistant plants to construction materials.

In this contribution basic strategies and structural properties of frost-resistant plants are introduced. Furthermore, an accompanying customised modelling strategy is proposed. Since plant tissues are formed by highly ordered arrangements of single cells with prescribed sizes, shapes and cell-wall properties, they represent graded (anisotropic) natural porous materials. These structural traits are involved in dealing with frost events. The presented continuum-mechanical modelling approach, based on the Theory of Porous Media (TPM), allows for the description of multicomponent and multiphasic materials. First numerical simulations show the influence of structural traits on the water transport within plant tissues.

6.1 Introduction

Porous materials are ubiquitous in technology, being used for vastly different purposes, such as filters, catalysators, insulation or biomedical applications. Building materials are usually also of a porous nature, including natural building materials. Moreover, biological structures are mostly porous, since all living materials are produced by cells, and, therefore, consist of cells and/or substances that are

L. Eurich • A. Wagner • W. Ehlers
Institute of Applied Mechanics (Civil Engineering), University of Stuttgart, Pfaffenwaldring 7, 70569 Stuttgart, Germany
e-mail: Lukas.Eurich@mechbau.uni-stuttgart.de; Arndt.Wagner@mechbau.uni-stuttgart.de; Wolfgang.Ehlers@mechbau.uni-stuttgart.de

R. Schott • A. Roth-Nebelsick (✉)
Stuttgart State Museum of Natural History, Rosensteintraße 1, D-70191 Stuttgart, Germany
e-mail: rena.schott@smns-bw.de; anita.rothnebelsick@smns-bw.de

J. Knippers et al. (eds.), *Biomimetic Research for Architecture and Building Construction*, Biologically-Inspired Systems 9, DOI 10.1007/978-3-319-46374-2_6

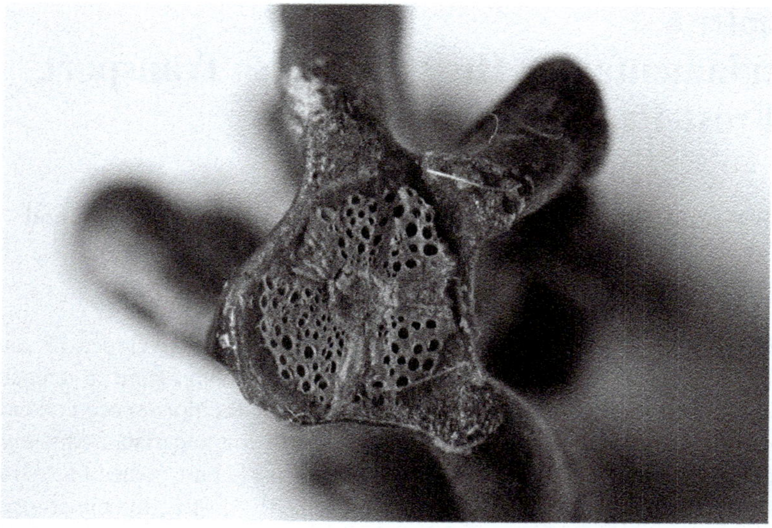

Fig. 6.1 Stem of the tropical liana *Adenia lobata*, in cross section. The diameter of the stem is about 1.5 cm

exuded by cells. The cellular origin of biological materials allows the construction of complex structures that can be intricately patterned down to the nanometer scale. The porous structure is particularly prevalent in many plant tissues.

Plant stems cut open often reveal pores that can be identified by the naked eye (Fig. 6.1). These belong to the xylem, the water-conducting tissue, consisting of dead cells or cell chains arranged parallel to each other, together with a certain part of living xylem parenchyma. The diameter of wood cells usually ranges from 20 μm to about 200 μm. Wood consists almost entirely of xylem and owes its excellent mechanical properties to the lignified cell walls of the conduits.

Less conspicuous are the other tissues that make up the plant stem (or any other plant organ) and whose cells are normally smaller than xylem conduits. Even at first glance, the stem cross section shown in Fig. 6.1 reveals the arrangement of diverse tissues. A cross section, however, does not reveal the three-dimensional structure of the various tissues. Xylem conduits are exceedingly long, reaching a length of several meters for ring-porous trees, while showing a width of about 200 μm. The walls of the wood cells are porous themselves, but with much narrower pores, in the nanometer range. In this way, highly anisotropic porous materials are built. Plant tissues thus represent porous materials with a graded structure. This strongly anisotropic water-conducting material shows high water conductance in the longitudinal direction and low conductance in the horizontal direction.

We can expect that these structural traits affect and control not only water flow, but also heat transport and ice formation. In the course of their evolutionary history, plants have developed an amazing resistance to various weather conditions. This is particularly evident with regard to frost-resistant plants, which are able to withstand many freezing and thawing cycles without any damage. In addition to physiological processes, another crucial factor of frost hardiness is the graded structure of plant materials. Indeed, ice nucleation starts within wood cells and subsequent ice formation proceeds along the longitudinal axis of conduits (Hacker and Neuner 2007).

In contrast, a change in the physical state of the aggregation of contained pore water within standard construction materials yields frequently to damage. In particular, this is typically caused by repeated thawing and freezing processes associated with volume changes of the pore water. Since plants have developed individual strategies in order to adapt to such circumstances, our vision is to transfer, at least in part, these properties to construction materials. The transfer of essential properties of plants to industrially feasible, porous construction materials should allow an optimal building physics involving frost resistance, thermal isolation and moisture transport to be obtained in heterogeneous porous structures.

In order to achieve this goal, several steps and basic considerations are necessary. In this contribution, basic strategies and structural properties of frost-resistant plants are presented. Furthermore, a modelling tool based on the TPM is introduced, which enables us to describe multicomponent and multiphasic materials.

6.2 Freezing Plant Tissues

Controlled "ice management" during freezing has repeatedly been reported for plant tissues, with special porous structures being involved, often associated with xylem (McCully et al. 2004, Roden et al. 2009, Lenné et al. 2010). During freezing, the water potential of the water body decreases, leading to a net flux of as yet unfrozen water towards the freezing site, followed by the continuous formation of ice crystals. Plant tissue is composed of "apoplastic" and "symplastic" regions. The apoplast encompasses all spaces belonging to non-living parts of the plant, i.e. cell walls and intercellular spaces. The symplast includes all living parts, i.e. the living content of cells. Ice crystal formation within the symplast is a critical process, threatening the integrity of the living structures, and has to be avoided.

The phenomenon of decreasing water potential during freezing can be utilized by plants to redirect water towards special apoplastic "freezing sites", while simultaneously dehydrating the symplast. This process has been repeatedly demonstrated. The first impressive demonstration dates back to the 19th century. Prillieux (1869, cited by McCully et al. 2004) showed, in frost-tolerant herbaceous plants, that ice accumulates in the intercellular spaces of petioles, whereas the living parenchyma dehydrates and is compacted to a much smaller volume. Upon thawing, the process is completely reversible. Another example is provided by the needles of frost-resistant pines. Here, water freezes within special sites that are prone to deformation (Roden et al. 2009).

Since the transport of heat and water are coupled in this process, the structure of the considered tissues should be essential, since it dictates the spatial conductances for both quantities. Within this project, these physical aspects of plant tissue freezing have been studied in various organs. Of course, physiological processes are essential, particularly the adaption of cell membranes. In this regard, interference with metabolic processes need to be avoided.

6.2.1 Stems and Wood

Wood or xylem shows a strong degree of porous anisotropy, with the additional property of the absence of a living content in conduits. This also means the absence of osmotic processes that require cell membranes. Consequently, freezing in xylem conduits is comparable with freezing within capillaries: the ice front progresses along the longitudinal axis of the conduits with high velocity (Hacker and Neuner 2007). Additionally, ice nucleation starts normally in xylem conduits, because the ionic content is here much lower than that in living cells. In this way, the xylem body freezes first, in a directed manner, along the longitudinal axis of the plant, while simultaneously dehydrating the surrounding living cells. Ice-filled conduits have been reported on various occasions, for example, by taking samples from winter wood and observing cross sections by cryo-scanning electron microscopy (cryo-SEM) (Utsumi et al. 2003).

The dwarf birch *Betula nana* is a richly branched copse with a height around 60 cm. Its origin in arctic alpine regions makes it a good example of the freezing tolerance of woody stems.

As an example of frozen regions, a cross section of a fresh *B. nana* stem and a frozen one were prepared (Fig. 6.2a, b). The frozen parts, mostly in the bark and the xylem, can be determined visually. The mechanical importance of the intercellular spaces shown in Fig. 6.3 as possible spaces for growing ice crystals becomes evident, since they allow ice formation without damaging living cells. These mechanisms need to be reversible every year.

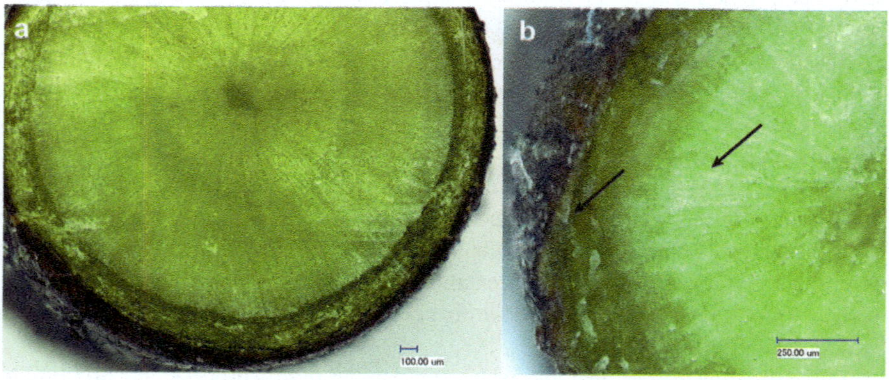

Fig. 6.2 (**a**) Fresh and unfrozen stem of *Betula nana*, in cross section. (**b**) Twig of *Betula nana* in the frozen state. Ice is visible within the xylem and in "pockets" close to the bark (*black arrows*)

Fig. 6.3 Cryo-SEM image of a stem of *Betula nana*, in cross section

Particularly frost-resistant is coniferous wood. Its structure differs from that of angiosperm wood by showing only tracheids. These are single cells that lack living content, are strongly elongated and have lignified cell walls. In conifers, the water-transporting capillaries are thus represented by single cells with a maximum length of several millimeters and a diameter of about 20 μm. Another special property of coniferous tracheids is the structure of their pits, which are pores within the tracheid walls. They show a central thickened lens attached to an elastic spoke-like structure acting as valve when the pressure between two adjacent tracheids changes, for example, because of embolism. This special structure together with the narrow tracheids is viewed as a major element of the frost-resistance of coniferous wood. Upon thawing, air bubbles will be present in the water, as products of the freezing process. The small dimensions of the tracheids keep the bubble radii low, facilitating the dissolution of the air in the thawing water.

6.2.2 *Leaves*

Evergreen leaves and needles have to be frost resistant, for example, the leaves of *Buxus sempervirens Arborescens*. In contrast to the stem, leaves contain much

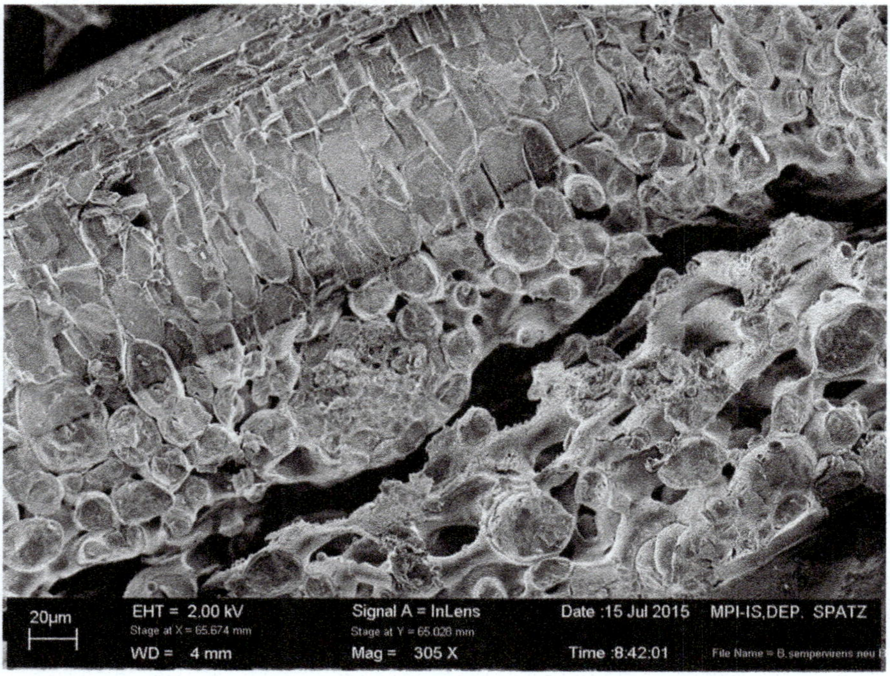

Fig. 6.4 Cryo-SEM image of a leaf of the evergreen *Buxus sempervirens Arborescens*, in cross section

less woody material and the assimilating cells have to be protected during the freezing process. They therefore undergo a special physical adaptation to frost during their normal aging process: the leaves begin to split internally, along their longitudinal axis, which is an irreversible process that provides space for ice crystals to grow. This physical adaptation was investigated by Hacker and Neuner (2007) and Hatakeyama and Kato (1965).

Observed from the outside, these leaves appear thicker and more inelastic during this aging process. During ice formation, the expanding ice widens the split, a process that is reversible in the spring, upon thawing. Figure 6.4 shows a Cryo-SEM image of this crack in an older leaf of *B. sempervirens*.

6.3 Multicomponent Modelling of Plant Tissues

The structure and biological properties of plants indicate that a description solely within either solid mechanics or fluid mechanics is not sufficient. Instead, a combined theoretical modelling approach of the mutually interacting solid–fluid aggregate is crucial. Because of the highly complex porous geometry of plants, a detailed micromechanical and, thus, locally single-phasic model requires the specification of all geometrical and physical transition conditions of the individual components. However, this is for the desired studies, coping macroscopic domains, numerically not feasible. Therefore, we have instead proceeded from a macroscopic (and multi-constituent) continuum-mechanical model based on the TPM. The TPM extends the Theory of Mixtures (Bowen 1976) by the concept of volume fractions, indicating the local amount of a constituent with the volume fraction. In particular, the constituents at the microscale are initially "smeared out" (volumetrically averaged) over a representative elementary volume (REV), resulting in a homogenised macroscopic model with superimposed and interacting constituents. This allows the application of continuum-mechanical theories for the individual constituents. The TPM provides a well-suited framework for the class of materials that are composed of a porous solid skeleton filled with a pore-space content. Within the pore spaces, fluids (liquids or gases) can be present that can themselves be (chemically miscible or immiscible) mixtures of further liquids and/or gases. For a comprehensive description of the TPM and its application fields, the interested reader is referred to the works of Ehlers (2002, 2009), de Boer (2005) or Bluhm et al. (2011) and citations therein.

Within the TPM, the definition of the number and type of constituents that reflect the complexity of the problem appropriately is crucial. In the present case of plant tissues, Fig. 6.5 indicates the consideration of the following constituents: the tissue cells as the structural and flow-guiding component and two pore fluids, namely, materially compressible air (gas) and materially incompressible water (liquid), whereby the latter can be subjected to a phase transition and may turn into ice. This leads to the model

$$\varphi = \varphi^S \cup \varphi^G \cup \varphi^L \cup \varphi^I,$$

where φ is the overall aggregate with its respective constituents.

The definition of the volume fractions $n^\alpha = \mathrm{d}v^\alpha/\mathrm{d}v$ relates the volume element $\mathrm{d}v^\alpha$ of a constituent φ^α to the volume element $\mathrm{d}v$ of the overall aggregate. Furthermore, the saturation condition

$$\sum\nolimits_\alpha n^\alpha = n^S + n^G + n^L + n^I = 1$$

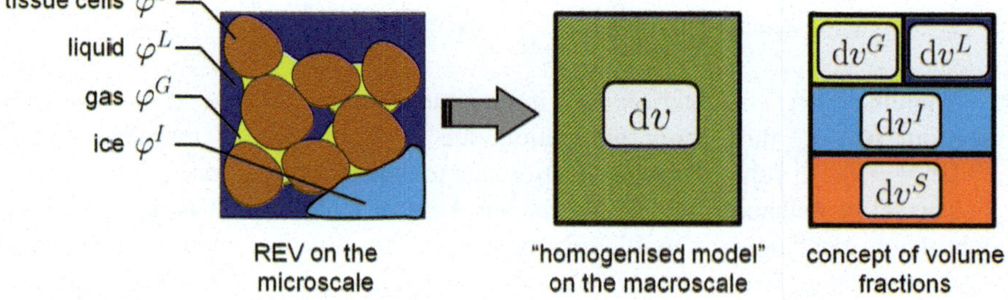

Fig. 6.5 Basic constituents of the TPM model for plant tissues

is fulfilled at any time, since no vacant space occurs within the considered domain. For the biological system of the plant tissue, the TPM proceeds from continuum-mechanical balance equations. These have to be fulfilled not only for the overall aggregate φ, i.e. the homogenised model from Fig. 6.5, but also for each constituent individually. According to Ehlers (2002), these are given for the particular constituents via

$$(\rho^\alpha)'_\alpha + \rho^\alpha \text{div}\mathbf{x}'_\alpha = \widehat{\rho}^\alpha\text{: mass balance}$$

$$\rho^\alpha \mathbf{x}''_\alpha = \text{div } \mathbf{T}^\alpha + \rho^\alpha \mathbf{b}^\alpha + \widehat{\mathbf{p}}^\alpha\text{: momentum balance}$$

$$0 = \mathbf{I} \times \mathbf{T}^\alpha + \widehat{\mathbf{m}}^\alpha\text{: balance of moment of momentum}$$

$$\rho^\alpha (\varepsilon^\alpha)'_\alpha = \mathbf{T}^\alpha \cdot \mathbf{L}_\alpha - \text{div}\mathbf{q}^\alpha + \rho^\alpha r^\alpha + \widehat{\varepsilon}^\alpha\text{: energy balance}$$

Therein, $(\cdot)'_\alpha$ denotes the material time derivative with respect to the motion of constituent φ^α. Furthermore, $\rho^\alpha = n^\alpha \rho^{\alpha R}$ is the partial density, with the effective density $\rho^{\alpha R}$, and $\widehat{\rho}^\alpha$ is the density production. Using this quantity, the phase transition from water to ice and vice versa can be described. The vectors $\mathbf{x}'_\alpha$ and $\mathbf{x}''_\alpha$ describe the velocity and the acceleration of constituent φ^α, respectively. The partial stress is given by $\mathbf{T}^\alpha$, and $\mathbf{b}^\alpha$ is the vector of the body forces. Furthermore, the second-order tensor $\mathbf{L}_\alpha$ describes the spatial velocity gradient, which is defined via $\mathbf{L}_\alpha = \text{grad } \mathbf{x}'_\alpha = (\mathbf{F}_\alpha)'_\alpha \, \mathbf{F}_\alpha^{-1}$, where $\mathbf{F}_\alpha = \text{Grad}_\alpha \mathbf{x}$ is the deformation gradient. The quantities ε^α, r^α and $\mathbf{q}^\alpha$ describe the internal energy, the radiation and the heat flux, in the given order. The remaining quantities $\widehat{\mathbf{p}}^\alpha$, $\widehat{\mathbf{m}}^\alpha$ and $\widehat{\varepsilon}^\alpha$ represent the direct production in terms of momentum, moment of momentum and internal energy, respectively. These production terms are, as is the density production $\widehat{\rho}^\alpha$, subject to the following constraints, since they have to vanish for the overall aggregate:

$$\sum\nolimits_{\alpha=1}^{n} \widehat{\rho}^\alpha = 0$$

$$\sum\nolimits_{\alpha=1}^{n} (\widehat{\mathbf{p}}^\alpha + \widehat{\rho}^\alpha \mathbf{x}'_\alpha) = \mathbf{0}$$

$$\sum\nolimits_{\alpha=1}^{n} \widehat{\mathbf{m}}^\alpha = \mathbf{0}$$

$$\sum\nolimits_{\alpha=1}^{n} \left[\widehat{\varepsilon}^\alpha + \widehat{\mathbf{p}}^\alpha \cdot \mathbf{x}'_\alpha + \widehat{\rho}^\alpha \left(\varepsilon^\alpha + \frac{1}{2} \mathbf{x}'_\alpha \cdot \mathbf{x}'_\alpha \right) \right] = 0$$

Therein, $\widehat{\rho}^\alpha \mathbf{x}'_\alpha$ characterises the momentum production due to density production, $\widehat{\mathbf{p}}^\alpha \cdot \mathbf{x}'_\alpha$ the energy production due to production of momentum, and $\widehat{\rho}^\alpha \left(\varepsilon^\alpha + \frac{1}{2}\mathbf{x}'_\alpha \cdot \mathbf{x}'_\alpha\right)$ the energy production due to density production.

The set of balance laws is valid for any kind of material that is treated within continuum mechanics of multicomponent materials. Specific material behaviour, e.g. elastic or inelastic material behaviour, isotropic or anisotropic fluid flow, etc., are described with constitutive laws. In order to propose reasonable constitutive laws, they have to be in accordance with the second law of thermodynamics, which poses restrictions to the direction of thermo-mechanical processes. The interested reader is referred to Coleman & Noll (1963) for the basic process and to Ehlers (2002) for a detailed discussion of the way to find these restrictions and to use them in the continuum-mechanical context of the TPM.

6.4 Numerical Simulations

Within a continuum-mechanical modelling of porous materials, the solution strategy for the acquired system of coupled partial differential equations proceeds from the formulation of the governing equations for the chosen primary variables. In the case of plant tissues, this is the displacement of the solid component, the fluid pressures, and the temperature of the overall aggregate. Additional equations, which can be derived from the balance equations, often in combination with constitutive assumptions, close the continuum-mechanical problem. In order to derive a form that can be solved numerically by means of the Finite Element Method (FEM), the strong formulations of the governing equations have to be transferred into weak formulations, inter alia by considering the boundary conditions of the initial-boundary-value problem. A monolithic solution scheme is then applied by using mixed finite elements. As the water-flow management was identified as one crucial factor for the frost resistance of plants, one numerical example examines the basic water transport in a stem of a tree (section 6.4.1), and another example discusses the water supply of a leaf (section 6.4.2). These two examples are simulated by a simplified continuum-mechanical model consisting of a single pore fluid without phase transitions within an anisotropic solid skeleton.

6.4.1 Basic Water Transport in a Stem

The first study aims to describe the basic properties of water transport in a stem. The structure of such a stem is a highly anisotropic. In particular, the permeability in the axial direction is significantly higher than in the transversal direction, as was described in the Introduction. In Fig. 6.6, the simulation set-up including the boundary conditions and the resulting perfusion within the solid skeleton are presented.

The perfusion can be characterised by the seepage velocity of the liquid $\mathbf{w}_F = \mathbf{x}'_F - \mathbf{x}'_S$ as the flow of the liquid through the porous material. For the present model, a Darcy-type equation (cf., e.g. Ehlers (2009)) is obtained by thermodynamical considerations, viz.:

$$\mathbf{w}_F = -\frac{\mathbf{K}^{SF}}{n^F \mu^{FR}} \left(\operatorname{grad} p^{FR} - \rho^{FR}\mathbf{g}\right).$$

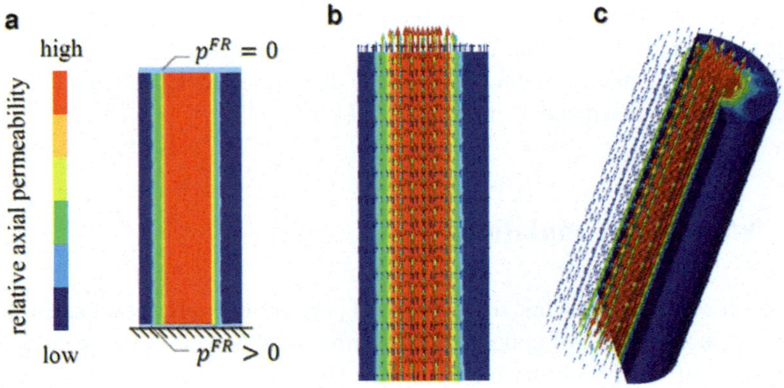

Fig. 6.6 (**a**) Problem description with axial permeability conditions, (**b**) seepage velocity (vertical cut), *large arrows* indicate a high velocity, *small arrows* indicate a low velocity, (**c**) seepage velocity (3D-perspective)

Therein, the intrinsic permeability $\mathbf{K}^{SF}$ is a second-order tensor, which describes, in general, anisotropic perfusion. This tensor can have preferred directions, as in case of the present model. Furthermore, the axial permeability in the centre of the stem is higher than that in the bark, as is shown in Fig. 6.6a. The viscosity μ^{FR} of the fluid and the gravitational acceleration $\mathbf{b}^{\alpha} = \mathbf{g}$ is included. The representation of the seepage velocity shows that the crucial driving force is given by the pressure gradient, which causes perfusion as is shown in Fig. 6.6b, c.

6.4.2 *Simulation of Water Supply in a Leaf*

With a locally varying permeability, the water transport in a leaf can be simulated, which is schematically shown in this example. The permeability within the veins of a leaf is distinctly higher than that in the rest of the leaf. The actual allocation of the regions of high permeability is schematically shown in Fig. 6.7, where the surrounding curve represents the shape of a leaf, and the inner black lines the veins with a high permeability. Since the numerical simulation is carried out by means of the FEM, as it is represented by the mesh in Fig. 6.7, an evaluation has to be carried out at each Gauss point. At each Gauss point, the permeability can be assigned, resulting in a locally varying permeability.

Applying the boundary condition to the leaf, i.e. a pressure gradient from the stem to the distal end of the veins, the resulting flow in terms of seepage velocity is shown in Fig. 6.8.

The attuning seepage velocity field shows high perfusion within the main vein in the centre of the leaf, a moderate perfusion in the remaining veins and low perfusion everywhere else in the leaf. This is also reflected by the experimental findings for the corresponding flow within a leaf in which the pressure gradient is mainly caused by the evaporation of the liquid on its surface.

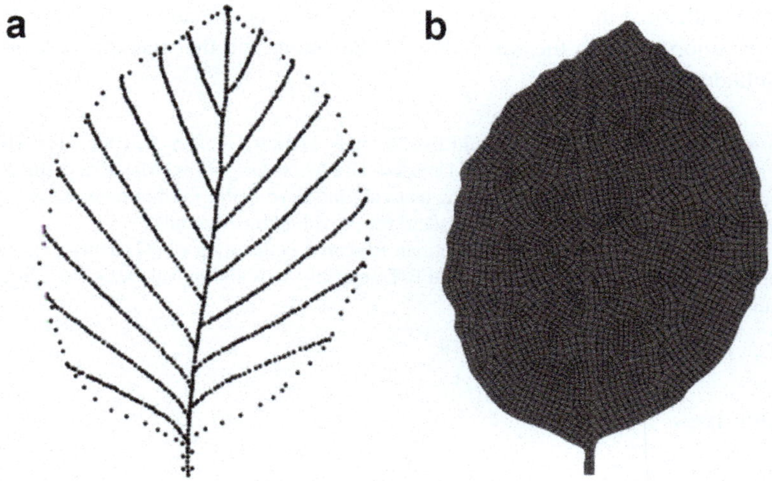

Fig. 6.7 (a) Shape and regions of high permeability (*black lines*), (b) FEM mesh

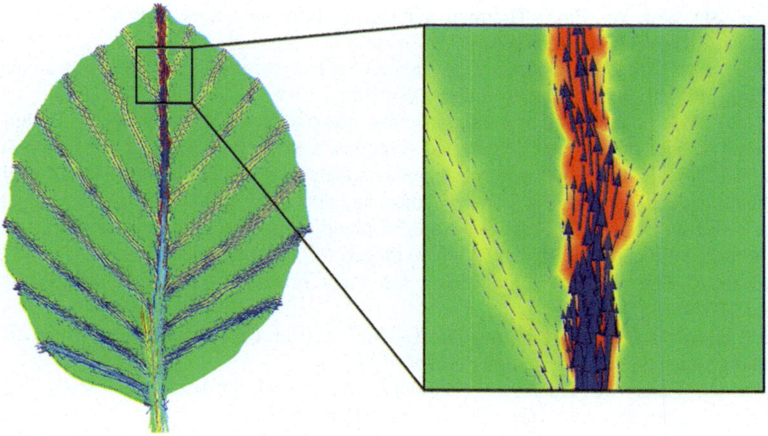

Fig. 6.8 Simulation results of the seepage velocity (*blue arrows*)

6.5 Conclusions

Within the scope of this publication, the presented article is aimed to introduce the issue of transferring mechanisms and structural properties of frost-resistant plant tissues towards engineering constructions and architecture. Therefore, the biological processes and structural properties of frost-resistant plant tissues and a modelling tool, which enables us to describe multicomponent and multiphasic materials, are introduced. The presented numerical examples illustrate the feasibility of this method with regard to plant tissues and describe flow properties, which is one factor that may influence the formation of ice by water-flow management. Nevertheless, the interaction between the water-flow management and the freezing process itself is still under investigation.

Acknowledgements The authors would like to thank Markus Günther, Dagmar Voigt (Institute of Botany at the Technical University of Dresden) and Ioanis Grigoridis (Max Planck Institute for Intelligent Systems, Stuttgart) for allowing us to use the Cryo-SEM. Furthermore, we are grateful to Barbara Ditsch (Botanical Garden, Dresden) for providing fresh samples.

This work has been funded by the German Research Foundation (DFG) as part of the Transregional Collaborative Research Centre (SFB/Transregio) 141 'Biological Design and Integrative Structures'/A01.

References

Bluhm J, Ricken T, Bloßfeld M (2011) Ice Formation in porous media. In: Markert B (ed) Advances in extended & multifield theories for Continua, vol 59, LNACM. Springer, Berlin, pp 153–174

Bowen RM (1976) Theory of mixtures. In: Eringen AC (ed) Continuum physics, vol III. Academic, New York, pp 1–127

Coleman BD, Noll W (1963) Thermodynamics of elastic materials with heat conduction and viscosity. Arch Ration Mech Anal 13:167–178

de Boer R (2005) Trends in continuum mechanics of porous media, vol 18, Theory and applications of transport in porous media. Springer, Dordrecht

Ehlers W (2002) Foundations of multiphasic and porous materials. In: Ehlers W, Bluhm J (eds) Porous media – theory, experiments and numerical applications. Springer, Berlin, pp 3–86

Ehlers W (2009) Challenges of porous media models in geo- and biomechanical engineering including electro-chemically active polymers and gels. Int J Adv Eng Sci Appl Math 1:1–24

Hacker J, Neuner G (2007) Ice propagation in plants visualized at the tissue level by infrared differential thermal analysis (IDTA). Tree Physiol 27:1661–1670

Hatakeyama I, Kato J (1965) Studies on the water relation of Buxus leaves. Planta (Berl) 65:259–268

Lenné T, Bryant G, Hocart CH, Huang CX, Ball MC (2010) Freeze avoidance: a dehy-drating moss gathers no ice. Plant Cell Environ 33:1731–1741

McCully ME, Canny MJ, Huang CX (2004) The management of extracellular ice by petioles of frost-resistant herbaceous plants. Annals Bot 94:665–674

Prillieux E (1869) Sur la formation de glaçons a l'intérieur des plantes. Ann Sci Nat 12:125–134

Roden JS, Canny MJ, Huang CX, Ball MC (2009) Frost tolerance and ice formation in Pinus radiata needles: ice management by the endodermis and transfusion tissues. Funct Plant Biol 36:180–189

Utsumi Y, Sano Y, Funada R, Ohtani J, Fuhikawa S (2003) Seasonal and perennial changes in the distribution of water in the sapwood of conifers in a sub-frigid zone. Plant Physiol 131:1826–1833

Chapter 7
Plants and Animals as Source of Inspiration for Energy Dissipation in Load Bearing Systems and Facades

Katharina Klang, Georg Bauer, Nicu Toader, Christoph Lauer, Kathrin Termin, Stefanie Schmier, Daria Kovaleva, Walter Haase, Christoph Berthold, Klaus G. Nickel, Thomas Speck, and Werner Sobek

Abstract From the manifold strategies that nature offers to materials under overload conditions, we describe two: the fibrous and multi-layered system of the bark of the Giant Sequoia, which possesses an impressive damping mechanism, and the spines of pencil and lance sea urchins. The latter introduce a new concept to energy dissipation in brittle construction materials, namely quasi-ductility by multiple local fracturing. The potential for transfer as bioinspired technical solutions is high as the biological role models combine several advantages such as lightweight, recyclability and high protective efficiency. We demonstrate that, in principle, the concepts found in the biological role models can be transferred to concrete-based building materials.

7.1 Introduction

For many applications in the sector of building construction, a combination of lightweight construction with high energy dissipation capacity and (locally adapted) puncture resistance is of increasing interest. This holds true not only for regions

K. Klang (✉) • C. Lauer • K. Termin • C. Berthold • K.G. Nickel
Applied Mineralogy (AMIN), University of Tübingen, Wilhelmstraße 56, 72074 Tübingen, Germany
e-mail: katharina.klang@uni-tuebingen.de

G. Bauer • S. Schmier • T. Speck
Plant Biomechanics Group, Botanic Garden, Faculty of Biology, University of Freiburg, Schänzlestraße 1, D-79104 Freiburg, Germany

Freiburg Centre for Interactive Materials and Bioinspired Technologies (FIT), University of Freiburg, Georges-Köhler-Allee 105, D-79110 Freiburg, Germany

N. Toader • D. Kovaleva • W. Haase • W. Sobek
Institute for Lightweight Structures and Conceptional Design (ILEK), University of Stuttgart, Pfaffenwaldring 7 + 14, 70569 Stuttgart, Germany

J. Knippers et al. (eds.), *Biomimetic Research for Architecture and Building Construction*, Biologically-Inspired Systems 9, DOI 10.1007/978-3-319-46374-2_7

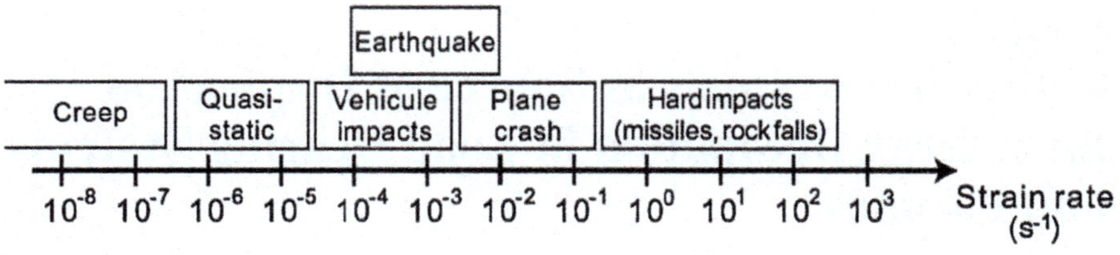

Fig. 7.1 Regimes of strain rates (Bischoff and Perry 1991; Hentz et al. 2004)

having a high earthquake or thunderstorm frequency or being endangered by rockfall events or avalanches, but also for endangered buildings covering sensitive technologies such as those in the chemical industries or in power plants.

Energy dissipation in a strictly physico-chemical sense is defined as the result of an irreversible process, in which entropy and ultimately heat is generated from other forms of energy (Zinoviev and Ermakov 1994). In our context, the term has a more general form, whereby coherent or directed energy is transformed into a more isotropic energy, as is known from processes involving friction. By analogy, the energy transmitted by a blow or directed stress to a component can be converted to heat, friction or surface energy and, in this way, becomes lost for any reversible deformation process or for continued stressing of the component. Figuratively speaking, we are searching for the "crumple zone" of natural constructions, where sacrificing some parts often protects the core life functions of an animate being. In terms of buildings, this is the quest to cope with temporary overload situations by allowing local damage without total collapse.

A building that has to withstand impact loads (car crashes, derailed trains, earthquakes etc.) can dissipate energy through plastic deformation or through damping. As a consequence of such actions, building components are exposed to dynamic loading at a wide range of strain rates (Fig. 7.1). The dependence of the concrete strength and its deformability with the strain rate is an important feature of concrete dynamic behaviour.

Energy dissipation attributable to damping can be obtained by using either active or passive energy dissipation systems. Active systems can be fully active (e.g. active bracing systems, active mass dampers, etc.) or semi-active (e.g. smart materials: shape memory alloys, piezoelectric materials, electrorheological fluids). Fully active systems are control systems in which the structural protection is ensured through controlling or modifying the motion of the structure with an external energy supply. The semi-active systems only need a nominal amount of energy to adjust their mechanical properties and can modify the motion of the structure only by dissipating energy, unlike fully active systems, which are also able to add energy for building motion control (Constantinou et al. 1998; Teuffel 2004).

Passive systems such as metallic dampers, friction dampers, viscoelastic dampers, viscoelastic or viscous fluid dampers are used for structural applications as devices incorporated into a building to modify the structural response. These devices have the basic function of absorbing or consuming a portion of the input energy, thereby reducing energy dissipation demand on primary structural elements and minimising possible structural damage (Constantinou et al. 1998).

In structures, energy dissipation attributable to plastic deformation can be obtained by proper structural compliance. In this case, the load-bearing structure of the building consists of dissipative and non-dissipative areas. Whereas the dissipative areas are characterised by a large deformation capacity, the non-dissipative areas have to remain in the elastic domain until the ultimate limit state is reached. Therefore, high amounts of energy have to be absorbed through plastic deformations of the dissipative areas without affecting the strength and stability of the main bearing structural system. The location of the dissipative areas is designed in correlation to the type of the used structural system (the end of beams – frame type structures, the base of the walls – wall type structures; bracings in tension – braced frame structures etc.) (Constantinou et al. 1998).

Nature may teach us some lessons to reduce these drawbacks, because biology offers a multitude of potential concept generators possessing one or several of the required properties. Biological lightweight constructions acting as energy-dissipating protective systems need to be investigated in order that their form-structure-function relationship can be quantitatively understood, abstracted, adapted and transferred to bioinspired built structures.

We have chosen the bark of the Giant Sequoia and sea urchin spines as starting points for quite differing strategies that nature uses to cope with mechanical stresses and impacts. The bark was chosen to represent a system in which an organic fibre-based layered structure is the damping unit. This is therefore a strategy akin to the passive damping endeavours in the building industry. Other possible role models are impact- and puncture-resistant fruits and seeds, such as the endocarp of the coconut, which will be presented in Chap. 14.

Sea urchin spines have been chosen, because they represent the concept of energy dissipation by quasi-plastic deformation based on multiple fracturing, which is rarely considered in classic architectural reasoning. Furthermore, the constituting material is very simple. It is made of calcite ($CaCO_3$) with a very low content of organic matter and yet is a hierarchically organised construction with amazing properties (Presser et al. 2009).

Progress has also been made in the direction of technical implementation by manufacturing graded/structured porous concrete constructions (Herrmann et al. 2013). Even though concrete is a material having a rather complex microstructure compared with the monomineralic natural materials, it is certainly the obvious choice for building constructions.

7.2 Energy Dissipation in Natural Materials

7.2.1 Impact-Resistant Tree Bark

Among tree barks, the bark of the Giant Sequoia (*Sequoiadendron giganteum* Lindl., Fig. 7.2) stands out because of its thickness, low density and high external and internal structuring. These properties are advantageous, for example, with regard to the highly frequent forest fires (Bauer et al. 2010) that occur in the Sierra Nevada (USA) in which the Giant Sequoia is native. However, a thick bark is also beneficial during rockfall events, which occur regularly on the slopes of the Sierra Nevada (Wieczorek 2002). The vascular cambium is located subjacent to the bark; this sensitive tissue comprises only a few cell layers and is crucial for the formation of secondary xylem (wood) and phloem (bast). As it cannot be regenerated from any other tissue after injury, the protection of the cambium is essential for the survival and growth of the whole tree. Because of high energy impacts that may occur during rockfall events, the bark of a tree might take considerable harm. Thus, the cambium might also be injured during impacts, especially in tree species with thin barks. Even if the immediate damage to the cambium during such a high energy rockfall event is not lethal for the tree, subsequent infections by pathogens 'entering' the tree through the injury are a threat for the whole organism.

In the literature, the barrier effect of trees during rockfall events is mainly considered in order to answer the question as to what advantage afforestation will have in hindering rockfall events from affecting populated areas (Stokes et al. 2005; Volkwein et al. 2011).

The effect of a rockfall event on a tree itself is mainly neglected. However, Stokes et al. (2005) have suggested examining the influence of bark thickness on the wound damage to a stem for various species in future studies. First results dealing with the question as to whether the impact-damping properties of the thick bark of the Giant Sequoia are outstanding compared with the bark of other tree species, and thus render the basis to protect the cambium from damage during rockfall events and/or subsequent pathogen infestation, will be presented in this chapter. Additionally, the possible contribution of the prominent outer and inner structuring (Figs. 7.2b and 7.3) of the bark of the Giant Sequoia to the dissipation of energy during rockfall

Fig. 7.2 Giant Sequoia (*Sequoiadendron giganteum*). (**a**) Giant Sequoia in Sequoia National Park, California (USA). The outer structuring of the thick bark is clearly visible. (**b**) A sample of Giant Sequoia bark showing the layered inner structuring of the bark (the front surface represents the longitudinal radial surface of the bark)

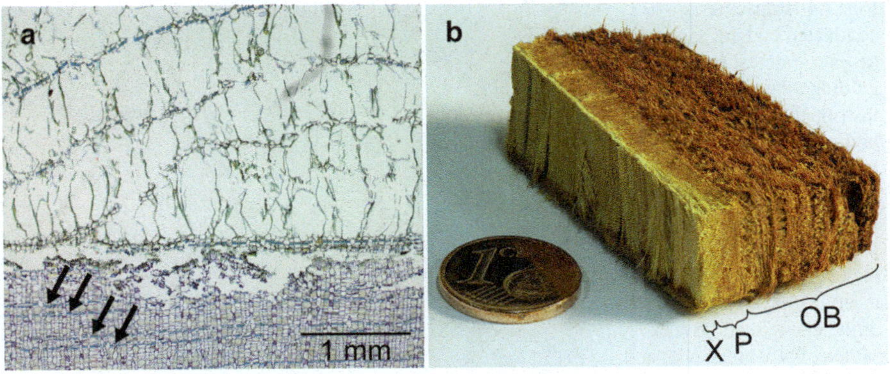

Fig. 7.3 Inner structuring of Giant Sequoia bark. (**a**) Thin section stained with toluidine blue. The inner bark or phloem (*lower part* of figure) consists of densely linked, thin-walled cells, horizontally traversed by rows of thick-walled cells (fibres, stained blue and marked by *black arrows*). These rows are also present in the outer bark (*upper part* of figure), whereas the cell walls of the interjacent cells are ruptured. This results in a still-layered but markedly less-dense outer bark. (**b**) Visualisation of the fibrous structure of the outer bark using a sawed out block of bark (upper surface represents the cross-sectional plane). Pulled out fibres are numerous in the outer bark (OB), but are absent in the inner bark (Phloem, P). Furthermore, the fibrous structure of the wood (Xylem, bright inner layer, X) becomes visible

events, will be elucidated. Finally, the potential of this biological role model for a transfer into graded/structured and fibre-reinforced porous concrete constructions will be outlined.

7.2.1.1 Quantification of Bark Energy Dissipation Using Impact Tests

In the following paragraph, a quantification method for the energy dissipation potential of different tree barks is presented. Therefore, impact tests of the Giant Sequoia bark were conducted and compared with identical tests carried out on the bark of the Tree of Heaven (*Ailanthus altissima* (Mill.) Swingle). This tree species is native to China and northern Vietnam, in flat rural areas without rockfall events. Its bark is much thinner and denser compared with the bark of the Giant Sequoia and thus was selected as a reference.

One individual Giant Sequoia (age: approx. 90 years and height: approx. 42 m) and one individual Tree of Heaven (age: approx. 70 years and height: approx. 28 m) were logged in the Botanic Garden of Freiburg and slices were cut from the tree trunks at various heights above ground, each slice comprising the whole cross section of the stem including the bark. The bark, defined here as comprising all tissues outside the vascular cambium, was cut off the wood and dried (DIN Norm EN 13183-1) before further preparation. For the sample preparation, the bark was cut into pieces with a quadratic basal area (3 × 3 cm). The thickness of the tree bark (representing the third dimension, i.e. the height, of the samples) was not cut into

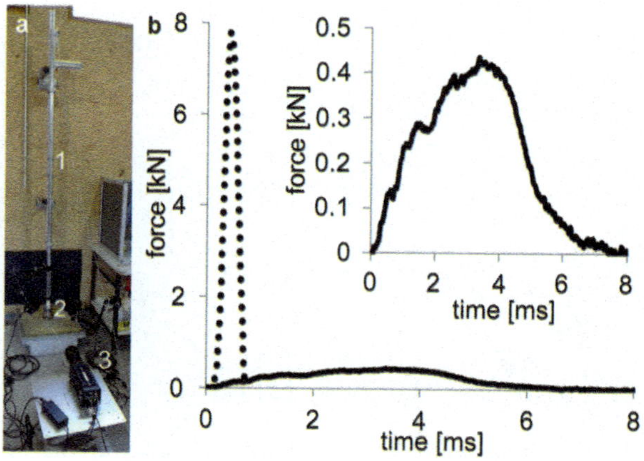

Fig. 7.4 Impact tests conducted on bark samples. (**a**) Free-fall test rig with (1) a guiding tube, (2) an instrumented anvil, and (**3**) a high-speed camera. (**b**) Exemplary force-time curves of an impact test of a Tree of Heaven bark sample (*dotted line*) and of a Giant Sequoia bark sample (*solid line*). The insert displays the same curve of the Giant Sequoia bark sample, but with an adjusted scale of the ordinate

shape to keep the inner and outer structuring of the bark sample unaltered. A free-fall test rig and a high-speed camera (model MotionPro Y4, Integrated Design Tools, Inc., Tallahassee, FL, USA) were used for the impact tests. The bark samples were placed on top of an instrumented anvil (equipped with a 20 kN force sensor, model 8402–6020, Burster Präzisionsmesstechnik GmbH & Co KG, Gernsbach, Germany) and a drop weight with a mass of 200 g was dropped through a guided tube onto the bark sample. The drop height was set to 1.99 m (distance of the lower surface of the drop weight to the upper surface of the ambos at test begin), resulting in an average impact energy of 3.62 J (calculated from drop weight velocity at impact). The impact was recorded with the high-speed camera (for a detailed description of the test setup, we refer the reader to Thielen et al. 2015). The test setup and exemplary force-time diagrams of impact tests with bark from the two selected tree species can be seen in Fig. 7.4.

The narrow force peak of the Tree of Heaven corresponds to the impact behaviour of a dense material with a high transmitted force. However, the bark of the Tree of Heaven does not represent an ideal elastic system in which the entire initial kinetic energy is released after impact. In contrast, the wide and shallow force-time curves found for the Giant Sequoia are less smooth, especially before reaching the force maximum (see insert in Fig. 7.4b). This indicates other mechanisms such as marked internal friction, revealing the potential to store and dissipate significantly higher amounts of energy during an impact event.

That tree barks do not represent ideal elastic systems becomes even more obvious when tracking the speed of the drop weight immediately before and after impacting the bark samples by using recorded high-speed videos of the impact test.

Converting the speed v of the drop weight (with the mass m) into its kinetic energy E_{kin}

$$E_{kin} = \frac{1}{2} \cdot m \cdot v^2 \tag{7.1}$$

immediately before and after the impact allows the calculation of the dissipated energy E_{diss} for each bark sample tested:

$$E_{diss} = E_{kin}^{init} - E_{kin}^{fin} \tag{7.2}$$

with the kinetic energy of the drop weight immediately before (E_{kin}^{init}) and after (E_{kin}^{fin}) the impact. For comparison, the percentage of the dissipated energy $E_{diss\,\%}$ was calculated using

$$E_{diss\,\%} = \frac{E_{diss}}{E_{diss}^{init}} \cdot 100 \tag{7.3}$$

However, this test does not define how much of the energy is simply transmitted as a shock wave and becomes dissipated elsewhere in the system and how much is actually absorbed in the bark. Nonetheless, these values and the shapes of the force-time curves allow an assessment of the processes involved to a good first order approximation.

The two groups of bark collected at two different heights from the Tree of Heaven show a median value of energy dissipation of 85.9 % and 82.8 %, respectively, and differ significantly ($p < 0.01$), whereas the values from the different heights of the Giant Sequoia do not differ significantly with the exception of bark samples taken from 19 m height when compared with the samples from 11.1 m ($p < 0.05$) and 14.5 m ($p < 0.05$) (Fig. 7.5). All values found in the Giant Sequoia show a significantly higher energy dissipation than the values measured for the Tree of Heaven with median values of energy dissipation ranging from 89.0 to 91.8 %.

The pattern of energy dissipation during an impact event is currently quantified in detail by relating the force to the deformation (compaction) measured during the

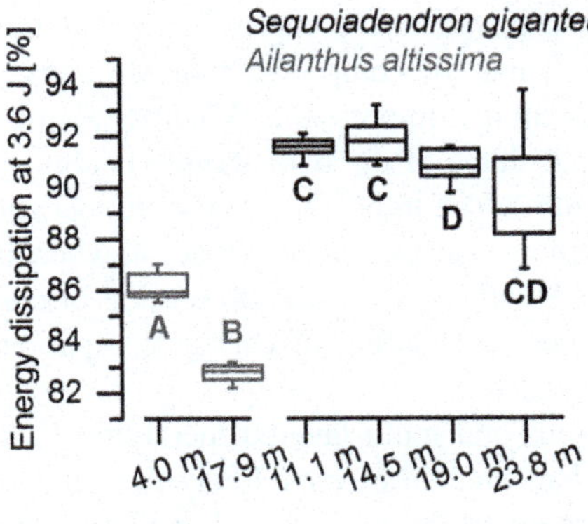

Fig. 7.5 Energy dissipation of bark samples taken from the two tree species at different heights above ground ($n = 8$ for both groups of the tree of Heaven, $n = 10$ for all four groups of the Giant Sequoia). Groups indicated with a same capital letter show no statistically significant difference, whereas groups indicated with different capital letters show a significantly different energy dissipation (pairwise Mann-Whitney U-test with Holm adjustment)

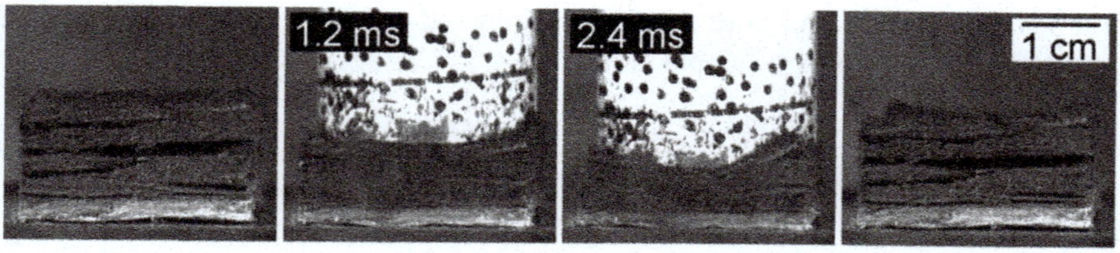

Fig. 7.6 High-speed video excerpts of an impact test of the bark of the Giant Sequoia. From left to right: bark sample before impact, bark sample 1.2 ms and 2.4 ms (maximum compression) after first contact with the impactor during the impact test and bark sample after the impact test being only slightly compressed

impacts. Our results show that a thin, not layered and less fibrous bark, like the one of the Tree of Heaven, results in a more direct transfer of the impact force through the sample. Thus, a bark type such as that present in the Tree of Heaven is markedly less capable of protecting underlying living tissues such as the sensitive cambium from damage caused by impacts.

Our results further establish that the energy dissipation capacity of the bark generally decreases with tree height mainly because of decreasing bark thickness in the increasingly younger upper tree parts. This holds true for both tree species but is less pronounced in the Giant Sequoia (Fig. 7.5). From an ecological point of view, this developmentally 'unavoidable' fact represents no shortcoming as the probability of facing an impact caused by rockfall decreases markedly with increasing tree height.

The high-speed videos reveal several stages of compression for bark samples of the Giant Sequoia (Fig. 7.6). In the uncompressed state prior to the impact, the division into the outer and inner bark and the layered structure of the bark are clearly visible. During the initial phase of an impact, the air-filled spaces between the layers are compressed first (1.2 ms after the impact). In later phases of the impact, the layers themselves are also compressed, most likely by (a) compression of the network of fibres and (b) compression of the cells, as is visible 2.4 ms after the impact, at maximum compression. When removing the force applied by the impactor, the original shape of the bark sample is almost completely restored (Fig. 7.6, rightmost picture). The impact tests indicate that energy is mainly stored and dissipated by friction or viscoelastic deformation, whereas plastic deformation is of minor importance. However, the exact proportion of elastic, viscoelastic and plastic deformation to the contribution of energy dissipation needs to be further examined in ongoing studies.

In conclusion, a fibrous and multi-layered hierarchical structuring of the bark is, in addition to sheer bark thickness, most likely the structural basis for a higher amount of energy dissipation in this biological damping system.

The hierarchical structuring of the Giant Sequoia tree bark thus represents a promising role model for transfer into energy dissipating graded technical damping constructions with hierarchical inner structuring. A more detailed analysis of the bark structure and of the deformation behaviour of the Giant Sequoia bark during impact will yield a wide array of ideas for such a biomimetic transfer, for example, in fibre-reinforced graded lightweight concrete.

7.2.2 Sea Urchin Spines

Sea urchins are members of an ancient group that is hundreds of million years old from the class Echinoidea in the Phylum Echinodermata and represent one of the taxonomically most diverse groups of marine invertebrates in the oceans of the world today. They are prevalent in all over the marine world and live in both high-energy tidal environments and low energy abyssal depths. Depending on their ecological niche and habitat, the morphology of their spines differs in terms of size, thickness and surface properties.

The investigated species, namely *Heterocentrotus mammillatus* (pencil sea urchin) and *Phyllacanthus imperialis* (lance sea urchin), are both reef inhabitants with thick, large and blunt spines adapted to strongly hydrodynamic environments (Fig. 7.7). *Heterocentrotus mammillatus* is native to reef crests, whereas *Phyllacanthus imperialis* occurs in backreef areas where lower energy conditions prevail (Nebelsick 1992). The spines of both species are used for anchorage within the reef and for protection from predators and the energy released by high tides (Amemiya et al. 2005).

Spines of these sea urchin species are made of brittle magnesium calcite ($Ca_xMg_{1-x}CO_3$) forming a highly porous three-dimensional mesh of trabeculae (stereom). The porous structure has several functions. As a lightweight construction, it helps the animal to achieve agility and limits the expenditure of energy concurrently. The cavities of the stereom are partly filled with collagen fibres

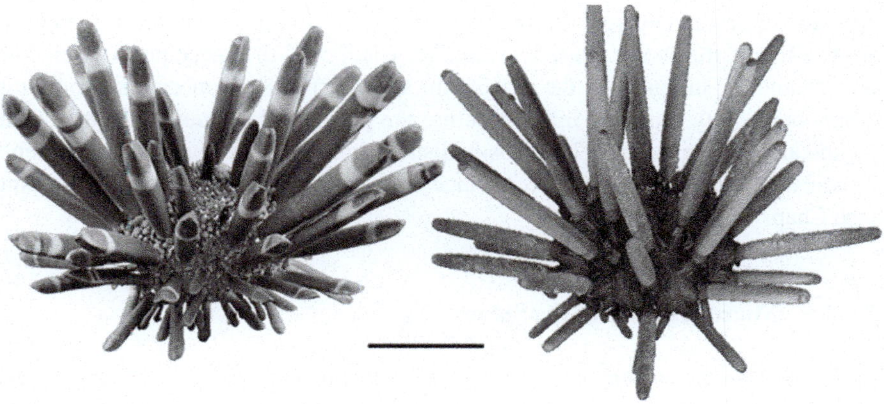

Fig. 7.7 The investigated sea urchins: *Heterocentrotus mammillatus* (*right*), *Phyllacanthus imperialis* (*left*) (*scale bare* = 5 cm)

and are permeable to fluids, possibly to enable the regeneration and growth of the spine (Großmann and Nebelsick 2013). As outlined below, the structure also achieves a 'graceful failure' mode under compressive overload. The stereom of the spine can vary in its porosity, pore size/shape and constructional design from species to species. Although differences can also appear in one single animal, for instance, when comparing primary and secondary or young and adult spines, the differences between the species probably reflect evolutionary adaptations to differing habitats.

Chemical analyses have shown that the $MgCO_3$-content varies between 2 to 12 mol% with an organic content of approximately 0.1 mol% (Magdans and Gies 2004). Despite their porous structure, these spines appear in X-ray diffraction or polarised light as single crystals. Indeed, they consist of highly ordered nanoparticle-sized domains (10–40 nm, mesocrystals), oriented parallel along the c-axis (Eiberger 2007; Oaki and Imai 2006). The misalignment angle between the individual domains is below 2° (Aizenberg et al. 1997). Amorphous regions and macromolecules are embedded in and between the magnesium calcite domains. Thus, the substitution of Mg for Ca (Astilleros et al. 2010) and the slight misalignment of the individual domains helps to increase fracture toughness and to avoid the typical cleavage of pure calcite along the cleavage plane. Instead, the spines show conchoidal fracture behaviour.

Studies at Applied Mineralogy in Tübingen (Presser et al. 2011) have shown that the spines of *Heterocentrotus mammillatus* and *Phyllacanthus imperialis* possess a high initial crushing strength σ_c and that their mechanical behaviour under uniaxial compression involve high energy dissipation by graceful failure in compression (mean σ_c of 47 ± 17 MPa for the spines of *Heterocentrotus mammillatus* and for spines of *Phyllacanthus imperialis* $\sigma_c = 61 \pm$ MPa). Currently, we are investigating the details of the relationship between stereom structure, strength and energy dissipation. The complexity of the failure process involving buckling, cracking and spallation are explored in uniaxial compression tests monitored by video extensometry. Local pin indentation tests have been developed to quantify stresses, strains and energy consumption as described in greater detail in Chap. 14, Sect. 14.1 of this book. Based on the knowledge of the mechanical behaviour and geometric structure of the sea urchin spines, studies upon concrete building components with a graded inner structure are performed as starting point for the technical implementation. The porosity distribution inside the concrete components is adjusted and optimized in accordance to the stress distribution under loading (see Chap. 4).

7.2.2.1 Microstructure of the Spines

Spines of *Heterocentrotus mammillatus* and *Phyllacanthus imperialis* vary locally in their stereom architecture. Figures 7.8a and 7.9a show the macrostructure of both

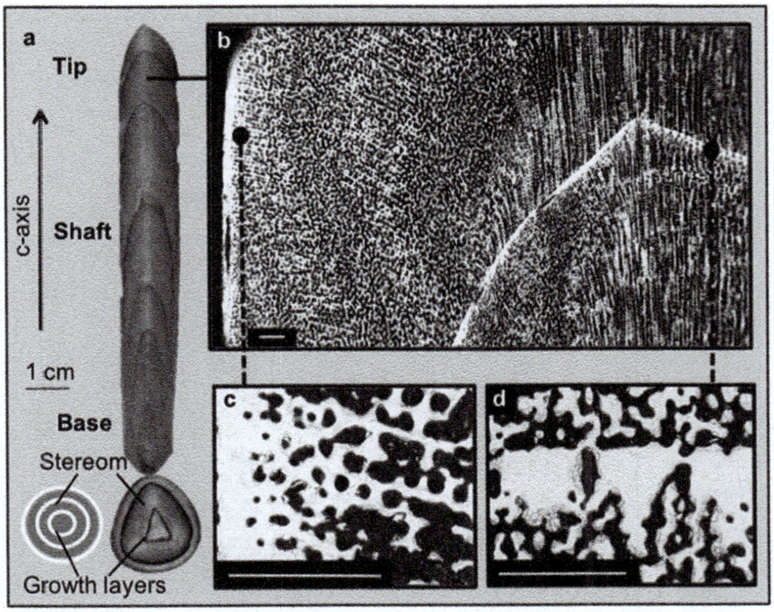

Fig. 7.8 (**a**) Microphotograph and (**b–d**) backscattered electron (BSE) images of *Heterocentrotus mammillatus* spines (*white scale* = 200 μm). (**a**) Longitudinal and cross sectional views of *Heterocentrotus mammillatus* demonstrating the massive growth layers. A BSE image (**b**) reveals the internal microstructure of the upper part of the spine including the medulla (*right*), a growth layer (*white area*) and the trabeculae that fan out perpendicular to the growth rings (*left*). A detailed view of the outer edge and of the growth layer can be seen in (**c**) and in (**d**), respectively

spines in longitudinal and cross sections. The spine can be divided into the collar, shaft and tip. A ball-and-socket joint with muscle fibres attached to the surface of the sea urchin skeleton allows the spine to be moved (Großmann and Nebelsick 2013). In the living organism, the spines of *Heterocentrotus mammillatus* are coated by a thin organic protective layer (epithelium) that inhibits incrustation. In contrast, the spines of *Phyllacanthus imperialis* lose their external epithelium during growth, with incrustation by small calcified crustaceans being often seen as a consequence in this species.

The internal structure of *Heterocentrotus mammillatus* spines is characterised by concentric stereom layers separated by layers of higher density interpreted as former shell surface or growth layers (Ebert 1988). Backscattered electron (BSE) images of spines infiltrated with low-viscosity resin (Fig. 7.8b) reveal details of the stereom structure with a highly porous region (porosity of >50 %) in the centre of the spine (medulla) surrounded by a denser stereom with pore sizes commonly between 10 μm up to 30 μm. Even though the latter portion of the stereom appears in ordinary scanning electron microscope (SEM) images as being disordered (Fig. 7.10a), or labyrinthic in the nomenclature of Smith (1980), Fig. 7.8a

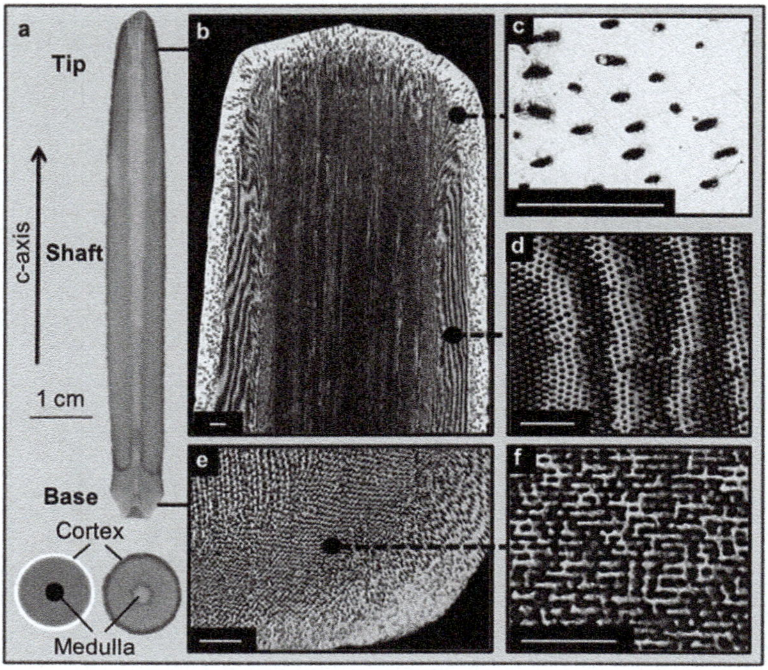

Fig. 7.9 (**a**) Microphotograph and (**b–f**) BSE images of *Phyllacanthus imperialis* (*white scale* = 200 μm). (**a**) Longitudinal and cross sectional views of *Phyllacanthus imperialis* showing the tabular structure of the spine. The internal microstructure of the upper part of the spine is demonstrated in (**b**). The outer edge and the zebra-like structure of the spine in (**b**) is displayed in detail in (**c**) and (**d**). At the spine base, a nearly imperforate stereom can be observed in (**e**). A detailed view of the base is displayed in (**f**)

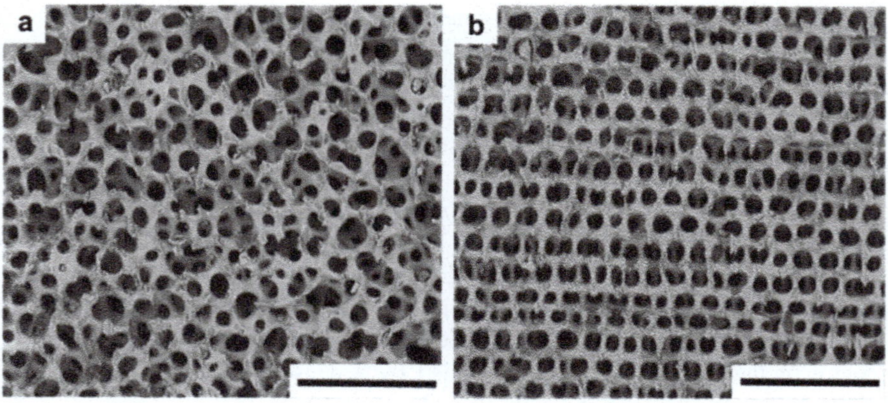

Fig. 7.10 SEM images of (**a**) *Heterocentrotus mammillatus* and (**b**) *Phyllacanthus imperialis* displaying the stereom microstructure (*scale bar* = 50 μm)

suggests that trabeculae preferentially fan out perpendicularly to the surface of the growth rings. The compact growth layers also contain pores, probably ensuring fluid permeability through the entire spine (cf. Fig. 7.8c, d).

A simpler tubular core-shell macrostructure is present in the spine of *Phyllacanthus imperialis* (Fig. 7.9a, b). It comprises the medulla, a less porous stereom towards the edge and a highly dense outer edge (cortex). However, the stereom structure of *Phyllacanthus imperialis* spines is clearly ordered (Figs. 7.9d, and 7.10b) and can be described as a galleried (Smith 1980) stereom (Fig. 7.9b, c). This layer is missing in the spine of *Heterocentrotus mammillatus*. A nearly imperforate stereom occurs, in both species, at the base of the spine, as displayed for the spine of *Phyllacanthus imperialis* in Fig. 7.9e, f.

7.2.2.2 Failure and Energy Dissipation in Uniaxial Compression

Spines of *Heterocentrotus mammillatus* and *Phyllacanthus imperialis* were cut for the uniaxial compression experiments into axial segments (i.e. along the c-axis) with coplanar ends to achieve uniform loading. To avoid bending and limit buckling, a 2:1 length/diameter ratio for the axial segments was chosen (Bargel and Schulze 2012). Uniaxial compression tests were conducted in a universal testing machine, namely the Instron 3180 (Instron Deutschland GmbH, Pfungsstadt, Germany) (Fig. 7.11a). The samples were placed on a Si_3N_4 plate on the crosshead and pressed against a tungsten carbide compression die. The force was measured simultaneously by a force transducer (Fig. 7.11b). The experiments were performed with a crosshead movement speed of 0.5 mm·min^{-1}. During the tests, the fracture behaviour was monitored with a video extensometer (LIMESS Meßtechnik GmbH, Type RTSS_HS) from one direction (Fig. 7.11a).

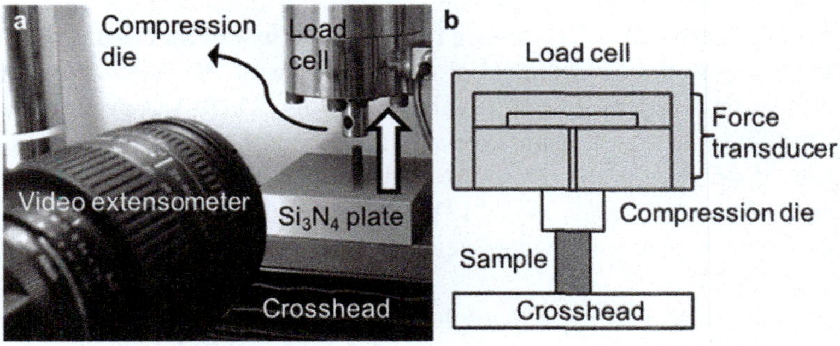

Fig. 7.11 (**a**) Photograph and (**b**) schematic view of the measuring principle of the universal testing machine. The *white arrow* in (**a**) indicates the crosshead movement direction. The spine segment is located between a compression die and an Si_3N_4 plate. The length of this specific sample shown here is 1.4 cm

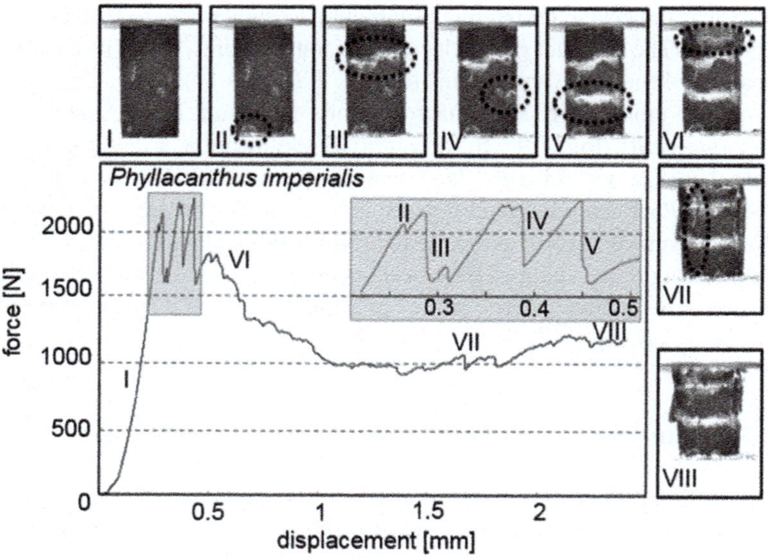

Fig. 7.12 Force-displacement diagram for compression tests of *Phyllacanthus imperialis* segments. The photographs on top and side show effects like initial spalling (*II*), ring crack formation (*III* to *VI*) and vertical crack development (*VII* and *VIII*). The corresponding force jumps are marked accordingly in the diagram and the enlarged section in the insertion

Force-displacement curves for the uniaxial compression of the spines of *Phyllacanthus imperialis* and *Heterocentrotus mammillatus* are shown in Figs. 7.12 and 7.13. The force-displacement curves were correlated with the images from the video monitoring system showing characteristic facture phases. The propagation of cracks is displayed in the images from one perspective during the uniaxial compression. Therefore, the correlation of cracks with the force-displacement can be considered only from one perspective. The drop in force during a first linear elastic compression is associated with the development of cracks. Spines of *Phyllacanthus imperialis* develop several independent horizontal cracks (Fig. 7.12II–VI). After each crack event, the force increases again to levels that reflect the compressive strength of the material/structure. Later, the horizontal cracks act as points of origin for the formation of vertical cracks (Fig. 7.12VII). The developing multiple crack system separates the spine into individual segments, which eventually buckle and spall (Fig. 7.12VII–VIII).

The latter phase requires a lower load than the peak values of phase II to V. However, during both periods, a substantial force is needed further to compress the spine. In this way, energy is dissipated by the continued development of cracks during the whole test. Notably, a displacement of 1 mm translates into a compression of approximately 5 %.

The spines of *Heterocentrotus mammillatus* also initially develop horizontal cracks (Fig. 7.13II–III) but concurrently spalls off smaller fragments at an

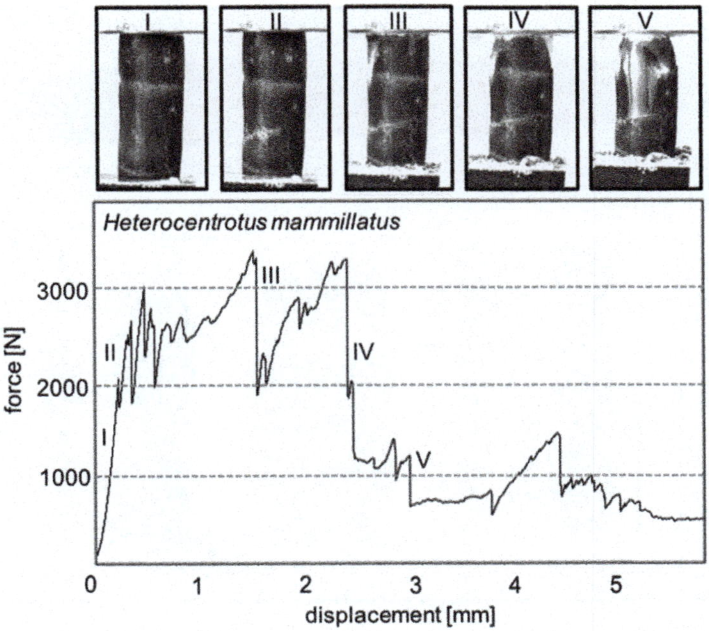

Fig. 7.13 Force-displacement diagram for compression tests of *Heterocentrotus mammillatus* segments. The photographs on top show effects like ring crack formation (*II*), oblique spallation (*III* and *IV*) and vertical crack development (*V*). The corresponding force jumps are marked accordingly in the diagram

oblique angle (Fig. 7.13III–IV). In many cases, the load needed to sustain continued compression is increased above the level of the pre-failure of the structure. However, at later stages, vertical cracks also develop in these spines (Fig. 7.13IV–V) significantly reducing the required force for continued compression. Nonetheless, the energy is dissipated, because of crack formation, all along the experiment.

In both cases, the failure behaviour is quasi-ductile, because segmental spallation and localised destruction relax the stress on the structure and allow the structural integrity of the undamaged parts of the spines to be kept and used over almost the entire displacement.

To link the failure mode with microstructural features, we used computer tomography (X-ray CT) in advance of the compression tests. The link of the failure mode with the microstructure is demonstrated for a spine of *Heterocentrotus mammillatus*. Figure 7.14 shows the macroscopic structure of the growth lines as revealed by CT (Fig. 7.14a). Figure 7.14b indicates that both the horizontal cracks (dashed line) and the spalled top parts are related to the geometry of the growth layers. Proper material parameters such as compressive stresses throughout the whole process cannot be gained from such measurements. Instead, we strive to

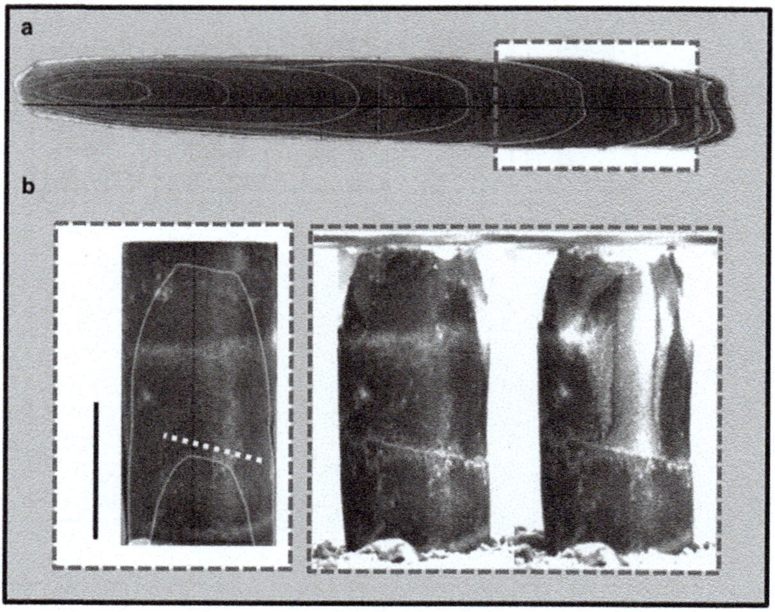

Fig. 7.14 (**a**) X-ray CT scan of a spine of *Heterocentrotus mammillatus*. (**b**) The spine segment after uniaxial compression (*scale bar* = 10 mm)

derive such parameters by a procedure that is described in detail in Chap. 14 of this book and combine it with the structural features and the findings described here.

7.3 Energy Dissipation in Building Materials

The idea of gradient concrete was first put forward by Werner Sobek (Herrmann and Haase 2013). The development of graded building materials, with a continuous variation of the porosity in the inner structure of a construction component, is driven by the idea of adapting material properties to the stresses that occur under load. In construction components subjected mainly to bending stresses, the variation of the density of the concrete from 2160 kg/m^3 down to 830 kg/m^3 allows local mass savings of up to 60 % to be generated. Through the targeted positioning of lighter concrete mixtures in lower stress areas, weight optimised beams can be obtained such as the one in Fig. 7.15.

Purely mineral, functionally graded exterior walls, presented in Fig. 7.16, simultaneously fulfil the specifications for load-bearing capacity, durability, architectural appearance, mass and CO_2 reduction, in addition to saving energy during the production process. Such walls are constructed as a monolith with high-strength concrete on both sides and a core zone made of lightweight concrete. This allows material consumption to be minimised and creates slimmer cross sections. Wall elements with a heat transfer of 0.275 W/m^{2}K and a thickness of 400 mm were

Fig. 7.15 Functionally graded concrete slab (100 mm thickness) with weight optimisation for load-bearing gravitational forces by using lightweight aggregates (Herrmann et al. 2013)

Fig. 7.16 Functionally graded exterior wall, U-value 0.275 W/m^2 with wall thickness of 400 mm (Herrmann et al. 2013). The dark nodules consist of expanded porous glass

achieved by using this pattern. Lightweight aggregates of expanded glass were used for these elements. By employing aerogels as lightweight aggregates, the wall thickness was decreased down to 220 mm and maintained the same insulation properties. In addition to wall elements, functionally graded concrete enables an important weight reduction to be made in building components subjected mainly to bending moments, such as slabs and beams, and for columns that are subjected to compressive forces and lateral impact loads.

7.3.1 Experimental Procedure

For the production of concrete specimens, two mixtures were used: one for normal density concrete and the other for lightweight concrete. The normal density concrete mixture consisted of cement (CEM I 52.5 R), sand (aggregate size 0–2 mm) and water, whereas the lightweight concrete mixture also contained lightweight aggregates (expanded glass granules), partially replacing the sand. The samples were prepared by two production technologies: concrete casting and automated dry spraying (Fig. 7.17).

Quasi-static and dynamic compression tests were carried out by using a compression test machine with a maximum capacity of 500 kN (CH-8224 Löhningen,

Fig. 7.17 Concrete samples obtained by casting (*left*) and by automated dry spraying (*right*) with an edge length of 100 mm

Switzerland) and having the possibility of running path-controlled tests with speeds between 0.01 mm/s and 2.50 mm/s. The samples are positioned between two parallel steel plates with the polished sides facing the plates. During the tests, the displacements and the corresponding force were monitored by using a displacement transducer and a force transducer, respectively. In order to make an observation of the fracture behaviour, videos and photographs were taken during each test.

The investigations were carried out on two main groups of functionally graded concrete (FGC) specimens. The first group consisted of cubes extracted from prefabricated beams that previously were optimised for their bending and shear bearing capacities. Therefore, each beam layout was the result of the load-bearing analysis for a 4-point bending test. In total, 50 concrete cubes were investigated, divided into sets of 2 and 4 pieces, according to their layout and the direction of the load applied. The layout of a specimen was characterised by the concrete layer sequence, the density of each layer and its thickness. For the second group, the layout of the cubes was inspired by the alternating porosity in the sea urchin spines of *Heterocentrotus mammillatus* and *Phyllacanthus imperialis*. For this second group of specimens, each layout was produced by casting and afterwards by dry spraying, leading to 10 sets of paired cubes. Each pair of sets had a particular layout consisting of concrete layers with up to five different densities. In total, 150 cubes were produced and investigated.

7.3.2 Energy Dissipation in Graded Concrete Under Compression

The authors started the analysis of the first biological concept generator (sea urchin spines) used for transfer into biomimetic building structures from the microstructural point of view (density, composite type, layer structures) and linked it to mechanical and physical properties (Young's Modulus, compressive strength,

Table 7.1 Mechanical and physical properties of the spines of *Heterocentrotus mammillatus (HM)*, calcite and the two concrete mixtures used in the production of the FGC specimens

	ρ_m [kg/m^3]	σ_{cm} [MPa]	E_m [GPa]	σ_t [MPa]
Spines of HM	900–1220[a]	40–80[b]	–	≈29[c]
$CaCO_3$	2710[d]	n.a.	≈85 (c-axis)[e]	n.a.
Lightweight concrete	830	2.9	9.9	1.1
Normal weight concrete	2160	43.0	35.2	5.0

Data are mean values for density (ρ_m), compressive strength (σ_{cm}), Young's modulus (E_m) and flexural strength (σ_t) of the samples tested
[a]Applied Mineralogy at the University of Tübingen
[b]Weber et al. (1969), Vecchio et al. (2007) and Presser (2009)
[c]Großmann (2010)
[d]Myers (2007)
[e]Chen et al. (2001)

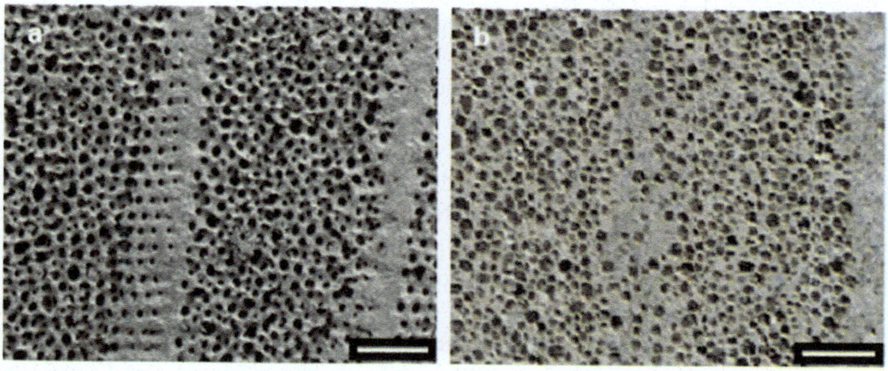

Fig. 7.18 (**a**) Changing density in *Heterocentrotus mammillatus* aboral spine, fully grown Großmann 2010) (*scale bar* = 200 μm) and in FGC specimen (**b**), concrete layers being shown in a vertical position (*scale bar* = 7 mm)

bending strength). These properties were then compared with the properties of the concrete samples that were produced at the Institute for Lightweight Structures and Conceptual Design (ILEK). Broadly speaking, the normal weight concrete can be considered as the dense "material" for the FGC specimens, just as calcite is for the *Heterocentrotus mammillatus* spine. On the other hand, the lightweight concrete can be considered as the porous "material" correlated to the highly porous calcite regions. Furthermore, the values in Table 7.1 indicate that the two extreme concrete mixtures have mechanical properties comparable with the dense pure calcite and with the porous calcite structure inside the spine. In the sprayed concrete specimens, we were able to reproduce the gradual transition of the material from a reduced porosity level to a high level. As in the spines of *Heterocentrotus mammillatus*, we can see a change of the porosity from the growth rings towards the inner side. Replicating this feature by spraying concrete of different densities in thin layers, we obtained the graded material distribution normal to the layer surface (see Fig. 7.18).

Spraying concrete with different densities in several layers was carried out by mixing lightweight concrete (LWC) with normal density concrete (NDC) in various ratios.

Previous investigations of FGC components showed favourable behaviour to bending moments and shear forces, and the tests proved that the components were able to withstand quasi-static compression forces with strain rates between 10^{-6} s^{-1} and 10^{-5} s^{-1} (Noack 2010; Schmeer 2015). In the current investigations, we studied the possibility of extending the applicability of FGC components to withstanding impact forces such as vehicle impacts and earthquakes, with strain rates of up to 10^{-2} s^{-1}.

Strain rates can decisively influence the bearing capacity of building components. For this reason, we first investigated the way that specimens of graded concrete cubes react to the velocity of applied compression force. Strain rates in the range of 0.05 x 10^{-2} s^{-1} to 1.50 x 10^{-2} s^{-1} gave surprising results, with the maximum force varying by only $\pm$ 10 %, whereas the deformability capacity was not affected. In other words, according to our study (which included 50 cubes tested for different strain rates), the compression strength and the capacity of energy absorbance of the graded concrete remained approximately the same for quasi-static loads, for vehicle impact loads and for earthquake loads.

In order to avoid local failure of the concrete, the top and bottom of the samples had added layers of NDC. Because the compression strength of the expanded glass granules themselves is 18 times smaller than the strength of the sand + cement + water (normal weight concrete), we considered them as substitutes for pores inside the concrete specimens. The ratio between the maximum pore size in the *Heterocentrotus mammillatus* spines and the maximum diameter of the used expanded glass granules is approximately 100. At the meso-structural level, the scaling factors between tested spine segments (diameter: approximately 10 mm, height: 20 mm) and concrete cubes (edge length: 100 mm) was 5 for the height and 10 for the width.

For each group of the FGC cubes, a representative specimen can be seen in Fig. 7.19 for which the deformation and degradation is illustrated corresponding to strains of 0.0 %, 2.5 %, 5.0 %, 7.5 % and 10 % (picture sequence from left to right). The fracture behaviour of the FGC cubes is characterised by unexpectedly large deformations for both casted and automatically dry sprayed samples. The first group consisted of dry sprayed FGC cubes having the compression force applied parallel to the layers. After the plastic domain on the strain-stress curve was reached, vertical cracks opened along the lightweight concrete layers. As the compressive strains increased, the number of cracks grew and, simultaneously, the crack openings enlarged. Finally, each specimen turned into a group of vertical slices equal to the total number of layers. The bursting forces, which appeared as a consequence of the applied compression force, caused the concrete layers to split at the sections at which low strength concrete was used. The second group of dry sprayed FGC cubes was tested against compression forces perpendicular to the layers. This time, the first cracks were horizontal and, as predicted, they appeared in the lightweight concrete layers. When the compressive strains exceeded 0.50 %, another type of

Fig. 7.19 Characteristic stages of the fracture behaviour for dry sprayed FGC cubes with an edge length of 100 mm tested parallel to the layers (*top*), dry sprayed FGC tested perpendicular to the layers (*middle*) and casted FGC (*bottom*)

material degradation appeared, namely spallation. For the casted FGC cubes, failure occurred more punctually and we observed that the lightweight aggregates were crushed. Beyond a deformation of 1.00 %, spallation of the concrete cube edges occurred, in addition to the punctual failure of the expanded glass aggregates. This behaviour continued until the total collapse of the test specimens.

A comparison in terms of stress-strain curves was made for the three types of concrete cubes: NDC, casted graded concrete and automatically dry sprayed graded concrete. Figure 7.20 reveals the way that the deformation capacity changed once we switched from NDC to a graded concrete. The peak stress f_c corresponded to a compressive strain ε_{c1} of 1.00 % for NDC, compared with the graded cubes that experienced strains of 3.50 % or even higher. On the other hand, NDC cubes experienced a final strain of 1.60 %, whereas the strains for FGC cubes went beyond the 20.00 % limit. The authors defined the final strain as corresponding to a drop of the compression force below 15 % of the maximum compression strength for the respective test specimen, considering the area of the undamaged cross section. We should also mention that the Model Code gives values for ε_{c1} in the range of (0.20 ... 0.23) % and final strains of (0.35 ... 0.60) % for concrete strength classes between C20/25 and C40/50. In the Model Code (CEB 1991) and in the Eurocode 1992 (CEN 2004) for normal density concrete under compression forces, the strains are indicated as being approximately 3 times smaller than those that our tests have shown. This difference might have occurred because the specimen production was performed under better controlled laboratory conditions than those on site and because the norms are conservative.

Figure 7.20 reveals that the NDC samples require a high load but fail at relatively low compressive strain, i.e. they fail as expected for a brittle material. The cast

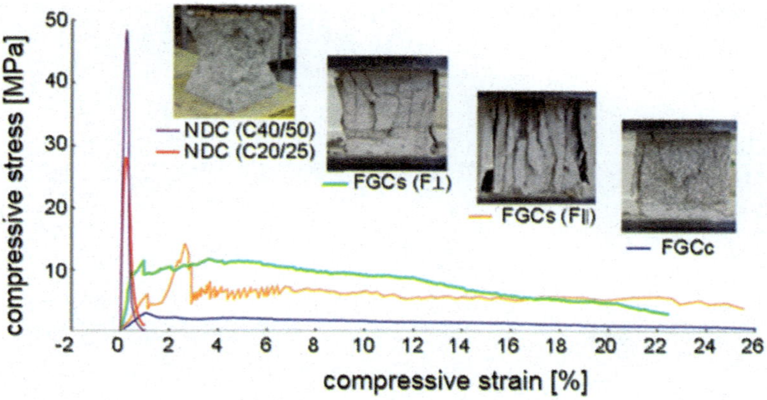

Fig. 7.20 Stress-strain diagram for uniaxial compression test of the investigated concretes. Photographs show the specimen at the end of the test

samples (FGCc) do not withstand high stresses, because the porous inner part has a low intrinsic strength. However, the stress level is held at an almost constant level for a long strain period. Hence, it behaves like a so-called "brittle foam", where quasi-ductility is obtained, and the energy absorbed (evaluated from the area under the stress-strain curves) is twice as high in NDC samples (C40/50).

The biomimetically structured concrete loaded parallel to its layering (FGCs F||) behaves similarly but at a much higher stress level. Thus, the dense interlayers give the structure an increased strength that it retains over the long straining period. Hence, this material can be considered as a self-reinforced brittle foam representing a passive damper system structured like *Phyllacanthus imperialis*. The amount of absorbed energy per unit of volume is 7 times higher than in NDC samples (C40/50).

When loaded perpendicular to its layering (FGCs F_⊥_), the stress level during compression is further increased and the energy dissipated in the process is strongly enhanced over the other variants. However, it still behaves like a brittle foam. A mechanism to enhance the stress bearing capacity, once cracking has started locally, is not implemented in the structure. Thus, changes have to be made until a structure akin to *Heterocentrotus mammillatus* spines can be achieved. The amount of absorbed energy per unit of volume is 10 times higher than in NDC (C40/50). Nonetheless, the experiments establish that extremely strong enhancements in terms of energy dissipation are achievable by bioinspiration.

7.4 Conclusions and Outlook

Nature is a good source of inspiration when it comes to energy dissipation in overload situations. In the case of tree bark, it shows us that fibre-based systems can absorb energy and slow down impact processes, both of which may be of the utmost importance for the protection of crucial components in technical structures.

Future approaches will extend the focus on tree bark as a role model for improved energy dissipation in concrete components. These investigations will involve fibre-reinforced lightweight concretes in which the fibre arrangement and composition of the layers are inspired by the bark of the Giant Sequoia.

A new concept for energy dissipation in buildings is emerging from the structure of specific sea urchin spines, because they absorb energy from impacts and similar situations by controlled local failure. At the same time, they are examples for very strong lightweight self-reinforcing materials. A range of reinforcing mechanisms including crack deflection, crack blunting and geometrical strutting seems to be involved.

The presented results indicate that our first approaches to the transfer of these findings to functionally graded concrete, in the form of mixtures of normal weight concrete and lightweight concrete with alternating porosities, can significantly improve the ability of concrete components to absorb energy when subjected to compression forces. In addition, the same concrete components maintain an overall graceful failure behaviour, thus overcoming an important drawback of regular concrete specimens subjected to compression: brittleness. Consequently, by using a brittle material, we are mimicking the biological role model by only changing the macro- and meso-structure of the concrete sample.

The concept of functionally graded concrete is a promising approach for obtaining weight-optimised building components with similar load bearing capacity to those of normal concrete components and also components with higher energy absorption capacity. Our FGC specimens exhibit high energy absorption capacity and graceful failure behaviour, unlike any normal concrete components subjected to compression. Such a combination of properties for concrete elements opens new possibilities for designing and constructing buildings able to withstand extreme loads.

Our objective is to develop graded and/or fibre-reinforced concrete components, which, when integrated into load-bearing structures, will help buildings, even if partially damaged, to survive extreme action such as vehicle impact, derailed trains and earthquakes. Structures with graded concrete components will be able to absorb energy in their dissipative regions not only because the reinforcement will yield, but also because the concrete will undergo large quasi-plastic deformations.

Acknowledgements This work has been funded by the German Research Foundation (DFG) as part of the Transregional Collaborative Research Centre (SFB/Transregio) 141 'Biological Design and Integrative Structures'/project A02 'Plants and animals as source of inspiration for energy dissipation in load bearing systems and facades'.

References

Aizenberg J, Hanson J, Koetzle TF et al (1997) Control of macromolecule distribution within synthetic and biogenic single calcite crystals. J Am Chem Soc 119(5):881–886. doi:10.1021/ja9628821

Amemiya CT, Miyake T, Rast JP (2005) Echinoderms. Curr Biol 15(23):R944–R946
Astilleros JM, Fernández-Díaz L, Putnis A (2010) The role of magnesium in the growth of calcite: an AFM study. Chem Geol 27(1–2):52–58. doi:10.1016/j.chemgeo.2009.12.011
Bargel HJ, Schulze G (eds) (2012) Werkstoffkunde. Springer, Berlin/Heidelberg/New York
Bauer G, Speck T, Blömer J et al (2010) Insulation capability of the bark of trees with different fire adaptation. J Mater Sci 45(21):5950–5959. doi:10.1007/s10853-010-4680-4
Bischoff PH, Perry SH (1991) Compressive behaviour of concrete at high strain rates. Mater Struct 24(6):425–450
CEB (Comité Euro-International du Béton) (1991) CEB-FIP Model code 1990. Bulletins d'Information. Thomas Telford, London, pp 203–205
CEN (Coomite Europeen de Normalization) (2004) Eurocode 2: Design of concrete structures – Part 1–1: general rules and rules for buildings (EN 1992-1-1). CEN, Brussels
Chen CC, Lin CC, Liu LG (2001) Elasticity of single-crystal calcite and rhodochrosite by Brillouin spectroscopy. Am Mineral 86:1525–1529
Constantinou CM, Soong TT, Dargush GF (1998) Passive energy dissipation systems for structural design and retrofit. In: MCEER Monograph serie no. 1. Multidisciplinary Center for Earthquake Engineering Research, Buffalo
Ebert TA (1988) Growth, regeneration, and damage repair of spines of the slate-pencil sea urchin Heterocentrotus mammillatus (L.) (Echinodermata: Echinoidea). Pac Sci 42(3–4):160–172. doi:10.2307/1539799
Eiberger J (2007) Röntgenographische Spannungsmessung durch Mikrodiffraktion. Diploma thesis, University of Tübingen
Großmann JN (2010) Stereom differentiation in sea urchin spines under special consideration as a model for a new impact protective system. Dissertation, University of Tübingen
Großmann JN, Nebelsick JH (2013) Comparative morphological and structural analysis of selected cidaroid and camarodont sea urchin spines. Zoomorph 132(3):301–315. doi:10.1007/s00435-013-0192-5
Hentz S, Donzé FV, Daudeville L (2004) Discrete element modelling of concrete submitted to dynamic loading at high strain rates. Comput Struct 82(29):2509–2524
Herrmann M, Mittelstädt J, Wörner M et al (2013) Precast components made from functionally graded concrete. CPI (Concrete Plant Int) 6:60–64
Herrmann M, Haase W (2013) Tragverhalten biege- und querkraftbeanspruchter Bauteile aus funktional gradiertem Beton (Load bearing behaviour of functionally graded concrete components under flexural and shear stress). Beton- und Stahlbetonbau 108(6):382–394. doi:10.1002/best.201300017
Magdans U, Gies H (2004) Single crystal structure analysis of sea urchin spine calcites: Systematic investigations of the Ca/Mg distribution as a function of habitat of the sea urchin and the sample location in the spine. Eur J Mineral 16(2):261–268. doi:10.1127/0935-1221/2004/0016-0261
Myers RI (2007) The 100 most important chemical compounds: a reference guide. Greenwood Press, Westport
Nebelsick JH (1992) Echinoid distribution by fragment identification in the Northern Bay of Safaga, Red Sea, Egypt. Palaios 7(3):316–328. doi:10.2307/3514976
Noack C (2010) Untersuchungen zu Betonsandwichdecken mit funktional gradiertem Leichtbetonkern hinsichtlich des Tragverhaltens und des Masseneinsparpotentials (Investigations upon the load bearing capacity and the mass savings potential of functionally graded concrete sandwich slabs). Diploma thesis, University of Stuttgart
Oaki Y, Imai H (2006) Nanoengineering in echinoderms: the emergence of morphology from nanobricks. Small 2(1):66–70
Presser V, Schultheiß S, Berthold C et al (2009) Sea urchin spines as a model system for permeable, light-weight ceramics with graceful failure behavior. Part I. Mechanical behavior of sea urchin spines under compression. J Bionic Eng 6(3):203–213. doi:10.1016/S1672-6529(08)60125-0
Presser V, Schultheiß S, Kohler C et al (2011) Lessons from nature for the construction of ceramic cellular materials for superior energy absorption. Adv Eng Mater 13(11):1042–1049. doi:10.1002/adem.201100066

Schmeer D (2015) Experimentelle Untersuchung automatisiert hergestellter Bauteile aus funktional gradiertem Beton (Experimental investigations upon automated manufactured functionally graded concrete components). Master thesis, University of Stuttgart
Smith A (1980) Stereom microstructure of the echinoid test. Spec Pap Palaeontol 25:1–81
Stokes A, Salin F, Kokutse AD et al (2005) Mechanical resistance of different tree species to rockfall in the French Alps. Plant Soil 278(1–2):107–117. doi:10.1007/s11104-005-3899-3
Teuffel P (2004) Entwerfen adaptiver Strukturen (Design of adaptive structures). Dissertation, University of Stuttgart
Thielen M, Speck T, Seidel R (2015) Impact behaviour of freeze-dried and fresh pomelo (Citrus maxima) peel – influence of the hydration state. R Soc Open Sci 2(6):140322. doi:10.1098/rsos.140322
Vecchio KS, Zhang X, Massie JB, Wang M (2007) Conversion of sea urchin spines to Mg substituted tricalcium phosphate for bone implants. Acta Biomater 3:785–793
Volkwein A, Schellenberg K, Labiouse V et al (2011) Rockfall characterization and structural protection – a review. Nat Hazards Earth Syst Sci 11:2617–2651. doi:10.5194/nhess-11-2617-2011
Weber J, Greer R, Voight B (1969) Unusual strength properties of echinoderm calcite related to structure. J Ultrastruct Res 26:355–366
Wieczorek GF (2002) Catastrophic rockfalls and rockslides in the Sierra Nevada, USA. In: Evans SG, DeGraff J (eds) Catastrophic landslides: effects, occurrence and mechanisms, Geological Society of America reviews in engineering geology XV. Geological Society of America, Boulder, pp 165–190
Zinoviev PA, Ermakov YN (1994) Energy dissipation in composite materials. Technomic Publishing Co. Inc, Lancaster

Chapter 8
Adaptive Stiffness and Joint-Free Kinematics: Actively Actuated Rod-Shaped Structures in Plants and Animals and Their Biomimetic Potential in Architecture and Engineering

Oliver Betz, Annette Birkhold, Marco Caliaro, Benjamin Eggs, Anja Mader, Jan Knippers, Oliver Röhrle, and Olga Speck

Abstract Plants and animals have evolved a variety of rod-like or tube-shaped systems capable of adapting their stiffness and concomitantly achieving deformations without the need of classical engineering-like joints. In architectural concepts, such adaptable systems would provide the technical basis for the construction of building structures and facades with the capacity for continuous adaptive actuation. The present contribution provides an overview of a literature screening for general principles of the deployment of continuous kinematics and adaptive stiffness of rod-like structures in animals and technical systems. With respect to the performance of

O. Betz (✉) • B. Eggs
Evolutionary Biology of Invertebrates, Institute of Evolution and Ecology, University of Tübingen, Auf der Morgenstelle 28, 72076 Tübingen, Germany
e-mail: oliver.betz@uni-tuebingen.de

A. Birkhold
Institute of Applied Mechanics (CE), University of Stuttgart, Pfaffenwaldring 7, 70569 Stuttgart, Germany

M. Caliaro • O. Speck
Plant Biomechanics Group, Botanic Garden, Faculty of Biology, University of Freiburg, Schänzlestraße 1, 79104 Freiburg, Germany

Freiburg Centre for Interactive Materials and Bioinspired Technologies (FIT), University of Freiburg, Georges-Köhler-Allee 105, 79110 Freiburg, Germany

A. Mader • J. Knippers
Institute of Building Structures and Structural Design (ITKE), University of Stuttgart, Keplerstraße 11, 70174 Stuttgart, Germany

O. Röhrle
Institute of Applied Mechanics (CE), University of Stuttgart, Pfaffenwaldring 7, D-70569 Stuttgart, Germany

Cluster of Excellence for Simulation Technology, Continuum Biomechanics and Mechanobiology Research Group, University of Stuttgart, Pfaffenwaldring 5a, D-70569 Stuttgart, Germany

J. Knippers et al. (eds.), *Biomimetic Research for Architecture and Building Construction*, Biologically-Inspired Systems 9, DOI 10.1007/978-3-319-46374-2_8

plants, the initial results of a case study of anatomical properties and wilting modes and of their interrelationship are presented.

More specifically, plants are characterized as anatomically heterogeneous and mechanically anisotropic fibre-reinforced materials. In herbaceous plants, movements can be classified into hydraulic movements caused by water transport and movements caused by mechanical instabilities. With our focus on water-dependent adaptive stiffness, the plant tissues of special interest are those that change their properties significantly according to water availability, such as parenchyma and collenchyma. In invertebrate animals and "protozoans", adaptive stiffness is found in hydroskeletons in which a change in internal hydrostatic pressure interacts with crossed-fibre reinforced external walls and is also achieved by other mechanisms, such as (1) amoeboid cell crawling, (2) muscular hydrostats and (3) slide-lock mechanisms. Slide-lock mechanisms are present, for instance, in insects that have a mechanism for releasing needle-like elongated cuticle structures interlocked along their entire length, thereby achieving joint-free movement (e.g. in needle-like mouthparts responsible for piercing and sucking fluids or in ovipositors used for egg-laying).

The transfer of these biological concepts of adaptive stiffness and continuous kinematics to technical applications in architecture might lead to improved architectural systems, for example, for sun-shading systems. As the proneness to failure is directly dependent on the number of individual movable components in these systems, a decrease in mechanical complexity is a guiding design criterion. Hingeless or compliant mechanisms transferred from biology offer this feature. However, abstractions are needed for technical application. Therefore, computational analysis is essential to identify the key characteristics and the underlying actuation principles of the biological role model. The identified principles from several biological role models can finally be implemented in a common demonstrator, thus providing completely novel bio-inspired technical actuation solutions.

8.1 Introduction

Many developments in current architecture and building technology are directed towards adaptive systems that adjust their geometry to changing external environmental conditions or internal comfort requirements. Their general aim is to contribute to a more sustainable architecture through adaptivity. These systems are envisaged in different sizes for a variety of applications, such as moveable panels for daylight-control or sun-shading, solar tracking for photovoltaic cells, flexible cladding for the adjustment of aerodynamic properties and noise reduction. Until now, these structures have been constructed out of rigid elements or soft textiles connected by hinges, bearings and other highly strained elements. They are actuated and locked in certain (often discrete) positions by external mechanical devices. However, proneness to failure is directly dependent on the number of individual movable components. A minimum level of mechanical complexity is therefore a guiding design criterion. Hinge-less or compliant mechanisms offer this feature. Ideally, these systems exhibit flexibility during movement and develop an increased stiffness and strength when needed. They are thus able continuously to adapt to various geometric configurations through elastic deformation and to resolve the potential conflicts of carrying external loads, mainly self-weight and wind. If these elements deploy a rod-like geometry, they can often be easily integrated in a variety of two-dimensional (2D) or three-dimensional (3D) configurations, similar to hydraulic or electric pistons.

Plants and animals have evolved a variety of linear systems capable of adapting their stiffness and of achieving deformations without the need of classical engineering-like joints. The aim of biomimetic research is to transfer these functional principles to the macro-scale of building constructions through fibre-composite elements with integrated actuators based on pneumatic or other principles.

The present contribution is structured in the following manner. First, general concepts and terms are defined that biology and engineering have in common under the framework of the current biomimetic project (Chap. 8.2). Chapter 8.3 provides the results of a screening of principles of adaptive stiffness and continuous kinematics of (1) a more specific case involving wilting processes in the genus *Gerbera* (botany) and (2) the more general literature of such principles as realized in zoology and engineering. Here, we also introduce two more specific insect examples that are to be investigated in greater detail. Chapter 8.4 exemplifies how the collaboration between biologists and engineers is structured in this concrete project including both data acquisition and engineer-like modelling. Finally, we outline the general way of elucidating their biomimetic potential resulting from the extracted functional principles (Chap. 8.5).

8.2 Definition of Terms

One precondition of the development of interdisciplinary research in the field of biomimetics is to reflect upon project-relevant terms and concepts in order to advance a common understanding of the underlying general functional principles and to ensure their correct use and comprehension (cf. Chap. 19 of this book). Thus, to begin with, we need to define the central concepts and terms used in the present project from the different involved disciplines, i.e. botanical and zoological functional morphology and engineering.

8.2.1 Adaptive Stiffness

The engineers' perspective: Adaptability is a major aim of current research and development in architecture. In general, it reflects the step from a steady and time-independent conception of architecture towards a built environment that constantly reacts to changing requirements. The current project focuses on the adaption of two selected and individual features: the elastic stiffness and the geometric configuration of a rod.

Stiffness defines the extent to which a body resists deformation in response to an applied force. It is a property dependent on the material, the shape and the applied boundary conditions of the body. In the context of this research, adaptive stiffness means that the stiffness of the system can be altered in such a way that the system is, on the one hand, stiff enough to withstand high external forces in a steady state and, on the other hand, the stiffness can be reduced far enough to allow the system to change its geometric configuration, at least for a defined period of time.

The biologists' perspective: The general capacity of plants and animals directly to adjust to changing habitat conditions is called adaptivity, a term that also refers to concepts such as physiological adaptation, phenotypical adjustment, acclimation or modification. Especially for immobile plants, which cannot escape from unfavourable environmental conditions and cannot hide themselves behind protective structures, adaptive traits confer some evolutionary advantage. Adaptivity, in general, and adaptive stiffness, in particular, mean a significant change of one or more features such as structures, materials and geometric configurations in a controlled fashion in answer to changing environmental conditions such as water availability, temperature or mechanical strains (Gibson 2012; Niklas and Spatz 2012; Fratzl and Barth 2009; Fratzl and Weinkamer 2007; Vincent 2007). In animals, adaptive stiffness occurs when single organs or body parts (usually reversibly) adapt their (flexural) stiffness in response to changing external loads.

In plants and animals, adaptive systems, which are capable of dealing with changing environments, are controlled by feedback loops (Brun et al. 2009). Key components of such open- or closed-loop control systems are (1) sensors in order to detect external or internal stimuli, (2) controllers in order to ensure positive or negative feedback in the system and (3) suitable actuators, which are controllable work-producing elements. Interestingly, this holds true for both natural and technical systems. In contrast to technical applications where sensors, controllers and actuators are realized in terms of block building constructions, biological solutions are often integrated systems. This is of special interest and is especially promising, if one considers biomimetic innovations.

8.2.2 *Joint-Free Movement*

The engineers' perspective: A joint or a kinematic pair is the connection between two (or more) bodies that imposes constraints on their relative movement. Movable technical systems are usually based on the conception of components deploying a maximum ratio of strength versus weight linked by individual elements deploying a minimum resistance to rotation or translation. These hinges and bearings are highly strained elements prone to wearing and failure. A joint-free movement offers a

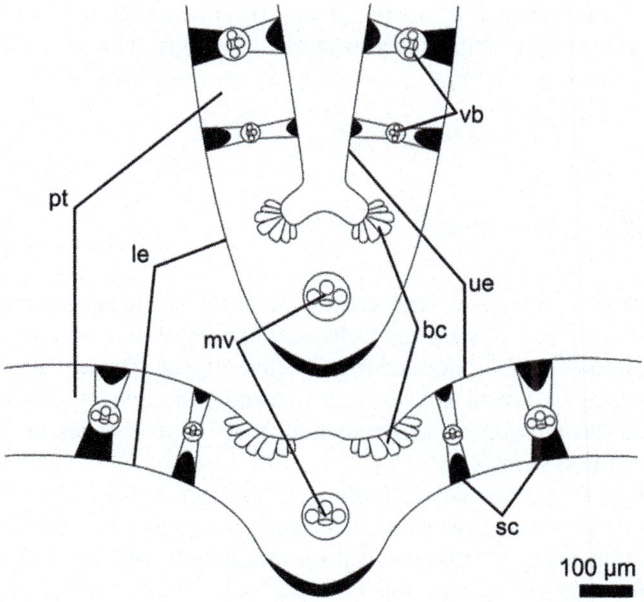

Fig. 8.1 Schematic drawing of the tissue arrangement in leaf blades of *Poa pratensis* L. allowing joint-free adaptive movement. By swelling or shrinking of the bulliform cells (*bc*) the leaf blades can change their conformation between closed (*drawing above*) and open (*drawing below*). Abbreviations: *mv* midvein, *vb* vascular bundles, *ue* upper epidermis, *le* lower epidermis, *pt* parenchymatic tissue, *sc* sclerenchyma strands

radically different approach. In this case, the compliant system consists of only one body. The movement is based on an elastic deformation of the material and is predetermined by a continuous distribution of structural or material properties.

The biologists' perspective: In plants, joints are non-existent according to the engineer's definition. Because of the geometrical arrangements of several tissues that have different mechanical properties and that can react to changing environmental parameters, plant organs, tissues or cells can adaptively – and often reversibly – take various shapes. An excellent example is the joint-free movement of leaves found in monocotyledonous plants such as grass species (Fig. 8.1). Close to the mid-vein, the anatomy is obviously asymmetric: some cells of the upper epidermis are large, highly vacuolated and bubble-shaped (so-called bulliform cells). During drought, the bulliform cells shrink because of decreasing turgor pressure with the result that the two edges of the grass blade fold up towards each other. As soon as adequate water is available, turgor increases and the cells enlarge once again resulting in an opening of the leaves.

In animals, joint-free movements of organs, body parts or entire bodies do not involve flexible links (joints = articulations) between rigid elements (e.g. bones in vertebrates or exoskeletal parts in arthropods) but function via reversible flexible shape-changes together with their enclosing walls. These movements require deformable body media (i.e. tissues, fluids, sheaths, materials) and might comprise (often localized) lumps, constrictions, dilations, flattenings, elongations, contractions, flexures, coilings, twists, evaginations, eversions and / or undulatory movements.

8.2.3 Active Actuation

The engineers' perspective: An actuator converts mechanical, chemical, flow, thermal or electrical energy into a mechanical motion. Active actuation means that adaption is initiated and controlled by an active trigger. The actuation is actively controlled based on sensor data or user demands. In contrast, passive actuation is initiated and controlled by an autonomous reaction to changes in the boundary conditions of the system.

The biologists' perspective: In biology, passive and active actuation is known. Passive actuation is a pure reaction to changing environmental conditions, whereas active actuation needs energy provided by metabolic processes. In botany, active hydraulic actuators are parenchyma or epidermal cells that induce movement by osmotic swelling or shrinking in terms of increasing or decreasing their turgor pressure, i.e. motor cells in carnivorous plants or groups of bulliform cells in leaves of grasses. Other water-driven active actuators include all kinds of growth-mediated movements. Active actuation is also possible by means of fluids such as latex being stored under high pressure in laticifers. Furthermore, mechanically pre-stressed tissues such as fibres under pre-tension and / or parenchyma under pre-compression can serve as actuators. Passive movement, however, is often driven by changes in the humidity of living cells and dead tissues (Dumais and Forterre 2012). A typically passive-actuated motion is the opening and closing of the seed capsules of *Delosperma nakurense* (Engl.) A.G.J.Herre. These capsules possess swellable tissue on the inside and, if saturated with rainwater, the tissue starts to swell and unfold the lids (Harrington et al. 2011). In zoology, active actuation refers to the biological engine-like mechanical effectors that confer motility. On the cellular level, these mechanisms are based on interacting proteins as achieved by the microtubules and microfilaments of single cells and the ratcheting thick (myosin) and thin (actin) filaments that constitute muscles. For the generation of force, muscles usually shorten, thereby performing work. In order to transmit the generated force into movement and support, muscles require interaction with a rigid or incompressible skeletal system.

8.2.4 Continuous Kinematics

The engineers' perspective: 'Continuous kinematics' describes a moveable system that displays similar features during all stages of movement. During this continuous movement, the system can take on any geometrical state between two specified states and is not limited to pre-defined discrete positions as, for example, membrane systems that are stable only in the fully opened or fully closed position. Thus, an indefinite number of geometric configurations can be achieved and no limitation is set to certain pre-defined discrete positions.

The biologists' perspective: Although present in many examples, 'continuous kinematics' has not as yet become a well-defined established concept in functional morphology. So, this term is lacking in the biological literature. Hence, we adopt its technical definition in this paragraph also for the biologists' perspective and, instead of providing a separate definition, just give illustrative examples for its occurrence in biological organisms. In-depth kinematic studies carried out by Dumais and Forterre (2012) as well as by Skotheim and Mahadevan (2005) revealed two categories of plant movement: (1) hydraulic movements caused by water transport and (2) movements caused by mechanical instabilities. This can be displayed graphically in terms of a plot that gives the duration of movement (in seconds) as a function of the tissue size (in meter), defined as the smallest macroscopic moving part. Hydraulic movements are limited by the poroelastic time of water diffusion through porous plant tissues. These include movements caused by growth, swelling and shrinking, which are generally slow. Movements by elastic instabilities are eventually limited by inertia. The release of stored elastic energy or rapid geometric changes can speed up movements beyond the limits imposed by simple hydraulic mechanisms.

In animals, conventional kinematics deal with systems that move via defined joints, whereas continuous kinematics address the versatile joint-free movements of organs, body parts or entire bodies that are continuously deformable (e.g. with respect to changing loads) and are thus less restrained by the rigid interconnection of joints.

8.2.5 Rod-Shaped and Tube-Like Structures

The engineers' perspective: A rod-shaped structure is defined by its geometry, which is typically straight, thin, much longer than wide, and has often a circular cross section. In mechanics, the term 'linear element' is more common.

The biologists' perspective: In biology, cylindrical structures can be massive (rod) or hollow (tube). Important biomechanical properties of a rod are its resistance to (1) bending referred to as flexural stiffness and (2) torsion referred to as torsional stiffness. Both these properties are modified by the cross-sectional shape of the rod and its material anisotropy (Vogel 1988, 2003). More specifically, with respect to their loading regime, rods can be distinguished into (bending) beams and columns. Whereas beams run horizontally and withstand transverse forces that would bend them, columns run vertically and withstand lengthwise forces that would compress them (Vogel 2003). In animals, rods might become elements of superordinate structural systems (frameworks) such as trusses, filled polymers and multibar linkages (Vogel 2003) that confer novel properties by combining static support with mobility.

In plants, rod-shaped structures (without pith) or tube-like structures (with pith) are very common. Examples include stems, pedicels, petioles and roots. With respect to the various bauplans of mono- and dicotyledons, specific patterns of tissue arrangement can be found in the respective plant organ. Bearing in mind that every single tissue has its specific 3D arrangement in the plant organ and that the tissue's mechanical properties are very different from one another, plant organs are characterized as anatomically heterogeneous and mechanically anisotropic fibre-reinforced materials. With regard to adaptive stiffness and because plant movements are caused by water transport and/or mechanical instabilities, the tissues that are of special interest are those that can change their properties significantly according to water availability, such as parenchyma and collenchyma (Caliaro et al. 2013a,b).

8.3 The Diversity of Functional Principles

8.3.1 Plants: Principles Found in Mono- and Dicotyledons

From the botanists' viewpoint, the main aim of this project is to investigate mechanisms of plants exhibiting adaptive stiffness and joint-free connections and to abstract the functional principles in order to translate these mechanisms into novel biomimetic actuators for built structures.

Water-dependent systems: Bearing in mind that plants have evolved a remarkable range of mechanisms to create adaptive stiffness in reaction to several external and internal stimuli and considering that water-driven mechanisms play a major role in adaptive design, we first focus on water-dependent adaptive stiffness. This is initially performed within the framework of a screening process of various herbaceous plants followed by the selection of one model plant in which quantitative analyses of the functional morphology and biomechanics with various relative water content are carried out. In future, an abstraction via computational models of variable detail, in particular by using principles of continuum mechanics and the finite element (FE) method, will be investigated.

Suitable model plants were selected on the basis of an extensive literature search and a screening process that has been underpinned by the following pre-defined criteria: (1) herbaceous plant organs, (2) unbranched rod-shaped or tube-like geometry, (3) active actuation, (4) adaptive stiffness dependent on water content and (5) reversible adaptation process.

On the basis of the diverse bauplans of mono- and dicotyledons and the specific pattern of various tissues, a variety of wilting modes and water-dependent recovery can be defined (Fig. 8.2). Whereas wilting is a passive mechanism, the energy-consuming recovery after full water supply can be regarded as an active process.

Table 8.1 shows the result of this screening process carried out on a variety of selected model plants. A relationship between the various wilting modes with or without full recovery and the specific anatomical conditions (Fig. 8.3) can

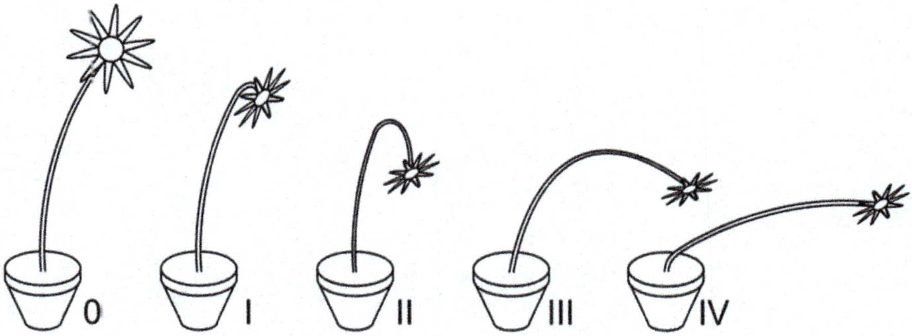

Fig. 8.2 Various wilting modes of rod-shaped or tube-like plant axes (e.g. stem, pedicel, petiole) with a top load (e.g. lamina, flower). *0* original position, *I* bending zone at the apical end of the axis close to the top load, *II* bending zone at a height of about half to two-third of the total length of the axis, *III* bending along the entire axis, *IV* bending zone at the basal end of the axis, thus tilting of the entire axis

be assumed and will be subject of future analyses. Especially promising results concerning the fulfilment of the pre-defined criteria could be found in pedicels of *Gerbera jamesonii* 'Nuance'. Therefore, this *Gerbera* cultivar was the first model plant to be analysed and initial findings are presented here (cf. Sect. 8.4.1).

8.3.2 Animals: Principles Found in Invertebrates

In animals, we distinguish between three fundamental ways of movement, i.e. (1) ciliate and flagellate via hair-like differentiated structures that beat in a pendulum-like manner, (2) amoeboid (also called cell crawling) via plasma protuberances (= pseudopodia) driven by contracting proteins and (3) muscular involving two sets of filaments (actin and myosin) that slip in between each other, such that muscles shorten. Whereas, in many taxa, muscular movements are regularly linked to jointed limbs, in this overview, we focus on the general principles of joint-free continuous kinematics involving flexible rod-like systems with adaptive stiffness. We limit ourselves to protistons and invertebrates, excluding vertebrates. Our screening of the literature has revealed four general principles of such movements, i.e. (1) amoeboid cell crawling, (2) hydroskeleton mechanisms, (3) muscular hydrostats and (4) slide-lock mechanisms (Fig. 8.4).

Amoeboid Cell Crawling (Fig. 8.4a)
Cell crawling occurs in a wide variety of cells including rhizopodean protistons and human white blood cells. To enhance the cell's grip with the substrate, the motile cells produce dynamic surface extensions (e.g. foot-like pseudopodia) of the gel-like ectoplasm containing concentrated networks of actin filaments forming a cell cortex. Forward movements are produced by actin polymerization beneath the cell's surface causing the leading edge to advance. The actin filaments grow at their plus

Table 8.1 Results of the screening process: plant species were characterized according to the pre-defined criteria

Plant species	Plant organ	Wilting mode	Reversible process	Anatomy
Monocotyledons				
Anthurium andraeanum Linden ex André 'Amalia Elegance'	Pedicel/petiole	Wilting shape IV: bending zone at the basal end of the pedicel/petiole thus slight tilting, no further change of shape of the pedicel/petiole	Fully reversible; recovery of original position	Atactostele: numerous scattered bundles (cf. Fig. 8.3a)
Caladium bicolor (Aiton) Vent. 'Candyland'	Petiole	wilting mode III followed by wilting mode IV: first, bending of the entire petiole; second, bending zone at the basal end of the petiole	Not reversible; permanent bending	Atactostele: numerous scattered bundles; collenchyma prevail in terms of strengthening tissue, nearly no lignified tissues (cf. Fig. 8.3b)
Dicotyledons				
Anemone sylvestris L.	Pedicel	Wilting mode I followed by wilting mode IV: first, bending zone at the apical end of the pedicel close to the flower; second, bending at the basal end of the pedicel and tilting of the entire pedicel	Fully reversible; recovery of original position	Eustele: broad closed ring of vascular bundles together with lignified tissue in the periphery, no pith (cf. Fig. 8.3c)
Armeria maritima (Mill.) Willd. 'Alba'	Pedicel	Wilting shape II: bending zone at a height of about half to two-third of the total length of the pedicel	Fully reversible; recovery of original position	Eustele: broad closed ring of lignified tissue in the periphery, inward following vascular bundles, pith (cf. Fig. 8.3d)
Begonia sanguinea Raddi	Pedicel/petiole	Wilting mode IV followed by wilting mode III: first, bending zone at the basal end of the pedicel/petiole, thus slight tilting; second, followed by a slight bending along the entire pedicel/petiole	Fully reversible; recovery of original position	Eustele: narrow closed ring of vascular bundles together with lignified tissue in the periphery, no pith (pedicel see Fig. 8.3e, petiole see Fig. 8.2f)
Bellis perennis L. 'Habanera'	Pedicel	Wilting shape I: bending zone at the apical end of the pedicel close to the flower	Fully reversible; recovery of original position	Eustele: narrow closed ring of vascular bundles together with lignified tissue in the periphery, pith (cf. Fig. 8.3g)

Pelargonium zonale (L.) L'Hér. ex Aiton	Pedicel	No visible wilting mode	–	Oval cross section; eustele: narrow closed ring of lignified tissue in the periphery, inward following vascular bundles, no pith (cf. Fig. 8.3h)
Gerbera jamesonii Bolus ex Hooker f. 'Nuance'	Pedicel	Wilting shape I: bending zone at the apical end of the pedicel close to the flower	Fully reversible; recovery of original position	Eustele: narrow closed ring of vascular bundles together with lignified tissue in the periphery, pith (cf. Fig. 8.3i)
Primula denticulata Sm. 'Red Select'	Pedicel	Wilting shape III: bending along the entire pedicel	Partially reversible; no recovery of original position	Eustele: narrow closed ring of lignified tissue in the periphery, inward following vascular bundles, huge pith (cf. Fig. 8.3j)
Primula obconica Hance 'Touch me'	Pedicel	Wilting shape IV: bending zone at the basal end of the pedicel, thus tilting of the entire pedicel	Fully reversible; recovery of original position	Eustele: narrow closed ring of lignified tissue in the periphery, inward following vascular bundles, no pith (cf. Fig. 8.3k)
Primula veris L. 'Cabrillo'	Pedicel	Wilting shape IV: bending zone at the basal end of the pedicel, thus tilting of the entire pedicel	Fully reversible; recovery of original position	Eustele: narrow closed ring of lignified tissue in the periphery, inward following vascular bundles, no pith (cf. Fig. 8.3l)

All plant organs possess an unbranched rod-like or tube-like geometry, active actuation and adaptive stiffness dependent on the water content.

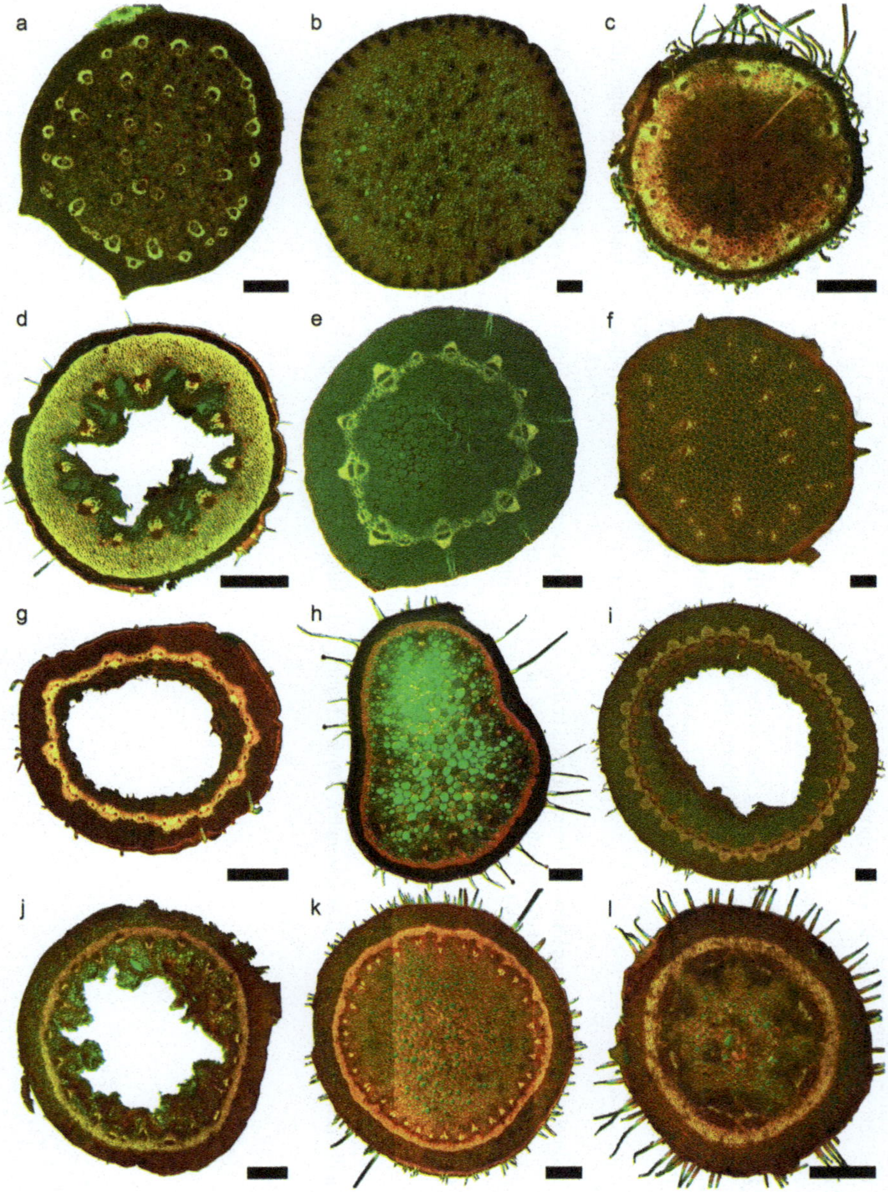

Fig. 8.3 Cross sections of plant organs stained with Acridine Orange highlighting lignified tissues in bright yellow. Monocotyledons: (**a**) *Anthurium andraeanum* Linden ex André 'Amalia Elegance' (petiole), (**b**) *Caladium bicolor* (Aiton) Vent. 'Candyland' (petiole); Dicotyledons: (**c**) *Anemone sylvestris* L. (pedicel), (**d**) *Armeria maritima* (Mill.) Willd. 'Alba' (pedicel), (**e**) *Begonia sanguinea* Raddi (pedicel), (**f**) *Begonia sanguinea* Raddi (petiole), (**g**) *Bellis perennis* L. 'Habanera' (pedicel), (**h**) *Pelargonium zonale* (L.) L'Hér. ex Aiton (pedicel), (**i**) *Gerbera jamesonii* Bolus ex Hooker f. 'Nuance' (pedicel), (**j**) *Primula denticulata* Sm. 'Red Select' (pedicel), (**k**) *Primula obconica* Hance 'Touch me' (pedicel), (**l**) *Primula veris* L. 'Cabrillo' (pedicel). *Scale bars* = 500 μm

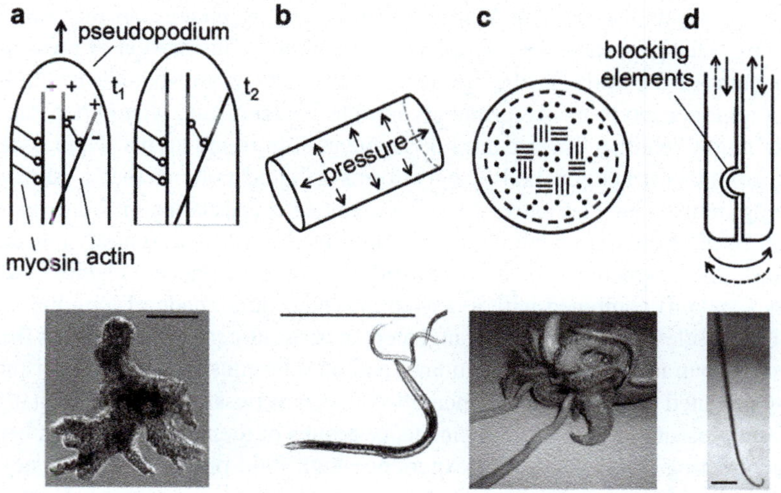

Fig. 8.4 Schemes and photographic examples of four general principles of joint-free movements with adaptive stiffness of rod-like structures in protistons and invertebrate animals. (**a**) amoeboid cell crawling, (**b**) hydroskeleton mechanisms, (**c**) muscular hydrostats and (**d**) slide-lock mechanisms. The *grey bars* in (**a**) are indicative of the outer actin elements to which new elements are added at their plus ends. The *drawing* in (**c**) shows a schematic cross section of a squid tentacle, in which longitudinal muscles are indicated as *dots*, and *circular* and crosswise muscles as *dashes*. The *straight arrows* in (**d**) indicate the relative lengthwise movements of the independent rod-like structures, and the *curved arrows* the resulting bending movement of the entire organ (see Fig. 8.9). The drawings were adapted and modified in (**a**) from Biewener (2003), in (**b**) from Barnes et al. (2002), in (**c**) from Vogel (2003). The *photos* show (**a**) the amoeba *Amoeba proteus* (Pallas) with protruding pseudopodia (*scale bar* = 200 μm); photo courtesy to Klaus Eisler, Tübingen, (**b**) the nematode *Pelodera strongyloides* (Schneider) (shown is one adult female and three L1 larvae) during its characteristic undulating body movements through a fluid medium (*scale bar* = 0.5 mm), (**c**) the arm of the squid *Sepia* sp. Linnaeus, and (**d**) the bending joint-free maxillae of the assassin bug *Dipetalogaster maxima* Uhler (Reduviidae: Triatominae) as recorded during their feeding on pig blood (*scale bar* = 0.6 mm). Abbreviations: + nucleation sites of actin filaments, − sites of disassembling of actin during the process of "treadmilling". t1, t2 two succeeding steps of pseudopodium protrusion

ends, while they become disassembled at their minus ends, a process that is called "treadmilling". Myosin is probably also involved in the generation of propulsive forces by pulling against the local actin networks. More detailed descriptions of this mechanism, which also involves hydrostatic pressure from the interior sol-like endoplasm of the cell and allows animal cells to adopt a variety of shapes, can be found, for example, in Biewener (2003) or Alberts et al. (2014).

Hydroskeleton Mechanisms (Fig. 8.4b)

Once muscles are involved in the production of movement, mechanical counter bearings are needed that are usually provided in the form of skeletons. The phylogenetically oldest skeleton type is the hydrostatic (= fluid) skeleton, whose working principle is a pressurized fluid-filled cavity that is surrounded by a tension-resisting fibre-reinforced skin or wall. In such systems, changes of the body shape and movements are caused by hydraulics through the transmission of pressure into other parts of the body. In the hydrostatic skeletons of soft-bodied invertebrates, the pressurized body cavity promotes muscle antagonisms, whereupon the body cavity might be either unstructured or segmented. In roundworms (Nematoda), for instance, an unstructured body cavity forms a liquid column that is surrounded by longitudinal muscles only. As a consequence, undulating movements result driven by the alternate contraction of opposing (dorsal and ventral) muscles. In contrast, the segmented worms (Annelida) show a segmental chambered body cavity (coelom) combined with a dermal muscular tunic made of longitudinal and circular muscles. From such an arrangement, a peristaltic movement results from the alternate contraction of the longitudinal and circular muscles. In comparison with non-segmented worms, such segmentation makes it possible to isolate changes in the fluid pressure to individual segments or groups of them and so regions between the contracting segments do not experience high fluid pressures and do not need to contract fully to maintain their body shape. This saves energy. Earthworms and other annelids use this principle to produce peristalsis; thus, instead of the body lengthening and shortening as a whole, waves of lengthening and shortening proceed backwards along the body (Gray 1968; Alexander 1983, 2003). In this way, the worms can produce both the propulsive forces (as exerted by contracting the circular muscles and elongating the segments) and, at the same time, the friction towards the substratum (as caused by the thickening of the fully shortening segments upon contraction of the longitudinal muscles) as necessary for an effective forward penetration through the substrate without slipping backwards.

According to Ruppert et al. (2004), the incompressibility of the fluid- (or, alternatively, parenchyma-) filled body cavity upon muscle pressure leads to two important properties, i.e. (1) the resulting hydrostats are constant in volume and (2) any localized increase in pressure as a result of muscular contraction is transmitted equally throughout the hydrostat. These properties are used by the animals (1) mechanically to support the body and adaptively to increase their stiffness and (2) to generate forces by displacing water in one region to do work in another. In addition to locomotion as described above, the dynamic interrelationship between the muscles of the body wall around a fluid skeleton is responsible for the variety of joint-free body movements shown by worm-shaped invertebrates in various biological contexts (e.g. Clark 1964; Gray 1968). Many animals with hydrostatic skeletons (e.g. such as of sea anemones, "worms") are characterized by more or less cylindrical bodies and, similar to the windings in the wall of a garden hose, their body wall is often reinforced by a mesh of inextensible (e.g. collagen) fibres (e.g. Clark 1964; Ruppert et al. 2004). Such fibrous mesh not only toughens the body wall, but also prevents uncontrolled bulges and aneurisms as the hydrostatic pressure rises. To prevent aneurisms, the meshwork fibres in natural hydrostats are oriented in a crossed-fibre array, in which the fibres wrap helically around the body in layers of parallel left-handed and right-handed fibres. Figure 8.5 presents a scheme (adopted from Vogel 2003 and originally suggested by Clark 1964) that shows the change in volume and fibre angle as the length of a helically wound cylinder is varied.

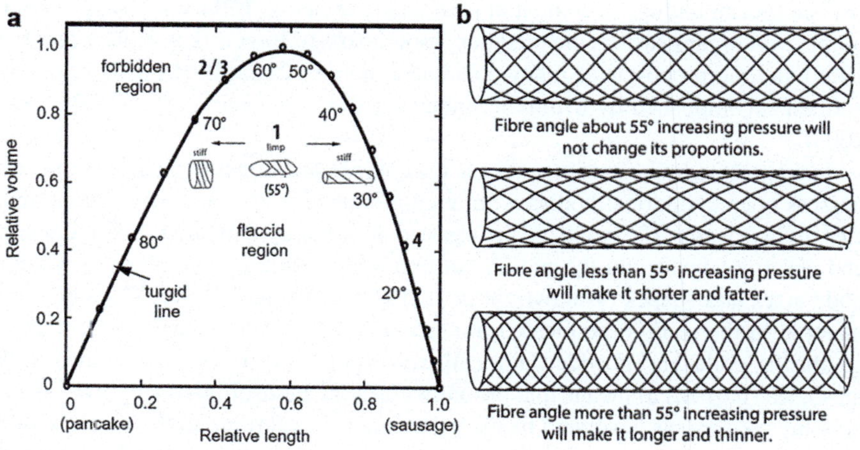

Fig. 8.5 (**a**) Relationship between the relative volume and the relative length as a function of the fibre angle of helically wound cylinders in natural hydrostats. (**b**) Three representative cylinders with different fibre angles and their consequences for shape changes. The **bold numbers** in the scheme represent different actual zoological examples. (*1*) Many "low-pressure" worms (e.g. (acoelomic) limp worms (Nemertini) and flatworms (Plathelminthes)) form flattened (not circular) cylinders, whose volume is not at its maximum ("flaccid"). By contracting the longitudinal muscles, these worms shift to the left towards the turgid line, whereby the mechanical capacity of the fibres is at its maximum. As a consequence, the worms become shorter, thicker, stiffer and more circular in shape. Upon contraction of the circumferential muscles, the worms shift to the right towards the turgid line, becoming longer, thinner and also stiffer and more circular. (*2*) "High-pressure" roundworms (Nematoda) possess longitudinal muscles only, have a strong external cuticle (with crossed collagen fibres) and show a circular cross section. Upon contraction of the longitudinal muscles, these worms shorten only a little but become stiffer (up to 30 kPa in *Ascaris* Linnaeus). They can increase their volume by relaxing the longitudinal muscles, thereby lowering the fibre angle towards 55°. Worms that possess fibre angles of 55° must be flat to be able to increase their volume by rounding their cross section; otherwise, shape changes would only be possible by further increasing the internal pressure. (*3*) Similar to roundworms, the small, soft and joint-free tube feet of echinoderms (e.g. starfish) have longitudinal muscles only and collagen fibres of 67° in the fully extended feet. Upon contraction of the longitudinal muscles, the foot shortens, increasing this angle further and making the tubes even stiffer. (*4*) The mantle of squids (Cephalopoda) possesses circumferential muscles only. Collagen fibres in the mantle make angles of 25° with the long axis. Upon contraction of the circumferential muscles, the fibre orientation resists lengthening, leading to a reduction in the volume of the mantle cavity, which is used to squirt water out of the siphon for jet propulsion. The relationship between volume and fibre angle are calculated as: volume = fibre length^3 ($\sin^2\theta \cos\theta$)/4π. Figure and examples taken from Vogel (2003); reproduced by courtesy of Princeton University Press

Worm-like invertebrates and other organisms under hydrostatic pressure can use this scheme predictably to adjust their stiffness. Whereas the area beneath the curve in the graph contains cylinders that are not fully inflated (and so not fully cylindrical), the turgid line represents so-called "high-pressure" worms that are under almost maximum pressure (and thus almost cylindrical). More detailed discussions can be found in Clark (1964), Wainwright et al. (1976), Vogel (1988, 2003) and Barnes et al. (2002).

One special biological context of hydrostatic skeletons is the effective protrusion of introvert organs such as the predatory mouthparts of insects (e.g. Kölsch and Betz 1998) and the proboscis of many coelomate worms (Clark 1964). This protrusion is often accomplished by hydraulic linkages and levers (Barnes et al. 2002; Vogel 2003).

In contrast to hydrostats of constant volume, sea anemones (Cnidaria: Anthozoa) possess open cylindrical bodies with circumferential, radial and longitudinal muscles in their walls. The cavity of their gastrointestinal system allows, by the ingress and egress of water via the mouth, an enormous range of hinge-less continuous body sizes and shapes (Wainwright et al. 1976; Alexander 1983). The supporting material in the wall is the mesoglea, i.e. a pliant composite visco-elastic material of fibrillar collagen imbedded in a cell-free matrix. The very low modulus of the mesoglea (0.01 MPa) means that the animal can inflate itself with very low internal pressures (<1 Pa) as generated by ciliary pumping. According to the viscoelasticity of the material, at higher strain rates, the mesoglea becomes more rigid (Wainwright et al. 1976).

Muscular Hydrostats (Fig. 8.4c)
According to Kier and Smith (1985) and Vogel (1988, 2003), muscular hydrostats are cylinders entirely composed of muscle in which the muscle fibres run in various directions, forming self-supported but fundamentally motile systems. Examples in invertebrates are the tentacles and arms of squids or the feet of snails. Like the fluid-based hydroskeletons, muscular hydrostats cannot change their volume, a feature that enables the muscles to antagonize each other without rigid skeletons or internal water compartments. These systems are able to perform an impressive range of motions and alter their stiffness without much shape change. Moreover, since the relationship between the radius R and the length L accounts for $R = 1/\sqrt{L}$ in cylinders of fixed volume, these motions imply the amplification of force and velocity. For example, by decreasing its diameter by only 20 %, a squid tentacle can reach a 70 % length extension, which is used by these animals for prey-capture (Vogel 2003).

Slide-Lock Mechanisms (Fig. 8.4d)
These mechanisms involve the lengthwise relative sliding of two or more rod-like elements. If the relative sliding movement between the elements is restrained in some way by some structural modifications at the interfaces, then joint-free bending movements result (similar to the principle of a bimetal strip in a thermostat). In interaction with the mechanical resistance of the penetrated medium, the direction and location of the bending deformations depend on (1) the direction of movement of the sliding elements, (2) the actual position and orientation of the blocking elements and (3) the material properties across the length of the rod-like structures. Examples of this mechanism occur in organisms as dissimilar as protistons and insects. Many protistons and other eukaryotic cells possess cilia and flagella for driving themselves through a fluid medium (e.g. spermatocytes) and / or producing currents for suspension feeding (e.g. collar cells of sponges). Whereas flagella produce sinusoidal motions across their longitudinal axis, cilia remain stiff and extended during their forward stroke and become flexible, bending near their base

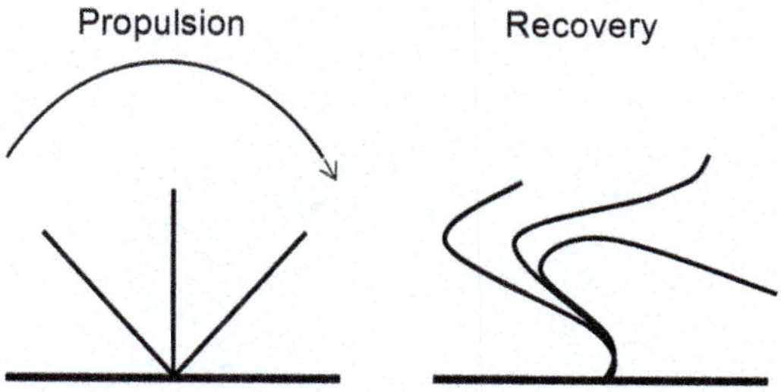

Fig. 8.6 Propulsion and recovery stroke of a cilium. During propulsion, its shaft remains stiff, whereas during the recovery stroke, it becomes more flexible and bends at certain locations along its shaft

during the backward stroke (Fig. 8.6) (Biewener 2003). This movement is understandable from the ultrastructure of a cilium / flagellum that, in transverse section, consists of a set of nine doublet microtubules (made of tubulin) surrounding a pair of single microtubules. Each doublet bears a pair of arms made of the motorprotein dynein, whereupon the dynein arms of one doublet and the tubulin of the adjacent doublet ratchet along each other. If this happens on one side only, the cilium bends. During forward movement, the sliding occurs across the entire length of the cilium shaft at the same time, whereas during the more flexible backward movement the sliding is restricted to certain regions of the shaft (cf. Alberts et al. 2014).

Several insects exhibit slide-lock mechanisms, which employ tongue and groove constructions that lack any joints in the direction of bending. Such mechanisms can be found within extended rod-shaped insect mouthparts responsible for piercing and sucking fluids or in the ovipositors used for egg-laying. Since such kind of functional systems are subjects of our current investigations, they are described in more detail in the following.

Mouthparts of blood sucking Triatominae (Hemiptera: Reduviidae): The haemotophagous heteropterans belonging to the subfamily of the Triatominae possess piercing mouthparts (a pair of mandibles that flank a pair of maxillae (Fig. 8.7b)). For feeding, the proboscis is folded outward by 90°-180° and the apex of the labium is pressed against the host's skin (Fig. 8.7a). The serrated mandibles penetrate the superficial layers of the host's skin by alternate movements but remain in this position to act as anchors. During the extension, the thin and bristle-like maxillae converge at the hypopharynx by a complex interlocking system but are still able to be continuously moved back and forth separately in the formed maxillary bundle (Wenk et al. 2010). They are projected forward as a single bundle and penetrate deeply into the tissue by permanent quiver-like movements. They are capable of a considerable degree of bending when searching for a blood vessel (Lavoipierre et al. 1959). The exact mechanisms behind these actively directed movements, however, remain unclear. When an appropriate blood vessel is tapped, the probing movement ceases. Hereafter, the left maxilla retreats (or the right maxilla further protrudes) suddenly, the thorn-like tip splays out and the connection between the two maxillary parts opens, forming the functional mouth opening (Lavoipierre et al. 1959; Wirtz 1987).

Ovipositors of parasitoid wasps (Hymenoptera): The vast majority of hymenopteran species are parasitoids of other insects, such as most members of the mega-diverse superfamilies of the Chalcidoidea and Ichneumonoidea.

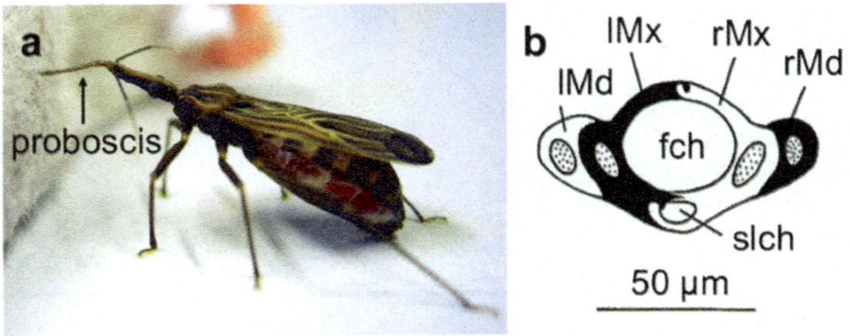

Fig. 8.7 (**a**) *Rhodnius prolixus* Stål (Reduviidae: Triatominae) feeding on an anaesthetized mouse. The proboscis (labium) is fully extended. (**b**) Its mandibles and maxillae in transverse section with the dorsally oriented food channel and the ventrally oriented salivary channel. Abbreviations: *fch* maxillary food channel, *lMd* left mandible, *lMx* left maxilla, *rMd* right mandible, *rMx* right maxilla, *slch* maxillary salivary channel. Adapted from Wenk et al. 2010

The ovipositor apparatus of parasitoids can serve several functions: navigating or penetrating the substrate (if the host is concealed), locating, assessing, marking and piercing the host, injecting venom, killing the competitor's eggs and oviposition (Fig. 8.8a), plus defensive stinging and forming a feeding tube (Quicke 2015). The ovipositor shaft (=terebra) consists of an upper valve (=2nd valvula) and a pair of lower valves (=1st valvulae). The upper valve is split into two asymmetric and overlapping halves that are fused at the apex in the Chalcidoidea (Fig. 8.8b), and is also completely divided, except at the apex, in some ophioniform ichneumonid taxa (Fig. 8.8d), although it is not divided in the majority of the Ichneumonoidea (Fig. 8.8c) (Oeser 1961; Quicke et al. 1994). The ventral surface of the upper valve interlocks with each of the lower valves by a longitudinal tongue-and-groove mechanism called the olistheter. The sublateral rhachis lies in a groove called the aulax along the dorsal surface of each lower valve. The olistheter mechanism tightly holds the valves together while allowing the three valves to slide independently relative to one another (Oeser 1961; Quicke et al. 1994). Despite the hymenopteran ovipositor completely lacking intrinsic musculature, many parasitoids are capable of actively bending and rotating them in various directions (Quicke et al. 1995; Quicke 2015). The movements of the ovipositor result entirely from the actions of muscles inside the metasoma. Several steering mechanisms have evolved in the various taxa as an adaption for reaching their hosts and permitting far greater control over egg placement. Several taxa have evolved mechanisms involving apposed stops on both the upper and the lower valves. Similar to a bimetal strip in a thermostat, the pushing of one valve beyond the location at which the stops have touched causes the ovipositor to bend. This is attributable to the tight interlocking of the upper and lower valves by the olistheter mechanism. This bending mechanism can be seen in several members of the Braconidae (cf. Fig. 8.8c): some taxa have a strongly swollen short region pre-apically on the rhachis of the upper valve with scale-like sculptures (Macrocentrinae) (Fig. 8.9a–b), others possess ancillary teeth on the lower valves next to the thinned outer wall of the aulax (Doryctinae) (Fig. 8.9c) (Quicke et al. 1995). A second type of ovipositor steering mechanism involves a largely longitudinally divided upper valve that is fused just at the apex (e.g.

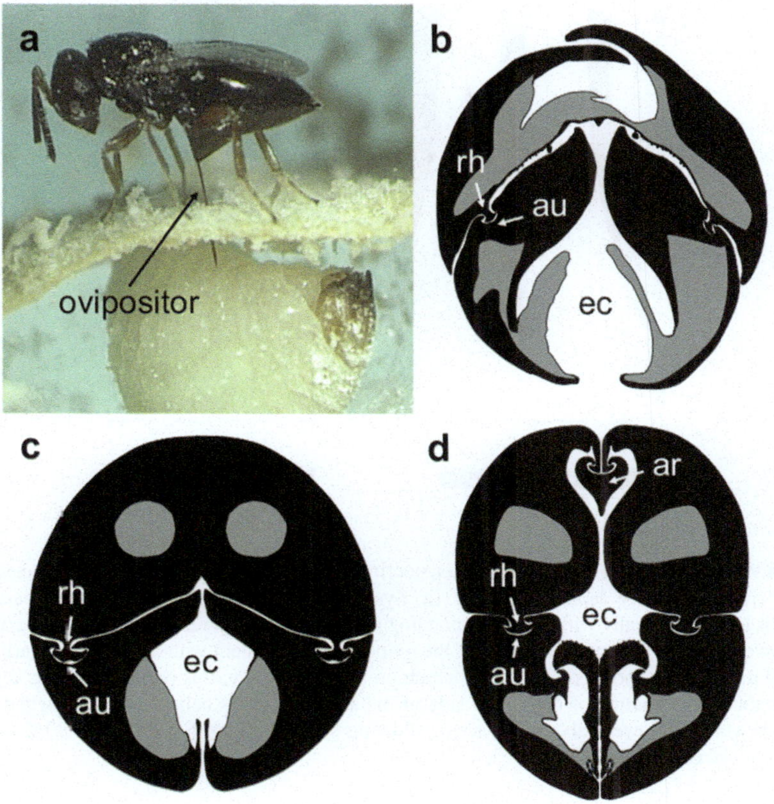

Fig. 8.8 (**a**) *Lariophagus distinguendus* (Foerster) (Pteromalidae) ovipositing on *Sithophilus granarius* (Linnaeus) (Curculionidae) larva in an artificial film chamber. The upper valve is oriented dorsally and the lower valves ventrally when at rest. (**b–d**) Diagrams of transverse sections through the medial part of the ovipositors of members of the genera (**b**) *Plutothrix* Foerster (Pteromalidae), (**c**) *Spathius* Nees (Braconidae: Doryctinae), (**d**) *Stethantyx* Townes (Ichneumonidae: Tersilochinae). Abbreviations: *ar* aulaciform rod, *au* aulax, *ec* egg canal, *rh* rhachis. (**a**) photo courtesy to Collatz et al. 2006, (**b–d**) provided by Quicke DLJ

in Chalcidoidea (cf. Fig. 8.8b) and several ophioniform Ichneumonidae in which the two parts of the upper valve are linked by the so-called aulaciform rod (cf. Fig. 8.8d)). The pulling of one part of the upper valve will cause the ovipositor to bend to the left or right. These wasps are able to bend their ovipositor both dorso-ventrally and laterally (Quicke 2015).[1]

[1]Another steering mechanism involves the formation of the distal part of the ovipositor into multi-arched and noticeably unevenly sclerotized regions: the intermodal arched sections are more heavily sclerotized than the thinner nodes (e.g. in the braconid genus *Zaglyptogastra* Ashmead). Therefore, a ventral protrusion of the lower valves will cause a flattening out of the nodal regions and hence a ventral flexing of the entire apex of the ovipositor (Quicke 1991).

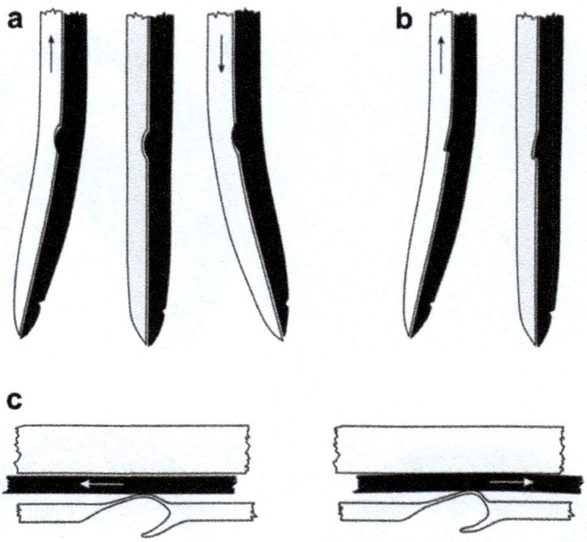

Fig. 8.9 Diagram of the various ovipositor steering mechanisms found in the Braconidae, with the upper ovipositor valve depicted in black. (**a**) System involving the swollen pre-apical part of the rhachis located centrally within a corresponding part of the aulax at rest. Retracting and extending the lower valves will cause a bending of the ovipositor. (**b**) System with a pre-apical scarped butt on the upper valve and a corresponding notch on the lower valve; this permits bending ventrally. (**c**) Flexible construction of the lateral side of the aulax, which is believed to be dragged by the rhachis when the lower valve is retracted, allowing some bending. From Quicke 2015, adapted from Quicke et al. 1995

8.3.3 Engineering and Architecture: Bio-inspired Mechanical Actuation and Adaptive Stiffness

Commercially available actuators include electric or fluidic actuators that operate by means of pneumatic or hydraulic pressure. Moreover, novel actuators based on so-called smart materials have been developed. Examples for these materials are shape memory alloys (SMAs), shape memory polymers (SMPs) or electro- and magnetorestrictive materials such as piezoelectric materials (Ham et al. 2009). Commercially bio-inspired actuators are rather rare. One of the few commercially available ones are pneumatic muscles and the Bionic Handling Assistant by Festo AG & Co. KG.

Recent advances in technology have led to many bio-inspired actuators in which knowledge gain from active biological systems has been transferred to energy-driven mechanical actuation, including electrospun climbing fibres and contractile polymers (Erb et al. 2013). Additionally, technical systems that take advantage of changes in mechanical behaviour, e.g. by changes in elasticity (Qin et al. 2012), in strength and toughness (Espinosa et al. 2011), or in stimuli responsiveness (Capadona et al. 2008) have also been inspired by biological role models (Egan et al. 2015). Rod-shaped adaptive actuators are commonly constructed as monolithic structures from compliant materials such as SMAs (Laschi et al. 2009), elastomers (Ilievski et al. 2011), electroactive polymers (Shi et al. 2012), hydrogels (Otake et al. 2002) or composites that undergo a solid-state phase transition (Brown et al. 2010). Actuation can be achieved by various mechanisms, including electrical charges in SMAs (Lin et al. 2011), piezoelectric drivers (Oldham et al. 2007) or micro motors (Yeom and Oh 2009), chemical reactions (Shepherd et al. 2013) and pressurized fluids (Roche et al. 2014; Polygerinos et al. 2015).

In particular, pressure-driven pneumatic- and hydraulic-powered flexible actuators are promising candidates for technical applications because of their high power-to-weight ratio, low material cost and "simple" fabrication process (Polygerinos et al. 2015). The widely used pneumatic artificial muscle actuators (for example, McKibben actuators) are compliant linear soft actuators consisting of elastomer tubes in fibre sleeves, inspired by the hydrostatic skeleton of worms (cf. Sect. 8.3.2). With pressurization, chambers embedded in the soft actuator expand in the directions associated with lower stiffness allowing extending or contracting motions (Tondu and Lopez 2000), bending (Ilievski et al. 2011) or twisting (Sun et al. 2013).

A plant-inspired bending actuation mechanism is based on the reversible adsorption and desorption of environmental humidity (Mazzolai et al. 2014). The combination of active and passive actuation by using a bilayer composite of an active water-absorbing polymer and a passive elastomeric layer allows humidity-dependent bending movements. An animal-inspired bending actuation mechanism has been, for example, adapted from ovipositing wasps, enabling a steering mechanism to be built for a multi-part probe that can undertake minimally invasive percutaneous interventions (e.g. Ko and Baena 2012) involving complex movements. As the probe is inserted into a compliant material, the bio-inspired bending mechanism of the device is influenced by the mechanical tissue-probe interactions and the interaction between the different parts of the probe.

However, most of the examples mentioned above are demonstrators as part of ongoing research.

8.3.3.1 Bending Actuators

The application of hinge-less actuators is often limited to small-scale applications in robotics or medical engineering in which only low forces occur. Research on large-scale applications is carried out in the field of aeronautics, e.g. morphing wings for airplanes (Sofla et al. 2010), or in maritime research, e.g. on fins for marine robots. So far, the principles for a joint-free actuator or for an adaptive stiffness have not been transferred to an architectural scale involving high external forces such as wind and snow. The compliant mechanism of the flectofin, a facade-shading system inspired by the opening mechanism of the flower of the Bird-of-Paradise (*Strelitzia reginae*) (Lienhard et al. 2011) is actuated and locked by external mechanical actuators. This is an example in which a rod-shaped joint-free actuator can be used to further reduce mechanical complexity and to achieve continuous adaptability of the system. When looking at achieved compliant systems such as the flectofin or the flectofold, we see that the bending of a stiff middle rib is the fundamental driving mechanism. Based on this, the focus of our project will be the development of a bending actuator. Bending describes the behaviour of a structural element subjected to an external load applied perpendicularly to a longitudinal axis of the body. The basic mechanism of bending is the extension of one side with respect to the other. This elongation or shortening can be triggered, for example, by the different thermal expansion of two materials such as in bilayers, SMAs or piezoelectric materials responsible for an elongation or shortening in a defined area of an actuator. Furthermore, fluids can be used to engender an extension. Several possible methods that create a bending motion in fluidic actuators exist. Usually, they require the use of extensible materials that can be elongated by fluid pressure. Limiting this elongation in different parts of the rod-shaped structure enables different actuation movements. Another possibility for creating a bending motion is the construction of an asymmetric inner structure of the actuator. For example, if pressure is applied at locations other than the centroid of the cross section of the structure, a bending motion is generated (Drimer et al. 2016). According to this principle, pneumatic cushions or elongating bellows built from a strong inextensible material can also be integrated into an actuation system. Here, the elongation is based on a shape change. By the arrangement of multiple off-centre elongations within the cross section of an actuator and by their separation in the longitudinal direction, complex motions can be generated. This is the case in several so-called continuous backbone robot manipulators, which are inspired by trunks or tentacles (Walker 2013). A bending actuator can also be built from two separate parts, e.g. with a spring-groove connection. If the two parts are fixed together at one end, a relative motion of the two parts can create the same effect as the elongation or shortening of one side. This principle is also found in biology and has been described for the ovipositors of parasitoid wasps or the feeding apparatus of assassin bugs (cf. Sect. 8.3.2).

8.3.3.2 Adaptive Stiffness

The stiffness of a system depends in general on the material and its geometry. These factors can be controlled in order to vary the stiffness. Adaptively changing the inherent stiffness by a change in material or by control strategies is an active field of research. Variability can be achieved, as for actuators, in many different ways. Compliant actuators in robotics, for example, exploit the possibility of changing their stiffness to allow safe human-machine-interaction. This is achieved through various control strategies: equilibrium-controlled, antagonistic-controlled, structure-controlled and mechanically-controlled strategies (Ham et al. 2009).

Especially in the context of morphing applications, materials of variable stiffness and various structural concepts have been studied. Some examples are the use of shape-memory materials, plant-inspired fluidic flexible matrix composites that derive their tuneable stiffness from the controlled pressurization levels of composite tubes embedded in a flexible matrix or pneumatic honeycomb structures in which the honeycomb cells are pressurized (Kuder et al. 2013). Materials also change their stiffness because of moisture or temperature changes leading, for example, to glass transition (Saavedra Flores et al. 2013) or phase changes (Schubert and Floreano 2013). The non-linear behaviour of materials can be used to create an adaptable stiffness through pre-tensioning. Based on this principle, a variable stiffness system has been developed inspired by the vertebral column. By compressing the elastic parts of a system consisting of bonded rigid and compliant segments arranged in an alternating way, one can increase the stiffness of the system (Huh et al. 2012). Furthermore, possibilities involving the variation of the moment of inertia by rotating a beam with a rectangular cross section around its axis or by increasing the effective thickness of a system by pressing multi-layered structures together, for example with the help of electrostatic forces or by means of establishing a vacuum, have previously been explored (Ham et al. 2009; Drimer et al. 2016).

8.4 Interdisciplinary Working Program

In this chapter, we exemplify how the collaboration between biologists and engineers is structured in the current project and how this will finally result in a demonstrator that combines major functional principles extracted from the biological role models.

8.4.1 Data Acquisition in the Biological Role Models

Within studies on the selected plant model *Gerbera jamesonii* 'Nuance', the water-dependent adaptive behaviour of the pedicels, which have access to more or to less water, are being investigated by using mechanical tests, e.g. three-point-bending tests for studying elastic properties such as the bending elastic modulus and for analysing visco-elastic properties such as the degree of elasticity or energy dissipation (Fig. 8.10).

In addition, morphological-anatomical studies are being carried out for every pedicel. From each pedicel, the corresponding relative water content (RWC) is determined in order to obtain evidence as to the way that the plant water status and the turgor pressure of the parenchyma cells associated therewith (Eastmond and Ross 1997) influence the mechanical properties. Because the mechanical properties also depend significantly on the degree of lignification of the pedicels, samples will be categorized in various age groups on the basis of their anatomical characteristics.

Morphological-anatomical studies of investigated pedicel samples ($n = 8$) reveal, to date, an almost perfect circular cross section (diameter 6.41 ± 0.75 mm, aspect ratio 0.99 ± 0.04), a total length of 45 cm, a varied pronounced taper (tapering mode 0.58 ± 0.18) along the later 3-point-bending-tested length of 280 mm (starting

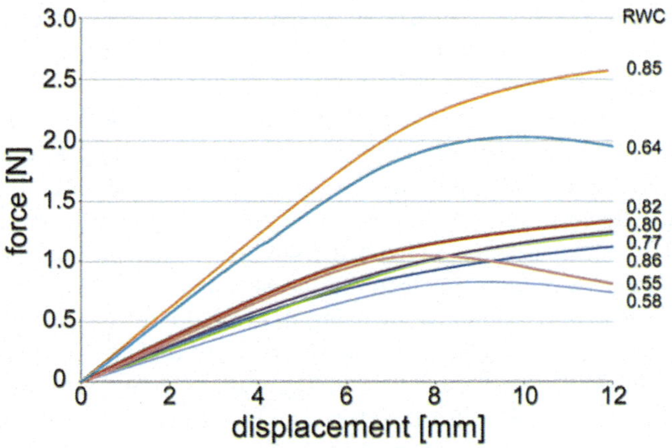

Fig. 8.10 Force-displacement diagram showing elastic behaviour measured during the three-point-bending test. Samples are pedicels of *Gerbera jamesonii* 'Nuance', which have different relative water contents (RWC). The relative water content can take values between 0 to 1, with 0 standing for a fully desiccated or dried plant organ and 1 for a fully hydrated or fully turgescent plant organ

5 cm below the inflorescence) and a density of approximately 0.909 g/cm^3. Hand-sections show a centripetal arrangement of four tissue layers: (1) an outer epidermis, (2) a peripheral cortex parenchyma with unlignified small cells, (3) a closed ring of lignified tissue (vascular bundles and sclerenchyma) and (4) an inner ring of unlignified parenchyma with larger cells than the cells of the cortex. The middle part of the pedicel often contains an irregularly shaped pith (Fig. 8.3i).

Video analyses show the water-dependent adaptive behaviour of the pedicels in detail. After the water supply is stopped, the plant wilts within a few days and the pedicels bend close to the flower. In this bending area, the flower stalks show pronounced ovalization events. After being re-watered, the stalks return totally to their original position within the time span of one day. This reversible mechanism and the comparison of the wilting phase relative to the rapid turgidity process are even more interesting, because, in addition to the weight of the pedicel, the weight of the flower head has to be elevated.

In zoology, it is necessary to combine morphological techniques at various scales from gross morphology to ultrastructural details in order to visualize and reconstruct the arthropod morphology under study in our project.

A wide range of histological and histochemical methods are available for the fixation and embedding of biological samples in order to make them processable for serial sectioning by ultramicrotomy. Histologically stained semithin sections (thickness of 500–1500 nm) provide information about the various types of tissue by light microscopic methods (LM, resolution of about 300 nm). Ultrathin sections (thickness of 50–60 nm), stained with uranyl acetate and lead citrate, provide information about the microstructure and chitin fibre directions under the transmission electron microscope (TEM, resolution of about 1 nm). Finally, all data of the section series are aligned and stacked by specialized 3D reconstruction software that also permits segmentation and 3D modelling. Fluorescence microscopy and confocal laser microscopy (CLSM) of both dissected samples and unstained semithin sections reveal differences in material composition, for example, by taking advantage of the cuticular autofluorescence of arthropod exoskeletons (Donoughe et al. 2011; Michels and Gorb 2012). In addition, scanning electron microscopy (SEM) allows the surface of dried specimens / samples to be viewed at a very high resolution (about 5 nm). Synchrotron X-ray microtomography (SR-μCT) is a non-invasive technique used to obtain complete sequences of virtual sections and can be applied in addition to classical histological serial sectioning. The tomographic dataset, consisting of voxels, is then visualized by using specialized volume graphics software that allows the 3D object to be sliced along any arbitrary axis (Betz et al. 2007).

A detailed analysis of both the functional morphology and the possible movements *in vivo* are necessary for the investigation of the underlying movement principles and biomechanics of a biological structure under study. Therefore, the quantification of the movements in the living animal is achieved by behavioural recordings with a high-speed video camera with movement tracking software.

All these techniques can be implemented in a detailed computational model.

8.4.2 Two-Step Modelling Approach for Translation from Biology to Architecture

The main purpose of biomimetic research is to solve technical problems by searching for solutions in biological organisms and transferring these into engineering applications. Thereby, a one-to-one reconstruction of the biological role model will not help to achieve a robust solution for technical or architectural applications. Abstractions are needed. To identify the key characteristics and the key underlying principles of the biological role model, a rigorous computational analysis is essential. For example, in order to gain deeper insights into the response of the biological and bio-inspired systems and, hence, to be able efficiently to design specific actuators, many different scenarios should be explored by employing FE simulations.

In living nature, small-scale features often influence large-scale features and, hence, cannot be ignored within FE models. However, because of the differences in scale (sometimes by several orders of magnitude), models resolving highly detailed small-scale features are often, from a computational point of view, not feasible. To achieve this translation, two-scale modelling approaches have been investigated (Table 8.2). Essentially, the two-scale modelling approaches can be categorized into two steps. First, precise micro-scale simulations are carried out to investigate the micro-scale features of the biological role models and to elucidate relevant biological mechanisms from experimental data. In a second simulation step,

Table 8.2 Differences between the detailed biomechanical finite element (FE) simulation and the reduced applied FE simulation

Detailed biomechanical FE simulation		Reduced applied FE simulation
Detailed morphology of biological role model	Geometry	Abstracted geometry
Experimental data & sensitivity analysis with respect to range of possible material parameters	Material properties	Well known technical materials; chosen to match technical requirements most closely
Scale of biological role model	Scale	Upscaling to desired technical application; independent of biological role model
Quantitative analysis of biological role model to analyse functional morphology	Aim	Simulation & analysis of functional principles for translation into technical applications

which aims to achieve the translation to architectural applications, an abstraction of the discovered key features is performed and the principles identified in the first simulations of biological actuation will be up-scaled to a technical actuation application.

By using computational models for investigating functional-morphological aspects of the selected biological concept generators, they can be distinguished into (1) geometrical aspects, (2) material modelling aspects and (3) homogenization techniques providing the link between detailed sub-models and the overall structure. This applies likewise for the use of computational models in order to understand and to abstract further the basic principles originating from plants or animals as biological concept generators. The challenge hereby is to construct models that exhibit the key features on the respective scale, while keeping the amount of data necessary for the respective computational model at a feasible level.

Most imaging techniques, such as light and electron microscopy, create 2D images, whereas micro-computed tomography (μCT) permits the direct 3D representation of the geometry of the micro-structure. For the computational models of the insect, a geometric image-based modelling approach is used based on high-resolution Synchrotron-μCT image data. Linking additional information from various 2D imaging techniques, such as high resolution microscopic images of the interlocking mechanism of maxillary or ovipositor parts, or complementary structural information, such as the arrangement of different composites, enables detailed information to be extracted from the biological sample. The behaviour of a bending movement, for example, can be captured in a realistic manner, while providing details about stress concentration points and generated strains. For plants, the computational models are based on 2D geometrical data derived from histological studies and mechanical properties of entire plant organs and single tissues.

Because of the complexity of the experimental data, a fully detailed small-scale model representing, for example, all the geometrical details of the interlocking phenomena between two mouthpart elements or the use of the developed detailed structural model for simulating the fibre-reinforced (micro) mechanical structure and function for an entire plant is unrealistic. Both the small-scale plant and animal models are employed to analyse and predict large-scale behaviour. Natural actuator systems are based on material systems with wide variability controlled by several parameters, such as the degree of crosslinking of the matrix or the variety of possible fibre arrangements of the chitin (arthropods) and cellulose fibres (plants) within the hierarchical structure of these composites (Fratzl and Barth 2009). Therefore, it is nearly impossible to distinguish between material and structure in these biological systems.

To identify the key features for the macro-scale, a sensitivity analysis with respect to the material parameters has to be carried out. Most likely, additional experimental measurements are necessary to verify constitutive laws qualitatively for the overall structures. This approach of using the output of simulations for the design of future experimental studies will also significantly increase the understanding of the biological concept generators (reverse biomimetics). In a second step, the identified underlying principles are verified by further reducing the FE model and focusing on the functional aspects of the biological role models. Therefore, abstracted large-scale FE models are developed. In these models, the geometry and the overall complexity has to be reduced to the necessary level. The material properties can be changed to support the underlying functional principles based on the range of available technical materials. The reduced large-scale FE models can be used for simulation-based analyses exploring the design variables and the potential of the system. To transfer and adapt the structures, material behaviour and methodologies to bio-inspired technical structures, the geometry model and the material parameters need to be scaled up and adapted to applications in building constructions in a further step. The reduced FE model will reveal the underlying principles of the biological role models and will contribute to a deeper understanding of the biological mechanisms and, at the same time, provide a proof-of-concept for technical transferability. As a final step in the development of a novel biomimetic actuator, functional demonstrators will show the feasibility of the developed system for large-scale applications. The development of a common demonstrator also allows the combination of different functional principles from several biological role models. The combination of principles found in plants and animals might lead to completely new bio-inspired technical solutions.

8.5 Multidisciplinary Functional Principles and Their Biomimetic Potential

The guidelines of the present biomimetic top-down approach, conducted in the framework of an interdisciplinary project, involve the bio-inspired application of linear (rod-like) elements in complex architectural solutions showing hinge-less kinematics driven by systemic and continuous actuation mechanisms and exhibiting adaptive stiffness (Fig. 8.11).

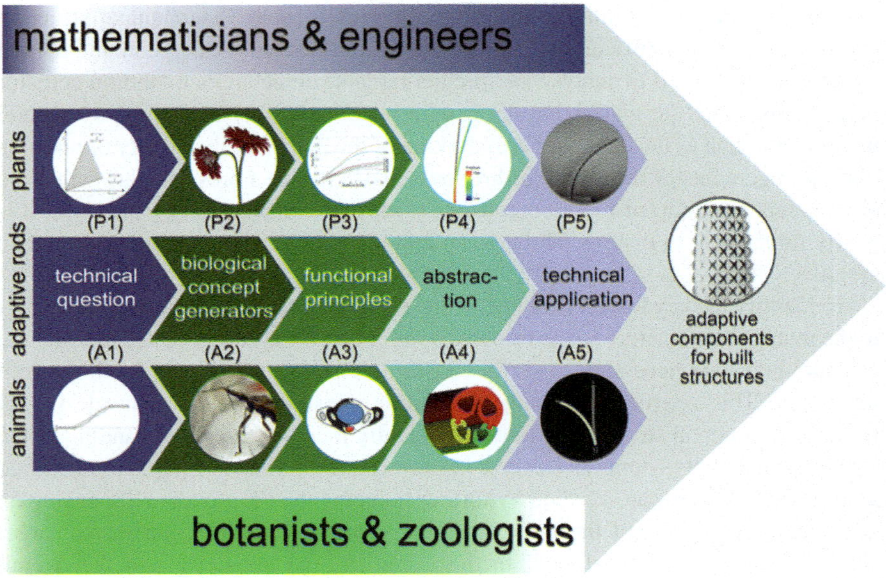

Fig. 8.11 Biomimetic top-down approach (technology pull process) of adaptive components for built structures. The technical challenge is to develop structures with adaptive stiffness and continuous hinge-less kinematics (P1 + A1). Biological concept generators are (P2) herbaceous plants and (A2) mouthparts and ovipositors of insects. The underlying functional principles found in plants (P3) and animals (A3) will be translated into computational models (P4 + A4). The first technical applications (P5 + A5) are feasibility studies for the subsequent demonstrator in which the functional principles from plant and animals will be combined

8.5.1 Extraction of the General Principles That Are of Technical Relevance

Most interestingly, in both plants and animals, similar principles and mechanisms are often deployed to consolidate both adaptive stiffness and joint-free continuous kinematics. From the biomimetic point of view, this largely facilitates the implementation of technical applications that concomitantly serve both of these functions. Many principles involve hydrostats that act against tension-resisting fibre-reinforced walls and thus encompass both movement and the possibility of adapting the stiffness to the current demands.

Both plants and animals exhibit examples of a pressure-dependent variable stiffness, whereby pressure is needed to maintain the stiffness of the system by pre-stressing the material or ensuring the transfer of stresses from one part to another. This principle can be transferred into a technical rod-like structure with integrated pneumatic cushions. The pressure of these cushions will be controlled and thereby allow different modes of operation. Pneumatic cushions are also a promising possibility for creating the actuation. This principle is also used by motor cells such as the bulliform cells that are responsible for the folding of grass leaves in response to drought conditions (cf. Fig. 8.1). The special shape of these cells and their groupwise arrangement and distribution over the cross section of the leaf might provide interesting insights into joint-free actuation with locally defined actuation areas. Discrete pneumatic actuators adapted from these cells can be integrated into fibre-reinforced composites. Fibre-reinforced composites are especially suitable for technical transfer as they allow the local adaptation of the mechanical properties. Thus, we can also learn, from the arrangement of elastic fibres in the insect cuticle and the special placement pattern of highly elastic resilin in the mouthparts or the ovipositors of insects, the way to control or limit deformation to locally defined areas. Furthermore, the prevention of material failure in highly stressed areas can be transferred to a technical application. The mouthparts and ovipositors of insects further show the way in which forces can be transferred between several parts by interlocking structures. The abstraction of these structures to a system that owns its variable stiffness to interlocking structures and, thus, the maintenance of a high stiffness independently of a certain pressure level in the system represents another interesting approach worth further exploration for its technical potentials. The principle of allowing the relative motion of two parts in order to create a bending motion might also be exploited. The potential of principles revealed in plants and animals for achieving a joint-free movement and an adaptive stiffness will be analysed carefully, compared and combined in the best suitable way in order to create a system matching the requirements for an application in load-bearing systems.

8.5.2 Biomimetic Potential from the Perspective of Natural Scientists, Engineers and Architects

With a focus on technical applications, our main interest is that adaptive stiffness, active actuation and joint-like movements of rod-shaped biological structures are not achieved in terms of single components but as a combination of multifunctional structures. Insights revealed in this context might lead to novel kinematic building envelopes and to a feasibility study of the way that markedly different methodologies based on various biological concept generators can be combined into an efficient and novel bio-inspired technical system. During the implementation of the functional principles discovered from biological solutions into biomimetic products (biomimetics), new findings will arise that, in turn, contribute to a deeper understanding of the functioning of the biological samples (reverse biomimetics).

8.6 Outlook

The long-term perspective of this project is to learn from biological systems in order to develop technical systems of higher complexity levels, i.e. multifunctional elements that autonomously adapt both their shape and stiffness according to changing environmental conditions such as temperature, humidity or wind load. This will provide the technical basis for highly efficient building structures and envelopes.

Acknowledgements This work has been funded by the German Research Foundation (DFG) as part of the Transregional Collaborative Research Centre (SFB/Transregio) 141 'Biological Design and Integrative Structures'/ project A03 'Inspired by plants and animals: actively actuated rod-shaped structures exhibiting adaptive stiffness and joint-free continuous kinematics'. We thank Heike Betz for drawing Fig. 8.4 and for proof reading.

References

Alberts B, Johnson A, Lewis J, Morgan D, Raff M, Roberts K, Walter P (2014) Molecular biology of the cell. Garland Publishing, New York

Alexander RMN (1983) Animal mechanics. Blackwell Publishing, Oxford

Alexander RMN (2003) Principles of animal locomotion. Princeton University Press, Princeton

Barnes RSK, Calow P, Olive PJW, Golding DW, Spicer JI (2002) The invertebrates. A synthesis. Blackwell Publishing, Malden

Betz O, Wegst U, Weide D, Heethof M, Helfen L, Lee W-K, Cloethen P (2007) Imaging applications of synchrotron X-ray phase-contrast microtomography in biological morphology and biomaterials science. I. General aspects of the technique and its advantages in the analysis of millimetre-sized arthropod structure. J Microsc 227:51–71. doi:10.1111/j.1365-2818.2007.01785.x

Biewener AA (2003) Animal locomotion. Oxford University Press, Oxford

Brown E, Rodenberg N, Amend J, Mozeika A, Steltz E, Zakin MR, Lipson H, Jaeger HM (2010) Universal robotic gripper based on the jamming of granular material. Proc Natl Acad Sci USA 107(44):18809–18814. doi:10.1073/pnas.1003250107

Brun Y, Di Marzo SG, Gacek C, Giese H, Kienle H, Litoiu M, Müller H, Pezzè M, Shaw M (2009) Engineering self-adaptive systems through feedback loops. In: Cheng BHC, de Lemos R, Giese H, Inverardi P, Magee J (eds) Software engineering for self-adaptive systems. Springer, Berlin/Heidelberg, pp 48–70

Caliaro M, Flues F, Speck T, Speck O (2013a) Novel method for measuring tissue pressure in herbaceous plants. Int J Plant Sci 174(2):161–170. doi:10.1086/668791

Caliaro M, Schmich F, Speck T, Speck O (2013b) Effect of drought stress on bending stiffness in petioles of *Caladium bicolor* (Araceae). Am J Bot 100(11):2141–2148. doi:10.3732/ajb.1300158

Capadona JR, Shanmuganathan K, Tyler DJ, Rowan SJ, Weder C (2008) Stimuli-responsive polymer nanocomposites inspired by the sea cucumber dermis. Science 319(5868):1370–1374. doi:10.1126/science.1153307

Clark RB (1964) Dynamics in metazoan evolution. The origin of the coelom and segments. Clarendon, Oxford

Collatz J, Steidle JJM, Wyss U (2006) Der Kornkäfer und sein natürlicher Feind *Lariophagus distinguendus*. Entofilm, Kiel

Donoughe S, Crall JD, Merz RA, Combes SA (2011) Resilin in dragonfly and damselfly wings and its implications for wing flexibility. J Morph 272:1409–1421. doi:10.1002/jmor.10992

Drimer N, Mendelson N, Peleg A (2016) A new type of hydraulic muscle. Actuators 5(3):1–12

Dumais J, Forterre Y (2012) "Vegetable dynamics": the role of water in plant movements. Annu Rev Fluid Mech 44:453–478. doi:10.1146/annurev-fluid-120710-101200

Eastmond PJ, Ross JD (1997) Evidence that the induction of crassulacean acid metabolism by water stress in *Mesembryanthemum crystallinum* (L.) involves root signalling. Plant Cell Environ 20:1559–1565

Egan P, Sinko R, LeDuc PR, Keten S (2015) The role of mechanics in biological and bio-inspired systems. Nat Commun 6:7418. doi:10.1038/ncomms8418

Erb RM, Sander JS, Grisch R, Studart AR (2013) Self-shaping composites with programmable bioinspired microstructures. Nat Commun 4:1712. doi:10.1038/ncomms2666

Espinosa HD, Juster AL, Latourte FJ, Loh OY, Gregoire D, Zavattieri PD, Zavattieri PD (2011) Tablet-level origin of toughening in abalone shells and translation to synthetic composite materials. Nat Commun 2:173. doi:10.1038/ncomms1172

Fratzl P, Barth FG (2009) Biomaterial systems for mechanosensing and actuation. Nature 462:442–448. doi:10.1038/nature08603

Fratzl P, Weinkamer R (2007) Nature's hierarchical materials. Prog Mater Sci 52:1263–1334. doi:10.1016/j.pmatsci.2007.06.001

Gibson LJ (2012) The hierarchical structure and mechanics of plant materials. J R Soc Interface, rsif20120341. doi:10.1098/rsif.2012.0341

Gray J (1968) Animal locomotion. Weidenfeld and Nicolson, London

Ham R, Sugar TG, Vanderborght B, Hollander KW, Lefeber D (2009) Compliant actuator designs. IEEE J Robot Autom 16(3):81–94

Harrington MJ, Razghandi K, Ditsch F, Guiducci L, Rueggeberg M, Dunlop JWC, Fratzl P, Neinhuis C, Burgert I (2011) Origami-like unfolding of hydro-actuated ice plant seed capsules. Nat Commun 2:337–343. doi:10.1038/ncomms1336

Huh TM, Park Y, Cho K (2012) Design and analysis of a stiffness adjustable structure using an endoskeleton. Int J Precis Eng Man 13(7):1255–1258. doi:10.1007/s12541-012-0168-2

Ilievski F, Mazzeo AD, Shepherd RF, Chen X, Whitesides GM (2011) Soft robotics for chemists. Angew Chem Int Ed Engl 50(8):1890–1895. doi:10.1002/anie.201006464

Kier WM, Smith KK (1985) Tongues, tentacles and trunks: the biomechanics of movement in muscular-hydrostats. Zool J Linn Soc Lond 83:307–324

Ko SY, Baena FR y (2012) Trajectory following for a flexible probe with state/input constraints: an approach based on model predictive control. Rob Auton Syst 60(4):509–521

Kölsch G, Betz O (1998) Ultrastructure and function of the adhesion-capture apparatus of *Stenus* species (Coleoptera, Staphylinidae). Zoomorphology 118:263–272

Kuder IK, Arrieta AF, Raither WE, Ermanni P (2013) Variable stiffness material and structural concepts for morphing applications. Prog Aerosp Sci 63:33–55. doi:10.1016/j.paerosci.2013.07.001

Laschi C, Mazzolai B, Mattoli V, Cianchetti M, Dario P (2009) Design of a biomimetic robotic octopus arm. Bioinspir Biomim 4(1):015006. doi:10.1088/1748-3182/4/1/015006

Lavoipierre MMJ, Dickerson G, Gordon RM (1959) Studies on the methods of blood-sucking arthropods: I. – The manner in which Triatominae bugs obtain their blood-meal, as observed in the tissues of the living rodent, with some remarks on the effects of the bite on human volunteers. Ann Trop Med Parasitol 53:235–250

Lienhard J, Schleicher S, Poppinga S, Masselter T, Milwich M, Speck T, Knippers J (2011) Flectofin: a hingeless flapping mechanism inspired by nature. Bioinspir Biomim 6(4):045001. doi:10.1088/1748-3182/6/4/045001

Lin H, Leisk GG, Trimmer B (2011) GoQBot: a caterpillar-inspired soft-bodied rolling robot. Bioinspir Biomim 6(2):026007. doi:10.1088/1748-3182/6/2/026007

Mazzolai B, Beccai L, Mattoli V (2014) Plants as model in biomimetics and biorobotics: new perspectives. Front Bioeng Biotechnol 2:2. doi:10.3389/fbioe.2014.00002

Michels J, Gorb SN (2012) Detailed three-dimensional visualization of resilin in the exoskeleton of arthropods using confocal laser scanning microscopy. J Microsc 245:1–16. doi:10.1111/j.1365-2818.2011.03523.x
Niklas KJ, Spatz H-C (2012) Plant physics. University of Chicago Press, Chicago
Oeser R (1961) Vergleichend-morphologische Untersuchungen über den Ovipositor der Hymenopteren. Mitt Zool Mus Berlin 37:1–119
Oldham K, Pulskamp J, Polcawich R, Ranade P, Dubey M (2007) Thin-film piezoelectric actuators for bio-inspired micro-robotic applications. Integr Ferroelectr 95:54–65. doi:10.1080/10584580701756482
Otake M, Kagami Y, Inaba M, Inoue H (2002) Motion design of a starfish-shaped gel robot made of electro-active polymer gel. Robot Auton Syst 40(2–3):185–191
Polygerinos P, Wang Z, Overvelde JTB, Galloway KC, Wood RJ, Bertoldi K, Walsh CJ (2015) Modeling of soft fiber-reinforced bending actuators. IEEE Trans Robot 31(3):778–789. doi:10.1109/TRO.2015.2428504
Qin G, Hu X, Cebe P, Kaplan DL (2012) Mechanism of resilin elasticity. Nat Commun 3:1003. doi:10.1038/ncomms2004
Quicke DLJ (1991) Ovipositor mechanics of the braconine wasp genus *Zaglyptogastra* and the ichneumonid genus *Pristomerus*. J Nat Hist 25:971–977
Quicke DLJ (2015) The braconid and ichneumonid parasitoid wasps: biology, systematics, evolution and ecology. Wiley Blackwell, Chichester
Quicke DLJ, Fitton MG, Tunstead JR, Ingram SN, Gaitens PV (1994) Ovipositor structure and relationships within the Hymenoptera, with special reference to the Ichneumonoidea. J Nat Hist 28:635–682
Quicke DLJ, Fitton M, Harris J (1995) Ovipositor steering mechanisms in braconid wasps. J Hymenopt Res 4:110–120
Roche ET, Wohlfarth R, Overvelde JTB, Vasilyev NV, Pigula FA, Mooney DJ, Bertoldi K, Walsh CJ (2014) A bioinspired soft actuated material. Adv Mater Weinheim 26(8):1200–1206. doi:10.1002/adma.201304018
Ruppert EE, Fox RS, Barnes RD (2004) Invertebrate zoology. A functional evolutionary approach. Brooks/Cole – Thomson, Belmont
Saavedra Flores EI, Friswell MI, Xia Y (2013) Variable stiffness biological and bio-inspired materials. J Intel Mat Syst Str 24(5):529–540. doi:10.1177/1045389X12461722
Schubert BE, Floreano D (2013) Variable stiffness material based on rigid low-melting-point-alloy microstructures embedded in soft poly(dimethylsiloxane) (PDMS). RSC Adv 3(46):24671–24679. doi:10.1039/c3ra44412k
Shepherd RF, Stokes AA, Freake J, Barber J, Snyder PW, Mazzeo AD, Cademartiri L, Morin SA, Whitesides GM (2013) Using explosions to power a soft robot. Angew Chem Int Ed Engl 52(10):2892–2896. doi:10.1002/anie.201209540
Shi L, Guo S, Li M, Mao S, Xiao N, Gao B, Song Z, Asaka K (2012) A novel soft biomimetic microrobot with two motion attitudes. Sensors 12(12):16732–16758. doi:10.3390/s121216732
Skotheim JM, Mahadevan J (2005) Physical limits and design principles for plants and fungal movements. Science 30:1308–1310. doi:10.1126/science.1107976
Sofla AYN, Meguid SA, Tan KT, Yeo WK (2010) Shape morphing of aircraft wing: status and challenges. Mater Design 31(3):1284–1292. doi:10.1016/j.matdes.2009.09.011
Sun Y, Song Y, Paik J (2013) Characterization of silicone rubber based soft pneumatic actuator. Proc IEEE/RSJ Int Conf Intell Robot Syst, pp 4446–4453
Tondu B, Lopez P (2000) Modeling and control of McKibben artificial muscle robot actuators. IEEE Control Syst Mag 20(2):15–38
Vincent JFV (2007) Adaptive structures – some biological paradigms. In: Wagg D, Bond I, Weaver P, Friswell M (eds) Adaptive structures – engineering approach. Wiley, Chichester, pp 261–286
Vogel S (1988) Life's devices. The physical world of animals and plants. Princeton University Press, Princeton
Vogel S (2003) Comparative biomechanics. Life's physical world. Princeton University Press, Princeton

Wainwright SA, Biggs WD, Currey JD, Gosline JM (1976) Mechanical design in organisms. Princeton University Press, Princeton

Walker I (2013) Continuous Backbone "Continuum" Robot Manipulators. ISRN Robotics 2013:1–19

Wenk P, Lucic S, Betz O (2010) Functional anatomy of the hypopharynx and the salivary pump in the feeding apparatus of the assassin bug *Rhodnius prolixus* (Reduviidae, Heteroptera). Zoomorphology 129:225–234. doi:10.1007/s00435-010-0115-7

Wirtz H-P (1987) Eindringen der Mundwerkzeuge von Raubwanzen durch eine Membran (Hemiptera: Reduviidae). Entomol Gener 12(2/3):147–153

Yeom S, Oh I (2009) A biomimetic jellyfish robot based on ionic polymer metal composite actuators. Smart Mater Struct 18(8):1–10. doi:10.1088/0964-1726/18/8/085002

Chapter 9
Compliant Mechanisms in Plants and Architecture

Simon Poppinga, Axel Körner, Renate Sachse, Larissa Born, Anna Westermeier, Linnea Hesse, Jan Knippers, Manfred Bischoff, Götz T. Gresser, and Thomas Speck

Abstract Plant movements can inspire deployable systems for architectural purposes which can be regarded as ideal solutions combining resilient bio-inspired functionality with elegant natural motion. Here, we first give a concise overview of various compliant mechanisms existing in technics and in plants. Then we describe two case studies from our current joint research project among biologists, architects, construction engineers and materials scientists where the aesthetic movements of such role models from the plant kingdom are analysed, abstracted and implemented in bioinspired technical structures for sustainable architecture. Both examples are based on fast snapping movements of traps of carnivorous plants. The Waterwheel plant (*Aldrovanda vesiculosa*) captures prey underwater and the Venus flytrap (*Dionaea muscipula*) snaps in the air. We present results on the motion principles gained by quantitative biomechanical and functional-morphological analyses as well as their simulation and abstraction by using e.g. Finite Element Methods. The

---

S. Poppinga (✉) • L. Hesse
Plant Biomechanics Group, Botanic Garden, Faculty of Biology, University of Freiburg, Schänzlestraße 1, D-79104 Freiburg, Germany
e-mail: simon.poppinga@biologie.uni-freiburg.de

A. Körner • J. Knippers
Institute of Building Structures and Structural Design (ITKE), University of Stuttgart, Keplerstraße 11, 70174 Stuttgart, Germany

R. Sachse • M. Bischoff
Institute for Structural Mechanics (IBB), University of Stuttgart, Pfaffenwaldring 7, 70550 Stuttgart, Germany

L. Born • G.T. Gresser
Institute for Textile Technology, Fiber Based Materials and Textile Machinery (ITFT), University of Stuttgart, Pfaffenwaldring 9, 70569 Stuttgart, Germany

A. Westermeier • T. Speck
Plant Biomechanics Group, Botanic Garden, Faculty of Biology, University of Freiburg, Schänzlestraße 1, D-79104 Freiburg, Germany

Freiburg Centre for Interactive Materials and Bioinspired Technologies (FIT), University of Freiburg, Georges-Köhler-Allee 105, D-79110 Freiburg, Germany

J. Knippers et al. (eds.), *Biomimetic Research for Architecture and Building Construction*, Biologically-Inspired Systems 9, DOI 10.1007/978-3-319-46374-2_9

Aldrovanda mechanism was successfully translated into a similarly aesthetic and functional technical structure, named Flectofold, which exists in a prototype state. The Flectofold can be used as a façade shading element for complex curved surfaces as existing in modern architecture.

9.1 Introduction

Compliant mechanisms are omnipresent in nature and, until now, have mostly been used for small technical devices, e.g. in the medical or packaging industries or in microelectromechanical systems (MEMS) but are widely unknown on the larger scale of building construction (Howell 2001). The urgent demand for more energy-efficient and sustainable architecture has led to a growing interest in adaptive building envelopes that can adjust themselves to changing external environmental conditions or internal comfort requirements. For kinetic systems in building construction, the criteria of "robustness" and "adaptability" are usually of major importance, whereas other aspects such as "accuracy" or "velocity" are of less relevance (Knippers and Speck 2012; Speck et al. 2015). Movements of many plant surfaces fulfil these criteria and are additionally interesting because of their often highly aesthetic movement patterns (Speck 2015). Petals and leaves do not only rotate around static axes (as it is the case in technical motions) but perform more complex, three-dimensional movement patterns based on countless robust principles important for many vital functions. These movements are based on the locally adapted stiffness of their components and avoid high stress and strain concentrations. An important issue for quantitatively understanding and describing these movements and for the transfer into bio-inspired technical solutions is the elucidation of movement patterns and of actuation principles and their interplay with the structural set-up of the mechanism because geometrical characteristics and material parameters are inseparably linked and similarly affect the motion behaviour of the compliant mechanism. In addition to quantitative studies of the biomechanics of plant movements and of the underlying structures, simulations of the movement and of the actuation mechanisms with kinetic FE models render especially valuable contributions for such analyses (Poppinga et al. 2013; Schleicher et al. 2015). A combination of these methodological approaches is not only mandatory for successful biomimetic transfer, but also facilitates an improved understanding of the biological role models in the process of reverse biomimetics (Speck and Speck 2008).

Despite an increasingly better comprehension of the functional aspects of the movements of plants since the appearance of the classic works of Charles Darwin (Darwin 1865, 1875), many questions concerning form-structure-function relationships remain to be solved, which, in addition to their biological interest, are also of great importance for successful transfer into biomimetic products. The idea that plant movements represent suitable role models for bio-inspired adaptive building envelopes has been established by the development and patenting (Knippers et al. 2011) of the biomimetic façade-shading system Flectofin®, which is composed of laminated glass-fibre reinforced polymers (GFRP). This case study covers the flapping mechanism inspired by the deformation principle found during the pollinator-induced deformation of the "perch" of the bird-of-paradise flower (*Strelitzia reginae*) (Lienhard et al. 2011; Knippers et al. 2012; Speck et al. 2015). Based on experiences with this successful biomimetic invention, further studies of selected especially promising biological role models, namely the Venus flytrap (*Dionaea muscipula*) and Waterwheel plant (*Aldrovanda vesiculosa*), have been performed and new biomimetic façade-shading systems have been and are currently being developed.

9.2 A Comparison of Technical Hinges and Flexible Joints in Plants

Deployability in technical constructions is typically achieved by joining stiff elements that have the ability to glide against each other (rigid-body mechanisms) (Howell 2001). With respect to adaptive building envelopes, Velasco et al. (2015) have classified certain types of mechanisms that are used in construction today and discerned, for example, deployable structures (folding grills, telescopic, tensegrity and tension structures, folding plates and pneumatic systems), various types of physical transformations (deformation, folding, deploying, retraction, sliding and revolving) and the involved direction of movements (spherical movement, circular tangential movement, radial movement, pivoting movement, monoaxial movement, biaxial movement and multiaxial movement). Most of the adaptive building systems such as façade-shading systems rely on rigid or membranous elements guided along linear and parallel translation axes or linear rotation axes, manufactured as standardised elements. This construction principle allows, in many cases, cheap and simple setups but the resulting structures also have the disadvantages of wear, backlash, high numbers of component parts and, hence, assembly costs and expenditure of time (amongst others). Furthermore, their adaptability to non-planar building geometries is limited. In contrast, compliant mechanisms show deployability because of their elastic deformation and rely on the flexibility of their components. Such mechanisms can be completely free of typical joints; hence, markedly reduced maintenance is needed and the whole structure can be constructed in a much lighter and less complex manner, as the amount of structural components (fasteners, springs, etc.) is greatly reduced.

As plant cells possess stiff cell walls (Burgert 2006), the implementation of contractile proteins, which are the basis for muscle-generated movement found in animals, is impeded. Therefore, plants cannot install "motors" for the actuation of rigid-body mechanisms but have instead evolved a multitude of flexible movement principles. Classically, one distinguishes between nastic motions, which follow morphologically predetermined movement patterns, and tropisms, which do not follow such patterns but are determined in movement direction and pattern by a stimulus. Nastic motions show a clear structure-function relationship and are therefore extremely interesting for biomimetic approaches (Burgert and Fratzl 2009). On the other hand, tropisms display a higher adaptability of the performed motions in terms of possible structural deformation as a reaction to triggering stimuli. Plant motions can be actuated hydraulically, which involves the swelling and shrinking processes of cells and tissues, and/or rely on a release of stored elastic energy (Skotheim and Mahadevan 2005; Forterre 2013). Both principles can be reversible (turgor change in cells, elastic organ deformation) or irreversible (growth, explosive fracture). Moreover, completely passive movements occur that are actuated by external mechanical forces, e.g. the pushing force or the body weight of pollinators (Skead 1975; Reith et al. 2006).

To our best knowledge, no analogues of rigid-body mechanisms exist in the plant kingdom, other than in fungi. The fruit bodies of *Macrolepiota* species (Agaricaceae), for example, are characterised by movable rings (annuli) around the respective stipes (cf. Ge et al. 2010) (Fig. 9.1a). During development of the fruit body, the cap-stipe-connection zone breaks because of cap expansion, leaving the ring, which behaves similar to an uniaxially moving prismatic joint thereby providing a linear sliding movement between two rigid bodies (one degree of freedom). The function of this remarkable structural feature remains to be investigated.

9.2.1 Mechanisms with Concentrated Compliance

Mechanisms with concentrated compliance ("lumped compliance") are characterised by the localisation of elastic deformation, during motion, only to a small structural component and not of the whole structure (Howell 2001). Such mechanisms behave similar to rigid-body mechanisms, but with the rigid hinge being replaced by a flexural joint. Small flexural hinges are sometimes referred to as "living hinges" when they consist of the same material as the two rigid parts that they connect. In plants, movability is often enabled by narrow/thin structures that allow motion of a "rigid" organ connected to a "rigid" plant body. For example, the pulvinus is the typical small joint of an actively moving leaf and functions via osmotically driven swelling and shrinking processes of antagonistically working motor cells (reviewed by Braam 2005; Dumais and Forterre 2012; Poppinga et al. 2013). In some plants, e.g. the nyctinastic silk tree (*Albizzia* spp.) and the thigmonastic sensitive mimosa (*Mimosa pudica*) (Fig. 9.1b), the pulvinus allows for a downwards and upwards bending of leaves (Weintraub 1952; Campbell and Garber 1980) and, hence, resembles a technical hinge joint with one degree of freedom.

Another uniaxially movable plant hinge, which instead works in a completely passive manner, is present in the flower lip (labellum) of the deceptive *Drakaea* orchid (Fig. 9.1c). Male wasps are attracted by the lip, which resembles a flightless

Fig. 9.1 Rigid-body-mechanism in fungi and mechanisms with concentrated compliance in the plant kingdom. Possible movement directions are indicated by dotted arrows, whereas previously performed motions are indicated by arrows with continuous lines. (**a**) The fruit body of *Macrolepiota* spec. is characterised by a movable ring (annulus) around the stipe. (**b**) The sensitive Mimosa (*Mimosa pudica*) folds its leaflets as a response to touch. Moreover uniaxial but, instead, completely passive motions are performed by the hinged labellae of the orchids (**c**) *Drakaea glyptodon* (image with kind permission from photographers Anja and Holger Hennern), (**d**) *Calopogon tuberosus* and (**e**) *Peristeria* spec. (**f**) The leaflets of *Desmodium motorium* move in elliptical circles. (**g**) *Lilium lancefolium* possesses dangling anthers that perform multiaxial passive motion induced by wind and pollinators. (**h**) The flowers of the Obedient plant (*Physostegia virginiana*) can be pushed and rotated in any direction, where they remain. In this image, they have been manually pushed to the right

female, grab the putative copulation partner and try to fly away with it (Peakall 1990). Because of the hinge, the wasp take-off results in an arch-like motion towards the flower and pollination takes place. Although this plant-insect interaction is well-investigated in an ecological context, the detailed functional morphology of the hinge, which allows the violent pollination/attempted copulation action, is still unknown. In other orchids, e.g. *Calopogon tuberosus* (Fig. 9.1d) and *Peristeria spec.* (Fig. 9.1e), the uniaxially movable labellum also aids in pollinator positioning (van der Cingel 2001) and the comparably large hinge lies along a strip-like connection line. A similar hinge mechanism is found in the staminal levers of sages (*Salvia* spp.) and secures pollen deposition on specific regions of the pollinator body (Reith et al. 2006, 2007).

Other plants possess hinged anthers or leaflets that can perform even more complex multiaxial motion, resembling technical structures with ball and socket joints with two degrees of freedom. The leaflets of the Telegraph plant (*Desmodium motorium*) (Fig. 9.1f), for example, perform ultradian motions in elliptical circles every few minutes (Antkowiak et al. 1991). The dangling anthers of *Lilium lancefolium* (Liliaceae) (Fig. 9.1g) help in stripping off the pollen onto the body of the pollinator. This noticeable anther flexibility is often initiated by the dehydration of the filament tip (Keijzer et al. 1987). Indeed, small, thin and, at the same time, flexible and mechanically resilient structures often enable organ deployability in the plant kingdom.

A special case is the movable flower of the Obedient plant (*Physostegia virginiana*) (Fig. 9.1h). The flower is connected to the plant body by a short stalk, which enables a movability that is, to our best knowledge, unique in the plant kingdom. The flower can be pushed and rotated in any direction, where it remains. This is in contrast to other passive plant motions whereby the organs flip back to their original position once the deflecting force stops acting. This hinge type can be compared with a lockable technical ball joint with three degrees of freedom. The biological meaning for this high movability and adaptive stiffness of the flower stalk remains unclear. Müller (1933) identified highly extensible cells that occur in a so-called twisting zone ("Drehzone") and that probably induce the observed flexibility.

9.2.2 Mechanisms with Distributed Compliance

A flexible structure that deforms over its entirety and not only at small flexible joints is considered to have a mechanism with distributed compliance. During the motion, stresses are not concentrated in small areas but are, rather, distributed over the whole body. Such a mechanism can be found in a multitude of moving plant organs that do not possess localised hinges.

Many uniaxial upwards and downwards bending motions are performed by the respective whole organs. For example, the seed scales of pine cones bend as passive reactions to changes in air humidity. This motion is based on the anisotropic swelling and/or shrinking processes of various cellular tissues (Dawson et al. 1997; Poppinga et al. 2013). Hence, the scale reacts similar to a bimetallic strip but responds to a change in humidity and not to heat. By this means, the whole pine cone opens when it is dry, enabling seed release, and closes into a protective state when it is wet (Fig. 9.2a). Multiaxial bending of entire rod-like structures is represented by the prey-induced bending motions of tentacles in carnivorous sundew plants (*Drosera* spp.) (Fig. 9.2b) (Williams and Pickard 1979).

Other cases of uniaxial motions are present in extendable/retractable plant structures. In horsetails (*Equisetum spec.*) (Fig. 9.2c) and bamboo (e.g. *Dendrocalamus giganteus*), growth and culm elongation can be compared with the extension of a tapering telescope (Niklas 1992). Interestingly, plants show a multitude of structures that can be regarded as an analogue to a technical spring, such as the cucumber

Fig. 9.2 Examples of mechanisms with distributed compliance from the plant kingdom. (**a**) A pine cone in the wet and closed (*left*) state and in the dry and open (*right*) state. (**b**) Sticky tentacles of the carnivorous Cape sundew (*Drosera capensis*) are bent towards captured prey. (**c**) Horsetail culms (shown: *Equisetum telmateia*) elongate by a telescope bar-like movement. (**d**) The Red bryony (*Bryonia dioica*) tendril is a natural spring featuring a reversal in its rotating direction. (**e**) Tropistic motions such as the leaf movement in the carnivorous Cape sundew (*Drosera capensis*) and (**f**) in butterwort leaves (*Pinguicula grandiflora*) around caught prey are complex mechanical responses depending on the direction of the triggering stimuli. (**g**) Processes of leaf unfolding in the European hornbeam (*Carpinus betulus*) follow a complicated origami pattern and (**h**) the opening movements of sympetalous flowers (*Ipomoea purpurea*) are important ontogenetic processes. (**i**) Explosive fracture of Touch-me-not (*Impatiens balsamina*) fruits

tendril (Fig. 9.2d) (Gerbode et al. 2012), which shows structural adaptations for strain-stiffening and for avoiding unwinding under tension. Even more complex three-dimensional motions of planar surfaces are performed by the sticky leaves of carnivorous sundews (Nakamura et al. 2013). Upon contact with caught prey, bending motions lead to the wrapping of the leaf around the animal and the formation of an "outer stomach" (Darwin 1875) (Fig. 9.2e). Similar motions are performed in butterworts (*Pinguicula* spp.) by their planar adhesive trapping leaves (Heslop-Harrison 1970), which either wrap around the prey or form depressions on their leaf surfaces for deterring escape (Fig. 9.2f). Mechanisms with distributed compliance also play a crucial role in many ontogenetic processes, for example, in unfolding leaves, which can follow a complex origami pattern (Kobayashi et al.

1998) (Fig. 9.2g) or in blooming processes of flowers (Liang and Mahadevan 2011; van Doorn and van Meeteren 2003; van Doorn and Kamdee 2014) (Fig. 9.2h).

Some plant organs change their shape very rapidly. The explosive bursting of fruits, e.g. in Touch-me-nots (*Impatiens* spp.) (Deegan 2012) (Fig. 9.2i), is irreversible because the whole structure is torn. Fast and reversible motions often rely on snap-buckling (snap-through) processes of doubly-curved shell-like structures. Because of the double curvature, tension and compression stresses occur when the organ is deformed and elastic energy is stored and eventually released when the curvature suddenly inverts. Hence, plant organs incorporating such elastic components first build up and store elastic energy by either passively or actively deforming a doubly-curved organ and achieve a very rapid and strong passive mechanical response. This movement principle is scalable and can be found in minute spores of eusporangiate ferns (diameter: 30 μm) (Hovenkamp et al. 2009), in ca. 500 μm long trapdoors of carnivorous bladderwort traps (*Utricularia* spp.) (Vincent et al. 2011; Poppinga et al. 2016a) and in up to 5 cm long snap traps of the Venus flytrap (*Dionaea muscipula*) (Forterre et al. 2005; Poppinga et al. 2013), which are described in detail in Sect. 9.3.2. Notably, some doubly-curved plant structures such as the false indusia of the Peruvian maidenhair fern (*Adiantum peruvianum*) (Poppinga et al. 2015) invert their curvature without performing a rapid snap-through transition; the underlying mechanism of this action remains to be investigated in detail.

9.3 Biological Role Models

*9.3.1 The Snap-Trap of the Waterwheel Plant (*Aldrovanda vesiculosa*)*

The aquatic carnivorous Waterwheel plant (*Aldrovanda vesiculosa*), which is distributed almost worldwide, possesses snap-traps typically 2.5–6 mm in length (Cross 2012) (Fig. 9.3) and captures small zooplankton prey. The trap is divided into two lobes that are connected by a midrib. Inside the trap and near to the midrib, several trigger hairs are located that are sensitive to touch by the prey. After being triggered, the fast trap closure lasts 100 ms (Poppinga and Joyeux 2011). Early investigations by Ashida (1934) showed that the trap lobe curvature does not change between the open and closed state but that the midrib curvature alters from straight (open trap) to strongly bent (closed trap). Much later, with the help of high-speed cinematography and computer-based mechanical modelling techniques, Poppinga and Joyeux (2011) were able to show that Ashida's observations are consistent with a trap closure mechanism that is markedly different from that of *Dionaea* (see Sect. 9.3.2). After being triggered by prey, volume changes attributable to turgor losses occur in motor cells situated next to the trap midrib (Ashida 1934), coupled to a bending of the midrib and movement of the lobes towards each other (Poppinga

Fig. 9.3 A closed *Aldrovanda vesiculosa* trap. Note the bent midrib (*arrow*)

and Joyeux 2011; Schleicher et al. 2015) (Fig. 9.3). A minute bending of the midrib suffices to initiate trap closure because of kinematic amplification deriving from the trap built-up.

The *Aldovanda* trap can be considered as an example of a mechanism with concentrated compliance (see Sect. 9.2). With the help of modern construction methods, we aim at implementing the *Aldrovanda* kinematic as a feature for inducing deployability in architecture, especially in façade elements for adaptive shading (see Sect. 9.5.1).

*9.3.2 The Snap-Trap of the Venus Flytrap (*Dionaea muscipula*)*

The *Dionaea* leaf is constituted of a flat petiolus and a leaf blade transformed into the (in)famous snap-trap, which is typically 2 cm in length (Bailey and McPherson 2012). Three trigger hairs sensitive to mechanical perturbations are situated on each of the two trap lobes. Several stimuli by prey (typically crawling arthropods) within a certain period of time are required for the trap to close (duration 0.3 s) (Forterre et al. 2005; Poppinga et al. 2016b). The rapid closure of the traps has fascinated scientists since Darwin's comprehensive works on carnivorous plants (Darwin 1875) and the mechanism by which they do so is still under debate. Active hydraulic processes at the cellular level are commonly accepted to drive the comparably slow first stage of the movement, although the exact mechanism remains to be elucidated. Either active changes in turgor pressure in cells of motor tissue or cell-wall loosening leading to an acid growth response drive this initial motion (Williams and Bennett 1982). Furthermore, the relaxation of pre-stressed tissue(s) has not been ruled out as supporting this process (Hodick and Sievers 1989; Colombani and Forterre 2011).

Fig. 9.4 The Venus flytrap wild type and two morphologically divergent cultivars used for investigation of the morphospace of functional (fast) snap-traps. (**a**) An open trap of a wild type plant. Note that the lobes are concave (as seen from the outside) when the trap is open and (**b**) convex when the trap is shut. Small prey can escape through the mesh of "teeth" protruding from the lobe margins. (**c**) Large prey, which cannot escape, triggers complete trap shutting and digestion. (**d**) The cultivar "Angel Wings" possesses trap lobes with extremely strong curvature and these do not shut after triggering (*right image*). (**e**) The cultivar "Korean Melody Shark" possesses trap lobes with little initial curvature and these shut very slowly

This initial motion leads to the deformation of the double-curved lobes, which are initially concave as seen from the outside. As described in Sect. 9.2, elastic energy is stored in the lobes and is released when they flip to a convex curvature (Fig. 9.4a, b). The trap snaps shut and the prey is caught. Hence, the *Dionaea* trap consists of two independent kinematical elements (the two independently, hydraulically and elastically moving lobes) and a midrib that does not actively take part in the motion. In contrast, the previously mentioned *Aldrovanda* trap comprises only one kinematic element (the midrib) whose bending deformation is coupled with lobe movement (see Sect. 9.3.1) (Poppinga et al. 2016b). The "teeth" on the closed trap lobe margins of the Venus flytrap interlock in such a manner that small prey unworthy of activating the energy-demanding digestion process can escape. If the prey is too large to escape and, hence, further stimulates the trigger hairs during its escape attempts, the trap closes completely and the digestion process is initiated (Böhm et al. 2016) (Fig. 9.4c). After digestion and nutrient uptake, the trap reopens after a few days by growth processes, hereby avoiding reverse snap-buckling (Poppinga et al. 2016b).

We have investigated various ontogenetic stages (Poppinga et al. 2016b) and diverse stable mutants (cultivars) with regard to their movement patterns during trap closing. The cultivars are characterised by different morphological variations in the trap form and structure, which hinders (fast) trap closure to a variable degree. By quantitatively analysing and simulating kinematics of the trap-closing motion of the diverse mutants, we aim to characterise a "morphospace" representing the morphological limits for successful trap action. In addition to an improved understanding of the biological structures and of the structural constraints that have to be fulfilled during trap evolution, this morphospace will also help to define the possible degrees of freedom in shape finding for biomimetic flapping systems inspired by the snap-buckling mechanism of the Venus flytrap.

Our first investigations dealt with the cultivars "Angel Wings" with its conspicuously more strongly curved traps (Fig. 9.4d) and "Korean Melody Shark", which is characterised by an apparent weak curvature of the lobes (Fig. 9.4e) compared with the wild type (Fig. 9.4a). By choosing the two "extremes" in terms of trap-lobe curvature, initial results concerning the possible boundaries of the snap-buckling morphospace could be gained. We performed 3D-scans with a laserscanner (DAVID-SLS-1 structured light 3D scanning, DAVID Vision Systems GmbH, Koblenz) of open and closed traps and recorded the respective, manually triggered motions with a high-speed camera (Motion Pro Y4, IDT, Tallahassee, USA, recording speed 1000 fps). We have found that the trap motion in the cultivar "Angel Wings" is blocked shortly after triggering. Because of the strong lobe curvature, the energy barrier between the concave and convex lobe states is apparently so high that pure hydraulics are not sufficient to cause trap closure. By manually pressing the trap lobes together, they can be brought into a closed state. On the contrary, in the trap of the cultivar "Korean Melody Shark", we have observed a slow and continuous motion (snapping duration: 2.2 s). Here, the lobe curvature and, consequently, the elastic instability is much less pronounced compared with the wild type. Hence, we hypothesize that little elastic energy can be stored and released during trap motion. The motion in a wild type trap shows a characteristic sequence of motion steps of highly different speeds (slow initial motion, fast snap-buckling, slow final closure) (cf. Forterre et al. 2005; Poppinga et al. 2016b). Assuming that the lobe anatomy and the physiology behind the hydraulic movement do not differ in general between the cultivars and the wild type, we speculate that the different trap behaviours (no snapping, normal snapping, slow snapping) can be explained by the different lobe curvatures. This indicates that geometric limitations are indeed present with respect to the mechanics and dynamics of *Dionaea muscipula* traps. A planned extension of the test specimens including cultivars with intermediate lobe curvatures and with trap shapes strongly diverging from the normal bauplan will lead to further insights into the structural requirements for functional snap-traps.

9.4 Simulation

Mechanical modelling of biological role models aims to provide a deeper understanding of their motion principles and may thus contribute to the field of reverse biomimetics. The Finite Element Method (FEM) provides new possibilities in quantifying the various parameters that have a significant influence on the overall deformation mechanism. Even though some information about material and geometry and the effect of the environment may still be lacking, assumptions can be made that allow the qualitative simulation of the principal mechanism.

Moreover, various simulation techniques enable the discovered kinetic mechanisms and involved geometrical principles to be abstracted, following the methodology of Schleicher et al. (2015). The process of modelling and simulating mechanisms is divided into three main categories. 1. The geometric model provides a parameterised representation of the underlying geometrical features responsible for the movements. It allows the development of topologically identical variations. 2. The kinematic model is used qualitatively to evaluate the way that the geometric parameters influence the actual movement. 3. Finite element analysis used in the kinetic model permits not only the evaluation of the geometric variations, but also the determination of the manner in which certain material distributions and stiffness gradients influence the energy needed and the related forces therein.

9.4.1 Simulation of the Aldrovanda *Trap*

The two trap lobes are connected by a midrib and each is divided into two parts differing in thickness and function (Fig. 9.5). The central portion is located next to the midrib and contains the motor zone that active-hydraulically drives the movement. The outer lobe part is composed only of two thin cell layers (Ashida 1934), leading to a (supposed) much smaller stiffness compared with the central portion consisting of three cell layers. The simulation of the trapping movement as a kinetic model, by using FEM, has been performed by Schleicher (2015). As a result, the principal mechanism, as described in Sect. 9.3.1, could be verified. The objective

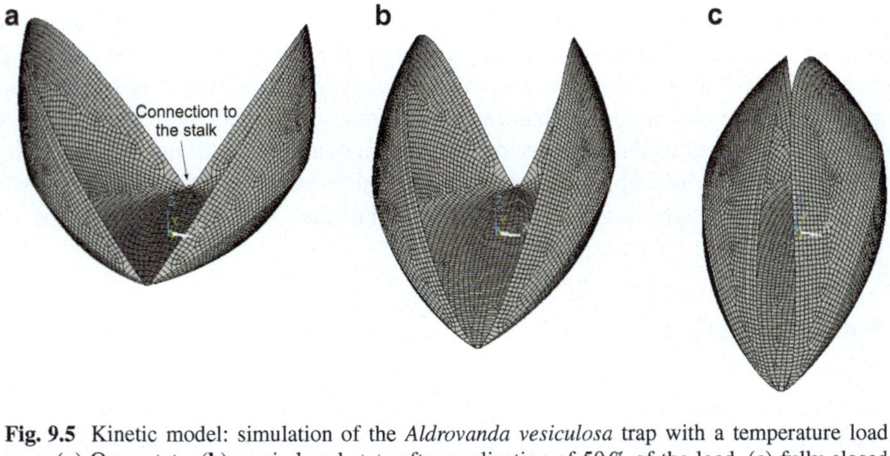

Fig. 9.5 Kinetic model: simulation of the *Aldrovanda vesiculosa* trap with a temperature load case. (**a**) Open state, (**b**) semi-closed state after application of 50 % of the load, (**c**) fully closed state

for future research is to simulate the mechanism in more detail and to quantify important values or limits constraining the closing mechanism. For this purpose, a different finite element model is constructed. Because much of the input data is hard to examine experimentally (e.g. mechanical parameters of tissues), various modelling assumptions for the geometry, material and even load cases need to be made.

Detailed images of the *Aldrovanda* trap from various angles serve as the starting point for the geometry model. The overall dimensions of the geometry match the real size of the trap but the geometric model itself is an idealised one. The base geometry has a midrib length of 6 mm and a width of 4.5 mm, leading to an opening angle of about 70°. The Young's modulus of the lobes is assumed to be 800 N/mm^2 and twice as high for the midrib. Further to emphasis the different levels of stiffness in the zones described in Sect. 9.3.1, thicknesses between 0.03 mm and 0.2 mm are allocated. The displacement boundary conditions are represented by the connection of the midrib to the rest of the leaf. In order to prevent rigid body movements, all degrees of freedom, three translations and three rotations are fixed at one end. To simulate the strain or volume change by turgor variation in the motor cells after trap triggering, a temperature load case is applied to the motor zone. In a geometrically nonlinear analysis of the model, we have found that an isotropic negative temperature load, meaning a contraction of the cells by 11 %, leads to a closure of the trap as illustrated in Fig. 9.5.

Schleicher et al. (2015) have focused on the translation of the trap mechanism into a kinetic curved-line folding model in which two flaps are connected to a stiffer middle part by areas of reduced bending stiffness (Fig. 9.6). The most intriguing aspect for technical use is the extensive amplification of a comparatively small bending deformation of the central part into a large closing movement of the two adjacent flaps. The relationship between bending and opening is determined by the radius of the curved-line fold. To evaluate this correlation, Schleicher et al. (2015) developed a kinematic model by using the "Rigid Origami Simulator" (Tachi 2009), which allows simulation of the folding mechanism of a discretised simplification of the curved-line fold. Although this kinematic model provides an overview of geometrical variations, it lacks precision and the consideration of material properties. Another possibility for approximating the folding behaviour is the method of reflection (Mitani and Iagarshi 2011), which requires that the curved-line fold remains entirely on one plane during the folding process. To simulate the

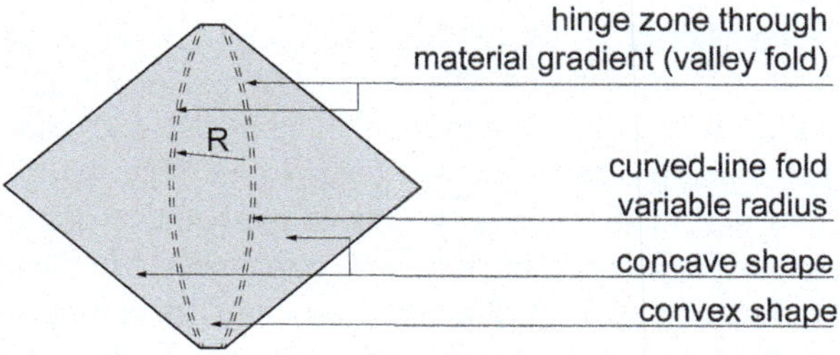

Fig. 9.6 Geometric model: abstracted curved-line folding model

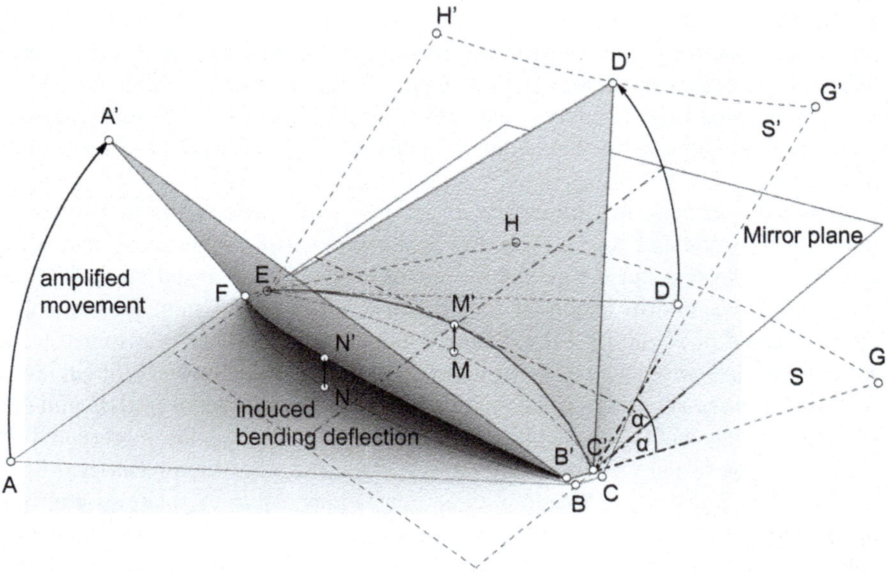

Fig. 9.7 Kinematic model of the abstracted mechanism based on the method of reflection

folding movement, which is controlled by the induced bending, we created a bent midrib geometry in the form of a lens-shaped surface based on the elastica curve in relation to the given translation of control points B and C (Fig. 9.7). The plane given by C′, E and M′ is used to mirror the surface S and the initial outline of the flap surface. Due to the amplification of the folding mechanism only a small translation of B to B′ and C to C′ is needed to result in a complete closure of the component. With the information of curvature and reflected boundary, the folded geometry can be rebuilt. This method not only allows real-time parameter manipulation, but also provides immediate information about the relationship between the displacement of the support points, folding angle and sensitivity of the mechanism and reduces the computational effort compared to spring based methods, especially when it comes to simulations of aggregations of a high number of elements.

The established geometric relationships were transferred into a kinetic model by using FEM software (SOFiSTiK, SOFiSTiK AG, Oberschleißheim, Germany). Several variations regarding the hinge zones in terms of width and stiffness gradient between the hinge, flaps and middle rib have been simulated and evaluated by Schleicher et al. (2015) and Schleicher (2015). When we continued the research with the main focus on the construction of a physical prototype, a rectangular shape with edge lengths of 520 mm and 420 mm was chosen because of the economic production of test objects (see Sect. 9.5.1). Furthermore, the integration of a pneumatic actuator to induce the bending deformation of the central portion was of importance and had to be implemented in the analysis model.

In a first series of digital experiments, an appropriate material gradient between the three zones was established, leading to the desired closing movement. The pneu-

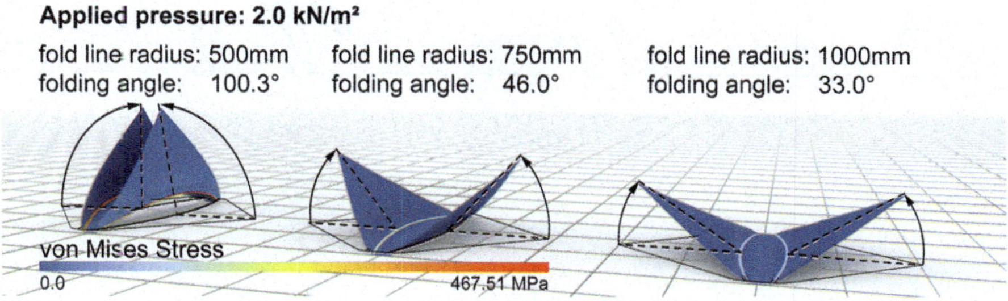

Fig. 9.8 Comparison of models exhibiting various curved-line fold radii to evaluate the required actuation force and corresponding folding movement

matic actuator was simulated by uniformly distributed pressure on the midrib. Even though this does not match accurately a pneumatic cushion, it gives a qualitative insight into the relationship between applied pressure and resultant movement. With this method and the material distribution of a 0.9 mm thick "midrib" and 0.8 mm flaps (Young's Modulus 14000 N/mm^2) and a 0.2 mm thick and 5.0 mm wide hinge zone (Young's Modulus 8000 N/mm^2), a series of different curved-line folds was simulated and evaluated in terms of the required pressure, resultant stresses, corresponding displacement and sensitivity to the actuator (Fig. 9.8). As a summary, it can be stated that a larger radius of the curved-line fold leads to a higher actuation force but less displacement in the midrib. With a smaller radius, the midrib exhibits higher curvature, which is transferred into the flaps, leading to higher geometrical stiffness in the folded position.

For implementation into technical systems, one of the most interesting results in the preliminary research by Schleicher et al. (2015) was the recognition that the basic rectangular shape can be scaled in size and distorted in geometry while remaining operative in these modified instances and boundary conditions. This makes it possible to translate it into various boundary conditions. To keep the movement as consistent as possible between several panels, we developed a method that subdivided a given surface, exhibiting positive and negative Gaussian curvature into adjusted quadrilateral patches of similar anticlastic curvature to ensure a certain pre-fold in the entirely closed condition. The geometric rules of curved-line folding permit the translation of each surface patch into the folded geometry (Fig. 9.9). Further investigations will be addressed to improve the basic geometry in terms of efficiency as shading device, including detailed climatic, physiologic and psychologic comfort studies.

9.4.2 *Simulation of the* Dionaea *Trap*

Simulation of the *Dionaea* snap-trap makes different demands as it functions with different mechanics than the *Aldrovanda* trap. As mentioned in Sect. 9.3.2, the closure of the *Dionaea* trap can be divided into three phases: a first slow movement

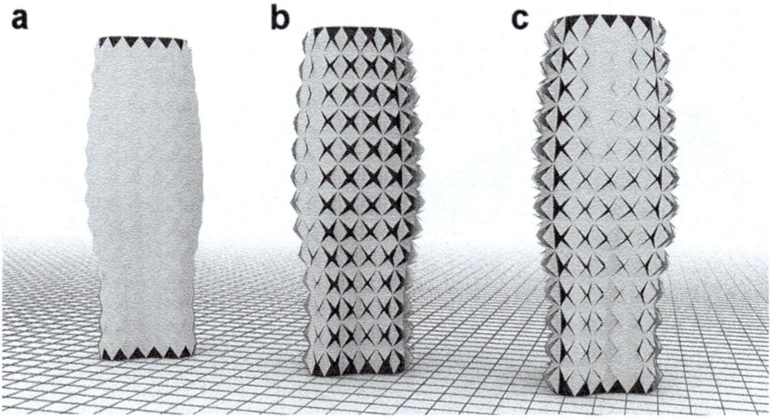

Fig. 9.9 The derived kinetic mechanism is adaptable to geometries of synclastic and anticlastic double curvature. (**a**) Closed postion. (**b** and **c**) Open configurations based on sun vector

driven by hydraulic pressure, a second phase characterised by a high deformation speed during snap-through (also called snap-buckling) and a third slow closing phase. The second process is characterised by a sign change in Gaussian curvature and by the bypassing of an unstable path with negative stiffness.

The geometry of the trap is described by a passive midrib and two active lobes on either side that perform the curvature inversion from convex to concave. These parameters can be directly transferred to a finite element simulation. Because the lobes act independently, in terms of their post-stimulatory movements, and in order to save calculation time, only one lobe is simulated. The geometry is given by a 3D scan that is idealised to a symmetric shape. For the approximate representation of the real cross section, the thicknesses vary between 1 mm at the connection to the midrib and 0.75 mm at the outer end of the lobe. Young's modulus and the Poisson's ratio are assumed to be 1400 N/mm^2 and 0.0, respectively. The boundary conditions can be described by fixing all translations and rotations along the midrib. The measured strain field described by Forterre et al. (2005) has been increased by 50 % and applied as a temperature load case to represent the hydraulic pressure. During a dynamic analysis, this load is gradually increased from zero up to the total load (load factor).

Figure 9.10 shows the load displacement curves of the mid-point of the outer edge of a trap lobe. The movement can be seen to start slowly and represents the change of hydraulic pressure in the initial motion phase. After application of approximately 85 % of the load, snap-through occurs and the deformation speed increases. In contrast to the wild type, the two cultivars behave differently in our simulations according to the different curvature of their lobes. The cultivar "Korean Melody Shark" is characterised by its very weak lobe curvature. The trap

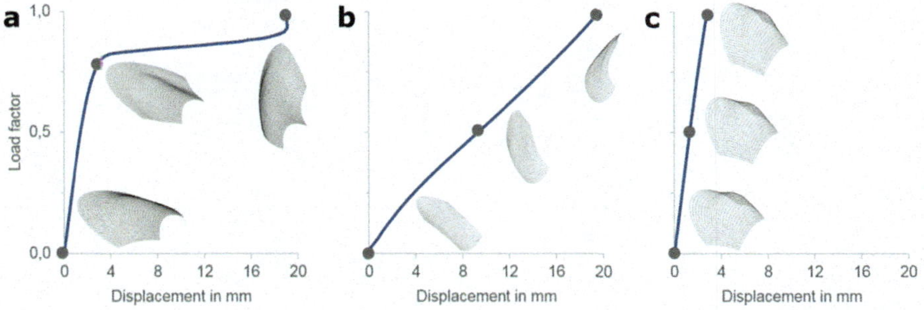

Fig. 9.10 Kinetic models of a Venus flytrap lobe. (**a**) Load-displacement curve and deformation of the normal shaped wild type. Snap-through occurs at 85 % of the load, corresponding to a load factor of 0.85. Load-displacement curve and deformation of the cultivars (**b**) "Korean Melody Shark" and (**c**) "Angel Wings"

closure is a smooth movement without any rapid deformation enhancement. By applying the same load as in the wild type, this behaviour is also visible in the load displacement curve, shown in Fig. 9.10. The opposite of the "Korean Melody Shark" type is presented by "Angel Wings", which exhibits strong curvature of its lobes. The shape of the curve is similar as that for "Korean Melody Shark" but the displacement values indicate that no closure occurs. The applied load is not sufficient to evoke snap-through. These initial simulations generally corroborate the observations described in Sect. 9.3.2.

With regard to technical applications, one of the most noteworthy aspects is the change from a concave to a convex geometry, associated with the bi-stability of open and closed configurations. This change of curvature direction could be advantageous regarding the adaptability to double curved geometries. The transition phase is difficult to control and involves a sudden release of stored elastic energy, which might be problematic for large-scale applications. The investigation of geometrical variations leads to the assumption that the released energy is dependent on the thickness and initial curvature of the lobes (cf. Forterre et al. 2005). Figure 9.11 shows a comparative study based on a parameterised surface with adjustable curvature (Poppinga and Joyeux (2011) and thickness. By using a certain range of expansion (0–0.06 %) uniformly over the whole surface, we can evaluate if and at which expansion rate the snap-through occurs.

To evaluate the influence of stiffness gradients on the released energy, we built a simulation model of about 700 mm/500 mm with constant curvature, actuated by the expansion of a central portion of increased thickness (5 mm, Young's Modulus 14000 N/mm^2). The adjacent surfaces varied in their Young's Moduli (10000 N/mm^2, 7000 N/mm^2 and 4000 N/mm^2) and possessed a thickness of 1 mm. Figure 9.12 shows the initial and deformed shapes and the curvature change along the central axis. The transition is smoother in the case of reduced stiffness of the adjacent surfaces.

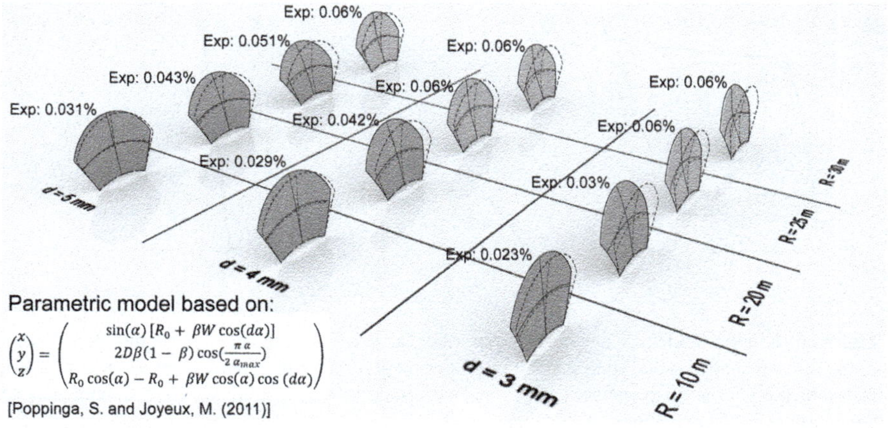

Fig. 9.11 Comparison of the snap-through behaviour of *Dionaea* trap lobe geometries with various initial curvature (based on R in m) and shell thickness (d in mm) with a maximum expansion (Exp) of 0.06 %

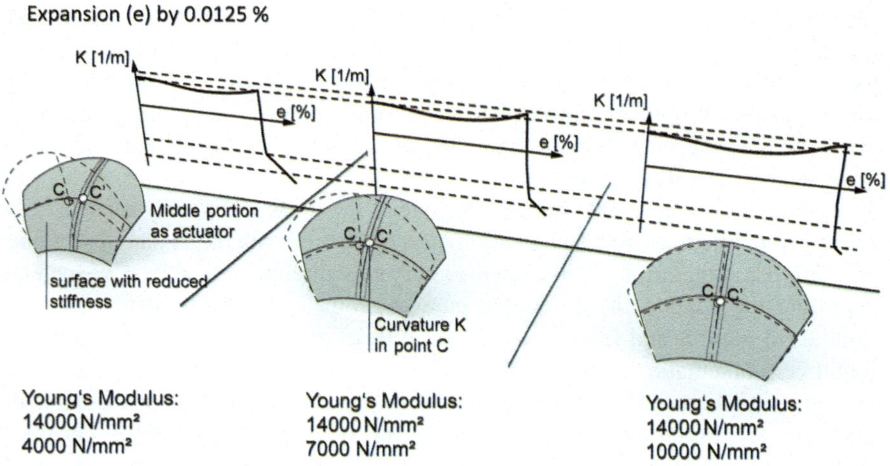

Fig. 9.12 Comparison of the snap-through behaviour of various stiffness gradients between actuator rib and adjacent surfaces. A higher gradient leads to a smoother transition

9.5 Technical Implementation

9.5.1 Technical Implementation of the Trap Movement Principle of Aldrovanda vesiculosa

The technical implementation of the *Aldrovanda* trap movement (Flectofold) is structurally based on defined areas with various mechanical properties. The midrib and the flaps are stiff but remain elastically deformable, while the curved hingezone (the "technical counterpart" of the elliptical bending zone in *Aldrovanda* trap) is very soft. These areas of the plant with huge differences in the mechanical properties are obtained by precisely defined gradient structures between them. The structure resembles a fibre-reinforced composite with anisotropic and tailored mechanical properties in which the matrix takes the role of load transmission and shaping, whereas the fibres absorb the forces and determine the strength of the component. By using fibre-reinforced plastics (FRP) it was it possible to transfer the structural design of the *Aldrovanda* trap into a first demonstrator and to meet the demanding material requirements in technical applications.

In FRP, mostly stiff fibres are embedded in a shaping matrix to generate tailored mechanical properties that depend on the fibre orientation: a high load-carrying capacity occurs in the fibre direction and a good load-transfer capacity exists perpendicular to the fibre direction (Schürmann 2007). By adaption of i.e. the fibre orientation and the numbers of layers, an anisotropic material can be created. The fibre orientations within one component depend on the textile processes and on the draping of the textile layers onto the form during the fabrication process. In order to obtain various fibre orientations, several textile layers are placed at a certain angle onto each other. Herewith, different fibre orientations can be generated within one component with the aim of defining anisotropic properties and to obtain components with designated bending-zones, i.e. by reduction of the number of textile layers, which are stiff in other zones. These bending-zones are necessary for the implementation of the curved folding principle.

For the technical implementation of curved folding into an FRP component, VAP® (Vacuum Assisted Process) was selected. By using a semipermeable membrane during the vacuum infusion it was possible to reduce air inclusions in the component. Details of this manufacturing process are provided in Henning (2011). During the manufacture of the demonstrator, the resin infusion was applied directly onto a laminating foil (Fig. 9.13), instead of the mould itself. By this means, no preparation of the mould with a release agent is necessary, which makes the process faster and less expensive. Furthermore, the laminating foil is directly integrated into the component in order to avoid potential finishing processes caused by the surface and especially by the colour of the FRP demonstrator.

The material set-up needed for the technical implementation (Fig. 9.13) of the curved folding was determined by a scientific material analysis. The tested resins

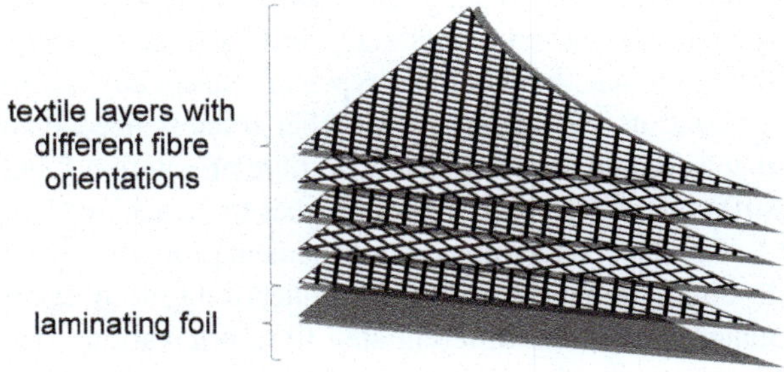

Fig. 9.13 Basic laminate set-up of one shading element of the Flectofold

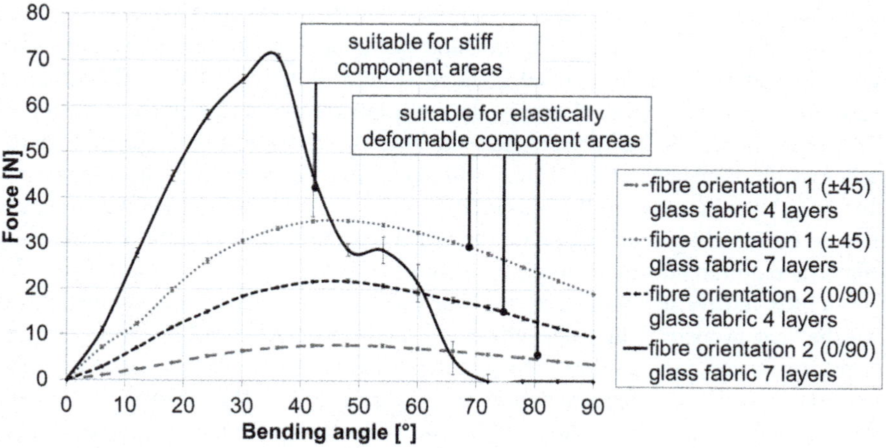

Fig. 9.14 Force-bending angle-diagram: various fibre orientations and laminate thicknesses, respectively

were analysed by Differential Scanning Calorimetry (DSC) to determine their glass transition point. The chosen resin fulfils the temperature requirements for outer façades, i.e. a glass transition temperature of a minimum of 80 °C. Additionally, the bending properties were tested by using a modified two-point-bending test based on DIN 53121 with regard to the curved folding movement. With respect to the established material and stiffness derived from gradients, which were derived from digital simulations concerning the different stiffness gradients within the component, various resin-fibre combinations, laminate thicknesses, fibre materials, fibre orientations and laminate set-ups were tested for their bending behaviour (max. bending angle 90°). An exemplary extract of the test results is given in Fig. 9.14. The tested samples consist of the same resin and the same woven glass fabric. This material combination was tested in two different laminate thicknesses (four and seven layers of fabric) and two different orientations of the fabric.

In Fig. 9.14, the forces necessary to induce various bending angles of the samples are indicated. The curves are based on a sample size of $n = 5$, the dimensions of the samples are 25x50 mm. Unless all samples consist of the same material combination, it is obvious that fibre orientation 2 with a laminate thickness of seven layers leads to a break. In contrast, samples with fibre orientation 1 and a laminate thickness of seven layers stay elastically deformable. In the cases of four-layer-thick laminae, both sample types are elastically deformable but the one with fibre orientation 2 is stiffer and a higher force is needed for bending. The results show that a combination of layers with fibre orientation 1 and 2 enables to achieve defined stiff and elastically deformable areas to be achieved in one component.

The data concerning the mechanical properties were fed back into the simulation to obtain more precise results concerning the possible stiffness range of the used materials and the differentiated fibre orientations. Thus, more information on the

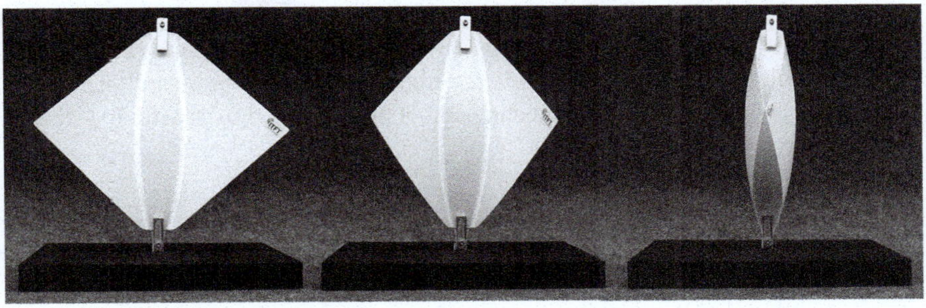

Fig. 9.15 Demonstrator Flectofold

stresses, the required actuation force and the behaviour under external loading can be gained. Eventually, a woven glass fabric was chosen for the building of a first prototype of the Flectofold (Fig. 9.15). By adaption of the fibre orientation and different laminate set-ups within the bending-zone, the backbone and the shading elements, it is possible to implement the *Aldrovanda* curved folding movement principle in the Flectofold demonstrator. In contrast to the biological role model, the backbone of the Flectofold is wider but the bending-zones are narrow. In the bending-zones, the fibres are primarily stressed with shear loads. In the stiffer regions of the Flectofold, as in the backbone and the shading elements, the fibres are stressed both in the fibre and shear direction.

9.5.2 *Technical Implementation of* **Dionaea muscipula's** *Trap Movement Principle*

The finding that a mechanical "trick" (snap-buckling) allows the Venus flytrap to snap fast was also beautifully brought into a wider context of physical limits of plant movement (Skotheim and Mahadevan 2005). Although the snap-through transition mechanism has long been widely known among construction engineers and is present, for example, in switches, caps, and even in children's toys, the idea that it is implemented as a plant movement speed-boost feature has resulted in a multitude of papers presenting energy-related and/or motion-related biomimetic applications of this principle (reviewed by Hu and Burgueño 2015; Guo et al. 2015). For example, Holmes and Crosby (2007) have fabricated arrays of micro-lenses, which, by snap-buckling, switch their focal points, Guo et al. (2014) have fine-tuned bistable helical ribbons for shape-shifting, and Lee et al. (2010) have made jumping microgels.

Because of the sudden release of energy and the barely controllable transition between the two stable states (see Sects. 9.2 and 9.4.2), no architectural application is, to our best knowledge, available so far, although the curvature transition from concave to convex offers potential for the geometrical adaptability of façade-shading systems. We believe that investigations into the geometric and mechanical properties of curvature inversions of doubly-curved shells without snap-buckling, as present in false indusia in *Adiantum* (Poppinga et al. 2015), some *Dionaea* cultivars, e.g. "Korean Melody Shark", and in *Dionaea* trap reopening (Poppinga et al. 2016b), will lead to new insights into the mechanisms at play and inspire transfer to technical structures.

9.6 Discussion and Perspectives

Movements in plants, for example, flower opening and closing, the unfolding of leaves or prey capture movements in carnivorous plants, are highly functional, robust and reliable. In addition, they are typically performed by elastic deformation, i.e. without localised hinges, and have therefore fewer regions of stress concentrations. These features make plant movements especially promising concept generators not only for a novel group of movable structures in architecture, but also for many other applications in various fields of technology, for example, automotive, air- and spacecraft or machine construction. Bioinspired or biomimetic solutions for movable structures in technical applications have the potential to be robust and both disturbance- and maintenance-free, which is of increasing interest for many applications, especially in architecture and building construction (Speck et al. 2016; Knippers and Speck 2012). In addition to these structural and functional features, which are the reasons for the increased interest in bioinspired solutions in building construction and many other field of technology observable during the last few decade(s), an additional feature has to be taken into account in order to understand the specific interest in architecture for bioinspired solutions: a different type of aesthetics based on the beauty of biological movements and functions in general. In the "typically" biomimetic approach, driven by natural scientists, material scientists and engineers, the predictable and reliable transfer of functionalities from the biological role model to the biomimetic product is central, although some of the "aesthetic value" of the biological role model is sometimes lost in this process (Speck 2015). On the other hand, in the art-based process of architectural design, the aesthetic values of biology are often the incentives for choosing a bio-inspired approach (Imhof and Gruber 2015). In the case of bioinspired movable structures, biomimetic products can be developed in which not only the physical functionality is transferred from biology, but additionally a functional elegance similar to that of the highly aesthetic biological role models can be achieved. These cases are still rare and can be considered as the "royal road", i.e. as ideal solutions combining bio-inspired functionality with the natural beauty of function (Speck 2015). The first biomimetic façade-shading system, named Flectofin®, represents such an example. The Flectofin® is a hinge-less resilient biomimetic façade-shading system inspired by the elastic deformation processes taking place during bird pollination on the perch of the Bird-of-paradise flower (*Strelitzia reginae*) (Lienhard et al. 2011; Knippers et al. 2011; Speck et al. 2015). This first successful development combining a "hard approach" to bio-inspired architecture and building construction (mainly driven by civil engineers, materials scientists and natural scientists) with a "soft approach" to bio-inspired architecture (mainly driven by artists, designers and architects) shows the great potential of biomimetic elastic architecture and allows a promising outlook, especially for bioinspired movable structures.

Acknowledgements We thank Anja and Holger Hennern for their kind permission to include Fig. 9.1c in this chapter. This work has been funded by the German Research Foundation (DFG) as part of the Transregional Collaborative Research Centre (SFB/Transregio) 141'Biological Design and Integrative Structures'/project A04.

References

Antkowiak B, Mayer WE, Engelmann W (1991) Oscillations of the membrane potential of pulvinar motor cells in situ in relation to leaflet movements of *Desmodium motorium*. J Exp Bot 42:901–910

Ashida J (1934) Studies on the leaf movement of *Aldrovanda vesiculosa* L. I. Process and mechanism of the movement. Mem Coll Sci Kyoto Imp Univ Ser B9:141–244

Bailey T, McPherson S (2012) *Dionaea*. The Venus's Flytrap. Redfern Natural History Productions, Poole

Böhm J, Scherzer S, Krol E et al (2016) The Venus flytrap *Dionaea muscipula* counts prey-induced action potentials to induce sodium uptake. Curr Biol 26:1–10

Braam J (2005) Plant responses to mechanical stimuli. New Phytol 165:373–389

Burgert I (2006) Exploring the micromechanics of plant cell walls. Am J Bot 93:1391–1401

Burgert I, Fratzl P (2009) Actuation systems in plants as prototypes for bioinspired devices. Phil Trans R Soc A 367:1541–1557

Campbell NA, Garber RC (1980) Vacuolar reorganization in the motor cells of *Albizzia* during leaf movement. Planta 148:251–255

Colombani M, Forterre Y (2011) Biomechanics of rapid movements in plants: poroelastic measurements at the cell scale. Comput Method Biomec 14:115–117

Cross A (2012) *Aldrovanda*. The Waterwheel Plant. Redfern Natural History Productions, Poole

Darwin C (1865) On the movements and habits of climbing plants. Bot J Linn Soc 9:1–18

Darwin C (1875) Insectivorous plants. Murray, London

Dawson C, Vincent JFV, Rocca AM (1997) How pine cones open. Nature 390:668

Deegan RD (2012) Finessing the fracture energy barrier in ballistic seed dispersal. Proc Natl Acad Sci U S A 109:5166–5169

Dumais J, Forterre Y (2012) "Vegetable dynamicks": the role of water in plant movements. Annu Rev Fluid Mech 44:453–478

Forterre Y (2013) Slow, fast and furious: understanding the physics of plant movements. J Exp Bot 64:4745–4760

Forterre Y, Skotheim JM, Dumais J et al (2005) How the Venus flytrap snaps. Nature 433:421–542

Ge ZW, Yang ZL, Vellinga EC (2010) The genus *Macrolepiota* (Agaricaceae, Basidiomycota) in China. Fungal Divers 45:81–98

Gerbode SJ, Puzey JR, McCormick AG et al (2012) How the cucumber tendril coils and overwinds. Science 337:1087–1091

Guo Q, Zheng H, Chen W et al (2014) Modeling bistable behaviors in morphing structures through finite element simulations. Bio Med Mater Eng 24:557–562

Guo Q, Dai E, Han X et al (2015) Fast nastic motion of plants and bioinspired structures. J R Soc Interface 12:20150598

Henning F (2011) Handbuch Leichtbau. Carl Hanser Verlag, München

Heslop-Harrison Y (1970) Scanning electron microscopy of fresh leaves of *Pinguicula*. Science 167:172–174
Hodick D, Sievers A (1989) On the mechanism of trap closure of Venus flytrap (*Dionaea muscipula* Ellis). Planta 179:32–42
Holmes DP, Crosby AJ (2007) Snapping surfaces. Adv Mater 19:3589–3593
Hovenkamp PH, van der Ham RWJM, van Uffelen GA et al (2009) Spore movement driven by the spore wall in an eusporangiate fern. Grana 48:122–127
Howell LL (2001) Compliant mechanisms. Wiley, New York
Hu N, Burgueño R (2015) Buckling-induced smart applications: recent advances and trends. Smart Mater Struct 24:063001
Imhof B, Gruber P (eds) (2015) Built to grow – blending architecture and biology. Birkhäuser Verlag, Basel
Keijzer CJ, Hoek IHS, Willemse MTM (1987) The processes of anther dehiscence and pollen dispersal. III. The dehydration of the filament tip and the anther in three monocotyledonous species. New Phytol 106:281–287
Knippers J, Lienhard J, Schleicher S et al (2011) Gelenkloser, stufenlos verformbarer Klappmechanismus (Hingeless, infinitely deformable folding mechanism). Eur Patent Off Filing 10013852(8)
Knippers J, Speck T (2012) Design and construction principles in nature and architecture. Bioinspir Biomim 7:015002
Knippers J, Scheible F, Oppe M et al (2012) Bio-inspirierte kinetische Fassade für den Themenpavillon One Ocean Expo 2012 in Yeosu, Korea. Bautechnik 90:341–347
Kobayashi H, Kresling B, Vincent JFV (1998) The geometry of unfolding tree leaves. Proc Roy Soc B 265:147–154
Lee H, Xia C, Fang NX (2010) First jump of microgel; actuation speed enhancement by elastic instability. Soft Matter 6:4342–4345
Liang H, Mahadevan L (2011) Growth, geometry, and mechanics of a blooming lily. Proc Natl Acad Sci USA 108:5516–5521
Lienhard J, Schleicher S, Poppinga S et al (2011) Flectofin: a hinge-less flapping mechanism inspired by nature. Bioinspir Biomim 6:045001
Mitani J, Iagarshi T (2011) Interactive design of planar curved folding by reflection. In: Pacific conference on computer graphics and applications, Kaohsiung, Taiwan, 21–23 September 2011
Müller L (1933) Über den Bau und die Entwicklung des Bewegungsmechanismus von *Physostegia virginiana*. Planta 18:651–663
Nakamura Y, Reichelt M, Mayer VE et al (2013) Jasmonates trigger preyinduced formation of 'outer stomach' in carnivorous sundew plants. Proc Roy Soc B 280:20130228
Niklas KJ (1992) Plant biomechanics: an engineering approach to plant form and function. The University of Chicago Press, Chicago
Peakall R (1990) Responses of male *Zaspilothynnus trilobatus* Turner wasps to females and the sexually deceptive orchid it pollinates. Funct Ecol 4:159–167
Poppinga S, Joyeux M (2011) Different mechanics of snap-trapping in the two closely related carnivorous plants *Dionaea muscipula* and *Aldrovanda vesiculosa*. Phys Rev E 84:041928
Poppinga S, Masselter T, Speck T (2013) Faster than their prey: new insights into the rapid movements of active carnivorous plants traps. BioEssays 35:649–657
Poppinga S, Haushahn T, Warnke M et al (2015) Sporangium exposure and spore release in the Peruvian maidenhair fern (*Adiantum peruvianum*, Pteridaceae). PLoS One 10:e0138495
Poppinga S, Weisskopf C, Westermeier AS et al (2016a) Fastest predators in the plant kingdom: functional morphology and biomechanics of suction traps found in the largest genus of carnivorous plants. AoB PLANTS 8:plv140
Poppinga S, Kampowski T, Metzger A et al. (2016b) Comparative kinematical analyses of Venus flytrap (*Dionaea muscipula*) snap-traps. Beilstein J Nanotechnol 7:664–674

Reith M, Claßen-Bockhoff R, Speck T (2006) Biomechanics in *Salvia* flowers, the role of lever and flower tube in specialization on pollinators. In: Herrel A, Speck T, Rowe N (eds) Ecology and biomechanics: a mechanical approach to the ecology of animals and plants. CRC Press, Boca Raton, pp 123–146

Reith M, Baumann G, Claßen-Bockhoff R et al (2007) New insights in the functional morphology of the lever mechanism of *Salvia pratensis* (Lamiaceae). Ann Bot 100:393–400

Schleicher S (2015) Bio-inspired compliant mechanisms for architectural design. Transferring bending & folding principles of plant leaves. to flexible kinetic structures. Dissertation thesis, University of Stuttgart

Schleicher S, Lienhard J, Poppinga S et al (2015) A methodology for transferring principles of plant movements to elastic systems in architecture. Comput Aided Des 60:105–117

Schürmann H (2007) Konstruieren mir Faser-Kunststoff-Verbunden. Springer, Berlin/Heidelberg

Skead CJ (1975) Weaverbird pollination of *Strelitzia reginae*. Ostrich 46:83–185

Skotheim JM, Mahadevan L (2005) Physical limits and design principles for plant and fungal movements. Science 308:1308–1310

Speck O, Speck D, Horn R, et al (2016) Biomimetic - bio-inspired - biomorph - sustainable? An attempt to classify and clarify biology-derived technical developments. Bioinspir Biomim (accepted)

Speck T, Speck O (2008) Process sequences in biomimetic research. In: Brebbia CA (ed) Design and nature IV. WIT Press, Southampton, pp 3–11

Speck T (2015) Approaches to bio-inspiration in novel architecture. In: Imhof B, Gruber P (eds) Built to grow – blending architecture and biology. Birkhäuser Verlag, Basel, pp 145–149

Speck T, Knippers J, Speck O (2015) Self-x-materials and -structures in nature and technology: bio-inspiration as driving force for technical innovation. Archit Design 85:34–39

Tachi T (2009) Simulation of rigid origami. In: Lang R, Peters AK (eds) Origami 4: the fourth international conference on origami in science, mathematics, and education, vol 4. A K Peters Limited, Natick, pp 175–187

van der Cingel NA (2001) An atlas of orchid pollination. A A Balkema, Rotterdam

van Doorn WG, van Meeteren U (2003) Flower opening and closure: a review. J Exp Bot 54:1801–1812

van Doorn WG, Kamdee C (2014) Flower opening and closure: an update. J Exp Bot 65:5749–5757

Velasco R, Brakke AP, Chavarro D (2015) Dynamic façades and computation: towards an inclusive categorization of high performance kinetic façade systems. In: Celani G, Sperling DM, Franco JMS (eds) Computer-aided architectural design futures. The next city, vol 527, New technologies and the future of the built environment. Springer, Berlin/Heidelberg, pp 172–191

Vincent O, Weißkopf C, Poppinga S et al (2011) Ultra-fast underwater suction traps. Proc Roy Soc B 278:2909–2914

Weintraub M (1952) Leaf movements in *Mimosa pudica* L. New Phytol 50:357–582

Williams SE, Bennett AB (1982) Leaf closure in the Venus flytrap: an acid growth response. Science 218:1120–1122

Williams SE, Pickard BG (1979) The role of action potentials in the control of capture movements of *Drosera* and *Dionaea*. In: Skoog F (ed) Plant growth substances 1979 – Proceedings of the 10th international conference on plant growth substances. Madison, Wisconsin, pp 470–480

Chapter 10
Branched Structures in Plants and Architecture

Larissa Born, Florian A. Jonas, Katharina Bunk, Tom Masselter, Thomas Speck, Jan Knippers, and Götz T. Gresser

Abstract In architecture and construction engineering, a vast number of connections and branched columns in frame structures exist that are exposed to high static and dynamic loads. The manufacture of many of these elaborate structures is both time-consuming and costly. Industry has no solution for cost-effectively producing aesthetic and mechanically stable branched columns. This challenge is addressed by the development of branched structures inspired by branched biological concept generators such as *Schefflera arboricola*. Here, we present methodological approaches allowing the reconstruction of the outer shape and inner structure of complex branching regions, such as in *S. arboricola*, by using and combining three-dimensional-image stacking of histological thin sections, micro-computer-tomography (μCT) imaging and laser scanning. Computer-aided design (CAD) and Finite Element (FE) models of such structures can then be produced that not only help to provide a better understanding of the functional morphology and biomechanics of the biological concept generator, but also render the basis for the intended biomimetic transfer to branched columns consisting of a braided hull filled with concrete. The current project results are mainly based on the analysis of *S. arboricola* branching and the results of a previous research project (SPP 1420) in which biomimetic branched fibre-reinforced plastic (FRP) columns inspired by the branching structure of *Dracaena* were produced. Currently a biomimetic hull geometry that can be manufactured industrially is developed. Initially, branched

---

L. Born • G.T. Gresser (✉)
Institute for Textile Technology, Fiber Based Materials and Textile Machinery (ITFT), University of Stuttgart, Pfaffenwaldring 9, 70569, Stuttgart, Germany
e-mail: goetz.gresser@itv-denkendorf.de

F.A. Jonas • J. Knippers
Institute of Building Structures and Structural Design (ITKE), University of Stuttgart, Keplerstraße 11, 70174 Stuttgart, Germany

K. Bunk • T. Masselter • T. Speck
Plant Biomechanics Group, Botanic Garden, Faculty of Biology, University of Freiburg, Schänzlestraße 1, D-79104 Freiburg, Germany

Freiburg Centre for Interactive Materials and Bioinspired Technologies (FIT), University of Freiburg, Georges-Köhler-Allee 105, D-79110 Freiburg, Germany

J. Knippers et al. (eds.), *Biomimetic Research for Architecture and Building Construction*, Biologically-Inspired Systems 9, DOI 10.1007/978-3-319-46374-2_10

FRPs based on triaxial braids with readily adjustable mechanical properties are filled with concrete and thus shall achieve sufficient mechanical properties for application and cost-effective fabrication in the building industry.

Abbreviations

FRP	Fibre-Reinforced Plastic
RC	Reinforced Concrete
SPP 1420	Priority Programme 1420 by the DFG
DFG	German Research Foundation
MRI	Magnetic Resonance Imaging
μCT	Micro-Computer-Tomography
FE-Model	Finite Element-Model
CAD	Computer Aided Design

10.1 Introduction and Target

In architecture and construction engineering, a vast number of connections and branched columns in frame structures are exposed to high static and dynamic loads. The manufacturing of many of these elaborate structures (Fig. 10.1a) is time-consuming and costly.

In a previous research project (SPP 1420), FRP-branched columns inspired by arborescent monocotyledons such as the genus *Dracaena* (Fig. 10.1b) were produced. These plants were chosen as biological concept generators as their inner arrangement, which consists of fibrous bundles embedded in a parenchymatous ground tissue, represents a natural fibre-reinforced composite so that the abstraction of functional principles and the implementation in technical branchings is simplified. At the Institute of Textile Technology and Process Engineering (ITV) Denkendorf and at the Institute of Lightweight Engineering and Polymer Technology (ILK), TU Dresden, hollow braidings of elementary forms were made

Fig. 10.1 (**a**) Airport Stuttgart; (**b**) *Dracaena reflexa*; (**c**) FRP-composite branch

(Fig. 10.1c) for use as construction elements in lightweight structures (Milwich et al. 2006, 2008; Müller et al. 2013). Within this research project, they are being employed as a first series for an outer FRP hull in concrete-filled columns. The disadvantages of low-torsional strength and low-compressive strength at the branching point have not been solved in previous research projects. Therefore the principal aim of this project is the development of novel, biomimetically-optimised branched columns for FRP-truss structures in architecture and construction engineering with a higher load-bearing capacity and reduced stress concentrations at the branching points compared with the state of the art.

The basis for the achievement of novel, biomimetically-optimised branched columns for FRP-truss structures is to find appropriate biological concept generators in plants. Therefore, plant species were selected that are lightweight and evolutionary adapted to bearing high static and dynamic loads for investigation.

A prerequisite for technical implementation in architecture is a detailed understanding of the basic mechanisms underlying the resilience of branched columns and their high load-bearing capacity, which can in turn also serve as a starting point for a detailed mechanical understanding of biological organisms with branched structural design (reverse biomimetics). This is approached via Finite Element (FE) simulations of models of the biological concept generators.

Within the project, new materials, new structures and new fabrication methods for bio-inspired branched columns are developed and combined. To date, the fabrication of branched columns requires complex expensive frameworks for concrete or for cast steel or the elaborate welding of steel members. The novel concept consists of a lightweight branched column based on FRPs with load-adapted fibre orientation that serves as the outer framework for a concrete filling and also as carrying element to bear torsion/tension forces. Since the middle of the 1990s, FRPs have been used for the confinement of reinforced concrete (RC) columns to prevent shear failure, lap splice failure, buckling of the longitudinal reinforcement and premature concrete failure (Mirmiran and Shahawy 1997; Saafi et al. 1999). However, the application of FRP-confined concrete for branching has not been investigated as yet. New fibre technologies are used to develop branched hollow structures in which fibres are placed mainly in the hoop direction and the longitudinal direction.

10.2 Branched Structures in Architecture and Biology

The aim of the quantitative and qualitative analyses of ramifications in plants is to gain insight into the mechanical behaviour of the biological concept generators and to abstract the underlying structural and functional principles for technical implementations in architecture. A deepened knowledge of the form-structure-function relationship in plant ramifications will contribute to the improvement of branched columns in general and especially in building construction. Additionally, the quantitative analysis and simulation of ramifications in plants will help to answer hitherto unsolved biomechanical questions.

Fig. 10.2 Typical bifurcations in architecture (**a**) Pedestrian bridge, Pragsattel Stuttgart; (**b**) Branched column, Airport Stuttgart

Branched columns in plants and architecture are subject to different challenges and constraints. The ability to grow and to adapt during growth is a central difference between plants and building constructions. Although the growth process itself is not transferable to building structures, tissue arrangements and functional principles can be abstracted for technical implementations. In addition to mechanical stability, nutrition and water supply and reproduction are essential for plants (Kull and Herbig 1987; Speck and Burgert 2011), of which the three last-mentioned can be ignored during a first biomimetic transfer process.

In building construction, not only large external loads have to be carried, but also deflections and vibrations need to be reduced and durability has to be ensured. Reliability, robustness, cost efficiency and aesthetic appearance are some of the main and often conflicting design criteria.

In spite of the optical resemblance to arborescent branched plant growth forms, branched columns as shown in Fig. 10.2 exhibit a fundamental difference in loading conditions and load transfer compared with trees or other biological ramified structures. In the artificial truss structure, the load is mostly transferred via normal forces (Fig. 10.3b). The geometry can be physically determined through hanging models (Otto 1982) but structural stability (i.e. buckling of the members) influences the design. Vertical external loads (snow, wind and self-weight of the roof or bridge) are horizontally distributed through the bending of the horizontal structure (i.e. the roof or bridge deck in Fig. 10.2) and induced into the columns (Fig. 10.3 N_b). The bending moments in the members are, in the latter case, very small and therefore not visible in schematic drawings showing internal force distributions (Fig. 10.3 M_b). In contrast, bending moments are the predominant type of load in branches of biological trees (Fig. 10.3 compare M_a with M_b)

Dynamic and static loads such as wind and snow act on both structures. However, trees are able to reduce loading by conspicuous deformations and the stream-lining of branches, twigs and leaves ('controlled flexibility', Speck and Burgert 2011). Because of serviceability requirements, these principles cannot be applied to building structures or are allowed only to a very limited extent.

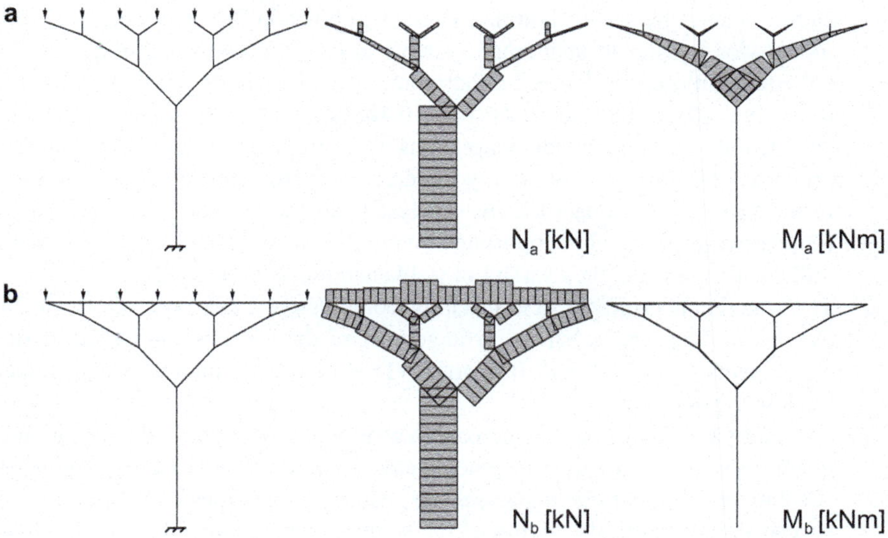

Fig. 10.3 Normal forces N and bending moments M of a branched structure (**a**) without horizontal tension elements (abstraction of a biological tree); (**b**) with horizontal tension elements (typical technical implementation as shown in Fig. 10.2)

Fig. 10.4 Different root morphologies in arborescent angiosperms (**a**) stilt roots (*Pandanus* sp.); (**b**) buttress roots (broad-leaved tropical tree); (**c**) adventitious roots (*Cocos nucifera*)

These considerations show that, even though many branched columns in architecture optically resemble ramifications in plants, adaptations to static and dynamic loads are markedly different. A fundamental analysis of the load adaptations in the ramifications of plants is necessary for a biomimetic translation into branched structures for the building industry. The current project has a special focus on adaptations to loading conditions in branched plant structures, for example, via (1) the formation of roots (compare Fig. 10.4), (2) the development of a load-specific branching architecture and (3) the avoidance of critical notch stresses:

(1) The wide-spread and branched root system of plants provides anchorage in the ground and is highly adapted to resist static and dynamic tension loads

caused mainly by self-weight and wind. The potential of such biological role models for transfer to anchorage systems in building construction has not as yet been explored in detail. Furthermore, specialised root types such as stilt roots (Fig. 10.4a), buttress roots (Fig. 10.4b) or adventitious roots (Fig. 10.4c) demonstrate morphological adaptations to soils with different mechanical properties (e. g. muddy or watery underground) (Kadereit et al. 2014). Such formations will be quantitatively analysed, abstracted and implemented in foundation structures for supporting large technical constructions, such as wind mills that are erected on a variety of stable and unstable subsoils.

(2) The functional principle of translating compressive loads into tensile loads by pre-stressing and a 'smart' arrangement of the (fibrous) biological tissue components (cf. Speck and Burgert 2011) is of special interest for technical implementation.

(3) Secondary growth leads to an increase in weight of stem, roots and branches and thereby also to potentially higher notch stresses in the stem-branch or stem-root attachments. Critical notch stresses are reduced by the adaptive formation of special curvatures in the region of the stem-branch-attachment. The development of form-optimised curvatures is enabled by adaptive secondary growth processes, i.e. the specific accumulation of wood material at areas of high stress and the resulting avoidance of critical stress peaks in notch regions (Mattheck 1998). An additional example of a structural adaptation is the formation of indentations at stem-branch junctions in columnar cacti (Schwager et al. 2010, 2013) and arborescent monocotyledons (Haushahn et al. 2012). FE simulations based on simplified shapes and taking into account the structural arrangement of the biological concept generators show that notch stress peaks are markedly reduced by these adaptations (Schwager et al. 2013). A technical transfer of critical notch stress reduction allows the construction of improved biomimetic branched lightweight structures with a high load-bearing capacity.

Because of the differences of mechanical constraints and adaptations between plants and the technical supporting structures discussed above, learning from biological concept generators for a biomimetic optimisation of branched pillars for building construction is a complex task. A careful selection of biological concept generators is a basic prerequisite, as is the abstraction of appropriate structures and functional principles. Despite its challenges, biomimetic research and development represents a highly promising approach for optimisations in lightweight engineering and technical load-adaptation thereby allowing for variety in constructional shape and pronounced building aesthetics.

10.3 Biological Concept Generators

Plants exhibit a broad diversity of branching morphologies that are highly adapted at several hierarchical levels to predominant physiological, climatic and mechanical constraints, to name but a few of the challenges that result in a high diversity of branching shapes and structures in plants. The identification of promising biological concept generators for optimising branched building structures in civil engineering requires careful selection and a detailed quantitative analysis of biological structures.

The genus *Dracaena* has been one focus for research into biological stem-branch-attachment over the last few years (Masselter et al. 2011; Haushahn et al. 2012). *Dracaena* is characterised by anomalous secondary growth in its stems, roots and branches (Zimmermann and Tomlinson 1970) and can form large, multiply branched trees with a complex arrangement and course of vascular bundles in the stem-branch-attachment region. These patterns have recently been visualised in living plants of *Dracaena* via magnetic resonance imaging (MRI), a method with a high potential for imaging intact living composite structures without damage from radiation and for differentiating between different tissue types (Masselter et al. 2015; Hesse et al. 2016).

Until now, biomimetic research into plant ramifications has mainly focused on the functional morphology and biomechanics of arborescent monocotyledons and columnar cacti for a technical implementation of lightweight structural elements in the automotive, aerospace or sports-equipment industries (Schwager et al. 2010, 2013). This 'narrow' focus was deliberately chosen as arborescent monocotyledons show a fibre-reinforced inner arrangement (see above) and columnar cacti develop stiff, yet lightweight hollow branchings that are excellent examples for 'smart' solutions in nature. Both groups of plants were therefore analysed in detail on their potential to be used as concept generators for the implementations in hollow bio-inspired technical fibre-reinforced branchings.

In the ongoing research project, the screening for biological concept generators has been expanded to selected dicotyledonous plants (other than columnar cacti) for implementation in architecture and structural engineering.

The growth habit of mature branched plants is established early in ontogenetic trajectory by the initiation of branch development, which often emerges from a lateral axillary bud meristem (Gola 2014; Kadereit et al. 2014). This type of side-branching results in a prominent main stem with subordinate lateral ramifications. In contrast, dichotomous (twofold) or trichotomous (threefold) branching is typically the result of a division of the apical meristem, resulting in several apical branches that might be equivalent (isotomous branching) or unequal in size (unisotomous branching) (Iwamoto et al. 2005; Gola 2014).

The diverse functional morphology and anatomy of side-branches compared with true dichotomous or trichotomous branching results in highly different structural adaptations for load-bearing. These adaptations are of great interest for optimising building structures that are also subjected to specific load conditions (compare Fig. 10.3).

The species *Edgeworthia chrysantha* and *Nerium oleander* have been identified as suitable biological concept generators with a trichotomous branching mode (Fig. 10.5a) as similar types of threefold branchings are often used for branched columns that support roof structures (see Fig. 10.1a, b). *Edgeworthia* is an arborescent plant of the family of Thymelaeaceae with a seemingly equal trichotomy as a result of a complex ontogenetic reorganisation of the apical meristem and

Fig. 10.5 Plant ramifications (**a**) *Edgeworthia chrysantha*; (**b**) *Dracaena marginata*

subsequent branching (Iwamoto et al. 2005). Even though the predominant loads acting on this plant are quite different from those acting on a roof-supporting technical structure, the branching shape of *Edgeworthia* highly resembles branched supporting pillars in architecture (compare Fig. 10.5a with Fig. 10.2).

A detailed anatomical and morphological analysis of the branched biological structure is a prerequisite for successful biomimetic abstractions that render the basis for optimising such technical constructions. Furthermore, the inflorescence of plants from the family of the Apiaceae (parsley family), such as *Angelica* or *Heracleum*, exhibit a sophisticated branching arrangement emerging from a single node into numerous ramifications, making these floral structures highly interesting for understanding and abstracting complex branching morphologies and for transferring them into technical applications.

Examples for conspicuous lateral branching structures are the flange-mounted species such as *Dracaena marginata* (Fig. 10.5b) (Masselter et al. 2011) or the finger-like branching morphologies of species of the Araliaceae family, such as *Fatsia*, *Polyscias* or *Schefflera* (Fig. 10.6) (Tomlinson et al. 2005).

The finger-like branching morphology of *Schefflera arboricola* and related Araliaceae species and genera has been selected as a first concept generator for a detailed quantitative and qualitative analysis. The *Schefflera* ramification is mainly characterised by individual woody segments that merge distally. These woody segments consist of highly lignified vascular bundles that extend from the main stem into the side branch (Fig. 10.6b) (Tomlinson et al. 2005). This arrangement allows for an abstraction of *Schefflera* into branched structures that emerge from a central pillar and form load-bearing struts which enable a high adaptivity (see Fig. 10.15) to various load conditions.

Although the anatomy of the woody branch-attachments of *Schefflera* has previously been elaborately analysed and various convincing scenarios have been proposed for its ontogenetic development (Tomlinson et al. 2005), the exact three-

Fig. 10.6 Plant ramifications (**a**) *Schefflera arboricola* living plant; (**b**) decorticated branching region of *Schefflera arboricola*

dimensional course and arrangement of vascular bundles through the branching area remains unclear. In order to obtain a detailed analysis of the anatomy of the three-dimensional stem-branch-attachment, sophisticated imaging methods are used that are highly invasive (histological sectioning and three-dimensional reconstruction of microscopic images; cf. Haushahn et al. 2014) or non-destructive (μCT).

Both methods have advantages and draw-backs: histological serial sectioning is time-consuming and destructive but provides high resolution down to the sub-cellular level and the reliable discrimination of different tissue and cell types by sophisticated staining procedures (Fig. 10.7a). Colour-filtered images of vascular bundles can be stacked in a z-direction via an image editing process that allows the observation of the course and arrangement of single lignified fibres (Fig. 10.7b). This type of analysis helps to understand the structural background of the mechanical and physiological functionality of the vascular system in the stem-branch-attachment region. The high resolution is essential for an in-depth understanding of plant ramifications and a scientifically sound abstraction as the basis for subsequent implementation into technical structures.

Complementary to the optical imaging and reconstruction method, μCT scanning is a fast and non-destructive method that currently allows the visualisation of the three-dimensional arrangement of fibrous material. Compared with the optical analysis, a much higher number of image layers can be gathered in a relatively short time, resulting in a much more refined volume reconstruction of the outer form (Fig. 10.8a) and of the inner structure (Fig. 10.8b, c). However, only small dried dead samples (in the case of *S. arboricola*, the decorticated woody core) can be scanned because of current limitations in the space of the measuring chamber and because of the necessary avoidance of drying artefacts that can occur in living material during μCT scanning. Additionally, the density-specific grey-value differences, which are the basis of the μCT images, exhibit less contrast and more background 'noise' signalling than the differential colour staining used for histological sections. Despite these disadvantages, μCT imaging is a valuable

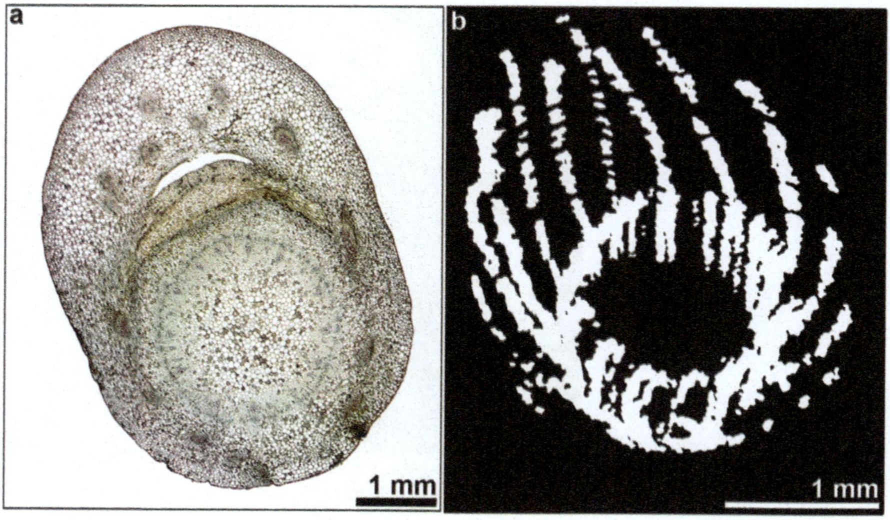

Fig. 10.7 Optical imaging and 3D-image stacking of a stem-branch-attachment in *Schefflera arboricola* (**a**) Light-micrograph of a thin section stained with safranin, methylene blue and acriflavin; (**b**) 3D-reconstruction of the vascular system via image z-stacking

Fig. 10.8 μCT scan of a *Schefflera arboricola* stem-branch-attachment region (**a**) 3D surface reconstruction based on μCT scans; (**b**) single μCT scan in longitudinal plane; (**c**) single μCT scan in transversal plane

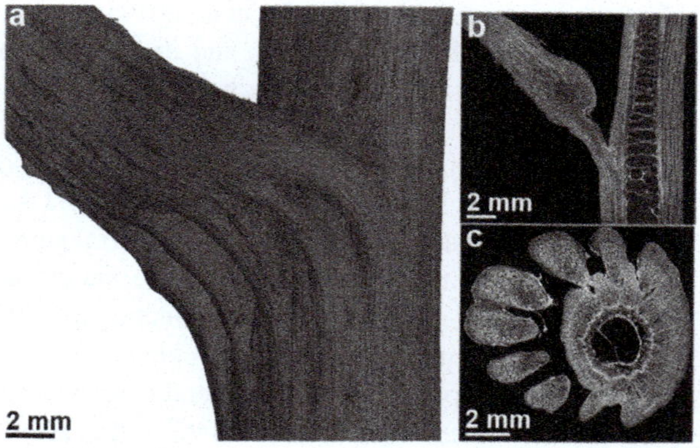

methodology for visualising the outer form and the internal arrangement of fibrous structures. In current projects, the visualisation of stem-branch-attachment regions with μCT scanning is to be intensified with an increased focus on the computational segmentation of the different tissue types and the use of scanned surface meshes as a geometrical input for FE simulations.

The transfer of a vast number of biological structures into distinct geometric models as a basis for simulations, abstractions and implementations into technical applications requires the development of a standardised computational methodology, which is capable of converting geometric information from a three-dimensional-imaging dataset into surface meshes. This is one of the main focuses in the ongoing investigations.

Furthermore, the high shape variety found in the diverse branched growth forms of plants will be analysed and categorised so that a 'shape-catalogue' of branched plant structures can be established. The categorisation will include data on the morphology and predominant stress-strain relationships of various branching shapes (such as in Figs. 10.5 and 10.6). This will ultimately allow the ready selection of an appropriate shape even in the planning phase of a technical branched column and for the establishment of a novel classification system for ramifications in botany. In a first phase, the well-established classification of Hallé et al. (1978) is used and will be adapted.

10.4 Modelling and Simulation in Biomimetics

The principle process from the biological role model to the FE simulation is illustrated in Fig. 10.9. This process follows the work flow that was established in the research project SPP 1420, in which a similar method was initially developed by the ILK of the TU Dresden together with the ITV Denkendorf as well as the Botanic Garden of the TU Dresden and the Plant Biomechanics Group of the University of Freiburg (Hufenbach et al. 2011b).

Initially, branchings of *Schefflera arboricola* (Fig. 10.9a) are decorticated and dried (Fig. 10.9b) as described in Section 3. Then, digital images of the stem-branch-attachment region are prepared by laser scanning and/or μCT (Fig. 10.9c, d). The potentials and challenges of both imaging methods for biomimetic research will be investigated and compared in ongoing studies. By overlaying several laser scans, a three-dimensional point cloud of the outer geometry of the plant sample is generated. A polygon mesh of the outer surface with no information about the inner structure can be created from this point cloud.

The remaining intensity of X-rays and their angle of deviation are evaluated in the μCT. This allows the visualisation of the internal 'material distribution' of the sample. Through the stacking of sections, a digital three-dimensional image with a resolution of 9 μm is created (Fig. 10.8). Every voxel, which represents a grid node in the volume grid, has a grey value and a distinct location. A light grey value correlates qualitatively with a high material density and therefore indicates dense

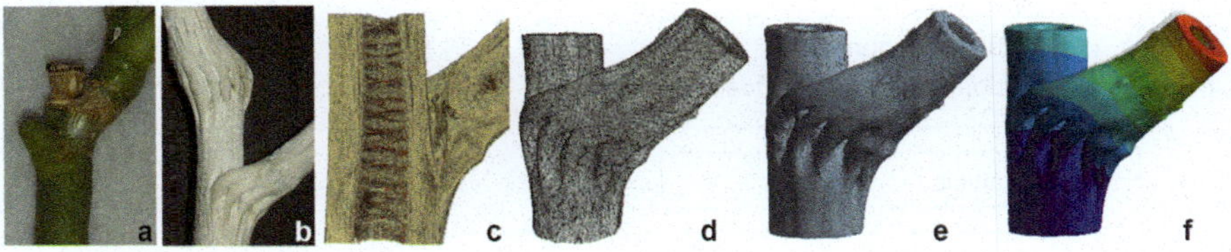

Fig. 10.9 Process chain of modelling (**a**) Preparation of a biological sample: stem-branch-attachment of a living specimen of *Schefflera arboricola*; (**b**) decorticated and dried woody core; (**c**) Digital imaging: μCT image; (**d**) CAD model; (**e**) Modelling: FE mesh; (**f**) FE postprocessing

regions of wood with a potentially higher material strength. However, quantification of the material density or the material strength of the wood is not possible based exclusively on differing grey values.

For a qualitative analysis, voxels up to a distinct grey scale threshold can be excerpted and used for the generation of a surface mesh (Amira or VG-StudioMax). In a next step, the arrangement of the vascular bundles has to be determined and integrated into a volume mesh of the model (Fig. 10.9e). This step differs for volume meshes that are generated for laser scans and that are based on μCT scans (see below). Thus, a (qualitative) anisotropic stiffness distribution in the branching can be incorporated into the FE model. The detailed simulation procedure is illustrated in Fig. 10.10.

The surface mesh generated by laser scanning or μCT is exported in STL-format and imported into a CAD program (Rhino) (Fig. 10.9d) for manual revision and reduction of the amount of data. The distal ends of the plant samples are virtually cut to ensure flat defined areas perpendicular to the stem and the branch for (virtual) load application.

In the laser scans, occlusions can be created by the complex geometries of the plant ramifications. The occlusions have to be manually eliminated by deleting superfluous polygon mesh parts and closing openings (Fig. 10.11a). Hollow regions of the plant have to be cut out manually as they cannot be detected by surface laser scanning.

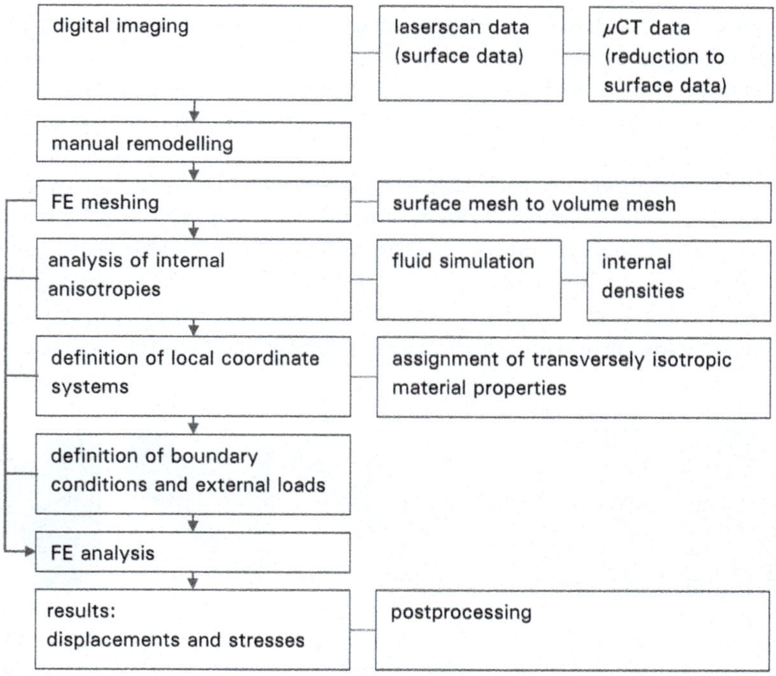

Fig. 10.10 Steps of the simulation process

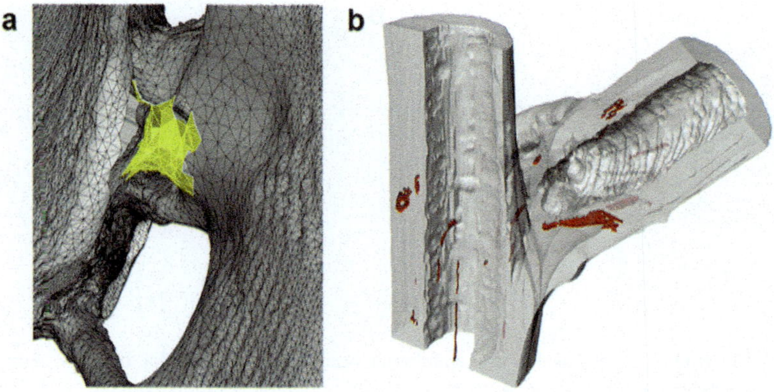

Fig. 10.11 Surface meshes based on stem-branch-attachment of *Schefflera arboricola* (**a**) Occlusions in the laser scan (*yellow* area) and (**b**) Gaps (*red* areas) in the μCT require manual remodelling

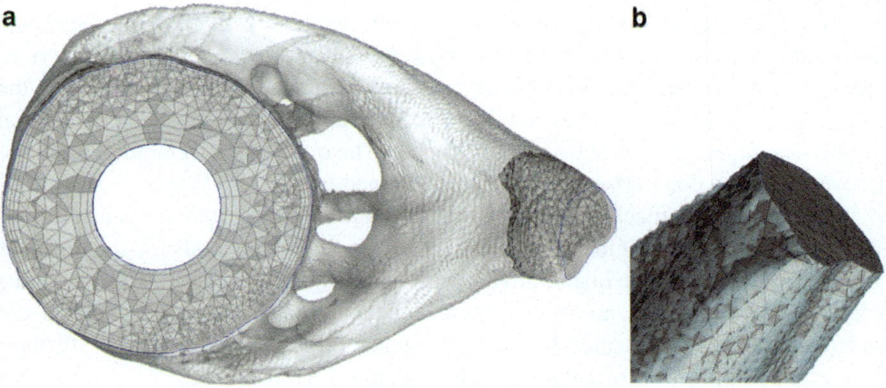

Fig. 10.12 FE volume meshes based on the biological concept generator (**a**) Mesh section; (**b**) Mesh detail

In the surface mesh generated by μCT data, gaps in the inner structures occur that are deleted to ensure proper meshing (Fig. 10.11b).

The CAD surface mesh is imported into commercial FE software (Ansys ICEM CFD). A smoothened volume mesh appropriate for FE calculations is generated (Fig. 10.12a). Because of the freeform geometry of the biological concept generators, a mesh with tetrahedrons is used. Figure 10.12b shows the difference between the finer geometry mesh of the three-dimensional model (Rhino, blue) and the coarsened FE volume mesh (Ansys, grey).

Realistic simulations that allow for a detailed visualisation of stress-strain relationships necessitate refined geometries and material parameters as input into the models. Therefore, two approaches are used for the transference of the anisotropy of the internal stiffness distribution to an FE-model.

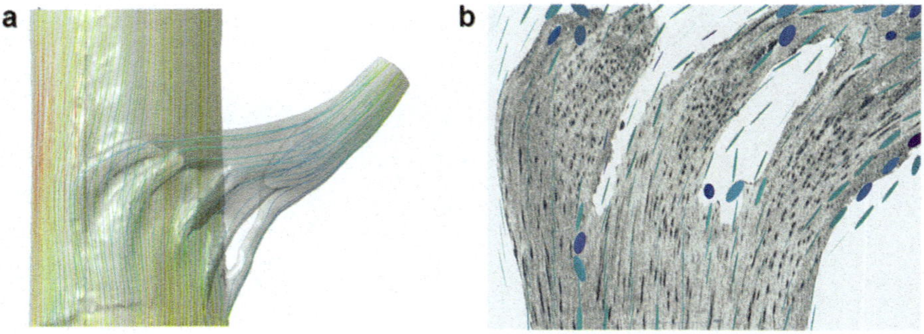

Fig. 10.13 Modelling approaches for simulating orientations of vascular bundles in the stem-branch-attachment region of *Schefflera arboricola* (**a**) Simulation of fluid dynamics with flow paths (*coloured lines*) of laminar stream; (**b**) Orientation tensors (*ellipses*) overlaying a μCT image

The first approach is based on a simulation of fluid dynamics starting with a volume mesh without information about the inner structure. Typically these meshes are generated from laser scan data and provided with prism layers at the outer and inner surfaces to minimise influences of boundary layer effects. A laminar flow from the stem base towards the branch and stem top is simulated. The orientation of the velocity vector is used as an estimation for the arrangement of anisotropic material properties. Figure 10.13a shows this flow path. The quality of this method has to be proved, i.e. by a comparison to directions obtained from μCT images.

The second approach gains information about internal material anisotropies directly from a three-dimensional μCT image. In the image analysis, the different grey scales that correlate quantitatively with material densities are rated per volume element. For each area, the directions between voxels of the same intensity are accumulated and represented through an orientation tensor. Figure 10.3b displays the results of the tensor orientation analysis graphically as ellipses.

The results of both approaches will be included in the FE models and compared. The local coordinate system of every volume element is reoriented according to the main direction of the vascular bundles. The material model is transversally isotropic, with a higher Young's modulus in the main direction. Material parameters of construction wood are being used initially (Ranz 2007) and will be replaced later by measured properties of the wood of *S. arboricola.*

In the FE model of the stem-branch-attachments, the stem base is considered to be fully fixed, the stem top is free and the branch is loaded with various actions (see Fig. 10.9f, deflections of the branch under a vertical load). Commercial FE programs (Ansys Workbench and Ansys Classic) are used to analyse displacements and stresses.

As the validation and implementation of the two approaches are ongoing preliminary results calculated with isotropic material properties are shown in Fig. 10.14. Both models are vertically loaded. The younger branching pattern (Fig. 10.14a) is based on laser scan image data and shows a high stress concentration at the innermost 'finger'. This stress peak probably results from a perturbation caused by the shortcomings of the laser scanning method because of unprecise overlays

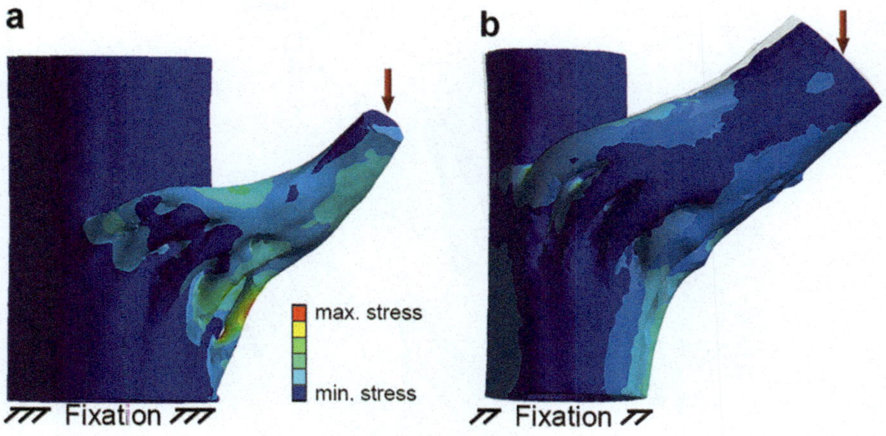

Fig. 10.14 Von Mises stresses under vertical loading for branchings with various growth habits

of single scans. The model based on micro-tomography data (Fig. 10.14b) indicates that the topmost and the lowest 'fingers' have to bear higher stresses than the middle ones. The highest utilisation occurs in the areas between the fingers and the stem.

The detailed results will allow verification of whether the simulations of fluid dynamics are a suitable tool for modelling the fibre orientation. If this is the case, laser scans that can be conducted faster and cheaper than μCT scans can be used to analyse the large variety of biological role models.

10.4.1 Abstraction

In order to understand the morphological adaptations in loaded ramifications of *S. arboricola* and to abstract the underlying functional principles, a parameter study is to be conducted in which the geometry of the ramifications is simplified, parametrised and varied (Fig. 10.15). A multitude of parameter variations have to be analysed for different load scenarios to evaluate the ratio of load-bearing capacity versus weight and to investigate the adaption of the biological role models to mechanical loading.

10.5 Technical Implementation in Fibre-Based Branched Structures

FRPs are used for technical components in which tailored anisotropic mechanical properties are required. In addition to their versatile mechanical characteristics, FRPs have the advantages of, for example, a high weight-related stiffness and strength, plus high fatigue strength against dynamic loads. The choice of the textile

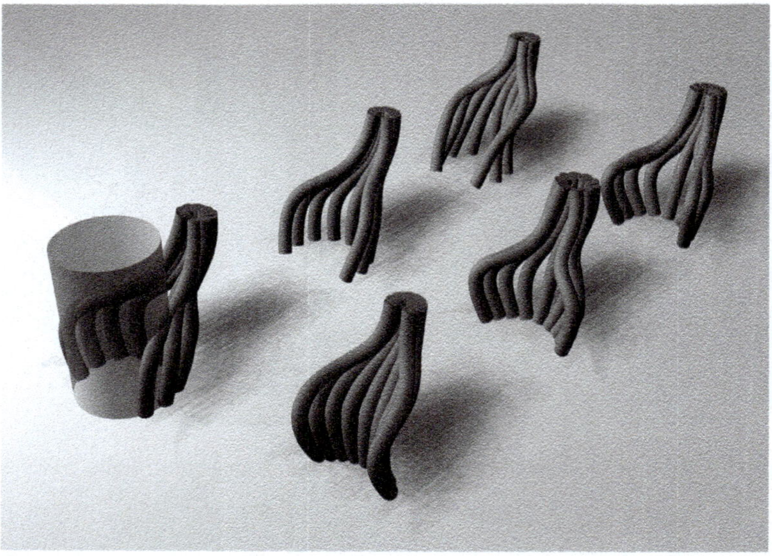

Fig. 10.15 Parameter study of various shapes of the branch-attachment region

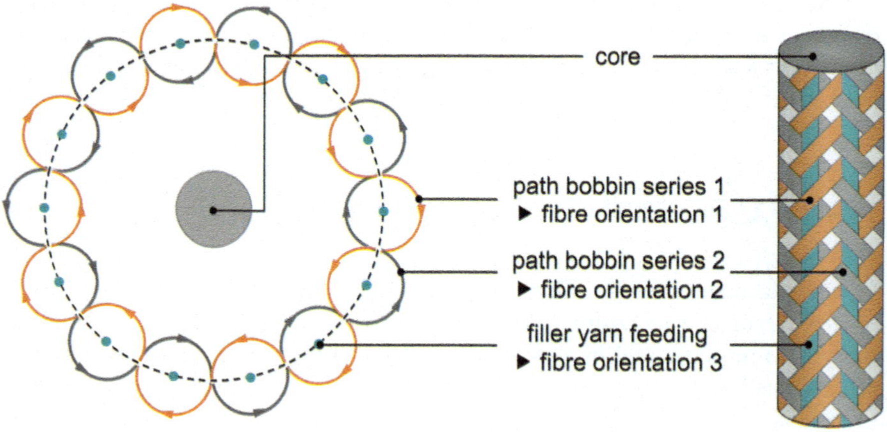

Fig. 10.16 Braiding principle of a triaxial braid (three fibre orientations)

process for manufacturing FRPs depends on the fibre orientations needed in the component. For maximum utilisation of the material properties, the fibres must be arranged according to the load paths. With regard to the manufacturing of a three-dimensional technical fibre structure for high load-bearing truss structures, braiding represents a suitable textile process (Yip and Ng 2008; Hufenbach et al. 2011a; Bilisik 2013; Mountasir et al. 2015a, b; Fazeli et al. 2016).

As shown in Fig. 10.16, a triaxial braid has three different fibre orientations: two defined by the braiding yarns and the third by unidirectional filler yarns. The two

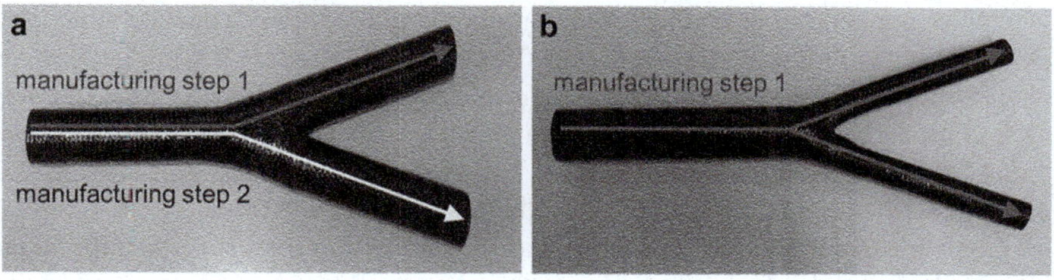

Fig. 10.17 Carbon-fibre-reinforced branched structures with (**a**) different fibre volume fraction in main stem and branches or (**b**) equal fibre volume fraction in main stem and branches

main directions of a braid (fibre direction 1 and 2 in Fig. 10.16) are generated by two yarn carrier groups that move on the machine bed of the braiding machine in opposite directions.

By the sinusoidal movement of the yarn carriers, an interlacing of the braiding threads is produced. Into the interlacing points, central filler yarns are fed (cf. Fig. 10.16), which are thus incorporated into the braid. Further explanations of the braiding principle are provided in Kyosev (2015). The external geometry of a fibre composite component is determined by the geometry of the core used for braiding (cf. Fig. 10.16). Complex three-dimensional structures can be manufactured depending on the core geometry. The braiding angle (fibre direction 1 and 2 in Fig. 10.16) is adjustable by altering the velocity of the yarn carriers and the throughput speed of the core. Thus, tailored mechanical properties of a branched FRP structure can be produced for technical applications in the building industry.

At present, two common methods can be used to produce fibre-based branched columns with one main stem and two boughs (Y-junctions, Fig. 10.17). The first method for manufacturing Y-junctions includes two braiding steps (Fig. 10.17a). In the first step, the main stem and one branch are braided. In the second step, the second branch and again the main stem are braided. The second method (Fig. 10.17b) is provided by a variation braiding machine. Here, the yarn carriers can change their position in the machine bed during the braiding process and the formation of two braiding centres up from the branching point is possible. Thus, the main stem and two branches can be braided with one fibre layer in one step (Müller et al. 2013). Other methods for braiding branchings in one step are known as for example the methodology of using a variable-braiding eye (Hufenbach et al. 2013). The selection of appropriate braiding processes for manufacturing suitable branched demonstrators is subject to intensive research at the Institute for Textile Technology, Fiber Based Materials and Textile Machinery (ITFT). For the intended technical applications, these braiding processes are of limited use as the Y-junctions are subject to production constraints such as that the fibre volume fraction always differs in the main stem and in the branches in the first production method (Fig. 10.17a) and the second production method only allows the braiding of rather small braids with necessarily differing diameters in the stem and branch (Fig. 10.17b).

For an application in building industry, concrete-filled FRP-branched structures with a greater diameter are envisaged (Fig. 10.1c). The next research objective therefore is to develop branched structures having a uniform fibre volume fraction at a constant diameter in the main stem and the branches. Furthermore, the new branches will integrate filler yarns (Fig. 10.16) to achieve, in accordance with the biological role models, a high tensile strength in the axial direction.

Concrete serves as a cheap filler to carry compression forces and stabilises the thin hull against buckling. In addition, it increases the fire resistance period, as exposed to fire the FRP hull loses its stiffness very quickly, whereas the concrete core can still carry loads. The FRP hull fulfils various functions simultaneously. It is the formwork for the wet concrete, it replaces steel reinforcement for tension forces and it confines the concrete core, which leads to an increase of its capacity. Furthermore, the smooth surface creates an aesthetic form.

Although the mechanical behaviour of confined concrete (Kupfer and Zelger 1973; Speck 2008) and FRP-confined concrete columns and beams have been the focus of many recent studies (e.g. Mirmiran and Shahawy 1996, 1997; Xiao and Wu 2000; Davol et al. 2001; Lam and Teng 2003; Eid et al. 2009; Kwan et al. 2015), no studies on branched concrete-FRP joints are as yet known.

In addition to the manufacturing of the hull, other challenges need to be addressed, such as the bond between the core and the hull to guarantee shear force transfer.

Furthermore, the calculation of the ultimate limit state is difficult, because of the complex geometry and the anisotropic properties of the FRP hull. In the next steps of the project, these challenges will be engaged and will then allow the production of promising alternatives for conventional high load-bearing truss structures in the building industry.

10.6 Summary and Outlook

The knot elements at the airport of Stuttgart (cf. Figs. 10.1a and 10.2b) were produced as cast steel parts and welded to adjoining steel pipes. This solution meets high mechanical and aesthetical demands but is very costly. Alternative designs such as the use of slotted plates or welded joints for the connection of steel tubes are optically less pleasing. In addition, durability, fatigue and notch stresses are issues that need to be addressed carefully. The use of reinforced concrete is challenging for such branching, as the reinforcement installation and the formwork are labour-intensive and difficult to produce.

Industry has no solution for an aesthetic and mechanically stable branched column with a cost effective assembly. This shall be addressed by the development of bio-inspired branched structures based on branched biological concept generators such as *Schefflera arboricola*.

The simulation of the finger-like branched structure of *S. arboricola* and an abstract implementation of its functional principles are the main objectives of the

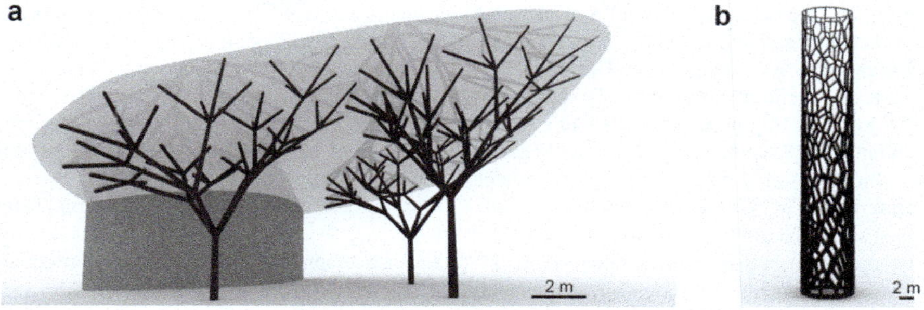

Fig. 10.18 Branched structures of concrete-FRP material in architecture (**a**) Advantageous application as replacement in buildings with existing design principles; (**b**) Innovative capabilities in structural design as supporting frame of a building

presented studies. Based on this analysis, a biomimetic hull geometry that can be manufactured industrially is developed. As a first step, branched FRPs based on triaxial braids with readily adjustable mechanical properties are filled with concrete and thus provide sufficient mechanical properties but based on a cost-effective fabrication method. Innovative and more automatable production methods might lead to new design options in architecture. The expansion of the design space inspires the development of appealing truss and frame structures with irregular and complex branching as building cores or as supporting structures (Fig. 10.18).

Acknowledgements We thank the two reviewers for their helpful comments and suggestions which have significantly improved the manuscript. This work has been funded by the German Research Foundation (DFG) as part of the Transregional Collaborative Research Centre (SFB/Transregio) 141 'Biological Design and Integrative Structures'/project A06.

References

Bilisik K (2013) Three-dimensional braiding for composites: a review. Text Res J 83:1414–1436. doi:10.1177/0040517512450766

Davol A, Burgueño R, Seible F (2001) Flexural behavior of circular concrete filled FRP shells. J Struct Eng 127:810–817. doi:10.1061/(asce)0733-9445(2001)127:7(810)

Eid R, Roy N, Paultre P (2009) Normal- and high-strength concrete circular elements wrapped with FRP composites. J Compos Constr 13:113–124. doi:10.1061/(asce)1090-0268(2009)13:2(113)

Fazeli M, Kern M, Hoffman G et al (2016) Development of three-dimensional profiled woven fabric on narrow fabric looms. Text Res J 86(12):1328–1340. doi:10.1177/0040517515606361

Gola E (2014) Dichotomous branching: the plant form and integrity upon the apical meristem bifurcation. Front Plant Sci 5:263. doi:10.3389/fpls.2014.00263

Hallé F, Oldeman R, Tomlinson PB (1978) Tropical trees and forests. Springer, Berlin/Heidelberg

Haushahn T, Schwager H, Neinhuis C et al (2012) Plant ramifications inspire branched lightweight composites. Bioinspired Biomim Nanobiomater 1:77–81. doi:10.1680/bbn.11.00011

Haushahn T, Speck T, Masselter T (2014) Branching morphology of decapitated arborescent monocotyledons with secondary growth. Am J Bot 101:754–763. doi:10.3732/ajb.1300448

Hesse L, Masselter T, Leupold J et al (2016) Magnetic resonance imaging reveals functional anatomy and biomechanics of a living dragon tree. Sci Rep 6:32685. doi:10.1038/srep32685

Hufenbach W, Böhm R, Thieme M et al (2011a) Polypropylene/glass fibre 3D-textile reinforced composites for automotive applications. Mater Des 32:1468–1476. doi:10.1016/j.matdes.2010.08.049

Hufenbach W, Gude M, Cichy F et al (2011b) Simulation of branched biological structures for bionic inspired fibre-reinforced components. Kompozyty/Composites 11:304–309

Hufenbach W, Gruhl A, Lepper M et al (2013) Verfahren für die Fertigung komplexer Faserverbund-Hohlstrukturen. Lightw Des 2:44–48

Iwamoto A, Matsumura Y, Ohba H et al (2005) Development and structure of trichotomous branching in *Edgeworthia chrysantha* (Thymelaeaceae). Am J Bot 92:1350–1358. doi:10.3732/ajb.92.8.1350

Kadereit J, Körner C, Kost B et al (2014) Strasburger. Lehrbuch der Pflanzenwissenschaften (37.Auflage). Springer Spektrum, Berlin/Heidelberg. doi:10.1007/978-3-642-54435-4

Kull U, Herbig A (1987) Pflanzen als natürliche Konstruktionen – oder das Prinzip Leichtbau. Arcus 1:11–16

Kupfer H (1973) Das Verhalten des Betons unter mehrachsiger Kurzzeitbelastung unter besonderer Berücksichtigung der zweiachsigen Beanspruchung. Deutscher Ausschuss für Stahlbeton 229:1–131

Kwan A, Dong C, Ho J (2015) Axial and lateral stress–strain model for FRP confined concrete. Eng Struct 99:285–295. doi:10.1016/j.engstruct.2015.04.046

Kyosev Y (2015) Braiding technology for textiles. Elsevier/Woodhead Publishing Limited, Cambridge

Lam L, Teng J (2003) Design-oriented stress–strain model for FRP-confined concrete. Constr Build Mater 17:471–489. doi:10.1016/s0950-0618(03)00045-x

Masselter T, Eckert S, Speck T (2011) Functional morphology, biomechanics and biomimetic potential of stem–branch connections in *Dracaena reflexa* and *Freycinetia insignis*. Beilstein J Nanotechnol 2:173–185. doi:10.3762/bjnano.2.21

Masselter T, Hesse L, Leupold J et al (2015) Using MRI for analyzing the anatomy and biomechanics of monocotyledons. In: The 8th plant biomechanics conference. Nagoya University, Nagoya, Japan, pp 230–234

Mattheck C (1998) Design in nature. Springer, Berlin/Heidelberg

Milwich M, Speck T, Speck O et al (2006) Biomimetics and technical textiles: solving engineering problems with the help of nature's wisdom. Am J Bot 93:1455–1465. doi:10.3732/ajb.93.10.1455

Milwich M, Speck T, Speck O et al (2008) The role of plant stems in providing biomimetic solutions for innovative textiles in composites. In: Ellison MS, Abbot AG (eds) Biologically inspired textiles. Blackwell, New York/London, pp 168–192

Mirmiran A, Shahawy M (1996) A new concrete-filled hollow FRP composite column. Compos Part B 27:263–268. doi:10.1016/1359-8368(95)00019-4

Mirmiran A, Shahawy M (1997) Behavior of concrete columns confined by fibre composites. J Struct Eng 123:583–590. doi:10.1061/(asce)0733-9445(1997)123:5(583)

Mountasir A, Löser M, Hoffmann G et al (2015a) 3D woven near-net-shape preforms for composite structures. Adv Eng Mater 18:391–396. doi:10.1002/adem.201500441

Mountasir A, Hoffmann G, Cherif C et al (2015b) Competitive manufacturing of 3D thermoplastic composite panels based on multi-layered woven structures for lightweight engineering. Compos Struct 133:415–424

Müller L, Gruhl A, Böhm H et al (2013) Biomimetisch optimierte verzweigte Faserverbundstrukturen mit hoher Tragfähigkeit. Melliand Textilberichte 2:88–93

Otto F (1982) Natürliche Konstruktionen. Dt. Verl.-Anst, Stuttgart

Ranz T (2007) Ein feuchte- und temperaturabhängiger anisotroper Werkstoff: Holz. Univ. der Bundeswehr München (Beiträge zur Materialtheorie)

Saafi M, Toutanji H, Li Z (1999) Behavior of concrete columns confined with fiber reinforced polymer tubes. ACI Mater J 96:500–509. doi:10.14359/652

Schwager H, Haushahn T, Neinhuis C et al (2010) Principles of branching morphology and anatomy in arborescent monocotyledons and columnar cacti as concept generators for branched fibre-reinforced composites. Adv Eng Mater 12:B695–B698. doi:10.1002/adem.201080057

Schwager H, Masselter T, Speck T, Neinhuis C (2013) Functional morphology and biomechanics of branch-stem junctions in columnar cacti. Proc R Soc B 280:20132244. doi:10.1098/rspb.2013.2244

Speck K (2008) Beton unter mehraxialer Beanspruchung. Ein Materialgesetz für Hochleistungsbetone unter Kurzzeitbelastung. Technische Universität Dresden

Speck T, Burgert I (2011) Plant stems: functional design and mechanics. Annu Rev Mater Res 41:169–193. doi:10.1146/annurev-matsci-062910-100425

Tomlinson PB, Fisher JB, Hallé F, Villalobos R (2005) Development of woody branch attachments in Schefflera (Araliaceae or Apiaceae). Am J Bot 92:1765–1773. doi:10.3732/ajb.92.11.1765

Xiao Y, Wu H (2000) Compressive behavior of concrete confined by carbon fibre composite jackets. J Mater Civ Eng 12:139–146. doi:10.1061/(asce)0899-1561(2000)12:2(139)

Yip J, Ng S-P (2008) Study of three-dimensional spacer fabrics: physical and mechanical properties. J Mater Process Technol 206:359–364. doi:10.1016/j.jmatprotec.2007.12.073

Zimmermann MH, Tomlinson PB (1970) The vascular system in the axis of Dracaena fragrans (Agavaceae) 2. Distribution and development of secondary vascular tissue. J Arnold Arboretum 51:478–491

Chapter 11
The Skeleton of the Sand Dollar as a Biological Role Model for Segmented Shells in Building Construction: A Research Review

Tobias B. Grun, Layla Koohi Fayegh Dehkordi, Tobias Schwinn, Daniel Sonntag, Malte von Scheven, Manfred Bischoff, Jan Knippers, Achim Menges, and James H. Nebelsick

Abstract Concrete double-curved shell constructions have been used in architectural design and building constructions since the beginning of the twentieth century. Although monolithic shells show a high stiffness as their geometry transfers loads through membrane forces, they have been mostly replaced by the more cost-efficient lattice systems. As lattice systems are covered by planar glass or metal panes, they neither reach the structural efficiency of monolithic shells, nor is their architectural elegance reflected in a continuous curvature. The shells of sand dollars' – highly adapted sea urchins – combine a modular and multi-plated shell with a flexible, curved as well as smooth design of a monolithic construction. The single elements of the sand dollars' skeleton are connected by calcite protrusions and can be additionally supported by organic fibres. The structural efficiency of the sea urchin's skeleton and the principles behind them can be used for innovations in engineering sciences and architectural design while, at the same time, they can be used to illustrate the biological adaptations of these ecologically important animals

---

T.B. Grun (✉) • J.H. Nebelsick
Department of Geosciences, University of Tübingen, Hölderlinstraße 12,
72074 Tübingen, Germany
e-mail: tobias.grun@uni-tuebingen.de; nebelsick@uni-tuebingen.de

L. Koohi Fayegh Dehkordi • M. von Scheven • M. Bischoff
Institute for Structural Mechanics (IBB), University of Stuttgart, Pfaffenwaldring 7, 70550 Stuttgart, Germany
e-mail: koohi@ibb.uni-stuttgart.de; mvs@ibb.uni-stuttgart.de; bischoff@ibb.uni-stuttgart.de

T. Schwinn • A. Menges
Institute for Computational Design (ICD), University of Stuttgart, Keplerstraße 11,
70174 Stuttgart, Germany
e-mail: tobias.schwinn@icd.uni-stuttgart.de; achim.menges@icd.uni-stuttgart.de

D. Sonntag • J. Knippers
Institute of Building Structures and Structural Design (ITKE), University of Stuttgart,
Keplerstraße 11, 70174 Stuttgart, Germany
e-mail: d.sonntag@itke.uni-stuttgart.de; j.knippers@itke.uni-stuttgart.de

J. Knippers et al. (eds.), *Biomimetic Research for Architecture and Building Construction*, Biologically-Inspired Systems 9, DOI 10.1007/978-3-319-46374-2_11

within their environments. The structure of the sand dollar's shell is investigated using modern as well as established imaging techniques such as x-ray micro-computed tomography (μCT), scanning electron microscopy and various optical imaging techniques. 3D models generated by μCT scans are the basis for Finite Element Analysis of the sand dollar's shell to identify possible structural principles and to analyse their structural behaviour. The gained insights of the sand dollar's mechanical properties can then be used for improving the state-of-the-art techniques of engineering sciences and architectural design.

11.1 Introduction

During the early twentieth century, iconic concrete shells were built with a double-curved geometry adapted to transferring loads through membrane forces. Today, monolithic shells are mostly replaced by grid shells consisting of cost-efficient lattice systems covered by planar glass or metal panes. These systems neither reach the structural efficiency of monolithic shells, nor is their architectural elegance reflected in a continuous curvature. The assembly of on-site segmented shells from prefabricated elements is a useful alternative. It reduces building time on-site and potentially lowers costs due to prefabrication and allows for a smooth surface curvature. Structural segment joining is necessary but can conflict with the desired membrane state for load bearing by reducing bending stiffness in the joints and lowering the capacity for the transfer of normal or shear forces. Heterogeneities in membrane stiffness can also trigger the bending of homogeneous shells in a pure membrane state. The properties and arrangement of the joints are thus of crucial importance for the load-bearing behaviour of segmented shells.

The shells of sand dollars, which are highly adapted sea urchins, are currently being studied as a biological concept generator. These shells consist of modular polygonal plates linked at their edges by calcite protrusions and organic fibres. Sand dollars show a variety of geometric variations with respect to overall shape and plate arrangements. The high load-bearing capacity of the sand dollar skeleton is well adapted to turbulent environments. These echinoids show morphological features that are also required from many shells in building construction: a mainly flat curvature, apertures and column-like connections between the upper and the lower half of the shell. The sand dollar thus serves as a suitable model for shells in building construction.

Analyses of the functional morphology of various sand dollars by focusing on the lightweight stereom of the plates, the connective joints and the role of organic and inorganic components provide insights into the biological principles of sand dollar adaptions. The examined material and structural properties can be used as a basis for Finite Element Analysis (FEA) leading to a better understanding of the functional morphology of the shell. The physical representation of the skeleton and joints by additive fabrication processes and the influence of additive manufacturing on the joints might represent innovations in building constructions of shells.

Various conditions for load transfer and stiffness at the joints and geometrical restrictions for the segments will be considered as variable parameters in the process. This design methodology can be used to investigate the various geometries of sand dollar species as the result of evolutionary adaptions to the mechanical conditions of their respective habitats and to develop design recommendations for joint pattern layouts for segmented shells in building construction. Investigations of possible ways to transfer the above-mentioned performative qualities and morphological principles of the sand dollar to building construction can be performed. Biological principles in modular shell systems allow high degrees of geometrical adaptability and structural efficiency attributable to component differentiation.

11.2 Clypeasteroid Echinoids as Biological Role Models

11.2.1 Hierarchically Organised Skeletons of Echinoids

Echinoids are hierarchically organised marine invertebrates featuring a multi-element skeleton. The animals exhibit five hierarchical levels (e.g. Nebelsick et al. 2015) including (1) the whole individual, (2) the multi-plated skeleton, (3) single skeletal plates, (4) a microstructure of various stereom types of which the single plates are constructed and (5) the nanostructure of the biomineralised stereom.

The highest organisational level is represented by the whole echinoid including all soft and hard parts of the individual. The skeleton is of mesodermal origin and is, in most cases, entirely covered by soft epidermis (e.g. Hyman 1955). Although it is an endoskeleton, the sea urchin test (calcareous skeleton without appendages) functions as an external skeleton (e.g. Goldberg 1992) protecting vital internal organs from both biotic (e.g. predators) and abiotic (e.g. strong water agitation) stress. The test also serves as a substrate for various surface appendages including spines, which are mounted on tubercles of the test surface, and pedicellariae, which are microscopic jaw-like structures that can remove particles and organisms from the echinoid surface. In addition, the echinoid skeleton features numerous apertures in the test including the peristome (mouth), periproct (anus), genital pores, ocular pores, the pore pairs for the respiratory tube feet of the ambulacral system and, in clypeasteroids, secondary unipores, which represent single holes for accessory tube feet. A further skeletal feature found inside the test is the complex jaw apparatus bearing self-sharpening teeth.

The skeleton itself consists of multiple rows of single plates that are interconnected by soft tissue such as collagenous filaments, as found in regular echinoids and, furthermore, interlocking skeletal protrusions in clypeasteroid echinoids (e.g. Seilacher 1979). Most regular echinoids, which show pronounced pentameral symmetry, possess similarly shaped plates throughout the skeleton, although these can be highly variable in some groups of irregular echinoids. Sea urchins feature two dominant types of plates in distinct double rows, the ambulacralia and the interambulacralia. Ambulacral plates bear the pores for the respiratory tube feet. In clypeasteroids, these pores are concentrated in a well-developed flower-like pattern on the dorsal surface. In addition, calcareous elements are present in spines, the jaw apparatus, and within diverse supporting structures on and within the test.

Every individual skeletal plate is constructed of a lightweight stereom, a sponge- or lattice-like three-dimensional trabecular system that is comparable with the microstructure of the bones of vertebrates (Millott 1967). The trabeculae are synthesised by specialised cells. Although biomineralisation occurs in various steps, each single plate behaves as a single crystal under polarisation, which combines both the advantages of a monolithic structure and a modular and multi-element construction (e.g. Raup 1959).

11.2.2 Morphological Details of Clypeasteroids

The echinoids are divided into two groups, the regular and irregular echinoids, based on their symmetry and position of the periproct (anus). In the mostly spherical regular echinoids, which show a generally pentamerous symmetry, the peristome (mouth) is oriented towards the oral (ventral) side, whereas the periproct is located on the aboral (dorsal) side. In the irregular echinoids, a prominent bilateral symmetry is developed and the periproct moves outside of the apical disc. In one group of irregular echinoids, namely the clypeasteroids, the periproct is positioned on the oral side of the test between the posterior ambitus (the "rim" of echinoids) and the centrally positioned peristome.

Clypeasteroids show a high degree of morphological variability (e.g. Müller 1854; Lang 1896; Mortensen 1948). Typical infaunal living individuals, such as the genera *Leodia*, *Mellita* and *Scutella* and some *Clypeaster* species, are highly flattened, whereas the epifaunal *Clypeaster rosaceous* is domed (Fig. 11.1). Exceptional features of the clypeasteroid test are the internal supports, buttress- or pillar-like structures that connect the oral and aboral plated surfaces of the test (Fig. 11.1d, e). Buttresses generally occur within the interambulacral fields, whereas pillars are restricted to the ambulacral fields (Mooi 1989). The morphology of these structures can be related to the overall test shape with flattened skeletons tending to form delicate needle-like (e.g. *Clypeaster subdepressus*) or labyrinthic (e.g. *Leodia sexiesperforata*) structures, whereas domed echinoids (e.g. *Clypeaster rosaceus*) are characterised by more robust and occasionally coalesced internal supports (e.g. Mortensen 1948; Mihaljević et al. 2011).

An additional important morphological feature of some especially flattened and discoidal sand dollars are the lunulae, which are round to slit-like holes (e.g. *Leodia*, *Mellita*) or indentations (e.g. *Rotula*, *Encope*) in the skeleton (Fig. 11.1b). Lunulae can occur in various numbers and can either grow during ontogenesis or arise by resorption of skeletal material (e.g. Hyman 1955). The clypeasteroids are further characterised by (1) the presence of respiratory tube feet in a flower-like petalodium

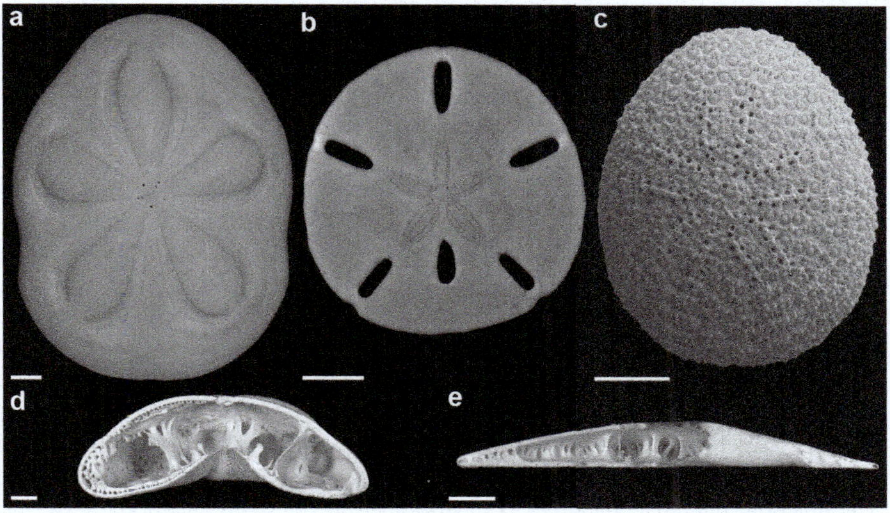

Fig. 11.1 Photographs of clypeasteroid echinoids. (**a**) *Clypeaster rosaceus*, (**b**) *Leodia sexiesperforata* and (**c**) *Echinocyamus pusillus* in aboral view. (**d**) Domed *Clypeaster rosaceus* featuring an inflated peristome. (**e**) Highly flattened *Mellita tenuis* in section with the anterior to the left. *Scale bars* (**a**, **b**, **d**, **e**) = 1 cm, *scale bar* (**c**) = 1 mm

on the aboral surface, (2) secondary tube feet distributed across the whole skeleton, (3) highly differentiated minute spines on the surface of the test, (4) a branching system of food grooves on the oral surface leading to the peristome, and (5) a highly modified jaw apparatus found inside the test.

The morphological variability of clypeasteroids is clearly shown by the species *Clypeaster rosaceus*, *Echinocyamus pusillus* and *Leodia sexiesperforata* (Fig. 11.1). The sea-biscuit *Clypeaster rosaceus*, one of the largest known clypeasteroids, is characterised by its inflated shape. It lives epifaunally on top of the sea floor and masks itself by using available materials such as algae and sediment particles. The test shape is moderately domed and the oral surface shows a bell-shaped indentation around the mouth. The oral and aboral surfaces are interconnected by massive pillars. Furthermore, a second internal wall is present that is partly perforated to supply the water vascular system between the two test layers (Mortensen 1948).

The minute sand dollar *Echinocyamus pusillus* is one of the smallest clypeasteroids rarely exceeding 15 mm in length. It is moderately domed and its oral surface is slightly indented. *Echinocyamus pusillus* features buttresses instead of pillars as internal supports. It lives infaunally in a wide variety of substrates and depths (Ghiold 1982; Grun et al. 2014; Grun and Nebelsick 2015).

The six-holed sand dollar *Leodia sexiesperforata* is a discoidal, highly flattened, medium-sized echinoid that burrows through the sediment just underneath the surface feeding on organic material. The distal areas of the skeleton near the ambitus are characterised by a dense meshwork of labyrinthic internal support structures.

11.2.3 Functional Morphology and Habitat Adaptation

Echinoids are highly adapted to their marine environments. Their habitats range from rocky areas with rough breaker zones and high water agitation to those characterised by soft bottom sediments and slow currents (see Mortensen 1948; Schultz 2006). Adaptations to the environment can be seen in the high variability of skeletal morphologies and behaviour in both regular and irregular echinoids.

Most clypeasteroids live shallowly buried within soft sediment, although some adopt an upright position stuck into the sediment (Timko 1976; Lawrence et al. 2004). *Clypeaster rosaceus*, in contrast, shows an epifaunal life style on the sediment surface (Kier and Grant 1965). These various life styles are accompanied by morphological adaptations of the echinoids shell including general shape and internal structures. Infaunal living genera are generally characterised by highly flattened skeletons that are discoidal. Numerous minute spines cover the surfaces of these echinoids and fullfil a variety of purposes (see Smith 1984; Mooi 1986). The function of lunulae, the hole- to slit-like apertures in some sand dollar tests, has been discussed in detail by Smith and Ghiold (1982) who concluded that lunulae, on the one hand, can increase the food gathering capacity of the animal by shortening the transportation distance from the aboral surface to the peristome (Goodbody 1960; Ghiold 1979; Seilacher 1979; Smith 1980; Alexander and Ghiold 1980) and, on the other hand, can improve the hydrodynamic stability of these near surface dwelling sand dollars (Telford 1981). Telford et al. (1985), however, argued that their observations conclusively showed that food material is collected by surface podia on the oral surface alone.

Most clypeasteroids possess internal supports connecting the oral and aboral plates of the test. These buttress- and pillar-like structures have been interpreted as increasing the overall test stability of the extremely flattened form of some sand dollars, which can inhabit high energy environments (e.g. Seilacher 1979). There are, however, no empirical studies that have investigated the role of internal supports on the stability of clypeasteroid sea urchins. Furthermore, the role of spines, as shown for regular echinoids (Nichols 1962; Strathmann 1981) has barely been explored. In addition, the significance of soft tissues, which not only cover the plates, but also interconnect them, for the stability of clypeasteroids is poorly understood.

Not only are macroscopic morphological features of interest with respect to the stability of the echinoids, but also their plate arrangements, their mode of jointing, and their microstructure. A high variation of stereom types in echinoid plates and spines is exhibited (e.g. Smith 1980; Grossmann and Nebelsick 2013). These diverse stereom types are highly variable with respect to arrangement and density. They are adapted for various tasks including support, the interlocking of muscle tissue and the provision of surfaces for spine articulations and space fillers (Smith 1980). A further feature seen at the microscopic level are skeletal extensions that reach from one plate into neighbouring plates, thus potentially increasing interlocking (e.g. Seilacher 1979). Again, few studies have revealed the distribution and morphology of these interlocking features or the way that they contribute to the general strength of the test.

Functional morphological aspects are present at all hierarchical levels (see Nebelsick et al. 2015) enabling the echinoid to adapt to various environments. Biomimetic investigations might reveal more detailed insights into these evolutionary adaptive strategies and such discovered functional principles can potentially be transferred from nature and used to inspire construction techniques in architectural applications. Furthermore, structural analysis and developmental strategies employed during building processes can be used in a reverse biomimetic approach to improve the knowledge of the adaptations of sea urchins and their skeletons to the specific environments in which they live.

11.3 Biomimetic Steps from Echinoid Role Models to Building Constructions

11.3.1 Modelling of Skeletons as a Tool for Structural Analysis

Computational simulations of organisms, such as echinoids, can be difficult because of their complexity and highly structured body plan. A detailed description of the model including every morphological detail is thus not practicable. To obtain a meaningful model, concrete questions and working hypotheses have to be tested in order to be verified or falsified. For the mechanical modelling of sand dollars envisaged here, the most important questions concern the role of the specific geometry and segmentation of the skeleton, the mechanical properties of the joints, the role of the trabeculae and the relationship of these components to the environments to which the sand dollars are exposed.

The simulations are based on three models of the sand dollar, a full model, a reduced model and an abstracted model. The aim of a full or detailed model, which is pursued at the beginning of this research project, is to strive for a best possible approximation of the actual geometric, topological and material properties of the sand dollar. Reduced and abstracted models are later derived from the lessons learned with the full model. The reduced model preserves the basic mechanical and structural features of the full model at significantly lower computational expense, whereas the abstracted model represents the principal underlying structural concept with the aim of implementation in architecture. Questions that are addressed here are whether certain mechanical functionalities are intrinsically related to the pattern layout, the properties of the joints and the overall geometric and topological features of the segmented skeleton, including the trabeculae.

One fundamental problem of modelling organisms, as opposed to engineering structures, is that no protocol is available specifying the geometry, constructional details and materials used. The same is true for loads and boundary conditions. Instead, two complementary methods gain such information by using measurements and making feasible assumptions. Such information forms the basis for the finite element modelling described in this section.

For the material behaviour, assumptions are made based on available data from the literature for the stiffness and strength of the base material and physical tests. In this context, we have to stress that the notions of "material" and "structure" are somewhat ambiguous, because neither of them can exist without the other. This means that, for instance, a certain porosity of the stereom might be advocated for material behaviour and related parameters specifying stiffness, strength etc. Porosity can be regarded as a structural feature, given a sufficient geometric fidelity. In this case, the different material parameters characterising the base material are relevant.

The skeleton of clypeasteroid echinoids consists of high magnesium calcium carbonate, for which material properties, such as Young's modulus (around $7 \cdot 10^7$ kN/m^2 for the solid base material, neglecting porosity), can be taken from the literature. The porosity of the stereom is partly represented in the identified geometry but needs to be extended for a proper reduction of the related moduli.

Today, mechanical problems in engineering are mostly solved with the finite element method, a numerical method that approximately solves partial differential equations. A major technical issue when deriving a finite element model is the generation of a feasible mesh. Powerful tools exist for mesh generation given the Computer Aided Design (CAD) data of the geometry. For the problem at hand, this means that, first, a CAD model has to be extracted from the raw data, which can be automatically processed by meshing software in a second step. Both steps are not trivial and, in general, cannot be performed in a fully automatic manner. The automatically generated mesh generally needs manual refinements. In addition to employing such a surface mesh, the use of voxel-based finite element meshes (Fig. 11.2) is possible, which can be created from the raw data. In this case, every finite element geometrically represents a regular hexahedron. The grey scale can be related to the density or porosity and can thus provide hints towards realistic material properties.

Such a voxel-based approach has both advantages and disadvantages when compared with a standard CAD-FEM transition. The voxel model is limited in smoothness because of the resolution of the CT-images and the available soft- and hardware. The obvious disadvantage is the poor geometric representation of the surface contour, which potentially compromises accuracy of the computational results. A decisive advantage is also the possibility of a direct transfer from raw data to a numerical model. Additionally, for linear analyses, the underlying element stiffness matrices have the same generic format. At this point, which of the advantages and disadvantages of both approaches will prevail is not yet clear. For the time being, both concepts are pursued in parallel.

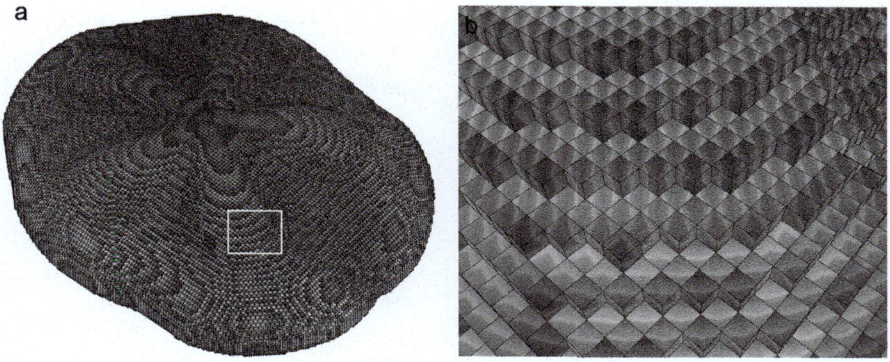

Fig. 11.2 Voxel-based mesh (Cartesian grid) of *Clypeaster rosaceus*. (**a**) Oblique view of the mesh. The region within the square shows (**b**) single voxels in detail containing material density information represented in different *grey* values

11.3.2 Morphospace and Optimisation of Skeletal Structures

In evolutionary biology, the morphological features of an individual are the observable instances of a theoretical solution space (Eble 2004). Morphological spaces or "morphospaces" constitute formal spaces defined by multiple dimensions that each correspond to a variable parameter of morphology. Morphospaces serve as computational and conceptual tools that allow the vast variation of organismal form in living nature to be described and associated (Mitteroecker and Huttegger 2009). In computational design, the generation of form is based on algorithmic processes that operate within specific variable ranges of selected parameters. The combined variance range of each parameter in the computational model can be conceptualised as delineating an n-dimensional space that is similar to the morphospaces of theoretical morphology in biology. In this case, individual building elements are represented as n-dimensional points.

In order to address the problem of the exploration into fabrication-related design, a method for evaluating design-spaces has already been developed in previous work and allows for the quantification of the available design space associated with a particular fabrication setup. The so-called *Machinic Morphospace Method* adopts and transfers the method from the field of evolutionary biology to architectural design in order to describe the morphological variance of building elements together with the machine constraints necessary for their fabrication within one methodological framework (Menges 2013). This transfer allows the representation of the variable parameter range of both form generation and materialisation in one theoretical morphospace. In this theoretical space, different regions can be distinguished such as the *Producible region of Possible Form* (PPF) as a subset of the *Geometrically Possible region of Form* (GPF) defined by the computational design process (Fig. 11.3) (Menges 2013). In the context of digital fabrication, the PPF

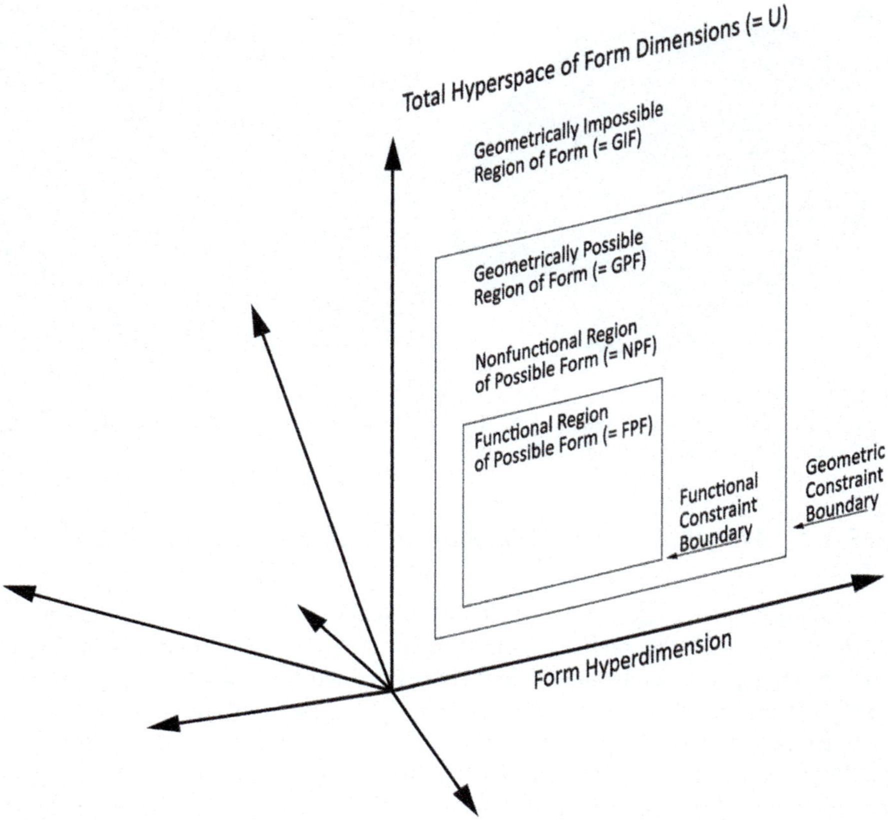

Fig. 11.3 Regions of a hyper-dimensional theoretical morphospace in which each dimension represents one parameter of form (Menges 2013)

relates a specific machine configuration to a space of possible producible outcomes, its machinic morphospace (Schwinn et al. 2012). While ongoing technological advances, such as the advent of robotics, increase the capacities of digital fabrication and, in consequence, expand the space of producible forms, the search space for the *Functional region of Possible Form* (FPF), as the subset of the GPF, increases proportionally. On the one hand, such correlations increase the possibility that a particularly functional region of form is contained within a producible region determined by a particular fabrication setup. On the other hand, the correlations also offer the possibility for the determination of fabrication requirements with respect to morphological parameters.

Based on the premise of biological evolution as an adaptive agent, a biomimetic design approach based on the analysis of biological role models and on the transfer of morphological features into the context of architectural and structural design provides a filter for identifying particularly functional regions of the producible form

(Krieg et al. 2011). In the context of the design of segmented shells, the machinic morphospace method in conjunction with a biomimetic design approach has proven to be a viable strategy as described in the case studies below.

11.3.3 Optimisation of Shell Structures in Architecture

Numerous strategies for shape finding and the optimisation of shell structures have been developed including the experimental work of Heinz Isler (Chilton 2000) and the analytical methods of Bletzinger and Ramm (1999) and Arnout et al. (2012). Insights into form finding and the optimisation of shell structures inspired by nature have been explored in previous research (e.g. Ramm et al. 1993). Most existing work on form finding deals with homogeneous smooth shell structures. Monolithic concrete shells are of less relevance for today's architecture mainly because of the extensive effort needed for individually customising complex reinforcement layouts. An alternative might be the erection of shells from prefabricated segments; this offers many advantages for prefabrication, transportation and on-site assembly. Only a few studies exist on the optimal joint layout for segmented shells. Some preliminary studies have, however, demonstrated possible applications (Fildhuth et al. 2012; Breitenberger et al. 2013). The major difficulties lie in the choice of a parameterised tessellation taking into account fabrication constraints, the choice of suitable optimisation algorithms, the coupling of optimisation algorithms with finite element software and the generation of a functional interface.

In the case of the structural optimisation of segmented shells, the objective functions are generally non-linear with respect to the design variables and are non-continuous. Evolutionary algorithms, which are metaheuristic search and optimisation processes, have shown promise in optimising such problems as they only require the evaluation of the objective function itself. Although not converging as rapidly as gradient-based methods, they tend to have fewer problems with discontinuities. In the context of segmented shells, it might be necessary to deal with multiple objective functions. This is true for the biological role model (e.g. mechanical stability, functional aspects and ontogenesis) as for the architectural counterpart (stiffness, weight, forces in the joints and stability). Evolutionary algorithms can deal with several objective functions at once without the need of preliminary reduction to one objective function; they can directly sample the Pareto-front and are therefore particularly suitable for multi-criteria optimisation. A comprehensive overview concerning evolutionary algorithms for multi-objective optimisation is given by Fonseca and Fleming (1995) and Deb (2011).

For the optimisation of grid shells, Dimcic and Knippers (2011) have used a standard evolutionary optimisation algorithm. An evolutionary optimisation algorithm has also been used by Fildhuth et al. (2012) to minimise tension forces in adhesive joints under various geometric conditions for glass panes and overall shell geometry. This has proved to be a suitable method as it handles discontinuities of a segmented shell and can be adjusted to various fitness criteria and penalty functions. The evolutionary methodology also requires the identification of essential geometric parameters and their constraints in order to describe the diversity of the studied skeletons.

The optimality of a sea urchin skeleton under various loading conditions has been analysed through parametric studies (Philippi and Nachtigall 1996). In this study, the mechanical response of a regular sea urchin shell to external loading was calculated by using Finite Element Analysis. The results were used to link shell shape to mechanical behaviour. The influence of the joints between the various shell segments was, however, not taken into account. Few studies of the mechanical behaviour of the shell are available for irregular sea urchins.

Today, new computational tools are available allowing an advanced optimisation of segmented shells in architecture. In order to study the biological role model with respect to its optimality, the data of the 3D model needs to be processed and reduced to a manageable data size. Thus, all relevant geometrical and mechanical characteristics need to be identified in order to generate a parametric design model. The definition of geometric parameters is established with the goal of modelling the biological diversity of the sand dollar's plate layout and morphological features, such as the length, width and height, plate topology, shell curvature, the quantity and position of skeletal apertures, and the quantity and position of internal supports. The 3D design model of the skeleton is therefore extended towards a parametric model that allows the description of natural diversity by variation of the design parameters. The goal is to keep the geometrical and mechanical model as simple as possible in order to reduce the amount of parameters and therefore computation time.

11.4 Examples for Echinoid Skeletons as Potential Models for Biomimetic Architecture

11.4.1 Potential for Building Construction

Continuous shells, if properly designed, can be highly efficient and aesthetic structures and have therefore a high potential in the field of architecture. In order to avoid disadvantages such as expensive and material-consuming formwork, which is presented by continuous shells during construction, segmented shells might be an interesting alternative to monolithic shells.

Segmented shells have been built from glass (Blandini 2005; Almegaard et al. 2007; Bagger 2010) or wood (La Magna et al. 2013) with only limited use at larger scales. Since connections between adjacent segments can hardly deal with forces occurring in the constructions, segmented shells are still restricted to an academic research level. The main reason for this is the lack of applicable joint constructions satisfying conditions of sufficient structural load transfer and the restricted adaptability to various geometric and static situations. Only a few studies on this subject can be found (Wester 1990; Veer et al. 2003), which either use adhesives for joining or apply combined mechanical and adhesive techniques. Joints between prefabricated segments nearly always weaken shell structures as they disturb the continuity of their stiffness. Joints can also reduce the capacity to transfer bending moments or membrane forces. The structural capacity of joints not only impacts the overall load-bearing behaviour of the shell, but also determines the efficient arrangement of the segments on the 3D surface. Their geometry and dimensions are usually restricted by specific fabrication conditions and transport and assembly options. Usually, size, aspect ratios, corner angles and curvature are limited. Simple engineering considerations for regular shell geometries are not possible for the complex interactions present in 3D shapes.

In order to profit from load-bearing behaviour via membrane forces, shells should (1) have a continuous smooth curvature, (2) be extremely thin compared with their span dimensions, (3) have continuous equal stiffness (no sudden changes in thickness), and (4) have membrane-specific support boundary conditions. The last-mentioned condition is often difficult to achieve as it might result in lateral bending and constraints in the case of settlement or temperature loads. Here, the arrangement of joints might reduce such interior constraints (Fildhuth and Knippers 2011).

11.4.2 Application in Demonstrators

In previous work, the Institute for Computational Design (ICD) and the Institute for Building Structures and Structural Design (ITKE) at the University of Stuttgart, Germany, in collaboration with the Department of Geosciences at the University of Tübingen, Germany, have examined the architectural and structural potential of transferring morphological principles from the skeletons of echinoids to lightweight construction and, specifically, to the design of segmented shells in architecture.

In a first study in 2011, a temporary pavilion was built that explored the biomimetic transfer of four essential morphological principles to architecture. These principles were identified in the skeleton of echinoids and are related to (1) hierarchical organisation, (2) plate arrangement, (3) differentiation of plate morphology, and (4) plate connection.

One of the most prominent features of the plate morphology of the sea urchin is the arrangement of polygonal plates in the shell, an arrangement that is based on a three-plate principle. This principle describes the fact that, on each node, three-plate edges intersect that allow the sea urchin's shell to work as a pure plate structure (Wester 2002), which consists of rigid plates interconnected by shear-resistant hinges. The three-plate principle potentially avoids instable structural configurations and has been transferred to the computational design model of the above-mentioned pavilion resulting in a modular system in which three edges are assured to meet at one point.

Similar to the biological role model in which plates are connected by microscopic calcite protrusions along the plate edges, primarily shear forces and normal forces

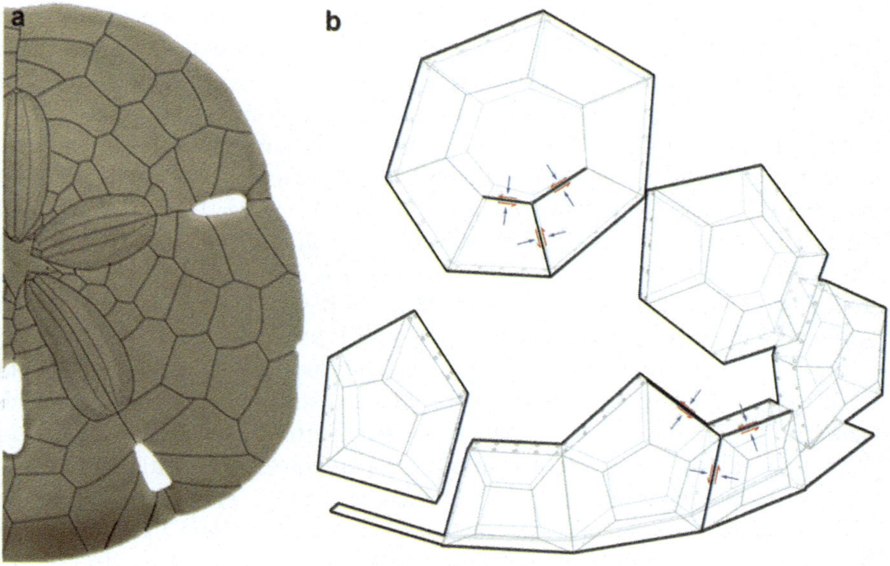

Fig. 11.4 ICD/ITKE Research Pavilion 2011 showing (**a**) the biological three-plate principle (Photo: Wolfgang Gerber, Department of Geosciences, University of Tübingen, Germany) and (**b**) the technological transfer in terms of plate arrangement and module assembly

are transferred at the connection level. Because of the three-plate principle, the transfer of bending moments globally is still possible but requires a certain bending stiffness of the segments, as can be provided by a double-layer structure of each element.

As with most biological structures, sea urchins and thus sand dollars exhibit a strict hierarchical organisation (e.g. Nebelsick et al. 2015). This principle has also been transferred to the research pavilion in which, at the plate level, finger joints have been glued together to form polyhedral modules, which, in turn, are assembled together with simple screw joints. At each hierarchical level, the three-plate principle has been respected (Fig. 11.4). Achieved by computational design methods and customised Computer Aided Manufacturing methods (CAM) (La Magna et al. 2013), the pavilion's overall form includes surface regions with strongly varying Gaussian curvature and various module and plate sizes in order to demonstrate the flexibility of the proposed prototypical building system. Whereas the arrangement and shape of the modules was further developed according to general design and engineering considerations, the transfer of these principles insured that, at the plate level, primarily in-plane forces occurred, which led to a highly efficient lightweight structure made entirely of only 6.5 mm thin plywood plates (Fig. 11.5).

In a follow-up project, the biomimetic transfer of the previously identified principles was further developed and also transferred from an academic to a building construction context resulting in the Landesgartenschau Exhibition Hall. Its main

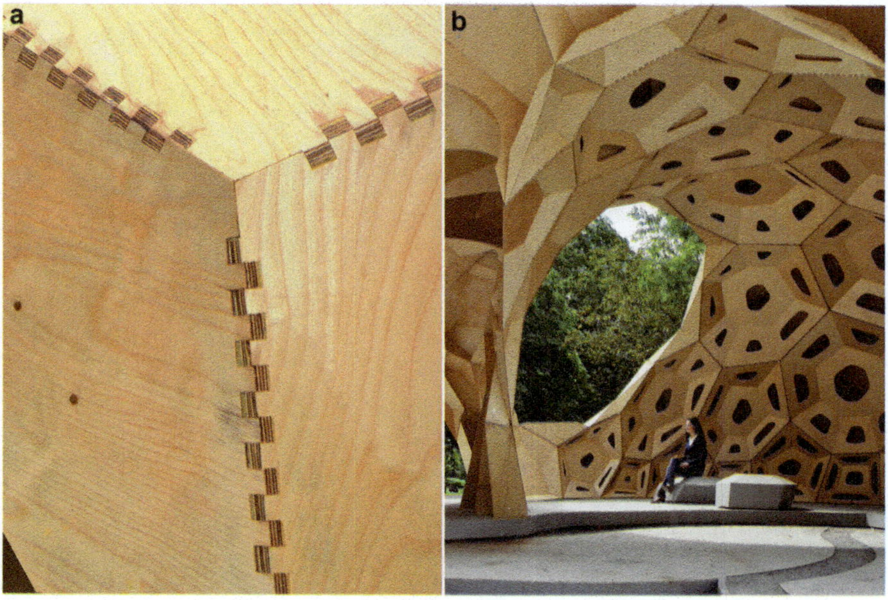

Fig. 11.5 ICD/ITKE Research Pavilion 2011 showing (**a**) plate connection details of finger joints and (**b**) the interior view (Knippers et al. 2013)

innovations compared with the previous research pavilion were (1) the single layered structural design, (2) the technical development of the joint detail with respect to structural performance, fabrication and assembly (Krieg et al. 2015), (3) the constructive development with regards to thermal insulation and weather proofing, (4) the integration of quality control into the fabrication and assembly process (Schmitt and Schwieger 2015) and, (5) the agent-based generation scheme for plate structures with its focus on material stock and producibility (Schwinn et al. 2014).

Completed in 2014, the complex plate structure consisting of 243 geometrically unique polygonal plates also complies with the three-plate principle. As a single layered structure made from 50 mm thin beech veneer plywood, the exhibition hall works as a pure plate structure spanning 11 by 17 m (Fig. 11.6). Although only connected by hinges, trivalent polygonal plate structures can transfer bending moments at the global level via out-of-plane shear forces (Li and Knippers 2015). This is also possible in planar configurations, leading, however, to high local force concentrations. Therefore, transition regions between synclastic and anticlastic regions are especially critical because of the lack of curvature and the resulting pattern.

In order to cover the double-curved design surface of the shell with only a single layer of planar stock material, a subdivision scheme involving the Tangent Plate Intersection (TPI) method (Wang and Liu 2009) was implemented as part of an agent-based modelling approach. In this multi-agent model, each agent corresponds

Fig. 11.6 Landesgartenschau Exhibition Hall in Schwäbisch Gmünd, Germany. Interior view showing the polygonal plate layout of the single-layered segmented shell (Krieg et al. 2015)

to exactly one input point for the TPI method and, correspondingly, to one plate in the segmented shell. Through the definition of parameterised behaviours on the level of each individual agent, the entire model can autonomously be steered towards producibility taking into account the planarity requirement, available stock material sizes, and workspace constraints of the fabrication setup (Fig. 11.7). Whereas the consideration of structural performance is technically possible in this kind of approach, the arrangement of the plates in this example primarily reflects geometrical and fabrication constraints, as the optimisation of the joint layout based on mechanical criteria had not previously been implemented.

In 2015, the ICD and the ITKE in collaboration with the work group Invertebrate Paleontology and Paleoclimatology at the Department for Geosciences, University of Tübingen (IPPK) and the work group for Evolutionary Biology of Invertebrates, Institute for Evolution and Ecology at the University of Tübingen, Germany started the development of a demonstrator in the form of a temporary research pavilion. Starting from the previous findings, the aim was (1) to revise some of the earlier biomimetic transfers with respect to the design of segmented shells, (2) to identify additional performative biological principles, and (3) further to improve the ratio of self-weight to loadbearing capacity of segmented timber shells.

The biological principles that have been further investigated include (1) the double-layer shell structure as found in the irregular echinoid species *Clypeaster rosaceus*, (2) the principles of hierarchical material organisation and geometric differentiation (Gruber and Jeronimidis 2012), which can be found in many

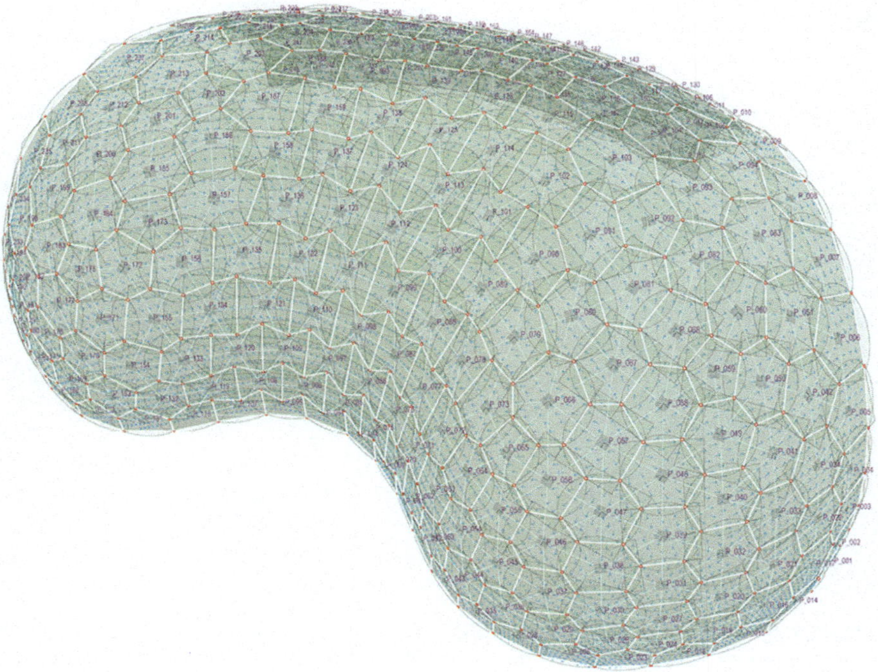

Fig. 11.7 Landesgartenschau Exhibition Hall in Schwäbisch Gmünd, Germany. Agent-based model of shell segmentation by using Tangent Plane Intersection (Schwinn and Menges 2015)

biological structures, and (3) the finger-jointing principle between building modules. New principles that have been identified as part of this research are based on (1) the differentiation of material composition of the echinoid shell, (2) the fibrous connections between shell segments, (3) the growth principles of plate addition and plate accretion (Raup 1968; Pearse and Pearse 1975; Abou Chakra and Stone 2011), and (4) morphological features such as plate morphology and shell openings, which can be observed in clypeasteroid echinoids and might be relevant in an architectural context.

Similar to the 2011 research pavilion, this demonstrator features a double-layered timber construction that is closely related to the constructional morphology of the sand dollar as a biological concept generator. Because of the double-layered nature of the structure, no bending moments need to be transferred between segments. Therefore, the joints between segments consist not only of finger joints, which transfer shear forces and compressive normal forces, but also of laces, which bear the tensile forces in between the segments.

Indeed, the joints of several irregular sea urchin species not only are formed of calcareous protrusions in form of finger joints, but also have often collagen fibres connecting the plates, which are able to withstand tensile forces (see Ellers et al. 1998). Although not necessary for overall stability, as the shell can work

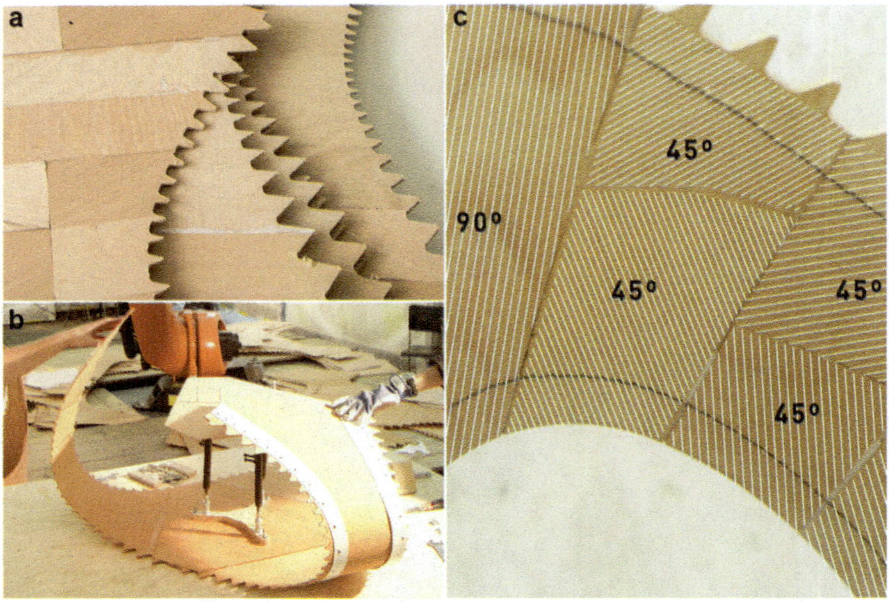

Fig. 11.8 6 ICD/ITKE Research Pavilion 2015. Process of active bending of (**a**) custom laminated veneer strips into (**b**) segments following (**c**) an angle pattern derived from stiffness requirements achieving desired curvature

as a pure plate structure, these fibres seem to be necessary for growth, assembly and robustness (Wester 2002). Echinoid shell growth occurs under maintenance of the structural integrity and mechanical stability of the test. This is a challenging task, as the increase in size of plates is often not possible without creating structural discontinuities. For pure plate structures, forces would be transferred mainly parallel to plate edges and thus perpendicular to the growth direction (Wester 2002). Moreover, during growth, fibres can be shown to play an important role in maintaining structural integrity (Ellers et al. 1998) with the flexible tissue allowing the creation of gaps into which new material can be inserted without the test collapsing. This principle of fibrous connection has been used as an inspiration when using textile connection techniques, such as sewing or lacing, for construction purposes.

The idea of transferring forces via textile connectors has been applied to construction at two levels of hierarchy. On a first level, thin sheets of beech veneer are laminated and bent to form extremely lightweight building modules of only 4 to 6 mm in thickness (Fig. 11.8). Traditionally, layers of veneer are bonded by using glue; however, in the case of bending glue-laminated veneer, delamination can be critical because the sheets need to be joined in a deformed position, thus the high pressures typically required for lamination cannot be achieved. Sewing layers of veneer together therefore presents an interesting alternative, as the yarn can

Fig. 11.9 ICD/ITKE Research Pavilion 2015. Robotic sewing of laminated veneer strips

continuously transfer forces between the sheets. The sewing only locally damages the material and, for timber constructions in general, many small connectors are preferable to a few large ones (Herzog et al. 2003). It is also at this local level that the principle of material differentiation with respect to mechanical requirements, as can be found in the calcite tests of echinoids, comes into play: the wooden strips are built up by laminating several veneer sheets, adjusting the thickness and grain orientation to stiffness requirements derived from the global segment layout and the resulting different local stresses that occur in each individual building module. Each module, in turn, is formed by three individually laminated strips, which are joined together by robotic sewing of the timber veneer laminates (Fig. 11.9).

The fibrous connective elements serve mainly for maintaining the structural integrity and for assembly, similar to the biological role models, but they can also transfer larger tensile forces in regions with low curvature. The laced connection is particularly suitable to variable connection angles, which arise from the module geometry. Furthermore, the connection allows flexible adjustment to tolerances during assembly, as they can be tensioned after assembly (Fig. 11.10).

Regular sea urchins grow by plate addition originating from the apical disc and by the incremental growth of individual plates by the accretion of calcite material (for an overview of echinoid growth, see Zachos 2009; Abou Chakra and Stone 2011). These principles have been abstracted and transferred by using a parametric circle packing approach as part of the computational design process (Fig. 11.11). In this process, circles are additively seeded from user-defined locations on a

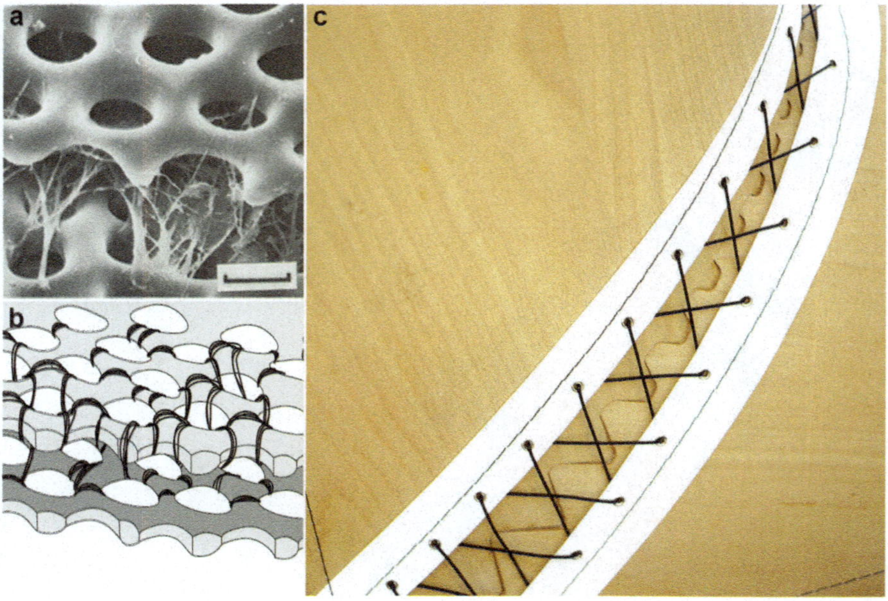

Fig. 11.10 ICD/ITKE Research Pavilion 2015. Principle of (**a**) collagen fibres connecting two adjacent plates of an echinoid test (Telford 1985), which has been (**b**) abstracted into a manufacturable system and transferred to (**c**) a connection detail between timber segments by using lacing

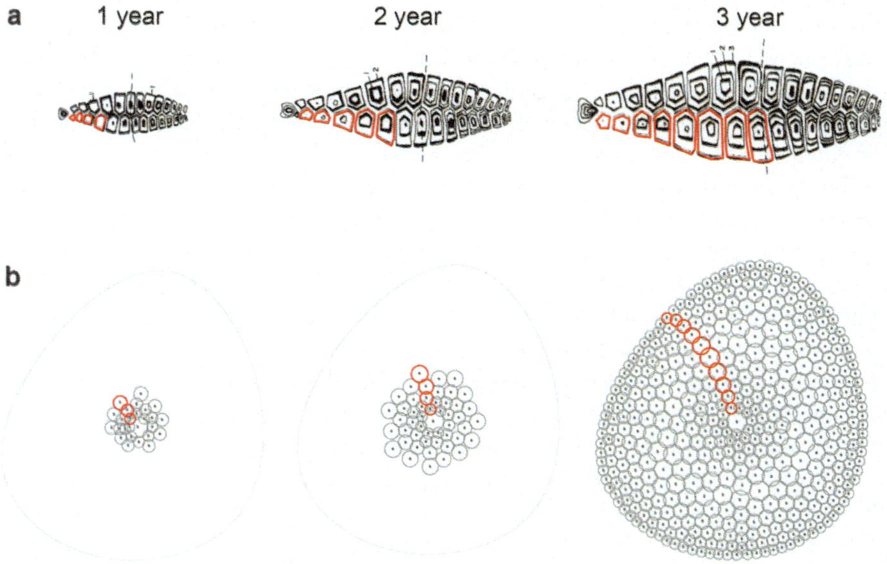

Fig. 11.11 ICD/ITKE Research Pavilion 2015. (**a**) Analysis of the growth process in echinoid plates (modified from Pearse and Pearse 1975) serving as (**b**) a form and pattern-generating algorithm for shell constructions

three-dimensional reference surface. After the circles have been seeded, they continue to grow, i.e. increase in radius, and to be pushed away from the seed origin until a previously defined surface area is filled with tangent circles. The diameters of the circles increase proportionally with time resulting in a configuration in which larger circles are located further away from the origin. The resulting topology of plate origins and shared edges forms the basis for the arrangement of the building modules and satisfies the requirements of the three-plate principles identified in previous case studies.

Morphological features such as lunulae and internal supports, which can be found in clypeasteroids, are particularly relevant in the context of architecture and shell design. Apertures are usually considered detrimental to the structural integrity of shells as they disrupt the continuity of the shell surface; however, they are useful from an architectural and daylighting point of view. Sand dollars, which incorporate lunulae, adjust the plate arrangement in order to accommodate the resulting discontinuity of the shell. This principle might be transferred to the demonstrator through the arrangement and geometry of the modules resulting in a characteristic arrangement of openings in the shell (Fig. 11.12). Supports as found in clypeasteroids are usually avoided in shell structures as they lead to force concentrations and require the thickening of the surface in the area around the support points. Clypeasteroids feature internal supports in the shell that obviously serves as a structural support of the plated skeleton. Transferred into the architectural context, openings and supports can significantly extend the solution for the space and range of applications for segmented shell structures. In the case of the research pavilion 2015, the double-layered shell gradually transforms into a plate-and-support system highlighting the flexibility of the proposed lightweight building system. Its capacity to transform between structural systems constitutes one of its most significant innovations, as this has not been achieved in any of the previous case studies (Fig. 11.12).

11.5 Conclusions

The hierarchically organised and segmented skeletons of the sand dollar feature a variety of adaptations with respect to their mode of life. These morphological and structural modulations are highly adapted to the specific conditions of the environment. The morphological response to mechanical stress caused by biotic and abiotic factors can be seen as form and principle generators in clypeasteroid construction. The adoption and abstraction of biological principles can improve building constructions, since buildings are possibly subject to similar loading conditions as echinoids.

The use of new methods and techniques such as high resolution x-ray micro-computed tomography, 3D imaging tools and advanced Finite Element Analyses enables the examination of the biological role model in detail. Computational simulations based on 3D data can provide insights into mechanical properties of

Fig. 11.12 ICD/ITKE Research Pavilion 2015. The front view showing the transition from shell to plate-and-pillar system (Photo: Roland Halbe)

morphological features and their structural adaptation. 3D models and specific material parameters are necessary for such advanced computational analyses. The model used here is based on x-ray micro-computed tomography records and has been post-processed producing both a surface polygon mesh and a voxel mesh that includes material density parameters. The mesh models can be additionally divided based on echinoid plate distribution and morphology including the connections between single plates.

Previous research on the echinoid skeleton has revealed various principles for segmented shells, such as the hierarchical organisation of the shell, which allows for a modular building concept by using prefabricated modules. The three-plate rule, meaning that three plates always meet in a single node, potentially avoids instable structural configurations. The development of a double-layered shell is the basis of a light-weight structure that not only results in an optimised weight of the building, but also in maximum material economisation. One of the most important developments in the clypeasteroid echinoids are the internal support structures that connect the oral and aboral surfaces. These structures vary in position and morphology with respect to the overall echinoid shape, which is correlated to the mode of life and ambient environmental conditions.

11.6 Outlook

The analyses of echinoids and, especially, the clypeasteroids have shown that these biological role models exhibit meaningful principles that can be transferred into building constructions and design. Mechanical and structural attributes of the echinoid skeleton, however, need to be investigated in more detail in order to obtain a more precise understanding of the revealed principles and the possible detection of new principles.

Further research will be performed by using more detailed models for simulations and material parameters that can be physically evaluated. Such focused analyses are predestined to reveal not only obvious principles on a macroscopic scale, but also latent principles that might be of major interest for structural engineering, form finding and functional morphological interpretations.

Acknowledgements This work has been funded by the German Research Foundation (DFG) as part of the Transregional Collaborative Research Centre (SFB/Transregio) 141 'Biological Design and Integrative Structures'/project A07. We also thank The Paleontological Society, The Gerace Research Centre, Hartmut Schultz (Scanning Electron Microscopy Lab, Department for Geosciences, University of Tübingen), Wolfgang Gerber (Photo Lab, Department for Geosciences, University of Tübingen), Ellen Struve (Applied Geosciences, University of Tübingen), Raouf Jemmali (German Aerospace Center, Stuttgart, Germany) and Rolf Pohmann (Max-Planck Institute for Biological Cybernetics, Tübingen, Germany). Thanks to the European Fund for Regional Development and the Cluster Forst und Holz Initiative. We also thank Theresa Jones for proof reading and Roland Halbe.

References

Abou Chakra M, Stone JR (2011) Holotestoid: a computational model for testing hypotheses about echinoid skeleton form and growth. J Theor Biol 285:113–125

Alexander DE, Ghiold J (1980) The functional significance of the lunules in the sand dollar *Mellita quinquiesperforata*. Biol Bull 159:561–570

Almegaard H, Bagger A, Gravesen J, Jüttler B, Šír Z (2007) Surfaces with piecewise linear support functions over spherical triangulations. Proc Math Surf XII 4647:42–63

Arnout S, Firl M, Bletzinger KU (2012) Parameter free shape and thickness optimisation considering stress response. Struct Multidiscipl Optim 45:801–814

Bagger A (2010) Plate shell structures of glass. Dissertation, University of Denmark

Blandini L (2005) Structural use of adhesives in glass shells. Dissertation, Universität Stuttgar

Bletzinger KU, Ramm E (1999) A general finite element approach to the form finding of tensile structures by the updated reference strategy. Int J Space Struct 14:131–145

Breitenberger M, Blenzinger KU, Wüchner R (2013) Isogeometric layout optimization of shell structures using trimmed NURBS surfaces. In: Proceedings of World Congress on Structural and Multidisciplinary Optimization, Orlando, 19—24 May

Chilton J (2000) Heinz Isler. The engineer's contribution to contemporary architecture. Thomas Telford Ltd, Reston

Deb K (2011) Multi-objective optimization using evolutionary algorithms. Kan Gal Rep 2011003:1–24

Dimcic M, Knippers J (2011) Structural optimization of grid shells. In: Proceedings of The International Association for shell and spacial structures, London, 20—23 September

Eble G (2004) The macroevolution of phenotypic integration. In: Pigliucci M, Perston K (eds) Phenotypic integration, studying the ecology and evolution of complex phenotypes. Oxford University Press, Oxford, pp 253–273

Ellers O, Johnson AS, Moberg PE (1998) Structural strengthening of urchin skeletons by collagenous sutural ligaments. Biol Bull 195:136–144

Fildhuth T, Lippert S, Knippers J (2012) Design and joint pattern optimisation of glass shells. In: Proceedings of The International Association for Shell and Spacial Structures, Seoul, 20—24 May

Fildhuth T, Knippers J (2011) Geometrie und Tragverhalten von doppelt gekrümmten Ganzglass-chalen aus kalt verformten Glaslaminaten. Stahlbau 80:31–44

Fonseca CM, Fleming PJ (1995) An overview of evolutionary algorithms in multiobjective optimization. Evol Comput 3:1–16

Ghiold J (1979) Spine morphology and its significance in feeding and burrowing in the sand dollar *Mellita quinquiesperforata* (Echinodermata: Echinoidea). Bull Mar Sci 29:481–490

Ghiold J (1982) Observations on the clypeasteroid *Echinocyamus pusillus* (O.F. Müller). J Exp Mar Biol Ecol 61:57–74

Goldberg WM (1992) The biology of reefs and reef organisms. The University of Chicago Press, Chicago

Goodbody I (1960) The feeding mechanism in the sand dollar *Mellita sexiesperforata* (Leske). Biol Bull 119:80–86

Grossmann JN, Nebelsick JH (2013) Stereom differentiation in spines of *Plococidaris verticillata*, *Heterocentrotus mammillatus* and other regular sea urchins. In: Johnson C (ed) Echinoderms in a Changing World. Proceedings of the 13th International Echinoderm Conference, Tasmania. CRC Press, London, pp 97—104

Grun T, Sievers D, Nebelsick JH (2014) Drilling predation on the clypeasteroid echinoid *Echinocyamus pusillus* from the Mediterranean Sea (Giglio, Italy). Hist Biol 26:745–757

Grun T, Nebelsick JH (2015) Sneaky snails: how drillholes can affect paleontological analyses of the minute clypeasteroid echinoid *Echinocyamus*? In: Zamora S, Rábano I (eds) Progress in echinoderm paleobiology. Publicaciones del Instituto Geológico y Minero de España, Madrid, pp 71–73

Gruber P, Jeronimidis G (2012) Has biomimetics arrived in architecture? Bioinspir Biomim 7:1–2

Herzog T, Natterer J, Schweitzer R (2003) Holzbau Atlas. Birkhäuser, Basel

Hyman LH (1955) The Invertebrates. Volume IV: Echinodermata. McGraw-Hill, New York

Kier PM, Grant RE (1965) Echinoid distribution and habits, Key Largo Coral Reef Reserve, Florida. Smithsonian Inst 149:1–62

Knippers J, Menges A, Gabler M, La Magna R, Waimer F, Reichert S, Schwinn T (2013) From nature to fabrication: biomimetic design principles for the production of complex spatial structures. In: Hesselgren L, Sharma S, Wallner J, Baldassini N, Bompas P, Raynaud J (eds) Advances in architectural geometry 2012. Springer, Wien, pp 107–122

Krieg OD, Schwinn T, Menges A, Li J, Knippers J, Schmitt A, Schwieger V (2015) Biomimetic lightweight timber plate shells: computational integration of robotic fabrication, architectural geometry and structural design. In: Block P, Knippers J, Mitra NJ, Wang W (eds) Advances in architectural geometry 2014. Springer, Cham, pp 109–125

Krieg OD, Dierichs K, Reichert S, Schwinn T, Menges A (2011) Performative architectural morphology: Finger-joined plate structures integrating robotic manufacturing, biological principles and location-specific requirements. In: Gengnagel C, Kilian A, Palz N, Scheurer F (eds) Computational design modelling: proceedings of the design modelling symposium berlin 2011. Springer, Berlin, pp 259–266

La Magna R, Gabler M, Reichert S, Schwinn T, Waimer F, Menges A, Knippers J (2013) From nature to fabrication: biomimetic design principles for the production of complex spatial structures. Int J Space Struct 28:27–39

Lang A (1896) Text-book of comparative anatomy, volume 2. MacMillan and Co, London

Lawrence JM, Herrera J, Cobb J (2004) Vertical posture of the clypeasteroid sand dollar *Encope michelini*. J Mar Biol Assoc UK 84:407–408

Li JM, Knippers J (2015) Pattern and form – their influence on segmental plate shells. In: Proceedings of The International Association for Shell and Spacial Structures, Amsterdam, 17—20 August

Menges A (2013) Morphospaces of robotic fabrication. In: Brell-Çokcan S, Braumann J (eds) Robarch 2012: robotic fabrication in architecture, art and design. Springer, Wien, pp 28–47
Mihaljević M, Jerjen I, Smith AB (2011) The test architecture of *Clypeaster* (Echinoidea, Clypeasteroida) and its phylogenetic significance. Zootaxa 2983:21–38
Millott N (ed) (1967) Echinoderm biology. Academic, New York
Mitteroecker P, Huttegger SM (2009) The concept of morphospaces in evolutionary and developmental biology: mathematics and metaphors. Biol Theory 4:54–67
Mooi R (1986) Structure and function of clypeasteroid miliary spines (Echinodermata, Echinoides). Zoomorphology 106:212–223
Mooi R (1989) Living and fossil genera of the Clypeasteroida (Echinoidea, Echinodermata): an illustrated key and annotated checklist. Smithsonian Institution Press, Washington, DC
Mortensen T (1948) A monograph of the Echinoidea IV. CA Reitzel, Copenhagen
Müller J (1854) Über den Bau der Echinodermen. Druckerei der Königlichen Akademie der Wissenschaft, Berlin
Nebelsick JH, Dynowski JF, Grossmann JN, Tötzke C (2015) Echinoderms: hierarchically organized light weight skeletons. In: Hamm C (ed) Evolution of light weight structures. Analyses and technical applications. Springer, Dordrecht, pp 141–154
Nichols D (1962) Echinoderms. Hutchinson and Co, London
Pearse JS, Pearse VB (1975) Growth zones in the echinoid skeleton. Amer Zool 15:731–753
Philippi U, Nachtigall W (1996) Functional morphology of regular echinoid tests (Echinodermata, Echinoida): a finite element study. Zoomorphology 116:35–50
Ramm E, Bletzinger KU, Reitinger R (1993) Shape optimization of shell structures. Revue Européenne des Éléments 2:377–398
Raup DM (1959) Crystallography of echinoid calcite. J Geol 67:661–674
Raup DM (1968) Theoretical morphology of echinoid growth. J Paleo 42:50–63
Schmitt A, Schwieger V (2015) Quality control of robotics made timber plates. In: Fédération Internationale Géometès, Sofia, 17—21 May
Schultz H (2006) Sea urchins I: a guide to worldwide shallow water species, 3rd edn. Heinke and Peter Schultz, Hemdingen
Schwinn T, Menges A (2015) Fabrication agency: Landesgartenschau Exhibition Hall. Archit Des 85:92–99
Schwinn T, Krieg OD, Menges A (2014) Behavioral strategies: synthesizing design computation and robotic fabrication of lightweight timber plate structures. In: Proceedings of the 34th annual conference of the Association for Computer Aided Design in Architecture, Los Angeles, 23—25 October
Schwinn T, Krieg OD, Menges A, Mihaylov B, Reichert S (2012) Machinic morphospaces: biomimetic design strategies for the computational exploration of robot constraint spaces for wood fabrication. In; Proceedings of the 32nd annual conference of the Association for Computer Aided Design in Architecture, San Francisco, 18—21 October
Seilacher A (1979) Constructional morphology of sand dollars. Paleobiology 5:191–221
Smith AB (1980) The structure and arrangement of echinoid tubercles. Philos Trans R Soc B 289:1–54
Smith AB (1984) Echinoid palaeobiology. George Allen and Unwin, London
Smith AB, Ghiold J (1982) Roles for holes in sand dollars (Echinoidea): a review of Lunulae function and evolution. Paleobiology 8:242–253
Strathmann RR (1981) The role of spines in preventing structural damage to echinoid tests. Paleobiology 7:400–406
Telford M (1981) Hydrodynamic interpretation of sand dollar morphology. Bull Mar Sci 31:605–622
Telford M (1985) Domes, arches and urchins: the skeletal architecture of echinoids (Echinodermata). Zoomorphology 105:114–124
Telford M, Mooi R, Ellers O (1985) A new model of podial deposit feeding in the sand dollar, Mellita quinquiesperforata (Leske): the sieve hypothesis challenged. Biol Bull 169:431–448

Timko PL (1976) Sand dollars as suspension feeders: a new description of feeding in *Dendraster excentricus*. Biol Bull 151:247–259

Veer FA, Wurm J, Hobbelman GJ (2003) The design, construction and validation of a structural glass dome. In: Proceedings of glass processing days, Tampere, 15—18 June

Wang W, Liu Y (2009) A note on planar hexagonal meshes. In: Emiris IZ, Sottile F, Theobald T (eds) The IMA volumes in mathematics and its applications. Springer, New York, pp 221–233

Wester T (1990) A geodesic dome-type based on pure plate action. Int J Space Struct 5:155–167

Wester T (2002) Nature teaching structures. Int J Space Struct 17:135–147

Zachos LG (2009) A new computational growth model for sea urchin skeletons. J Theor Biol 259:646–657

Chapter 12
Continuous Fused Deposition Modelling of Architectural Envelopes Based on the Shell Formation of Molluscs: A Research Review

James H. Nebelsick, Christoph Allgaier, Benjamin Felbrich, Daniel Coupek, Renate Reiter, Günter Reiter, Achim Menges, Armin Lechler, and Karl-Heinz Wurst

Abstract Land snails produce a highly structured functionally integrated self-supporting surface from a composite of organic and inorganic materials that posses a potentially rich source of aspects for possible transfer into the technical realm. These investigations form a basis for the concept of a machine setup harvesting potentials for architectural manufacturing. A computational design tool, incorporating the limitations of the production mechanism, the design intent and structural, architectural and functional aspects, has been established to integrate the fabrication in early stages of the architectural design.

The project presented here aims at developing a novel additive-manufacturing process and integrated computational design method for the construction of self-supporting lightweight architectural envelopes based on the investigation of the shell formation process in land snails with a strong focus on the production of

J.H. Nebelsick (✉) • C. Allgaier
Department of Geosciences, University of Tübingen, Hölderlinstraße 12, 72074 Tübingen, Germany
e-mail: nebelsick@uni-tuebingen.de; christoph.allgaier@uni-tuebingen.de

B. Felbrich • A. Menges
Institute for Computational Design (ICD), University of Stuttgart, Keplerstraße 11, 70174 Stuttgart, Germany
e-mail: benjamin.felbrich@icd.uni-stuttgart.de; achim.menges@icd.uni-stuttgart.de

D. Coupek • A. Lechler • K.-H. Wurst
Institute for Control Engineering of Machine Tools and Manufacturing Units, University of Stuttgart, Seidenstr. 36, 70174 Stuttgart, Germany
e-mail: Daniel.Coupek@isw.uni-stuttgart.de; armin.lechler@isw.uni-stuttgart.de; karl-heinz.wurst@isw.uni-stuttgart.de

R. Reiter • G. Reiter
Faculty of Mathematics and Physics, University of Freiburg, Hermann-Herder-Str. 3, 79104 Freiburg, Germany
e-mail: renate.reiter@physik.uni-freiburg.de; guenter.reiter@physik.uni-freiburg.de

J. Knippers et al. (eds.), *Biomimetic Research for Architecture and Building Construction*, Biologically-Inspired Systems 9, DOI 10.1007/978-3-319-46374-2_12

microscopically structured functional surface textures, the use and adaptation of composite material properties and the interplay between mechanical formation and molecular self-assembly at various scales.

In architecture, layer manufacturing processes are currently based on the upscaling of relatively small-scale 3D printing techniques. The inherent drawbacks are that these methods suffer from a trade-off between the degree of surface detail, the overall size of the building and its elements as well as production speed. Biological role models can serve both as conceptual idea generators and deliver tangible manufacturing setups. This can lead to an additive-manufacturing technology that not only satisfies the structural, spatial and functional requirements of architectural fabrication, but can also operate at larger scales.

12.1 Introduction

Current state of the art layer manufacturing techniques in architecture are primarily based on scaling up of relatively small-scale 3D printing techniques. They therefore inherently suffer from a trade-off between the degree of surface detail, the overall building or building element size and the speed of production. To generate an additive-manufacturing technology tailored to the structural, spatial and functional requirements of architectural fabrication and operating in a larger band width of scales, the investigation of biological role models can serve both as a conceptual idea generator and a supplier of tangible manufacturing setups.

In this context, the shell generation of land snails, specifically the building process of its periostracum, represents a formation process of particular interest. The snail produces a highly structured functionally integrated self-supporting surface from a composite of organic and inorganic materials that posses a potentially rich source of aspects for possible transfer into the technical realm. Furthermore, it performs this shell generation only by utilizing a limited area of productive tissue, the periostracal groove, a soft, freely movable, cellular band. This aspect represents a parallel to the principal of material extrusion through independently manoeuvrable extrusion nozzles in state of the art additive manufacturing, but with a much higher degree of detail.

The project presented here therefore aims at developing a novel additive-manufacturing process and integrated computational design method for the construction of self-supporting lightweight architectural envelopes based on the investigation of the shell formation process in land snails with a strong focus on the production of microscopically structured functional surface textures, the use and adaptation of composite material properties and the interplay between mechanical formation and molecular self-assembly at various scales. This chapter introduces novel concepts for bio-inspired additive manufacturing setups which will be showcased in a subsequent realization phase.

An elaborate screening of land snail shells with regard to shell form, surface texture, composite material structure and functional morphology has been performed to detect the relationship of these characteristics to the specific habitat of the snails and to identify relevant principles for technological transfer. In a next step, the shell growth mechanism located in the periostracal groove has been studied with regard to the transition of molecular material assembly towards a conscious mechanical formation process. These investigations form a basis for the concept of a machine setup harvesting potentials for architectural manufacturing. Furthermore, a computational design tool, incorporating the limitations of the production mechanism, the design intent and structural, architectural and functional aspects, has been established to integrate the fabrication in early stages of the architectural design.

12.2 Terrestrial Snails as the Biological Role Model

The mode of formation and the materials used in the shells of invertebrate animals are at the forefront of bio-material sciences (e.g. Hansell and Ruxton 2013). The ability of a biological system to drive the formation of a microstructure as complex and ordered as the molluscan shell is of immense interest for other science fields, such as engineering or medicine (McDougall et al. 2011). The natural mechanical design principles of shells have yet to be fully understood at both smaller and larger scales (Ortiz and Boyce 2008). Shell formation in molluscs has mostly been studied in bivalves (Jacob et al. 2011). In contrast, snail shells have received far less attention (Sud et al. 2002; McDougall et al. 2011). The shell of a snail consists of a composite material of structurally distinct layers and is an example for multi-scale mechanical design principles. In contrast to typical complex biological structures, snails assemble their shells extracellularly (Luchtel et al. 1997).

The soft body of the snail is firmly attached to the shell at a single point inside the tip of the shell. Shell growth is integrated with body growth as the animal adds material to the distal border of the shell (Fig. 12.1). The organ that forms the shell, the so-called mantle, releases organic macromolecules and $CaCO_3$ at the mantle edge. This secretion of the mantle epithelium is initiated at a single restricted groove of "productive" tissue, namely the periostracal groove. A thin proteinaceous layer (the periostracum), which is secreted into and assembled within the periostracal groove, is laminated to the shell margin and then extruded onto the dorsal surface of the shell (Jackson et al. 2006; de Paula and Silveira 2009; McDougall et al. 2011). The periostracum is moulded by inherent stereotypic movements of the snail body. This layer becomes hardened and is subsequently strengthened with inorganic building material from the inner side (ostracum). Consequently, the outer form of the shell is determined by the living tissue (the mantle edge) of the animal (e.g.

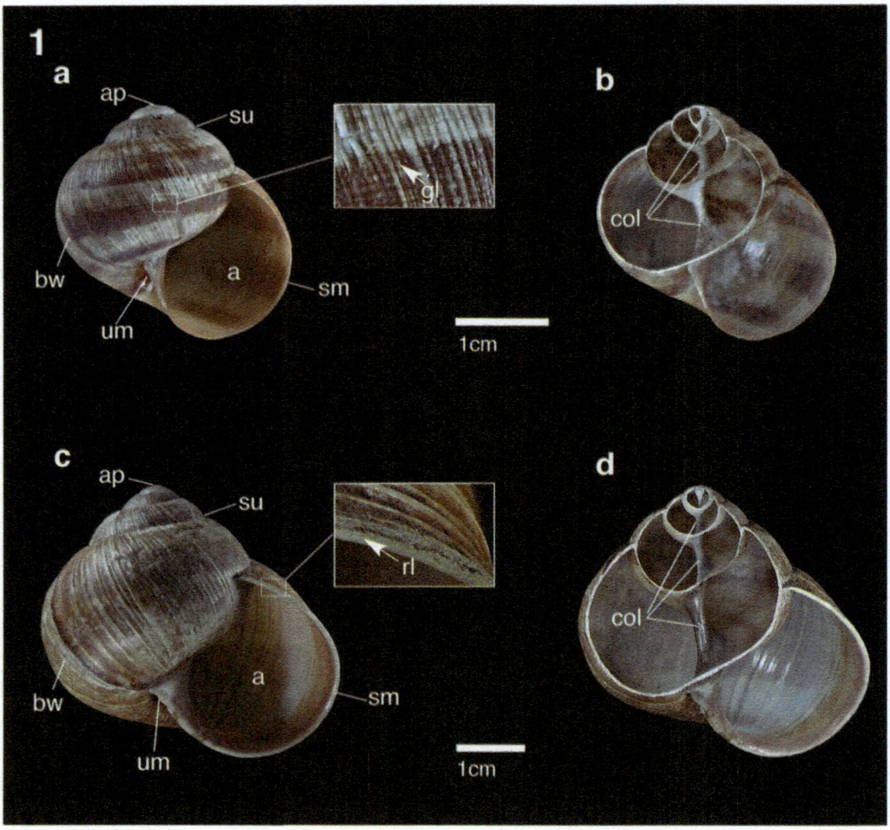

Fig. 12.1 A typical snail shell is shaped like a modified tube that spirals around its axis (columella). (**a**) Juvenile shell of the land snail *Helix pomatia*, outer view. On the surface of the shell, growth lines are visible (detail) but the shell margin is without a lip. (**b**) Shell cut open to reveal the columella. (**c**) Adult shell of the land snail *Helix pomatia*, outer view, the shell margin has a reflected lip (detail). (**d**) Shell cut open to reveal the columella. *a* aperture, *ap* apex, *bw* body whorl, *col* columella, *gl* growth line, *rl* reflected lip, *sm* shell margin, *su* suture, *um* umbilicus

Marin and Luquet 2004; Marin et al. 2008). During shell production, the snail takes up a defined position with respect to the mantle and the shell margin (Fig. 12.2). In this manner, the snail adds minute layers to subjacent layers (Fig. 12.3), a method that thus represents growth by increments (Luchtel et al. 1997).

The shells of the diverse snail species vary widely in shape and surface morphologies but are all based on this same construction principle. Few studies have explored in detail the morphology and exact mode of periostracum production, especially the interaction of the moulding and hardening of the polymer giving the shell its species-specific shape (Allgaier 2011; Salas et al. 2012). The periostracum can be of variable thickness and consists of one to four layers (Meenakshi et al. 1969).

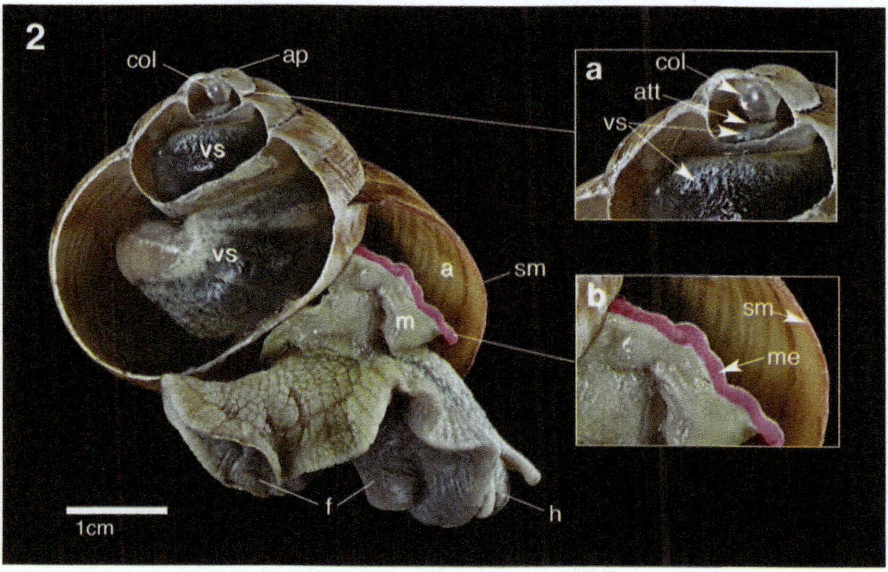

Fig. 12.2 Section of a juvenile *Helix pomatia* within its shell. (**a**) Attachment of the retractor muscle; (**b**) Mantle edge (highlighted in *pink*), detached from the shell margin. *a* aperture, *ap* apex, *att* attachment of the retractor muscle, *col* columella, *f* foot, *h* head, *m* mantle, *me* mantle edge, *sm* shell margin, *vs* visceral sac

Fig. 12.3 Thin section through the shell of a land snail. *p* periostracum, *o* ostracum, *ms* microstructure

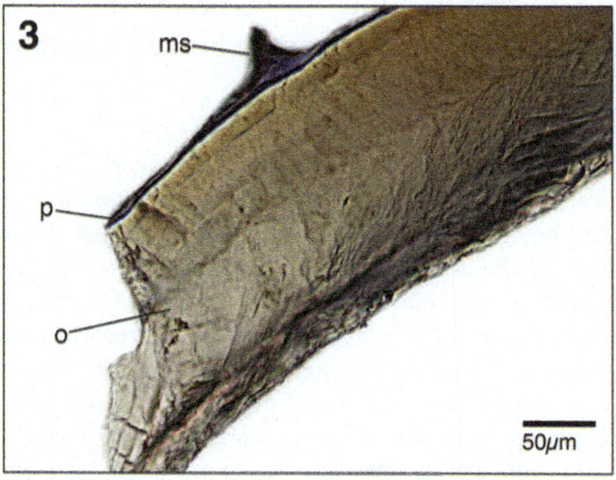

The surface structures of the periostracum are also highly varied ranging from microscopic spines to scale-like structures, grooves and ribs (Fig. 12.4). The functional significance of these surface structures is unknown, as is the manner that this surface variation is generated during the normal shell building process (Allgaier 2007, 2011; Boettiger et al. 2009).

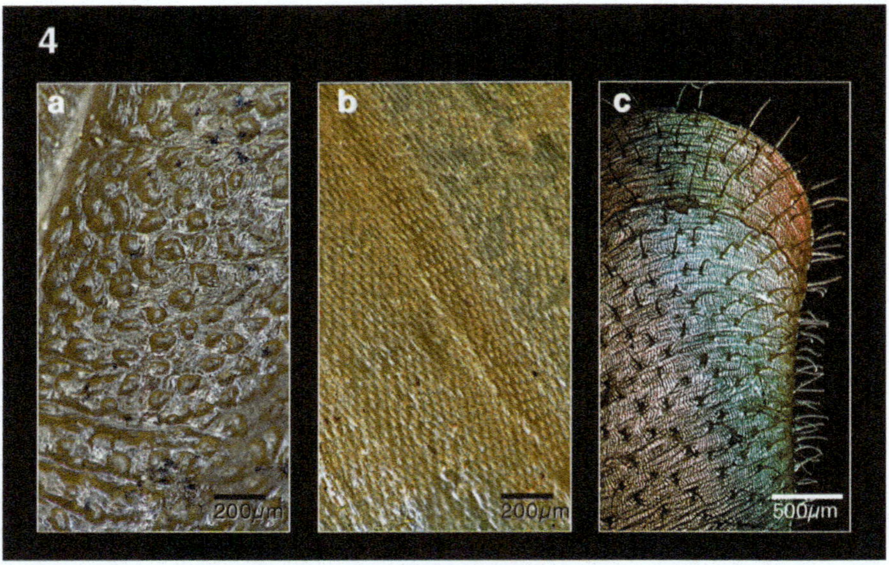

Fig. 12.4 Examples of shells with microstructure in land snails: (**a**) tubercles, (**b**) scales, (**c**) periostracal hairs

12.3 Investigating the Structure–Property Relationship of the Periostracum for the Assembly of Biomimetic Polymer Layers

One promising possibility for investigating the periostracum is to focus on mechanical properties by using Atomic Force Microscopy (AFM). The potential of AFM can be shown in preliminary measurements of an isolated piece of periostracum that has been lifted from the shell substrate (Fig. 12.5). The optical micrograph clearly shows the microstructural pattern of the shell surface. The first AFM results suggest differences in mechanical properties for the optically distinguishable regions. The elongated structures (blue arrow), which have a length of about 70 μm and protrude out of plain, show increased hardness with respect to the underlying surface with fibrillar structures (green arrow). Figure 12.6 shows preliminary AFM measurements of these two structural elements identified by optical microscopy (violet frame). On top of the elongated structure (blue arrow), the material appears considerably harder than in the underlying fibril-containing regions (green arrow). Based on the presented periodic microstructure, the bending rigidity along the elongated structures is considered to be higher than that in the orthogonal direction along the axis of the fibrils.

A possibility for studying appropriate polymeric components is by using the Langmuir trough to obtain laterally structured thin films at the air water interface

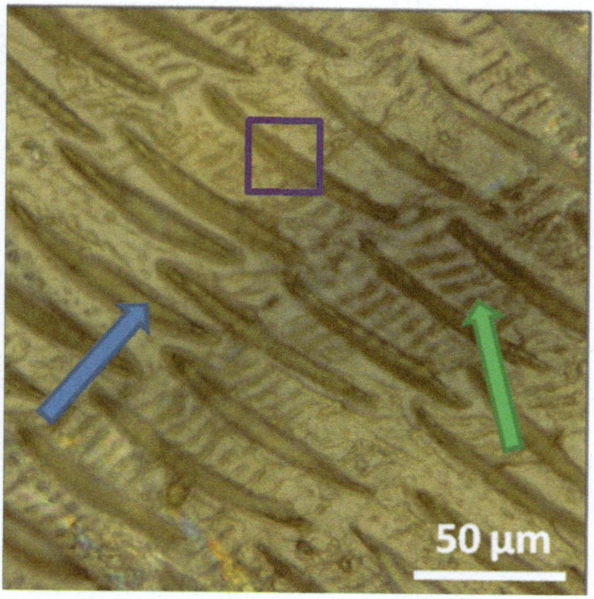

Fig. 12.5 Optical microscopic image of a piece of periostracum, peeled from the shell of a land snail. A periodic microstructure with elongated structures of an elevation of about 10 µm is revealed. *Image size: 286 µm × 212 µm*

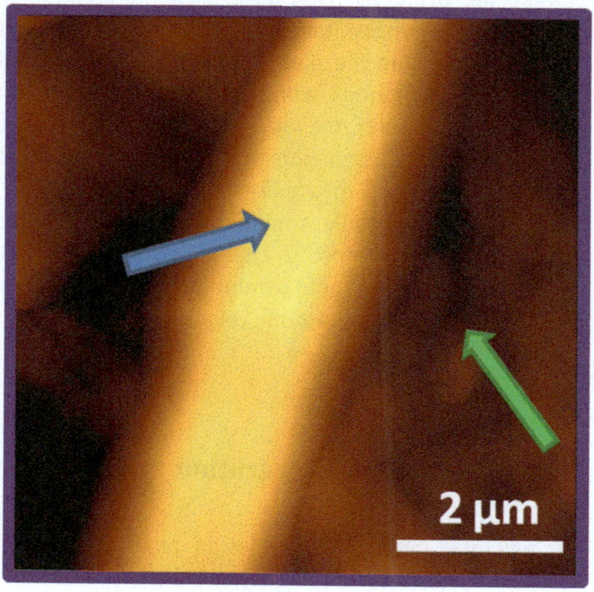

Fig. 12.6 AFM micrograph of the violet-rimmed region of the optical image (see Fig. 12.5) is shown. Mechanical properties were investigated on top of the elongated structure and the underlying organic film

by means of self-assembly. Poly-γ-benzyl-L-glutamate (PBLG) represents a biologically relevant system with controllable morphology (Chandran et al. 2015). Figure 12.7 shows an AFM height image of a monolayer of a mixture of short and long chain PBLG that adopts a helical conformation. Solid-like regions are imbedded in a fibrillar matrix.

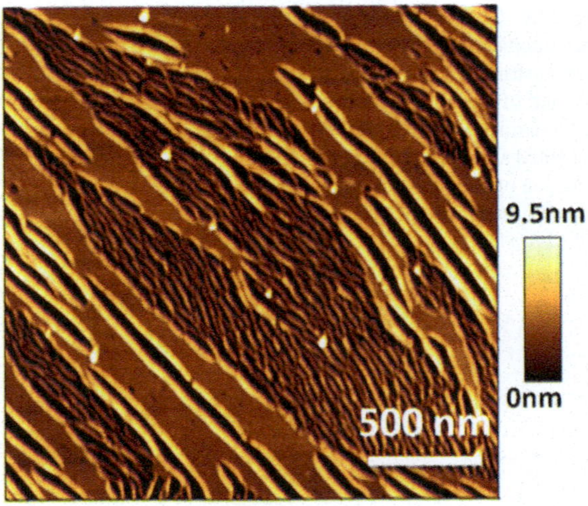

Fig. 12.7 AFM height image of a monolayer of a mixture of short and long chain poly-γ-benzyl-L-glutamate

With the Langmuir Schäfer technique, it is possible to withdraw these layers from the water surface and to stack them horizontally. Preliminary results reveal coalescence of the fibrils in those stacks in both the lateral and vertical directions. The assembly process takes place at various hierarchical levels, e.g. tuning the morphology within the plane of the monolayer followed by horizontal stacking. As the fibrillar regions within the stacks are thought to merge and form larger interconnecting fibres, the ratio of fibrillar areas to solidified areas in the monolayer plane is one of the parameters that determine the final structure in the stack. Another decisive parameter is the rotation of the monolayers in the stack with respect to the orientation of the fibrillar regions around the vertical axis in the stack. Currently, a device is being developed to enable these parameters to be controlled. Multilayers obtained in this way should exhibit differences in bending rigidity in the structurally different regions. Furthermore, they might serve as templates in three-dimensional (3D) printing processes.

12.4 Fabrication Integration in Computational Design

Architectural computational design aims at integrating given constraints, such as usability, structural integrity, aesthetics, material properties and fabricability, into a virtual geometric representation of architectural elements or entire building envelopes. To achieve this, designers employ a variety of strategies at the overlap of architectural design, engineering and computer science.

Approaches in rule-based generative design aim at assembling predefined architectural building elements into aggregated building structures in an attempt to provide a high degree of standardization in pre-fabrication. Research in this field is either performed in a purely exploratory bottom-up manner (Hansmeyer 2007), in which the final architectural design is hardly foreseeable and the incorporation of additional design criteria, such as structural integrity, becomes cumbersome. Other approaches attempt to gain control over the global design by defining target geometries and subdividing them into sets of available architectural elements, such as hexagonal (Jonas 2013) or triangular (Felbrich et al. 2014) panels. Thus, a variety of design criteria can be incorporated. The inherent contradiction between bottom-up growth and top-down target geometry design, however, poses a challenge.

Agent-based approaches employ the interaction of a virtual generative agent and its computational environment. Multi-agent systems extend this procedure by the interaction between diverse agents. The implementation of design constraints is carried out by defining favourable behavioural rules for the agent to consider when iteratively operating within its environment. In this manner, a variety of design drivers (behavioural rules) can be integrated in an open-ended fashion (Gerber et al. 2015). They extend from purely stress-driven multi-agent setups (Tsiliakos 2012) to behaviour-based agent systems informed by constraints of the fabrication tool (Baharlou and Menges 2013). The latter reveals an important, yet often neglected aspect: the integration of fabrication and computational design at early stages. To highlight the relevance of this issue, the conventional workflow and design constraint implementation in additive manufacturing needs to be discussed.

In a condensed way, the workflow can be separated into four distinct consecutive steps: (1) identification and evaluation of relevant design parameters; (2) mediation of design parameters in a computational model, with the goal of generating a single geometric solution that fulfils as many design constraints as possible; (3) post-processing of design geometry into machine instructions (tool paths); (4) production. The application of computational design tools solely takes place in step (2). Steps (3) and (4) are neglected by the designer. In fact, most state of the art additive manufacturing and design research considers the computational design of a product and the optimization of the subsequent production process as two disconnected consecutive steps. This top-down information flow, with a thoroughly designed object at its beginning and a separated fabrication process is common to most design and production processes. Thus, the potentials of computational optimization are limited to the static initial design phase.

The inspiration of this research, however, is the shell generation process as observed in land snails. The project attempts to derive a biologically inspired *production process*, rather than a biologically inspired *product*, based on the principles of material extrusion, material property transformation and molecular self-assembly. This opens up the opportunity to shift the focus of the computational design tool from an object-related geometric optimization towards process-related geometry generation, incorporating the characteristics of the fabrication tool and material properties at an early design stage. By overcoming the conventional sequential information flow in favour of a synthesized design and fabrication cycle, the benefits of computational integration can be utilized over the entire span of design and production.

To develop such a setup, a variety of diverse approaches can be utilized. In the agent-based approach of Schwinn et al. (2014), fabrication constraints are incorporated in early design stages by employing a machinic morphospace (Menges 2012). This multi-dimensional solution space provides structured information about physical limitations of the fabrication machine to the agent and thereby prevents it from generating design solutions that are not fabricable.

Vasey et al. (2015) have extended this approach by the immediate integration of physical fabrication during the design phase in real-time. The study employs a cyber physical agent that drives the fabrication machine, while constantly being informed about its environment, the building progress and relevant design parameters. Its virtual representation operates within a machinic morphospace and thereby integrates the machine limitations with the design requirements drawn from an internally stored 3D model in order to make decisions about the subsequent production steps. The physical representation of the agent collects information of the actual construction through sensors and executes the production. This study illustrates a way to merge design-time with construction-time, as observed in construction-performing animals such as the land snail. Obviously, it does not account for the multi-scalar material formation control, material property gradients and deformations which occur in the generation of the land snail's periostracum. Because of the openness and generality of agent-based behavioural strategies, however, such a setup extended to the nature of technological bulk deposition, dynamic material behaviour and the semi-controllability of its assembly at the micro-scale have proven most feasible and are conducted within the scope of this research. An overview of the desired workflow is shown in Fig. 12.8. This workflow is currently being elaborated and will include aspects of both hardware and software implementation.

12.5 Construction-Scale Additive Manufacturing

Additive manufacturing has gained much popularity in the design field over recent years. This is because, by using these techniques, a large variety of individual geometries can be produced easily and cost efficiently, thereby simplifying prototyping and model building, not only in architecture. Additive manufacturing at the architectural scale, however, is often based on attempts to scale up these known techniques. To illustrate the limitations of this approach, a few additive-manufacturing techniques existing in automated construction are discussed below.

State of the art concrete printing employs the principle of fused deposition modelling. Usually, a 9- to 20-mm-wide nozzle is used to extrude wet custom-engineered concrete with a layer thickness of 6–25 mm (Lim et al. 2009; Yingchuang 2014). Reinforcement is added subsequently. As opposed to the principle of extruding layers of material on top of each other in an additive manner, the D-shape process employs the approach of selectively fusing granular materials from a bed. By using fine sand or stone powder and a chlorine-based binder, a layer thickness of 4–6 mm can be achieved. Both these processes result in highly stable, yet massive, heavy-weight structures with little control over the surface texture. The dimensions of

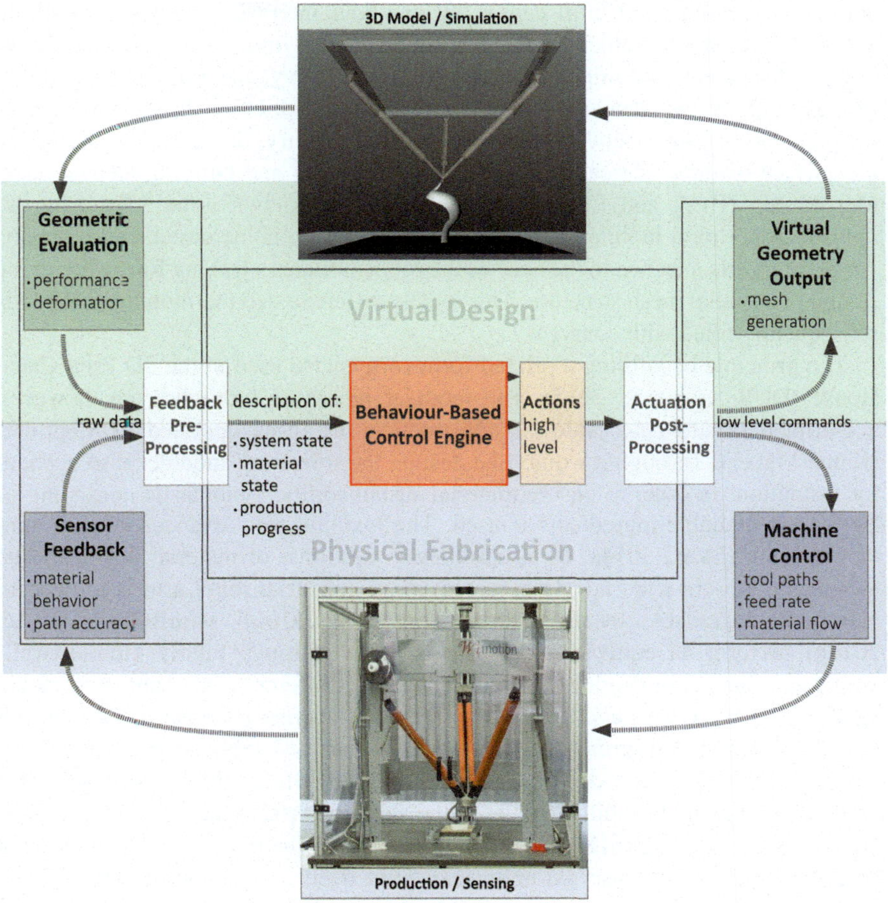

Fig. 12.8 Integrated design and fabrication workflow based on a feedback loop between sensor data/construction progress in physical fabrication and agent-based path planning in a virtual design model

the prints are limited to the size of the printing frame and only allow horizontal layering. Other approaches incorporate additional steps of post-processing. MineralJet combines 'near net-shape' additive manufacturing and subsequent 'net-shape' subtractive machining processes (Soar and Andreen 2012). It employs high-density calcium sulphate paste and achieves high tolerance and density but is limited to non-load-bearing interior applications.

A different approach is used in contour crafting. After initial extrusion of two outer form defining mortar contours in a layer-based manner, this mould is subsequently backfilled with concrete (Khoshnevis et al. 2006). The two-phase process enables both a relatively high resolution in the outer surface texture by using a trowel and a fast construction by extruding bulk quantities into the previously "printed" formwork. Contour Crafting, however, provides relatively weak bonding between the mould and infill caused by one-hour intervals between back-filling batches. It is, like the other examples discussed here, limited to horizontal layers and results in massive mineral structures. These heavy materials and the three-axis control of the deposition heads limits the freedom in design and causes reciprocal constraints between possible global geometry, detail and resolution in the surface texture, fabrication speed, bulk deposition and building element size. MeshMould (Hack and Lauer 2014) tackles these known issues. By using an industry robot on a mobile platform, the robot's reach is increased tremendously. An FDM nozzle attached to the head of the robot extrudes a leaking formwork in the design of a spatial mesh structure. This locally differentiated thermoplast scaffold is subsequently filled with concrete.

The principle of infilling a printed formwork is also used in the 3D Print Canal House (DUS Architects 2015). The modular pieces of formwork are, however, not printed in situ but beside the construction site in an upscaled FDM printer (KamerMaker). Through a womb-like design, the formwork modules also account for insulation. In order to ensure material sustainability, a bioplastic consisting of 80 % biodegradable ingredients is used. The Institute for Advanced Architecture of Catalonia (IAAC 2014a, b) also addresses the issue of material degradability. Whereas a 1.5-cm-wide nozzle mounted on an industrial robot arm is utilized to extrude biodegradable, locally collected soil (IAAC 2014a), MiniBuilders (IAAC 2014b) perform an equivalent material extrusion through highly mobile, wall-climbing robots. MiniBuilders are designed in a differentiated manner to tackle variable specified tasks and practically solve the inherent problem of limited tool reach. MX3D in Amsterdam focuses on the automated fabrication of steel construction. Although the widely known project M3XD Bridge (MX3D 2015a) is still under development, feasibility studies of the developed steel printing technique have proved very promising (MX3D 2015b). In contrast to the other projects introduced here, the concept is to use the unfinished print itself as a stable scaffold for the printing robot to follow the local construction site, an approach made possible by the high rigidity of printed metal and the incorporation of fabrication steps in the bridge design. The Free Form Construction Initiative with its focus on the climate management of termite mounds (Turner and Soar 2008) has investigated a biological shape in order to employ positive climate effects and to transfer them into the architectural realm. The biological formation process itself is little understood or investigated and a transfer of such biologically inspired construction techniques is not intended. This initiative has led to follow-up projects and is currently aiming at developing the "world's first commercial concrete-printing robot" in collaboration with concrete suppliers and architects (Buswell et al. 2014).

The focus of the research work described in this chapter, however, is on the examination and development of a biomimetic production process based on the shell formation of land snails and current additive processes such as Fused Deposition Modelling (FDM). This will lead to the development of an adapted FDM production process including machine-based technological components and a process-related approach to generating machine paths. Starting from possible material combinations, those combinations that partially resemble the biological process need to be analysed. The results have to feature necessary modifications and additional requirements in the layout and specifications for the adaptability of the components. An essential development is characterized by the fact that the guiding system for the print head has to have the functional properties of the moving biological model. An analysis of the model will show which kinematic arrangements and degrees of freedom are needed. The implementation of control algorithms for a production-specific process sequence will take place in the control module. The results of the work performed will be verified by a process simulation at the model scale, whereby the model size is adapted to requirements given by the biological model and to the potentials of an available guiding machine. Finally, the results have to be transferred to building scale for producing architectural envelopes by upscaling of the new processing unit and the kinematic system.

A promising kinematic concept for building scale manufacturing is that of cable robots consisting of a platform, wires and drive systems. In 1989, the RoboCrane system at NIST was presented, which seems to be the first prototype evaluated for large-scale handling (Dagalakis et al. 1989; Albus et al. 1992). Subsequently, several cable robots were developed differing in structure, scale and application domain. In addition to the well-known cable camera system at sporting events, however, only a few systems have been deployed to be used outside research laboratories. Since 2006, a family of robots named IPAnema has been developed for medium- to large-scale inspection, for handling and for assembly operations (Pott et al. 2010); these robots might serve as a starting point for large-scale additive-manufacturing systems by carrying the printing head.

12.6 From Biological Shell Formation to an Adapted Technical FDM Process

12.6.1 Building Direction and Production Area

In contrast to current FDM processes, snails extrude material not at a single location but through an expanded area of productive tissue, namely an array of glands resulting in a film extrusion of the periostracum. The newly extruded band is glued to the border of the existing shell. Thus, microscopic extrusion of the material for single bands is executed in the shell growth direction. Current FDM machines use one or two extrusion dies that are moved mainly orthogonally to the actual build direction. In a pre-processing step, the workpiece geometry is separated into single slices and nozzle paths are generated for each slice. After finishing one layer during fabrication, the print head is moved in the z-direction (build direction) to the next layer.

Based on the microscopic extrusion of the periostracum generation of the snail, new arrays of extrusion dies will be developed for the adapted FDM process. These arrays require novel strategies in control, path planning and collision avoidance. Multi-axis kinematic systems are needed as the orientation of a nozzle array requires more degrees of freedom than a single symmetric nozzle. The array will take the form of a flexible tool that can adapt its shape, thereby increasing the design possibilities of the workpieces.

12.6.2 Post-extrusion Forming and Material Properties

Most additive-manufacturing technologies are based on a specifically engineered temporary change in material properties. FDM for example employs a thermoplast (PLA, ABS, TPU, etc.); the print nozzle imposes a first transition in the material hardness and cohesion by melting it with heat. During the following molten phase, the extrusion nozzle is able to bring the material into place and connect it to previously extruded material (adhesion through melting). A second transition in material properties takes place when the thermoplast cools down and regains its initial hardness in its new geometric form. In order to avoid unforeseeable shape changes, it is useful to shorten the time span of the hardening phase as much as possible, meaning the active or passive cooling down of the molten material, after its deposition, to a state in which it no longer changes its form. Land snails, however, employ a different strategy. As opposed to FDM, the time span of the material hardening after its segregation from the productive tissue stretches over a long time span. This extended soft phase in the polymer, together with its stringy character, enables the snail to perform the muscular formation of the extruded material and thereby position it in its place. Translated into a technical setup, this post-extrusion formation can be thought of as extending the machine setup of a material depositing nozzle by the ability to mechanically form and subsequently shape a soft material; this adds another dimension of formation control to the setup.

12.6.3 Multiscalar Formation Process

The land snail shell and architectural envelopes can be split into several layers of different scales and functionality. The outermost layer of the periostracum is the surface layer that often bears microstructures potentially fulfilling various functions such as self-cleaning or water and dust repulsion. Mostly, a second uniform periostracum layer is present underneath the outer layer. The load-bearing calcium carbonate shell structure, which is much thicker than the periostracum, is subsequently added from below. The periostracum is generated by the production of periostracal material at the molecular level with the involvement of self-assembly and, then, the deposition as a film inside the periostracal groove. The formation of surface microstructures is included in this production step.

In an attempt to transfer these multi-scale aspects into the fabrication of architectural envelopes, the microstructures of the surface layer can either be produced by self-assembly of the used material or by the process of using fine nozzle diameters.

The macroscopic layout of the building or building segment defines the meso-scale, which is the internal structure of the formwork and the load-bearing structure. In a first step, the three layers will be produced by separated processes and process units. Then, combinations of micro-, meso- and macro-scale processes will be defined. The final goal is to create an adapted FDM process that is able to build surfaces with multi-scalar articulation in a single fabrication process step.

12.6.4 Product Representation and Fabrication Integration

In conventional FDM, the design of a printed geometry results primarily from the specific functional needs and predefined design constraints of the product, as implemented in a sequential top-down design and fabrication process (as discussed in Chap. 4). Once the design phase has been completed, the geometry is stored as a digital model. Subsequently, this virtual representation of the design is transformed into machine instructions and printed. The linear downstream information flow from a desired product target geometry leads to a second version of this geometry, which is post-processed to cohere with the limitations of the production machine, such as modularization because of limited work space, striation attributable to layer built-up, limited undercuts because of maximum material overhang and other constraints.

The biological role model exhibits a fundamentally different information flow and resulting control mechanism. The production of the periostracum in the land snail is an iterative process as there are periods of periostracum production and periods of rest; moreover, the snail can interrupt production when disturbed. Periostracum formation does not represent passive uncontrolled growth, but rather is part of a programmed active biological process that has developed during evolution. Furthermore, this active building process is enabled and framed by the physical production unit, namely the periostracal groove. Hence, the shell form can be understood as the product of an interplay between the snail's bauplan (the physical prerequisite for the formation of the periostracum) and behavioural aspects determining the timing of shell formation. If we further understand that the mechanisms of evolutionary optimization apply to the snail's building behaviour and the physical form of the periostracal groove, the necessity of a geometric shell design in a product-based context becomes obsolete in favour of a production-based procedural shell design. This is fundamentally different from state of the art computational design in which optimization procedures are solely applied to the geometric form of a product, rather than to its formation process. This bottom-up information flow, however, implements the production unit as an essential part of the design. It is furthermore highly adaptive towards changes in environmental conditions and therefore exhibits a high degree of procedural robustness. On the other hand, it still integrates the required design characteristics (enclosure, protection, position of openings) in the final product. For transfer into a technical control mechanism, agent-based behavioural strategies are the most promising paradigm in state of the art computational design.

12.7 Conclusion and Outlook

The periostracum is a polymer-like thin organic film built from the synthesis of macromolecules in specialized cells along the periostracal groove (e.g. Lowenstam and Weiner 1989). Macromolecules folded into structurally ordered units are then self-assembled into the periostracum. This sheet is malleable enough to bend towards the outside of the shell where it then forms the outermost layer of the multi-layered shell. Once extruded, the periostracum serves as a substrate for crystallisation of calcium carbonate that takes place at the inner side of the periostracum, which is directed towards the extrapallial space. An important objective in the future will be to discriminate between the scales of production involving the periostracum, i.e. the production of this polymer at the molecular level and its structural characteristics, to the formation of its surface morphologies by the snail itself after production. This is a prerequisite for developing reinforced polymer composite materials and self-assembled functional surfaces.

The described interdisciplinary approach will eventually lead to the implementation and application of generative design method within a surface modelling environment and the generation of larger surface geometries. This will subsequently lead to the development of adapted FDM process for large-scale 3D-printed architectural surfaces composed of polymer composites. Finally, reverse biomimetic consideration will help in the recognition of the specific adaptations of the surface structures of terrestrial snails to the environment in which they live.

Acknowledgements We thank the German Research Foundation (DFG) for supporting the Collaborative Research Center SFB-TRR 141: Biological Design and Integrative Structures – Analysis, Simulations and Implementation in Architecture, project A08.

References

Albus JS, Bostelman RV, Dagalakis N (1992) The NIST robocrane. J Robotics Syst 10(5): 709–724

Allgaier C (2007) Active camouflage with lichens in a terrestrial snail, *Napaeus* (*N.*) *barquini* Alonso and Ibáñez, 2006. Zool Sci 24:869–876

Allgaier C (2011) A hairy business – periostracal hair formation in two species of helicoid snails (Gastropoda, Stylommatophora, Helicoidea). J Morphol 272(9):1131–1143

Baharlou E, Menges A (2013) Behavioral protyping, rethinking prototyping. Proceedings of the Design Modeling Symposium, Berlin, pp 291–303

Boettiger A, Ermentrout B, Oster G (2009) The neural origins of shell structure and pattern in aquatic mollusks. PNAS 106:6837–6842

Buswell RA, Foster N, Skanska (2014) The world's first commercial concrete-printing robot. http://www.dezeen.com/2014/11/25/foster-partners-skanska-worlds-first-commercial-concrete-3d-printing-robot/. Accessed 26 Feb 2016

Chandran S, Dold S, Buvignier A, Krannig K, Schlaad H, Reiter G, Reiter R (2015) Tuning morphologies of Langmuir polymer films through controlled relaxations of non-equilibrium states. Langmuir 31:6426–6435

Dagalakis NG, Albus JS, Wang BL, Unger J, Lee JD (1989) Stiffness study of a parallel link robot crane for shipbuilding applications. ASME J Mech Design 111:183–193

de Paula SM, Silveira M (2009) Studies on molluscan shells: contributions from microscopic and analytical methods. Micron 40(7):669–690

DUS Architects (2015) 3D print canal house. http://3dprintcanalhouse.com. Accessed 14 Feb 2016

Felbrich B, Lordick D, Noennig JR, Wiesenhütter S (2014) Experiments with a folding multi-agent system in the design of triangle mesh structures. In: Proceedings of the 16th international conference on geometry and graphics ISGG 2014, Innsbruck, Austria

Gerber DJ, Pantazis E, Marcolino LS (2015) Design agency prototyping multi-agent systems in architecture. In: Proceedings of 'Next City' CAAD futures bi-annual conference, Sao Paulo, Brazil, July 2015, pp 214–235

Hack N, Lauer WV (2014) Mesh-mould: robotically fabricated spatial meshes as reinforced concrete formwork. Archit Design 84:44–53

Hansell MH, Ruxton GD (2013) Exploring the dichotomy between animals building using self-secreted materials and using materials collect from the environment. Biol J Linn Soc 108(3):688–701

Hansmeyer M (2007) L-systems in architecture. In: Liu Y-T (ed) Distinguishing digital architecture. Birkhäuser Verlag, Basel, pp 91–203

IAAC (2014a) Minibuilders. http://robots.iaac.net. Accessed 17 Feb 2016

IAAC (2014b) Pylos. http://pylos.iaac.net. Accessed 17 Feb 2016

Jackson DJ, McDougall C, Green K, Simpson F, Wörheide G, Degnan BM (2006) A rapidly evolving secretome builds and patterns a sea shell. BMC Biol 4(40):1–10

Jacob DE, Wirth R, Soldati AL, Wehrmeister U, Schreiber A (2011) Amorphous calcium carbonate in the shells of adult Unionoida. J Struct Biol 173(2):241–249

Jonas K (2013) From generic to specific: prototyping a computational growth model. In: Rethinking prototyping, Proceedings of the design modeling symposium, Berlin, pp 305–321

Khoshnevis B, Hwang D, Yao K-T, Yeh Z (2006) Mega-scale fabrication by contour crafting. Int J Ind Syst Engin 1(3):301–320

Lim S, Le T, Webster J, Buswell R, Austin S, Gibb A, Thorpe T (2009) Fabricating construction components using layer manufacturing technology, (GICC'09). Paper presented at the Global Innovation in construction conference, Loughborough University, Leicestershire, UK, 13–16 September 2009

Lowenstam HA, Weiner S (1989) On biomineralization. Oxford University Press, Oxford, 324 pp

Luchtel DL, Martin AW, Deyrup-Olsen I, Boer HH (1997) Gastropoda: Pulmonata. In: Harrison FW, Kohn AJ (eds) Microscopic anatomy of invertebrates, vol 6B, Mollusca II. Wiley-Liss, New York, pp 459–718

Marin F, Luquet G (2004) Molluscan shell proteins. Compos Rend Palevol 3:469–492

Marin F, Luquet G, Marie B, Medakovic D (2008) Molluscan shell proteins. Primary structure, origin, and evolution. Curr Top Dev Biol 80:209–276

McDougall C, Green K, Jackson DJ, Degnan BM (2011) Ultrastructure of the mantle of the gastropod Haliotis asinina and mechanisms of shell regionalization. Cells Tissues Organs 194:103–107

Meenakshi VR, Hare PE, Watabe N, Wilbur KM (1969) The chemical composition of the periostracum of the molluscan shell. Comp Biochem Physiol 29:611–620

Menges A (2012) Morphospaces of robotic fabrication – from theoretical morphology to design computation and digital fabrication in architecture. In: Brell-Cokcan S, Braumann J (eds) Robotic fabrication in architecture, Art and Design. Springer, Wien/New York, pp 28–61

MX3D (2015a) MX3D bridge, http://mx3d.com/projects/bridge. Accessed 13 Feb 2016

MX3D (2015b) MX3D metal 3D printing, http://mx3d.com/projects/metal/. Accessed 13 Feb 2016

Ortiz C, Boyce MC (2008) Bioinspired structural materials. Science 319(5866):1053–1054

Pott A, Meyer C, Verl A (2010) Large-scale assembly of solar power plants with parallel cable robots. In: Proceedings for the joint conference of ISR 2010 (41st International symposium on robotics) and ROBOTIK 2010 (6th German conference on robotics), pp 1–6

Salas C, Marina P, Checa AG, Rueda JL (2012) The periostracum of Digitaria digitaria (Bivalvia: Astartidae): formation and structure. J Moll Stud 78:34–43

Schwinn T, Krieg OD, Menges A (2014) Behavioral strategies: synthesizing design computation and robotic fabrication of lightweight timber plate structures, in design agency. In: Proceedings of the 34th annual conference of the Association for Computer Aided Design in Architecture (ACADIA), Los Angeles, pp 177–188.

Soar R, Andreen D (2012) The role of additive manufacturing and physiomimetic computational design for digital construction. Arch Design 82(2):126–135

Sud D, Poncet J-M, Saihi A, Lebel J-M, Doumenc D, Boucaud-Camou E (2002) A cytological study of the mantle edge of *Haliotis tuberculata* L. (Mollusca, Gastropoda) in relation to shell structure. J Shellfish Res 21(1):201–210

Tsiliakos M (2012) Swarm materiality: a multi-agent approach to stress driven material organization. In: Digital physicality – Proceedings of the 30th eCAADe conference – volume 1, Czech Technical University in Prague, Faculty of Architecture (Czech Republic), 12–14 September 2012, pp 301–309

Turner S, Soar R (2008) Beyond biomimicry: what termites can tell us about realizing the living building. In Hassan T, Ye J (eds) Proceedings of the 1st international conference on industrialised, integrated, intelligent construction (I3CON), Loughborough University, pp 221–137

Vasey L, Baharlou E, Dörstelmann M, Koslowski V, Prado M, Schieber G, Menges A, Knippers J (2015) Behavioral design and adaptive robotic fabrication of a fiber composite compression shell with pneumatic formwork. In: Combs L, Perry C (eds) Computational ecologies: design in the anthropocene, Proceedings of the 35th annual conference of the Association for Computer Aided Design in Architecture (ACADIA), University of Cincinnati, Cincinnati, OH, pp 297–309

Yingchuang (2014) 3D printed concrete house. http://www.dezeen.com/2014/04/24/chinese-company-3d-prints-buildings-construction-waste. Accessed 1 Mar 2016

Chapter 13
Analysis of *Physcomitrella* Chloroplasts to Reveal Adaptation Principles Leading to Structural Stability at the Nano-Scale

Pouyan Asgharzadeh, Bugra Özdemir, Stefanie J. Müller, Oliver Röhrle, and Ralf Reski

Abstract Proteins of the FtsZ (*filamentous temperature sensitive* Z) family establish complex polymeric spatial patterns in plastids of the moss *Physcomitrella patens*. These structures represent a "plastoskeleton" that might contribute to plastid shape and stability. Because additional forces, such as electrostatic interactions, are effective between molecules and atoms at the nano-scale, the applicability of the rules and principles of structural analysis to molecular structures (diameter of a moss plastid: 3–6 μm in the short axis and 4–8 μm in the long axis) is unknown.

The aim of this project is to develop mathematical models of FtsZ network connectivity and dynamics in order to investigate whether molecular structures of the plastoskeleton are evolutionarily optimised to withstand mechanical stresses. To our knowledge, this is the first study focused on the nano-scale characterisation of molecular features/ultrastructures found in plant organelles with the goal of developing new ideas and approaches for bioinspired architecture/building construction.

P. Asgharzadeh • O. Röhrle
Institute of Applied Mechanics (CE), University of Stuttgart, Pfaffenwaldring 7, D-70569 Stuttgart, Germany
e-mail: roehrle@simtech.uni-stuttgart.de

B. Özdemir • R. Reski (✉)
Plant Biotechnology, University of Freiburg, Schänzlestraße 1, D-79104 Freiburg, Germany
e-mail: ralf.reski@biologie.uni-freiburg.de

S.J. Müller
Faculty of Biology, Albert-Ludwigs-University of Freiburg, Schänzlestraße 1, D-79104 Freiburg, Germany

INRES—Chemical Signalling, University of Bonn, Friedrich-Ebert-Allee 144, 53113 Bonn, Germany

J. Knippers et al. (eds.), *Biomimetic Research for Architecture and Building Construction*, Biologically-Inspired Systems 9, DOI 10.1007/978-3-319-46374-2_13

13.1 Introduction

Eukaryotic cells are highly compartmentalised and can thus be compared with a building containing many different rooms. Cell compartments are separated from each other by lipid membranes and often show stability of shape, on the one hand, and movement and dynamic form changes, on the other (Osteryoung and Pyke 2014; Usami et al. 2012). Several protein families forming various types of cytoskeletons have been implicated in influencing cellular and organelle shape and dynamics. Currently, the underlying mechanisms of stability on the cellular scale are largely unclear (Ingber 2003a, b; Sultan et al. 2004), based on a lack of the ability to quantify, to monitor and to simulate the stability of the complex and dynamic cell system.

Plastids are endosymbiont-derived plant organelles that are surrounded by two membranes and that exhibit a characteristic lens-shape and a diameter of a few micrometers. Proteins of the FtsZ family have been implicated in chloroplast division and shaping (Martin et al. 2009a). This project utilises the plastid compartment of the model moss *Physcomitrella patens* to monitor stability-conferring proteins by fluorescent protein tags and confocal laser scanning microscopy and applies engineering-based algorithms to implement computational stability models. These models will be used in simulations (1) to derive hypotheses regarding the biological function of FtsZ protein isoforms for validation in the biological model generator (reverse biomimetics) and (2) to analyse and compare derived principles with standard structure networks and shell theory from architecture (in cooperation with the SFB-TRR141 project covered in Chap. 17).

The *Physcomitrella patens* genome (cf. Chap. 15, Zimmer et al. 2013) encodes five FtsZ isoforms in three different subfamilies. The *in vivo* interactions of these isoforms have been elucidated (Gremillon et al. 2007). Here, we generate separate models in order to reveal differences in function and polymer organization.

At the *in silico* level, the robustness of the mathematical models of FtsZ networks are tested (in cooperation with the SFB-TRR141 project covered in Chap. 17) with regard to the expected high robustness of complex biological multifunctional structures.

Analysed principles of adaptive stability at the nano-scale are then placed into an evolutionary (in cooperation with other SFB-TRR141 projects covered in Chap. 15) and architectural context and integrated with results from the micro-scale (in cooperation with other SFB-TRR141 projects covered in Chaps. 8 and 11) and macro-scale (in cooperation with the SFB-TRR141 project covered in Chap. 9). Thus, the scalability of structures will be characterised. Furthermore, the modelling results are constantly integrated into biological hypotheses of organelle evolution and FtsZ isoform functional diversification and are tested experimentally in the moss system.

Thus, the main aims of this project are to reveal functional and organisational properties of the plastoskeleton in moss by investigating the FtsZ polymer networks. The first aim is concerned with deriving from the FtsZ network the general nano-network principles that can be used to derive conceptual biomimetic ideas. The second aim focuses on revealing the underlying adaptivity of FtsZ network. Fulfilment of both aims will eventually provide the basis for performing the biomimetic transfer of the biological nano-scale principles to architectural structures.

13.2 Research Rationale

The protein FtsZ (*filamentous temperature sensitive Z*) was discovered as a component of the "bacterial cytoskeleton" and provides a scaffold for bacterial cell division. In many eukaryotes, FtsZ genes have been introduced by the bacterial ancestors of endosymbiont-derived organelles, namely the plastids and mitochondria (TerBush et al. 2013). The presence of FtsZ proteins in the model plant *Physcomitrella patens* (Reski 1998a) has been demonstrated and visualised *in vivo* (Kiessling et al. 2000; Strepp et al. 1998). Further research on this protein family has revealed the presence of five FtsZ isoforms, which can be divided into three sub-families (Rensing et al. 2004; Martin et al. 2009b). The basic principles of FtsZ localisation, interaction and function have been established by fusing FtsZ proteins to fluorescent reporter proteins and by the targeted knockout of FtsZ genes by homologous recombination (Gremillon et al. 2007; Kiessling et al. 2004; Martin et al. 2009a, b; Suppanz et al. 2007). Proteins of all three subfamilies are necessary for plastid division, although differences in plastid size and shape between the distinct mutant lines indicate functional diversification (Martin et al. 2009a). Visualisation of FtsZ by fusion to the fluorescent reporter *Green Fluorescent Protein* (GFP) has shown complex and isoform-specific patterns of FtsZ localisation and interactions (Fig. 13.1 and Gremillon et al. 2007). Concerning the molecular structure of bacterial FtsZ polymers, Lu et al. (2000) have suggested a double helix that has a diameter of about 23 nm with two protofilaments.

As the confocal microscopy of fluorescence-labelled FtsZ in moss reveals complex geometrical patterns and as chloroplasts in mutant lines show distinct shape defects (see Fig. 13.2), polymers of FtsZ might provide scaffolds ensuring the stability and structural integrity of plastids (Reski 2002, 2009). Thus, the term

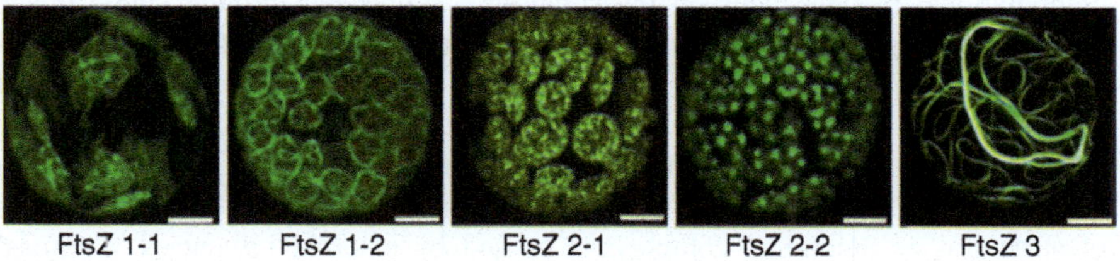

Fig. 13.1 FtsZ:GFP Network in Plastids (PhD thesis L. Gremillon, Universität Freiburg, modified). Localization of GFP-tagged FtsZ proteins in moss protoplasts. Protoplasts were transiently transfected with 35 *s::FtsZ::GFP* and imaged 3 days after transfection. Scale bars = 5 μm

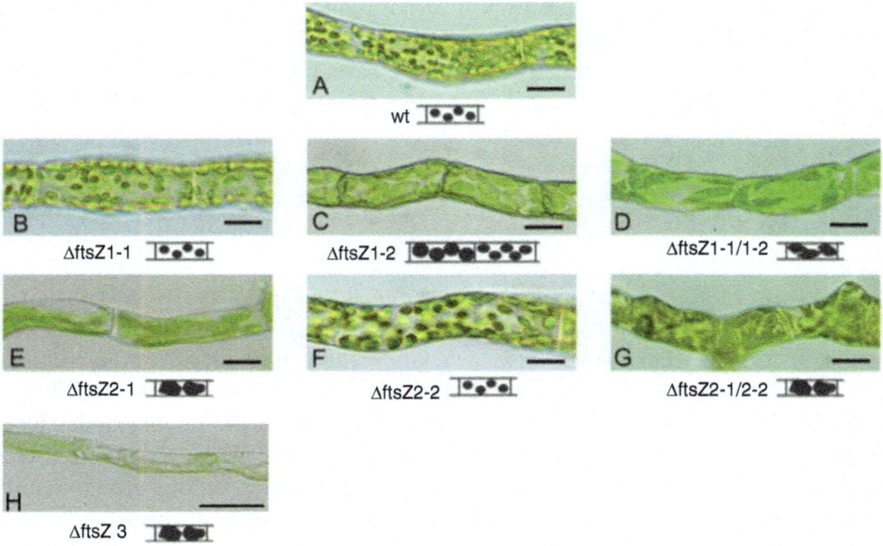

Fig. 13.2 Mutant macrochloroplasts (Martin et al. 2009a, modified). Knock-out of FtsZ genes affects chloroplast morphology. Moss protonema cells with mutated FtsZ isoforms show distinct defects in chloroplast division and shape. (**a**–**g**) scale bar = 25 μm, (**h**) scale bar = 50 μm

"plastoskeleton" has been coined for FtsZ polymers in plastids (Reski 2002). FtsZ networks are reminiscent of geodesic domes in architecture.

Although confocal or electron microscopy provides many new insights into geometrical patterns of various phenomena at the (sub)cellular scale, computational and geometrical models describing detailed functional and structural aspects at this scale remain rare. A few three-dimensional computational models exist in which the modelling focuses mainly on the cytoskeleton, cell morphogenesis or cell aggregates. Mechanical behaviour in these three-dimensional studies is modelled either by means of Tensegrity networks (Ingber 2003a, b; Sultan et al. 2004) or by three-dimensional continuum mechanical models either reduced to truss models (Muñoz et al. 2010) or discretised by the Finite Element method (Or-Tzadikario and Gefen 2011). Many of the three-dimensional structural or computational models focus on investigating a single cell through single image data sets or focus on determining material properties by simulating experimental setups, such as Atomic Force Microscopy (AFM) experiments. However, individual sets of images of particular cells or cell structures are not necessarily capable of revealing new insights into functional aspects. Further, such single-specimen image data sets do not allow the investigation of potential dynamic changes of structural or functional components caused by external influences. Confocal microscopy imaging is capable of investigating dynamic changes in living cells in three dimensions and over time. The challenge of acquiring the appropriate image data and automatically reconstructing computational models from image data have prevented researchers from using three-dimensional computational models for investigating functional aspects of cell mechanics. The large amount of data and the specific characteristics of the biological specimen require tailored non-standard image registration and segmentation algorithms, e.g. the manual selection of adaptive thresholding algorithms, noise reduction by means of deconvolution algorithms and structure reconstruction in an iterative process including biological characteristics of FtsZ isoforms. This approach of extracting computer models from 3D images of biological environments, which proposes truncating the resulting intricate geometry of the segmented confocal images into a simpler form of structure through a specifically developed process for confocal microscopy data, can be seen, for example, in Röhrle et al. (2011) for the case of segmenting biological image data to provide geometrical input to finite element simulation. Furthermore, challenges based on performing image processing on small-scale data need to be taken into consideration e.g. the lack of resolution in the z-direction (perpendicular to the imaging planes) vs in-plane resolution and the greater impact of the noise on the segmented image. In addition, micromechanical aspects for an extensive finite element analysis of the biological structures, i.e. the FtsZ network, individuates itself from regular treatment of larger scale structures as mentioned in Röhrle et al. (2012).

13.3 The Process of Developing a Well-Grounded Computer Model of Plastoskeleton Leading to Reverse Biomimetics

As mentioned above, the aim of this research project is to analyse nano-structures, i.e. FtsZ isoforms in moss chloroplasts, in order to find new principles of structural integrity and adaptation that can be efficiently transferred to larger scales. To do so, the structural components of a single fluorescence-tagged FtsZ isoform are first analysed by using imaging data of single chloroplasts (see Sect. 13.3.1). The specifically designed image-processing algorithms for confocal microscopy images are further extended to include multiple FtsZ isoforms and dynamic changes of the protein networks (Sects. 13.3.1.4 and 13.3.2). This development of methods is only possible when data sets that contain information of all five isoforms of FtsZ within the different cell types of moss are available. Therefore, the acquisition of stably transformed transgenic moss lines is essential (Sect. 13.3.2). Fundamental principles behind the adaptive stability of plastoskeleton are driven and used to establish a framework that provides the necessary details to test and investigate the biological hypotheses and to derive methodologies for structural stability and adaptation on larger scales (Sect. 13.3.3). Furthermore, the application of image-processing algorithms to the image data set allows us to use computational techniques to investigate FtsZ redundancy and interdependence in a reverse biomimetics perspective (Sect. 13.3.3.4).

Analysed principles of adaptive stability at the nano-scale are then placed into an evolutionary and architectural context and integrated with results from the micro-scale (in collaboration with projects mentioned in Chaps. 8 and 11) and macro-scale (in collaboration with projects mentioned in Chap. 9). Thus, the scalability of structures will be characterised. Furthermore, the modelling results are constantly integrated into biological hypotheses of organelle evolution and FtsZ isoform functional diversification and tested experimentally in the moss system.

13.3.1 From Imaging to Models of FtsZ-Isoforms

13.3.1.1 Monitoring FtsZ Networks by Using Fluorescent Proteins

In molecular biology, the characterisation and mutagenesis of naturally occurring fluorescent proteins from jelly fish and corals has resulted in the emergence of a palette of genetically encoded fluorescent reporters that can be fused to proteins of interest (Shaner et al. 2005). DNA constructs for all five FtsZ isoforms fused to GFP (Gremillon et al. 2007) are available. These constructs are under the control of the constitutive cauliflower mosaic virus (CaMV) 35S promoter and can be used for the transient transfection of moss protoplasts. In addition to these existing constructs, new transient expression vectors have been generated by C-terminally labelling each FtsZ isoform with different fluorescent protein variants (e.g. EGFP, mCitrine, mCerulean and mCherry) and placing these fusion constructs under the control of the PpAct5 promoter (see Fig. 13.3). PpAct5 is a *P. patens*-specific constitutive promoter that has also been used for the production of human proteins in moss bioreactors, as shown by Weise et al. (2006). Transient transfection of moss protoplasts with these constructs and subsequent confocal microscopy analysis

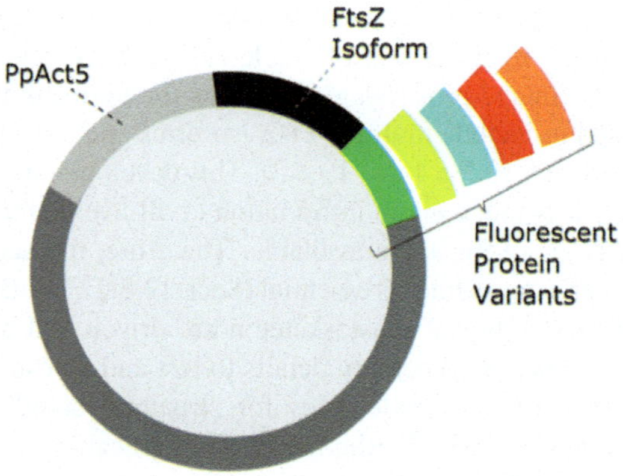

Fig. 13.3 Schematic illustration of the transient expression vectors: the plasmids were genetically engineered to carry the expression cassette that contains a fusion construct and a strong promoter (PpAct5) controlling its expression. The fusion construct consists of the coding sequence of a specific FtsZ isoform and the coding sequence of a specific fluorescent protein variant

enable the simultaneous imaging of two or more FtsZ isoforms within a single cell. This, in turn, permits a better understanding of the hierarchy and interplay of the different isoforms in the plastoskeleton structure.

In an initial phase, confocal microscopy of the networks formed by the various isoforms in single chloroplasts was employed to determine the necessary imaging set-up and data structure to provide a basis for generating computer models.

13.3.1.2 From Imaging to Model Geometry

Application-specific image analysis techniques for single FtsZ isoforms are being developed based on new image data sets that have been generated via the confocal microscopic analysis of moss protoplasts transfected with transient expression constructs (see Fig. 13.3). Within this work, the FtsZ1-2 isoform has been used to investigate and develop the basic mechanism for image analysis. Based on the images generated by confocal microscope, the diameter of the chloroplasts in moss is 3–6 μm in the short axis and 4–8 μm in the long axis. The resolution of the confocal microscopy of the preliminary data set is 0.5 μm between the single image slices (z-direction) and ~0.1 μm in-plane (xy-plane). Based on the voxel data set, diffusion-based image segmentation techniques are employed to detect the isoform and the three-dimensional arrangements. Digital models based on rod-like structures are developed to analyse the structural arrangements of the investigated FtsZ isoform. In cases in which the out-of-plane resolution is too coarse, i.e. the FtsZ isoform connections are located in the space between two consecutive confocal planes, a novel post-processing algorithm is being developed to detect such interactions and to proceed with the structural reconstruction of the FtsZ isoforms. Since the development of the digital models is closely related to the image preparation and imaging itself, the feedback from the modelling tasks is needed to adjust the imaging protocols to improve the overall image quality.

Extraction of the structural information of FtsZ isoforms from the three-dimensional images starts with the decision as to whether a voxel contains material or not. This requires the application of different thresholding algorithms to the images. First, the parameters of the local adaptive thresholding algorithms are chosen in an interdisciplinary manner, i.e. based on expectations of the geometry of the structure such as continuity of the structure inside the chloroplasts and the localisation of certain isoforms in the chloroplasts. Then, a machine learning mechanism is utilised to decide on the selection of the thresholding process based on the previous results as part of the learning set.

A predictable problem in the segmentation procedure is the eradication of the noise produced in the process of imaging. To do so, deconvolution processes are used that will produce results whereby their quality will strongly depend on the parameters of the deconvolution algorithm. Again, a close interaction between the biologists and mechanical engineers is essential. Deconvolution, however, often leads to images that are smoothed to too great an extent. This causes problems in further steps of structure reconstruction. Therefore, thoughtful precautions should be considered when dealing with deconvolution algorithms.

13.3.1.3 Structural Analysis of the Spatial Arrangement of the Digitised FtsZ Isoform Arrangement

In order to obtain an understanding of the structural features of the plastoskeleton via an analysis of computer models based on three-dimensional images, a representative model (structure) is needed that provides essential parameters of any plastoskeleton that has been subjected to imaging. Building such a computer model is by no means an easy task. However, with a statistical analysis of the structural parameters of the individually extracted models from numerous image samples, one can come as close to building a representative model as possible. These structural parameters include a variety of characteristics, namely the different types of connections and their occurrence, the angles between the different members in a connection and the thickness in the different parts of the structure. In order to be able to extract the mentioned characteristics from each image data set, a skeleton building method is utilised founded on the images that have gone through the previous image processing steps. This skeleton formation method is based on edge detection techniques and a cellular automata algorithm that then helps to reconstruct the missing parts of the structure. The outcome of this technique is a graph consisting of connections, segments and points that are used for the statistical analysis of the mentioned structural features. An example of the initial image processing process and skeleton formation is shown in see Fig. 13.4. On consideration of the results produced by this statistical analysis, visual inspection of the digital models reveals further structural features that are needed to be investigated as part of this research project.

The resulting algorithm leads to image-informed digital models. The developed algorithm can be applied not only to the images of the plastoskeleton, but also to any other three-dimensional fluorescent images of biological structures such as the cytoskeleton.

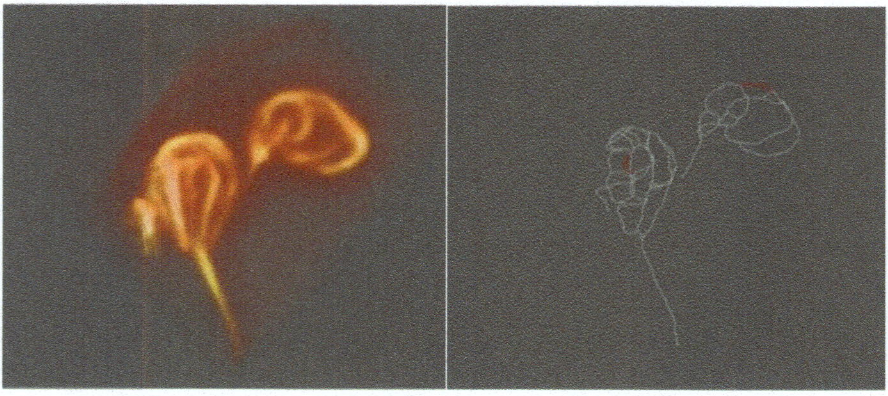

Fig. 13.4 The spatial graph as a result of applying the algorithm for extracting structural information from confocal images of FtsZ isoforms

13.3.1.4 Enhancing Image Analysis Techniques for Digital Model Creation Taking into Account Multiple FtsZ Isoforms

The methodologies and algorithms developed in Sects. 13.3.1.2 and 13.3.1.3 can be further extended to take into account the co-existence of multiple FtsZ isoforms. The image segmentation algorithms based on and reduced to two-colour images (fluorescence of the isoform and the background) are capable of being extended to a multi-colour FtsZ isoform image segmentation algorithm. Furthermore, the structural analysis algorithms need to include the ability to detect the interactions between various isoforms. The performance of a statistical analysis among the data sets between the different isoforms provides the basis for identifying concepts leading to structural stability and adaptation.

13.3.2 Integrating Dynamics: From Dynamics to Adaptivity

13.3.2.1 FtsZ Networks in Stably Transformed Transgenic Moss Lines

In order to reveal the structure and the dynamics of the plastoskeleton in distinct cell types and to visualise several FtsZ isoforms simultaneously, stable moss lines with tagged FtsZ proteins need to be established. The amenability of *Physcomitrella patens* to gene targeting (Strepp et al. 1998; Reski 1998b) can be used to integrate distinct fluorescent protein variants (e.g. EGFP, mCitrine, mCherry and mCerulean) into the coding sequence of the distinct FtsZ isoforms (a process called "knock-in", see Fig. 13.5). Under the control of the endogenous promoter, protein abundance is regulated as in wild-type plants and thus the differences between tissues and the influence of parameters such as the light conditions, growth medium and age become apparent, as evidenced by, for example, Mueller et al. (2014), Schuessele et al. (2016) and Horst et al. (2016). The use of different fluorescent protein variants enables us to perform the simultaneous imaging of the different FtsZ isoforms

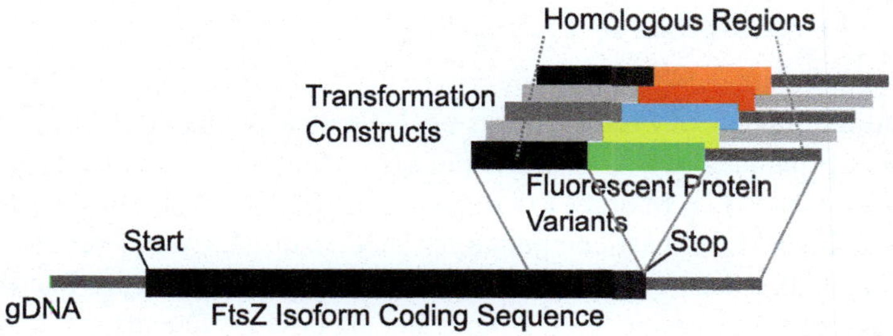

Fig. 13.5 The schematic illustration of the "knock-in" process. The coding sequence of various fluorescent protein variants are integrated in their target loci based on the homology between the flanking homologous regions and the corresponding sequences at the targeted locus. For each different FtsZ isoform, a specific fluorescent protein with a distinct colour was chosen

in a single cell. By stable transformation of the existing FtsZ mutant collection (Martin et al. 2009a) with the newly established knock-in constructs for fluorescent reporters, the effect of the lack of specific FtsZ isoforms can be characterised and integrated into the *in silico* network models. Hypotheses and predictions derived from the models can be tested *in vivo* in an iterative process. Confocal imaging of chloroplasts can be extended to various growth stages and conditions from which a subset of distinct plastoskeleton morphologies in well-defined cell types can be selected for modelling. By the simultaneous imaging of isoforms, problems such as the co-localisation of distinct isoforms and interactions (e.g. by Förster resonance energy transfer FRET, as shown in Gremillon et al. 2007) can be addressed.

13.3.2.2 Statistical Analysis of the Dynamics of the Plastoskeleton

In order to integrate dynamics into the analysis of FtsZ networks, time-lapse confocal imaging monitoring the dynamics of the plastoskeleton in various cell types and under diverse conditions will be assessed. Statistical distributions for specific measures, e.g. the number of connections or the angles at which the isoforms connect, will provide the basis for a statistical analysis of individual components and the way that these evolve over time.

Furthermore, to analyse the dynamic changes of the overall structure and for individual isoforms, a principal component analysis comparable with the statistical shape analysis methods frequently used for the segmentation of biological structures (Heimann and Meinzer 2009) will be employed to identify the most relevant variations within the time-lapse image data set. This information will then be used again in a statistical shape model representing FtsZ-isoform template structures.

The goal here is to use advanced imaging techniques and digital models qualitatively to identify the timing, reasons and manner of plastoskeleton changes. Thus, reverse biomimetics need to be employed to derive testable biological hypotheses concerning the plastoskeleton.

13.3.3 From Models to Principles

The resulting models of FtsZ networks enable us to investigate the robustness and underlying principles of FtsZ isoforms in cooperation with another SFB-TRR141 project (see Chap. 17). In order to use the methodologies developed in the project covered in Chap. 17, robust computational models of the overall structure need to be developed. A robust computational model in this context means that the model can be stably solved for a variety of input and material parameters.

13.3.3.1 Testing of Material Parameters

The establishment of realistic mechanical models requires either the approximation or the experimental determination of material parameters of the experimental system, i.e. the moss chloroplast.

Application of mechanical stress to an isolated chloroplast (Lang et al. 2011) with a micro-scale cantilever coupled to an Atomic Force Microscope yields force-deformation data describing the overall mechanical behaviour of the chloroplast. By replicating the experimental set-up with the help of computational models, one can estimate homogenised material parameters for the overall mechanical models of the chloroplast.

13.3.3.2 Mechanical Models for *In Silico* Testing

Complex biological systems are often more flexible and thus more resistant than simple optimised models suggest. The FtsZ models can be tested *in silico* as to whether they can adapt to and thus withstand mechanical stress and to what extent. For this purpose, the three-dimensional structural models are enhanced to three-dimensional continuum mechanical models. Based on these findings, a Finite Element modelling approach based on chain statistics as developed by Böl and Reese (2006) can be adopted. This modelling approach provides a methodology that takes into account, at the macro-level (the entire chloroplast), the statistical data of the various FtsZ isoforms at the micro-level (single FtsZ isoforms). The structural model obtained via image-processing-identified isoform structures will be utilised to construct the Finite Element mesh. The isoforms themselves will form hereby the element edges. This choice of mesh has the advantage that isoform-specific chain-associated parameters can be included as a reinforcement within the overall structure. Depending on the structural arrangements of the various FtsZ isoforms, the meshing or the modelling needs to be adapted, e.g. by considering volume fractions, as the different isoforms can be present within the same element.

The continuum-mechanical model will be implemented within the open-source software library OpenCMISS (www.opencmiss.org), which has specifically been designed as a mathematical modelling environment enabling the application of Finite Element analysis techniques to a variety of complex bioengineering problems. The research group of Oliver Röhrle is a key contributor to OpenCMISS. The advantage of OpenCMISS over other Finite Element simulation software is its markup-language-based data structures, such as CellML (www.cellml.org) or FieldML. In particular CellML, which is used to describe models in a system-biological and standardised way, will provide the potential for easy integration into the generic evolutionary tools developed in another SFB-TRR141 project (see

Chap. 15). The link between the use of a standardised description of processes at the subcellular level and the tools generated as part of the project covered in Chap. 15 needs to be further explored.

13.3.3.3 Abstracting Molecular Construction Concepts for Complex Structures

Following the extraction of the underlying principles from FtsZ network models, they can be investigated in simulations of structures in cooperation with another SFB-TRR141 project (see Chap. 17). Thus, the biological concept generator might provide the source for alternative construction options. However, the step from molecular structures to man-made structures is substantial. An intermediate step for analysing this potential of knowledge transfer will be the definition of general rules to generate generic but representative isoform network models. The rules are developed based on statistical analysis and shape analysis techniques (Sects. 13.3.1 and 13.3.2). In the same way as digital anatomical atlases are constructed, the principal component shape analysis of the overall chloroplast structure will reveal the geometrical structures (modes) that are the key contributors to their overall structure. To verify that the shape-analysis-identified modes are the key contributors to the mechanical stability of the overall chloroplast, the single modes will be mechanically tested by *in silico* experiments with the Finite Element modelling approach (Sect. 13.3.3.2) and will be compared with results from *in vivo* tests (Sect. 13.3.3.4).

A comparison of the results of single modes with the overall mechanical behaviour allows the separation of the essential modes from the redundant modes in an abstract way. Furthermore, whether the essential modes might directly translate to simplistic man-made structures in civil engineering and architecture can be analysed. If key structural modes can be identified, the remaining modes might be less essential.

13.3.3.4 Reverse Biomimetics: What Can Be Learned from *In Silico* Models

In silico predictions for FtsZ networks will be translated back into experiments in the moss system. The computational models of the distinct FtsZ isoforms allow predictions to be made regarding the behaviour of chloroplasts in the available loss-of-function mutant collection. The combination of the outputs enables us to design experiments that will provide insights into the biological meaning of FtsZ isoform functional diversification. Furthermore, specific perturbations of the plastoskeleton are feasible by synthetic biology approaches, e.g. by the fusion of a light-induced protease to an FtsZ isoform in the moss genome by homologous recombination. Thus, the spatial-temporal control of plastoskeleton perturbation can be achieved and used to verify digital-model-derived biological hypotheses.

Furthermore, results from transient transfection of protoplasts and stable moss lines expressing tagged FtsZ isoforms in specific mutant backgrounds will be used to integrate the functional diversification of FtsZ isoforms into the models. Changes in network architecture caused by the absence of a specific FtsZ isoform will give insights into the interdependence of the five isoforms. Differences between the distinct FtsZ isoforms, especially of the FtsZ1 and FtsZ2 subfamilies, are expected (TerBush et al. 2013), as FtsZ2 subfamily proteins are suggested to provide stability, whereas FtsZ1 subfamily proteins might enhance dynamics.

Moreover, the co-assembly of FtsZ isoforms into networks will be investigated in moss lines with several distinctly tagged FtsZ isoforms.

The key outcome is to obtain mechanical models of the chloroplast that can be analysed by using methodologies developed in another SFB-TRR141 project (see Chap. 17) in order for the computational results to be discussed in a molecular biological, an evolutionary biological and a constructional and design methodological context.

13.4 Outlook

Stability on a cellular scale (i.e. in the nanometer-micrometer range) is influenced by many forces and factors that do not play a role in building structures on the macro-scale. However, the investigation of cellular stability mechanisms by using molecular biology suggests a stability-conferring role of several protein networks. Chloroplasts as cell organelles with a defined lens shape and a reduced set of candidate proteins, namely the FtsZ family, can be used as a simplified cellular system to investigate dynamic stability mechanisms. The combination of engineering-based algorithms with fluorescence imaging will be key to unravelling the underlying principles.

In the future, this project will aim at the translation of any new insights gained from this methodological basic research project at the nano-scale to the building of structures. Hence, the extension of this research will see a shift from a purely methodologically driven approach to a more translational approach.

By analysing methodological principles of scalability, one can establish links between key structural features at the nano-scale and technical translations. Thus, the model principles define the technical limitations and hence provide the scope and potential for built structures. This can be achieved by integrating the findings of this project together with the results of some other projects in the SFB-TRR141 (see Chaps. 15 and 17) by investigating the boundaries of identified principles in relation to the scaling of the experimental system (nano-scale vs. micro- and macro-scale) and three-dimensional printing techniques in order to generate conceptual studies for demonstrators based on the computational models.

In addition, reverse biomimetics experiments can be taken to the next level by the overexpression and purification of FtsZ protein isoforms and *in vitro* studies on physical parameters, network assembly and self-organisation.

Acknowledgements This work has been funded by the German Research Foundation (DFG) as part of the Transregional Collaborative Research Centre (SFB/Transregio) 141 "Biological Design and Integrative Structures", project A09.

References

Böl M, Reese S (2006) Finite element modelling of rubber-like polymers based on chain statistics. Int J Solids Struct 43:2–26. doi:10.1016/j.ijsolstr.2005.06.086

Gremillon L, Kiessling J, Hause B et al (2007) Filamentous temperature-sensitive Z (FtsZ) isoforms specifically interact in the chloroplasts and in the cytosol of *Physcomitrella patens*. New Phytol 176:299–310. doi:10.1111/j.1469-8137.2007.02169.x

Heimann T, Meinzer H (2009) Statistical shape models for 3D medical image segmentation: a review. Med Image Anal 13:543–563. doi:10.1016/j.media.2009.05.004

Horst N, Katz A, Pereman I et al (2016) A single homeobox gene triggers phase transition, embryogenesis and asexual reproduction. Nat Plants 2:15209. doi:10.1038/nplants.2015.209

Ingber D (2003a) Tensegrity I. Cell structure and hierarchical systems biology. J Cell Sci 116:1157–1173. doi:10.1242/jcs.00359

Ingber D (2003b) Tensegrity II. How structural networks influence cellular information processing networks. J Cell Sci 116:1397–1408. doi:10.1242/jcs.00360

Kiessling J, Kruse S, Rensing S et al (2000) Visualization of a cytoskeleton-like Ftsz network in chloroplasts. J Cell Biol 151:945–950. doi:10.1083/jcb.151.4.945

Kiessling J, Martin A, Gremillon L et al (2004) Dual targeting of plastid division protein FtsZ to chloroplasts and the cytoplasm. EMBO Rep 5:889–894. doi:10.1038/sj.embor.7400238

Lang E, Mueller S, Hoernstein S et al (2011) Simultaneous isolation of pure and intact chloroplasts and mitochondria from moss as the basis for sub-cellular proteomics. Plant Cell Rep 30:205–215. doi:10.1007/s00299-010-0935-4

Lu C, Reedy M, Erickson H (2000) Straight and curved conformations of FtsZ are regulated by GTP hydrolysis. J Bacteriol 182:164–170. doi:10.1128/jb.182.1.164-170.2000

Martin A, Lang D, Hanke S et al (2009a) Targeted gene knockouts reveal overlapping functions of the five *Physcomitrella patens* FtsZ isoforms in chloroplast division, chloroplast shaping, cell patterning, plant development, and gravity sensing. Mol Plant 2:1359–1372. doi:10.1093/mp/ssp076

Martin A, Lang D, Heckmann J et al (2009b) A uniquely high number of fts Z genes in the moss *Physcomitrella patens*. Plant Biol 11:744–750. doi:10.1111/j.1438-8677.2008.00174.x

Mueller S, Lang D, Hoernstein S et al (2014) Quantitative analysis of the mitochondrial and plastid proteomes of the moss *Physcomitrella patens* reveals protein macrocompartmentation and microcompartmentation. Plant Physiol 164:2081–2095. doi:10.1104/pp.114.235754

Muñoz J, Conte V, Miodownik M (2010) Stress-dependent morphogenesis: continuum mechanics and truss systems. Biomech Model Mechanobiol 9:451–467. doi:10.1007/s10237-009-0187-9

Or-Tzadikario S, Gefen A (2011) Confocal-based cell-specific finite element modeling extended to study variable cell shapes and intracellular structures: the example of the adipocyte. J Biomech 44:567–573. doi:10.1016/j.jbiomech.2010.09.012

Osteryoung K, Pyke K (2014) Division and dynamic morphology of plastids. Annu Rev Plant Biol 65:443–472. doi:10.1146/annurev-arplant-050213-035748

Rensing S, Kiessling J, Reski R, Decker E (2004) Diversification of *fts*Z during early land plant evolution. J Mol Evol 58:154–162. doi:10.1007/s00239-003-2535-1

Reski R (1998a) Development, genetics and molecular biology of mosses. Bot Acta 111:1–15. doi:10.1111/j.1438-8677.1998.tb00670.x

Reski R (1998b) *Physcomitrella* and *Arabidopsis*: the David and Goliath of reverse genetics. Trends Plant Sci 3:209–210. doi:10.1016/s1360-1385(98)01257-6

Reski R (2002) Rings and networks: the amazing complexity of FtsZ in chloroplasts. Trends Plant Sci 7:103–105. doi:10.1016/s1360-1385(02)02232-x

Reski R (2009) Challenges to our current view on chloroplasts. Biol Chem 390:731–738. doi:10.1515/bc.2009.089

Röhrle O, Davidson J, Pullan A (2012) A physiologically based, multi-scale model of skeletal muscle structure and function. Front Physiol 3:358. doi:10.3389/fphys.2012.00358

Röhrle O, Köstler H, Loch M (2011) Segmentation of skeletal muscle fibres for applications in computational skeletal muscle mechanics. In: Wittek A et al (eds) Computational biomechanics for medicine: soft tissues and the musculoskeletal system, pp 107–117. doi: 10.1007/978-1-4419-9619-0_12

Schuessele C, Hoernstein S, Mueller S et al (2016) Spatio-temporal patterning of arginyl-tRNA protein transferase (ATE) contributes to gametophytic development in a moss. New Phytol 209:1014–1027. doi:10.1111/nph.13656

Shaner N, Steinbach P, Tsien R (2005) A guide to choosing fluorescent proteins. Nat Methods 2:905–909. doi:10.1038/nmeth819

Strepp R, Scholz S, Kruse S et al (1998) Plant nuclear gene knockout reveals a role in plastid division for the homolog of the bacterial cell division protein FtsZ, an ancestral tubulin. Proc Natl Acad Sci U S A 95:4368–4373. doi:10.1073/pnas.95.8.4368

Sultan C, Stamenović D, Ingber D (2004) A computational tensegrity model predicts dynamic rheological behaviors in living cells. Ann Biomed Eng 32:520–530. doi:10.1023/b:abme.0000019171.26711.37

Suppanz I, Sarnighausen E, Reski R (2007) An integrated physiological and genetic approach to the dynamics of FtsZ targeting and organisation in a moss, *Physcomitrella patens*. Protoplasma 232:1–9. doi:10.1007/s00709-007-0284-5

TerBush A, Yoshida Y, Osteryoung K (2013) FtsZ in chloroplast division: structure, function and evolution. Curr Opin Cell Biol 25:461–470. doi:10.1016/j.ceb.2013.04.006

Usami H, Maeda T, Fujii Y et al (2012) CHUP1 mediates actin-based light-induced chloroplast avoidance movement in the moss *Physcomitrella patens*. Planta 236:1889–1897. doi:10.1007/s00425-012-1735-6

Weise A, Rodriguez-Franco M, Timm B et al (2006) Use of *Physcomitrella patens* actin 5′ regions for high transgene expression: importance of 5′ introns. Appl Microbiol Biotechnol 70:337–345. doi:10.1007/s00253-005-0087-6

Zimmer A, Lang D, Buchta K et al (2013) Reannotation and extended community resources for the genome of the non-seed plant *Physcomitrella patens* provide insights into the evolution of plant gene structures and functions. BMC Genomics 14:498. doi:10.1186/1471-2164-14-498

Chapter 14
Developing the Experimental Basis for an Evaluation of Scaling Properties of Brittle and 'Quasi-Brittle' Biological Materials

Stefanie Schmier, Christoph Lauer, Immanuel Schäfer, Katharina Klang, Georg Bauer, Marc Thielen, Kathrin Termin, Christoph Berthold, Siegfried Schmauder, Thomas Speck, and Klaus G. Nickel

Abstract The development of lightweight structures exhibiting a high energy dissipation capacity and a locally adapted puncture resistance is of increasing interest in building construction. As discussed in Chap. 7, inspiration can be found in biology, as numerous examples exist that have evolved one or even several of these properties. Major challenges in this interdisciplinary approach, i.e. the transfer of biological principles to building constructional elements, are scaling (different dimensions) and (at least for the botanic examples) the fact that different material classes constitute the structural basis for the functions of interest. Therefore, a mathematical description of the mechanical properties and the scalability is required that is applicable for both biological and technical materials. A basic requisite for the establishment of mathematical descriptions are well-defined test setups rendering a reliable data basis. In the following, two biological role models from the animal and plant kingdoms are presented, namely, sea urchin spines and coconut endocarp, and two experimental setups for quasi-static and dynamic testing of biological and bio-inspired technical materials are discussed.

S. Schmier (✉) • G. Bauer • T. Speck
Plant Biomechanics Group, Botanic Garden, Faculty of Biology, University of Freiburg, Schänzlestraße 1, D-79104 Freiburg, Germany

Freiburg Centre for Interactive Materials and Bioinspired Technologies (FIT), University of Freiburg, Georges-Köhler-Allee 105, D-79110 Freiburg, Germany
e-mail: stefanie.schmier@biologie.uni-freiburg.de

C. Lauer • K. Klang • K. Termin • C. Berthold • K.G. Nickel
Applied Mineralogy (AMIN), University of Tübingen, Wilhelmstraße 56, 72074 Tübingen, Germany

I. Schäfer • S. Schmauder
Institute for Materials Testing, Materials Science and Strength of Materials (IMWF), University of Stuttgart, Pfaffenwaldring 32, 70569 Stuttgart, Germany

M. Thielen
Plant Biomechanics Group, Botanic Garden, Faculty of Biology, University of Freiburg, Schänzlestraße 1, D-79104 Freiburg, Germany

J. Knippers et al. (eds.), *Biomimetic Research for Architecture and Building Construction*, Biologically-Inspired Systems 9, DOI 10.1007/978-3-319-46374-2_14

14.1 Introduction

Intuitively, problems are bound to arise in the transfer of the principles and characteristics of natural materials, which come in millimetre or centimetre sizes, to architectural objects, which may span many metres. However, the analysis of this scaling problem is far from being trivial and is of particular importance in the realm of brittle materials, to which at least the matrix of common building materials such as cement and concrete belongs. A theoretical and experimental evaluation requires proper data sets. In the following, we describe our efforts to develop the methods by which to obtain such data.

In 1921, Griffith reported that the experimentally determined strength of homogeneous, isotropic and brittle solids is several times lower than can be expected from calculations based on their intrinsic properties such as elastic moduli and surface energies (Griffith 1921). Weibull characterised this phenomenon statistically, stating that the probability of failure depends on two factors: a given distribution of stresses (e.g. according to the test method) and, in particular, the volume of the tested specimen. Weibull was the first who took into account the volume of the tested specimen as a reason for the large scatter of experimentally obtained ultimate strength data (Weibull 1939; Bažant 2000).

Weibull's theory is based on the weakest link hypothesis, which assumes that a specimen fails at its weakest point (e.g. an internal flaw or imperfection). In a homogeneous, isotropic and brittle material, in which the failure events are independent of each other, the size of the flaws controlling the strength follows a statistical distribution function for a given volume. If the size of flaws in a volume is a matter of probability, then it is tantamount that the stress needed to break the material of a given volume is statistically distributed. The probability of failure P_{i,V_0} for a specimen i within a given volume V_0 is then, in its simplest form, defined by the two-parameter Weibull formula:

$$P_{i,V_0} = 1 - e^{-\left(\frac{\sigma_i}{\sigma_0}\right)^m}, \tag{14.1}$$

where σ_i is the critical stress at failure for specimen i, σ_0 is the characteristic strength and *m* the Weibull modulus. The Weibull modulus *m* is therefore a measure of the reliability of the material: the lower the Weibull modulus, the higher the scatter of critical stresses in a material and the lower the material's reliability. For this reason, technical ceramics are desired to have high Weibull moduli ($m = 10 - 20$) (Danzer 2014). The characteristic strength is the critical stress at which 63.2 % of the tested specimens with volume V_0 fail. Unlike the Weibull modulus *m*, the characteristic strength σ_0 is not a material constant as it depends on the volume of the investigated specimen and, therefore, the corresponding material constant is $\sigma_0 {V_0}^{1/m}$. The Weibull modulus and characteristic strength can be determined experimentally by using the empirical distribution function *F* for P_{V0}. By fitting a regression line to $ln\ (ln\ (1/(1 - F_i)))$ versus $ln\ (\sigma_i)$, the Weibull

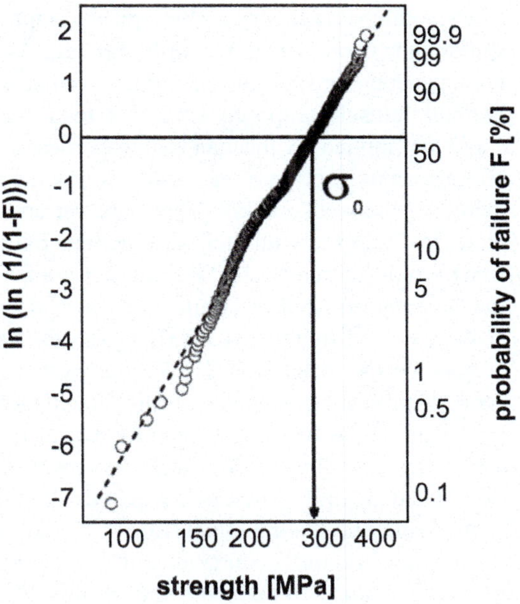

Fig. 14.1 Determination of characteristic strength σ_0 and Weibull modulus m of a set of 600 tested varistor ceramics ($\sigma_0 = 284$ MPa, $m = 5.8$). The regression line represents the related Weibull distribution. The distribution fits the data very well and the material is said to show 'Weibull behaviour' (Adapted from Danzer 2014, p. 3437, Copyright (2016), with permission from Elsevier)

modulus m can be identified as the slope and $\sigma_0 = exp\,(-c/m)$ with c being the intercept of the regression line (Fig. 14.1). Materials to which this method can be applied are said to show 'Weibull behaviour' (Weibull 1939, 1951; Danzer et al. 2007; Danzer 2014).

Weibull defined the characteristic strength as depending on the respective volume of the specimen. This is because it is more likely to find a critical flaw in a large specimen than in a small one. Thus, a large specimen will statistically fail at lower stresses (i.e. at lower σ_0) than small ones. This effect is called the 'size effect' and is expressed for specimens of the same material having the same probability of failure by:

$$\left(\frac{\sigma_1}{\sigma_2}\right) = \left(\frac{V_2}{V_1}\right)^{\frac{1}{m}} \tag{14.2}$$

where σ_1 is the strength at failure of the object with volume V_1, σ_2 is the strength at failure of the object with volume V_2 ($F_{1,V_1} = F_{2,V_2}$), and m is the Weibull modulus of the material (Danzer et al. 2007).

For brittle components (e.g. ceramics), the procedure of defining the Weibull modulus as a material constant is well established (e.g. Lawn 1993). Since samples at their final volume often cannot be tested because of costs and testing procedures, Eq. (14.2) can be applied for scaling up to the size used under realistic operational conditions of the respective material. This automatically leads to a strong decrease in characteristic strength.

Extending this theory to biological samples and role models (e.g. sea urchin spines and coconut shells) is fairly difficult. As biological samples rarely constitute homogeneous or isotropic materials, verification of Weibull's theory is not easy. Weibull himself proposed empirical tests for whether his distribution function could be applied to mechanical properties of other materials than the standard homogeneous, isotropic and brittle solids, as demanded as preconditions for the theory of the weakest link. Therefore, an application of the Weibull approach to biological samples with their specific material properties seems to be justified. For some biological materials, 'Weibull behaviour' has previously been proposed, e.g. for the determination of failure strength of sea urchin spines (Presser et al. 2009), abalone shells (Menig et al. 2000), pea seeds (Łysiak 2007), teeth (Chan et al. 2009) or bone (Wagermaier et al. 2015). Many other distributions in nature have recently been identified as being Weibull-distributed (Krumbholz et al. 2014).

For the bones of various animals, Taylor (2000) could prove a fatigue strength behaviour according to Eq. (14.1). However, interestingly, bones from larger animals failed at similar stresses compared with bones of smaller animals. This seems to contradict the scaling effect (Eq. 14.2) and Taylor assumed that an improvement in the bone structure counteracted it. Detecting, analysing and understanding such structural features influencing the scaling effect would be of great interest for the development of building construction involving scaling problems and would also help us to comprehend the specific failure behaviour of the biological role models used as concept generators for biomimetic materials in greater detail (reverse biomimetics).

14.2 Sea Urchin Spines – Biological Role Model from the Animal Kingdom

A more detailed description of the spines of *Heterocentrotus mammilatus* (HM) and *Phyllacanthus imperialis* (PI) is given in Chap. 7 "Plants and animals as source of inspiration for energy dissipation in load bearing systems and facades" and elsewhere (Grossmann and Nebelsick 2013; Presser et al. 2009). However, the most relevant properties for scaling purposes will be briefly described here: Sea urchin spines

- are lightweight constructions because of their highly porous microstructure.
- have a hierarchical structure (Seto et al. 2012).
- are composed of a highly sophisticated composite material comprising calcite (>99 wt.%) and organic macromolecules (<1 wt.%, e.g. Seto et al. 2012; Su et al. 2000) that overcome the brittleness of calcite and make it a good building material for undertaking the tasks of such a spine (Moureaux et al. 2010; Presser et al. 2009).

- of PI and especially of HM show a 'graceful failure' behaviour (Presser et al. 2009), which is a quasi-ductile partial failure under uniaxial compression without losing the capacity to withstand the load.
- have a remarkable capability to absorb energy (cp. 14-2c).

All these characteristics make the sea urchin spine an object of great scientific interest and of even more interest for biomimetics. In order to scale these properties, the exact figures of merit of the sea urchin spines need to be measured.

Presser et al. (2009) determined the Weibull modulus *m* by uniaxial compression tests for HM and PI. Su et al. (2000) determined *m* with 4-point bending tests for *Heterocentrotus trigonarius* (HT), which belongs to the same genus as HM.

PI exhibits the highest Weibull modulus ($m = 6.0$), whereas HM and HT show a significantly lower reliability with $m = 3.4$ and $m = 3.2$, respectively. Presser et al. (2009) attributed the higher reliability of PI to its regular structure (Fig. 14.2a), compared with the fairly heterogeneous structure of HM. PI is characterised by dense dome-shaped growth layers, which induce crack deflection (Presser et al. 2009). Although the strength in PI is more reliable, considerable variations exist in the spine structure depending on spine size (Fig. 14.2a). The highly porous medulla has, in all three spines of Fig. 14.2a, a comparable size (as indicated by the black bar

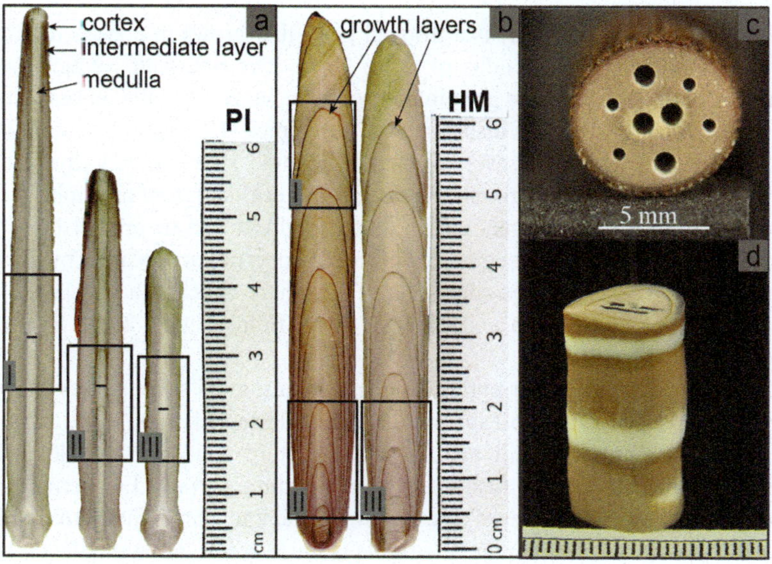

Fig. 14.2 Photographs of longitudinal spine sections of (**a**) PI (*Phyllacanthus imperialis*) and (**b**) HM (*Heterocentrotus mammilatus*). *Black boxes* represent sample dimensions for uniaxial compression tests. (**c**) Detailed view of a spine segment of PI after 8 longitudinal indentation tests (5–10 mm deep). Although the spine is locally damaged, the whole spine is still intact highlighting its capacity to absorb energy. (**d**) Detailed view of a spine segment of HM typically used in the indentation tests

in the boxes I-III, length = 1.5 mm), so that the highest proportion of very porous and low-strength material occurs in the smallest spine, presumably resulting in lower strength.

In addition to the above-mentioned heterogeneities in the structure leading to a lower reliability, the determination of *m* itself is challenging with regard to all these biological samples. A good estimate of the critical stress (σ_c, MPa) can only be found with precise geometric information of the spine. The spines of HM and PI are not perfectly cylindrical (Fig. 14.2c, d), as is assumed for the quantitative assessment of uniaxial compression experiments, and the spines of HT do not have two perfectly parallel surfaces as assumed by Su et al. (2000) for the 4-point bending tests.

Hence, the geometry of the spines and the heterogeneity of their structure challenge the determination of the Weibull modulus with uniaxial compression or 4-point bending tests. A method is thus needed that does not depend on the geometry of the spine and that has a high spatial resolution for an easier determination of the Weibull modulus and for easier scaling.

14.3 Coconut Endocarp – Biological Role Model from the Plant Kingdom

The coconut (*Cocos nucifera*) is a tropical palm that grows in coastal regions and riverbanks and can reach heights of up to 30 m. The ripe fruits, which have the ability to float, remain able to germinate even after several months in seawater and, therefore, the palm is widely distributed in the tropics. The fruits reaching weights of 3.7 kg have to withstand severe impacts of up to 1 kJ when they drop from the crown of the palm onto the ground. To assure the germination of the embryo inside, they have to absorb the impact energy and remain closed to protect the embryo from fungal and bacterial infection. All of these properties are ensured by the triple-layered fruit wall of the drupe consisting of a leathery exocarp, a fibrous mesocarp and a tough endocarp (Chan and Elevitch 2006; Franke et al. 2012; Bauer et al. 2015) (Fig. 14.3).

The tough endocarp is currently under investigation. Its shape varies from elongated and egg-shaped to almost spherical (Chan and Elevitch 2006). The tricarpelary structure of the ovary is still visible in the mature fruit, as three longitudinal ridges exist representing the fusion marks of the three carpels. Furthermore, three micropyles or 'germinating eyes' are present in the endocarp. These are regions in which the endocarp is thinner and the tissue is much softer, allowing the embryo to 'escape' the endocarp during sprouting (Winton 1901).

The endocarp consists predominantly of red to brownish-coloured, isodiametric or strongly elongated stone cells (Fig. 14.4a), comprising several layers of secondary cell walls (Fig. 14.4d, e). Dark channels can be observed that serve as a connection between the still young and living cells and can be interpreted as lignified plasmodesmata (Fig. 14.4e). The elongated cells are mostly oriented with their long axis parallel to the outer surface of the endocarp (Fig. 14.4b, c). The vascular

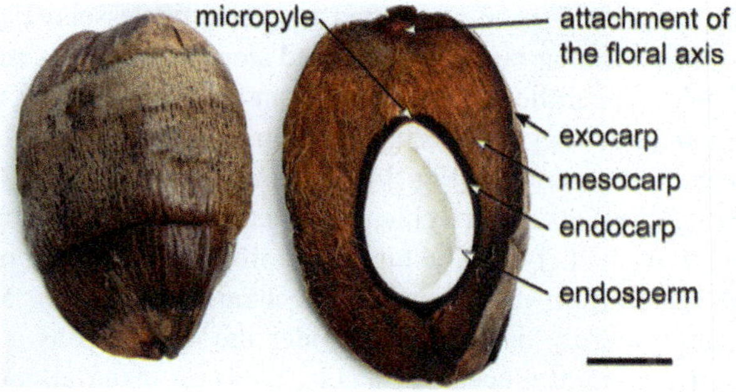

Fig. 14.3 Cut open ripe fruit of the coconut palm (*Cocos nucifera*). *Left.* Exterior view. *Right.* Interior view: The pericarp consisting of exocarp, mesocarp and endocarp lies over the endosperm, in which the embryo is embedded. The embryo is located under one of the three micropyles at the side at which the fruit is attached to the palm via the floral axis. Not visible is the testa or seed wall surrounding the endosperm. Scale bar: 5 cm

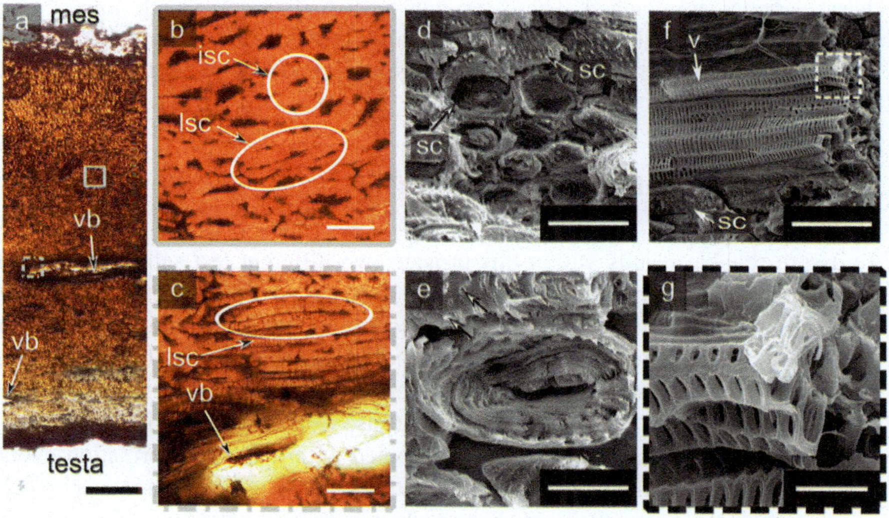

Fig. 14.4 (**a–c**) Light micrographs of a polished cross section and (**d–g**) SEM micrographs of a fracture surface through the coconut endocarp. (**a**) The structure mainly consists of densely packed, thick walled stone cells and embedded vascular bundles (vb). The diameter of the stone cells becomes gradually smaller in the direction from the outer side of the endocarp facing the mesocarp (mes) towards the inner side adjacent to the testa. (**b**) Detailed view of stone cells, which can be either (nearly) isodiametric (isc) or elongated (lsc). (**c**) Detailed view of a vascular bundle surrounded by longitudinally elongated stone cells. (**d**) Image of the fracture surface showing individual stone cells (sc). Whereas some cells are still intact, others are fractured, revealing the layered inner structure of the cell walls. (**e**) Detailed view of a fractured stone cell showing the multi-layered cell wall. The roundish structures (*arrows*) within the cell walls are plasmodesmata. (**f**) Pulled out vascular bundle with several rigid tracheids/vessels (v) surrounded by densely packed aligned stone cells (sc). (**g**) Detailed view of some tracheids/vessels from (**c**). The rigidity of the scalariform tracheids/vessels is caused by the ladder-like design and lignin deposition in the cell walls. Scale bars: *a*: 500 μm, *b*, *c*, *d*, *f*: 50 μm, *e*: 20 μm, *g*: 9 μm

bundles mainly consist of spiral and pitted tracheids or vessels (Fig. 14.4f, g). In close vicinity to the vascular bundles, elongated stone cells are found arranged with their longitudinal axis parallel to the bundles, an arrangement that we interpret as ensuring a smooth structural and mechanical transition between the stone cells and vascular bundles in the endocarp (Fig. 14.4c, f) (Winton 1901).

The endocarp represents a hierarchically structured material comprising various types of densely packed, typically highly lignified cells in an inhomogeneous agglomeration, resulting in its anisotropic mechanical behaviour. Since botanical materials are rarely purely brittle, we consider the endocarp as a "quasi-brittle" material. This term was introduced by Bažant (2004) for materials such as concrete, bone and biological shells. In the fracture process zone of quasi-brittle materials, microcracks occur that lead to a softening of the material (in brittle materials this zone is negligible). The coconut endocarp was chosen as the botanical role model for quasi-brittle materials.

Once the mechanical tests are well established, comparative studies with coconut dwarf varieties are planned. These varieties have smaller fruits with thinner fruit walls, features that will enable us to test smaller sample sizes of the endocarp and therewith to contribute to the question of the size dependency of fracture behaviour.

14.4 Mechanical Tests

To assure reliable results within this project, two different test setups have been developed. Since the behaviour of materials differs in static and dynamic loading situations, both situations have been taken into account. For the quasi-static loading situation (penetration tests), a constant feed motion has been applied to the samples, whereas in the dynamic situation (impact tests), the sudden impact is caused by an impact pendulum. With the mechanical tests, we aim to identify the Weibull behaviour of the biological materials. Combined with detailed information about the inner structure of the materials, we hope to identify structural parameters (such as cell arrangement or porosity gradients) which have an impact on $\sigma_0 V_0^{1/m}$.

14.4.1 Penetration Tests on Sea Urchin Spines

The method of choice is known to metallurgists as the 'impression test' (Yang and Li 2013). This test is used, for example, to characterise highly porous ductile metallic foams (e.g. Olurin et al. 2000; Kumar et al. 2003). A cylindrical indenter penetrates with a constant velocity several millimetres into the sample. Various geometries of the indenter tip have been reported in the literature, such as a flat-ended cylindrical indenter tip (FEI) or a spherical-ended indenter tip (SEI) (Olurin et al. 2000; Kumar et al. 2003) (Fig. 14.5a). Based on the first measurements by Presser et al. (2009), this method has been transferred to brittle porous materials and is called the 'penetration test' to avoid any possibility of confusion with the 'impression test'.

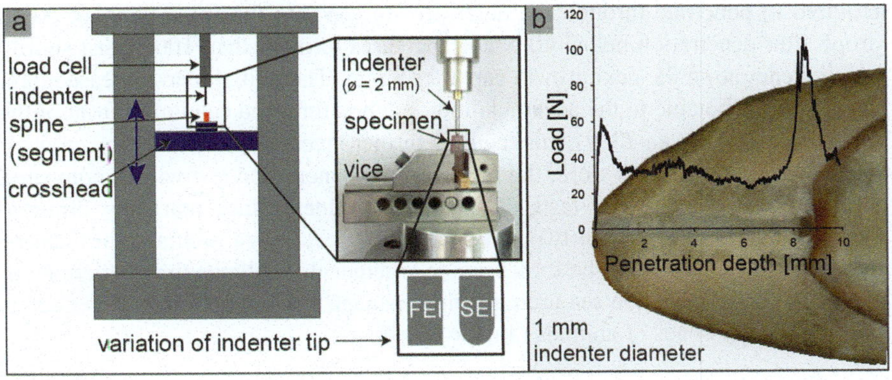

Fig. 14.5 (**a**) Representation of the universal testing machine Instron 4502. The first magnification is a photograph of a fixed spine on the crosshead and the second magnification sketches the shape of a flat-ended and a spherical-ended indenter tip, FEI and SEI, respectively. (**b**) Correlation of the load penetration curve of HM with its internal structure. The spine was cut longitudinally afterwards. The force is highest when penetrating through the dense growth layers

Hereby, we eliminate some difficulties of the uniaxial compression tests (Chap. 7) and provide:

- Simple and low cost sample preparation: a highly precise geometric sample is not required.
- High spatial resolution, which is of special interest for the growth layers of HM and characterisation of the individual intrinsic material properties of the substructures medulla, intermediate layer and cortex.
- The specimens are not destroyed entirely after measurements (Fig. 14.2c, d), which is advantageous for further investigations.
- Multiple measurements can be made on a single specimen (Fig. 14.2c).

Nevertheless, the overall principles of indentation are still similar to uniaxial compression tests (Kumar et al. 2003; Yu et al. 1985; Olurin et al. 2000; Andrews et al. 2001).

For our measurements, cylindrical tungsten carbide indenters (d = 0.6, 1, 2 mm) were used with flat-ended indenter tips and spherical-ended indenter tips (Fig. 14.5a). Sea urchin spines were cut into segments (Fig. 14.2c, d) and indented from the top of the spine (= crystallographic c-axis). The indentations were performed with an Instron 4502 universal testing machine (Instron Deutschland GmbH, Pfungstadt, Germany) with a constant cross head speed of 0.5 mm/min. The indentation depth varied from 5 to 20 mm and was mostly chosen to be 10 mm.

14.4.1.1 Spatial Resolution of the Penetration Test

Figure 14.5b highlights the spatial resolution of this method by showing the load/penetration depth curve of an SEI (ø = 1 mm) in a HM spine section compared with its interior structure (the spine was cut longitudinally afterwards). The force required to penetrate through the outer growth cap into the spine is about 60 N, drops after penetration below 40 N and increases to more than 100 N just shortly before penetrating the next growth cap after 8 mm. This initial load increase might in part be attributable to the accumulation of debris (crushed calcite) in front of the indenter (Presser et al. 2009); this requires further investigation.

In a later stage of this work, the results of the penetration test will be compared with uniaxial compression tests of sea urchin spines. Since markedly different volumes will be used in the two tests, the size effect in sea urchin spines can be evaluated. Of interest will be to determine whether this follows the predictions of Eq. (14.2) or whether the sea urchin spines manage to overcome the size effect as observed in the bones of animals (Taylor 2000).

14.4.1.2 Influence of Indenter Geometry

Figure 14.6a shows the influence of the shape of the indenter tip on the measurements. Representative measurements for FEI (d = 0.6 mm, solid grey line) and SEI (d = 0.6 mm, black dotted line) are given. These measurements were conducted in the intermediate layer of the same spine segment of PI (Fig. 14.2c). For FEI, the load increases sharply right after the touchdown of the indenter on the spine, whereas for SEI, the load increases gradually. This is attributable to the shape of the SEI tip, whereby the cross sectional area increases in the first 0.3 mm of penetration (Fig. 14.6a on the top left: shape of SEI tip). The load/penetration depth curves of the FEI and SEI intersect after 0.3 mm of penetration, exactly at the point at which the cross sectional areas of these indenters are identical (indicated by the black line in Fig. 14.6a). Subsequently, similar curves highlight the independence of the indentation response and indenter geometry. These observations agree with Kumar et al. (2003) who have found the same results for the impression of metallic foams. Because calculations with the FEI are simpler and because the deformation zone is less complicated (Mouginot and Maugis 1985), future measurements will be conducted with an FEI.

The effect of the indenter diameter on penetration stress (strength) is demonstrated in Fig. 14.6b. Indenter diameters of 0.6, 1 and 2 mm) were used for the intermediate layer (blue colours) and medulla (red colours) of PI. The shaded background colours highlight the range of stresses obtained in the 31 experiments. For clarity, only six stress/penetration depth curves are included in Fig. 14.6b. The strength of the intermediate layer ranges between 100–200 MPa and is generally higher than the strength of the highly porous medulla with 30–100 MPa, concentrating around 40–80 MPa. The scatter in both structural units of the PI spine is attributable to the heterogeneity of biological materials. Although the statistic is not exhaustive, the 31 measurements suggest that the strength data obtained with various indenter sizes do not vary systematically with indenter sizes.

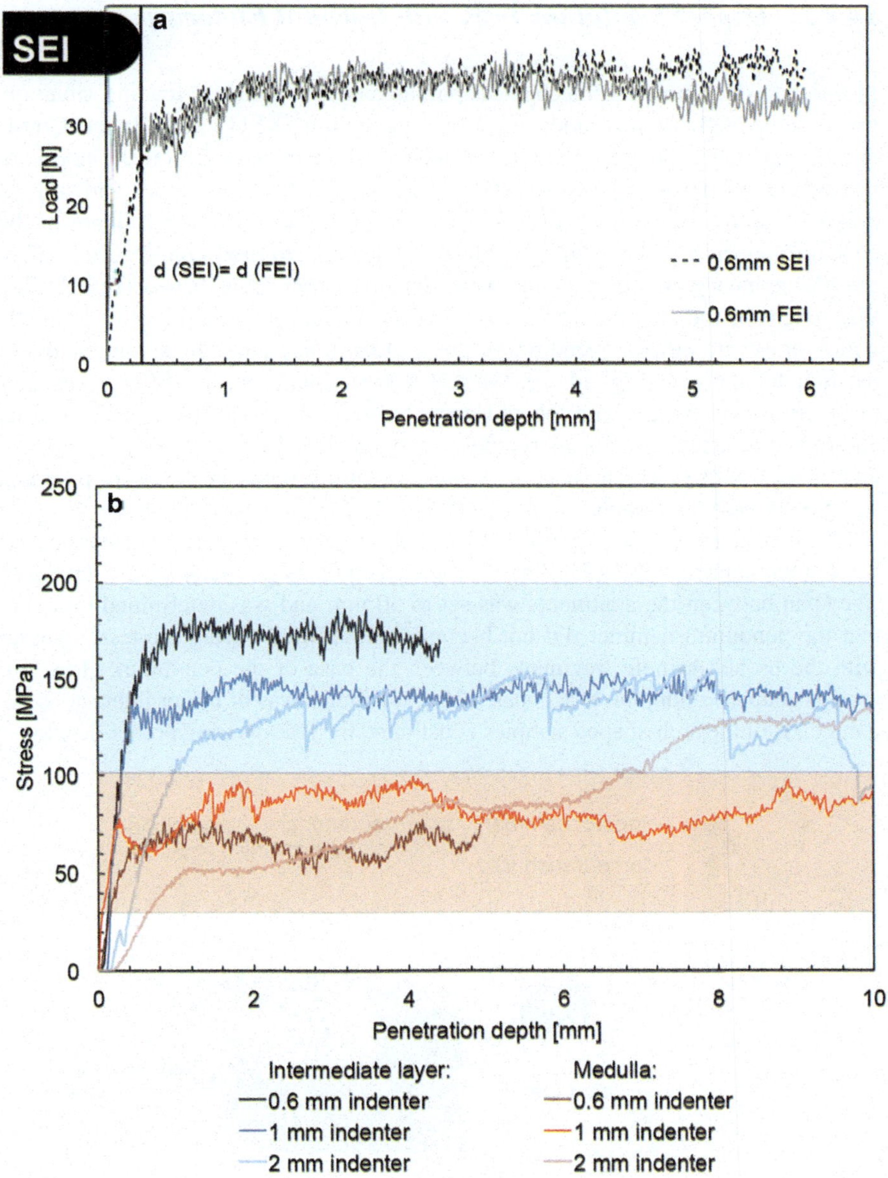

Fig. 14.6 (**a**) Load/penetration depth curve for SEI and FEI of PI. (**b**) Stress/penetration depth curves for various indenter diameters conducted on 31 PI spine segments; for clarity, only six representative measurements are shown

14.4.2 Impact Pendulum Tests with Coconut Endocarp

In order to determine the mechanical parameters in a dynamic loading situation, a test setup with an instrumented impact pendulum (B5113.303, Zwick GmbH & Co. KG, Ulm, Germany) was developed. The setup is based on the norm for instrumented Charpy impact tests (DIN EN ISO 179-2) of unnotched and straight polymer samples. Our setup differs from that norm because the curvature of the coconut endocarp makes it impossible to obtain the required cuboid and straight sample geometry, as our samples were unnotched but arch-shaped (Fig. 14.7a). The height of the samples was approximately 10 mm; the curvature and sample width, however, varied depending on the geometric and anatomical parameters of the individual fruits (Fig. 14.7c). We tested two sample groups differing in their respective orientation within the endocarp: termed equatorial and meridional. The end of the coconut with the micropyles was defined as the north pole, the opposite end as the south pole. The equator was set centrally between the north pole and south pole, with the meridians running through the two poles (Fig. 14.7b).

The described tests were performed with an instrumented pendulum hammer of 7.5 J work capacity (BRA5113.3145, Zwick GmbH & Co. KG, Ulm, Germany). The span between the abutments was set to 50 mm and was determined to assure that the pendulum hammer did not become immobilised during the test, together with the broken sample fragments between the base of the pendulum. However, for this adapted setup, the exact span of the abutments was of minor interest, as the inner ends of the arch-shaped samples constituted the load-bearing points. The inner

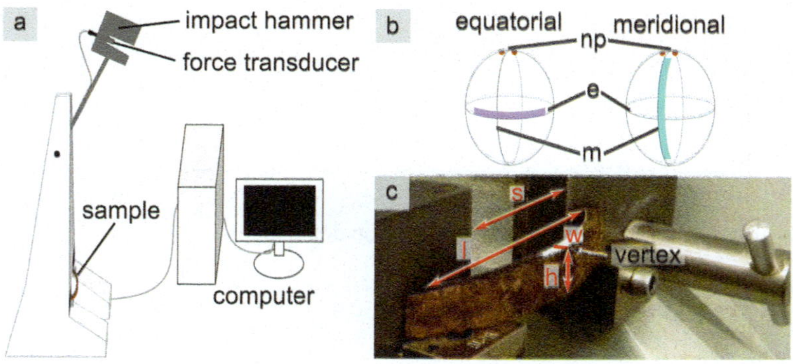

Fig. 14.7 Instrumented impact pendulum tests. (**a**) Sketch of the test setup. (**b**) Orientation of the test specimen as cut from the coconut endocarp. Abbreviations: *np* north pole, *e* equator, *m* meridian. (**c**) Detailed view of the mounted sample with geometric sample parameters. The sample geometry is arch-shaped because of the fruit morphology. The impacting hammer hits the sample at the vertex. Abbreviations: *s* span between abutments, *l* inner length = distance between the inner edges of contact areas of the sample with the abutments, *w* specimen width at sample centre, *h* specimen height at sample centre

sample length is essential for evaluating the comparability of the biological samples, which inevitably differ (slightly) as to their size and shape (Fig. 14.7c). The samples were centred, so that the force transducer hit the sample in its vertex. This was also the point at which the width and height of the samples was measured (Fig. 14.7c).

The tests were performed with 41 equatorial and 58 meridional samples. The equatorial samples had a smaller radius of curvature (median: 43.69 mm) than the meridional samples (median: 63.48 mm), because of the ovoid shape of the endocarp. Moreover, the radius of curvature of the meridional samples had a larger scatter. The radius of curvature showed no significant influence on the impact resistance ($r_s = -0.098$). For all other parameters, no significant differences between the equatorial and meridional samples were found (Table 14.1) and no significant relationship could be observed between the sample dimensions and the impact resistance.

The impact resistance of the equatorial samples (median = 4.46 kJ/m^2) was significantly higher than the impact resistance of meridional samples (median = 2.36 kJ/m^2) (Fig. 14.8). This reveals the anisotropy of the biological material, reflected by structural differences within the endocarp, such as the different course of the vascular bundles or the variable cell orientation and cell distribution. A detailed analysis of the structural variations at various hierarchical levels is of special interest not only for a transfer to biomimetic products, but also for characterising quasi-brittle biological materials and is currently being performed.

Table 14.1 Dimensions of the 99 tested coconut endocarp samples

		Height [mm]		Radius of curvature [mm]		Width [mm]		Inner length [mm]	
Sample orientation	n	Median	IQR	Median	IQR	Median	IQR	Median	IQR
Equatorial	41	10.22	0.76	43.69	3.02	3.64	0.60	72.18	17.18
Meridional	58	10.09	0.20	63.48	20.78	3.82	0.48	75.08	14.05

n number of samples, *IQR* interquartile range

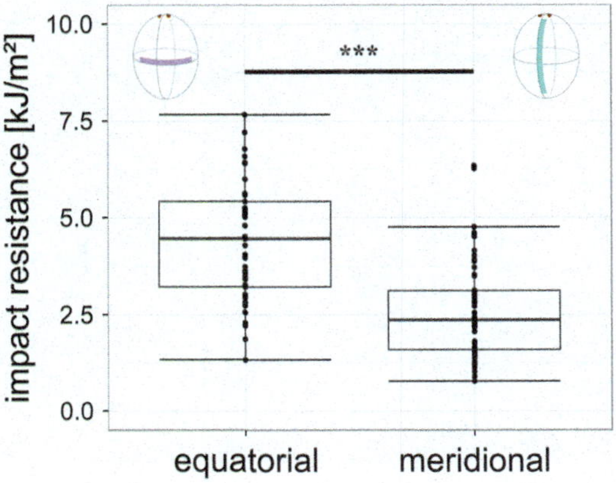

Fig. 14.8 Comparison of the impact resistance (absorbed energy per crossectional area) of equatorial (41 specimens) and meridional samples (58 specimens). The equatorial samples had a significantly higher impact resistance (median = 4.46 kJ/m^2) than the meridional samples (median = 2.36 kJ/m^2) (Mann-Whitney-*U* test; W = 434, $p < 0.001$)

14.5 Simulation: Support for Measurement and Scaling Challenges

The description and analysis of the material properties of biological materials is a great challenge. Natural materials show non-uniform directed properties because they often consist of an aggregation of various cell types (such as stone cells and vascular bundles within the coconut endocarp) or have other microstructures, for example, pores of various sizes (such as in the sea urchin spines), which result in inhomogeneous properties. Finite Element (FE) simulations of the materials and of the experiments might support the measurements and reveal the influence of specimen geometry and microstructure on the results of the experiments.

The failure behaviour during the impact of the coconut endocarp in the revised Charpy impact tests is highly complex, being the result of the inhomogeneous and anisotropic structure of the material and the arch shape of the sample. To generate comparable results and to analyse the influence of the shape of the coconut endocarp (and artefacts connected with it), we have used flat and arch-shaped PVC samples to determine the influence of the geometry of the probe. When the influence is known, a comparison of the properties of the natural arch-shaped samples with samples of other materials tested by Charpy impact tests can be performed. Variations not only of the shape, but also of the material itself can be easily taken into account in the simulation and can help to provide a better understanding of the influence of the geometries.

Simulations can only be as realistic as the input data and boundary conditions. Important input parameters in the case of the PVC samples are the density, elastic modulus and plastic behaviour. All these data can be found in the data sheet provided by the producer, except for the plastic material behaviour. For a first simulation (Fig. 14.9), literature values of the plastic behaviour of common PVC materials are used, but tensile tests will be performed in the near future to obtain the required information directly from the real tested specimens. The explicit linear

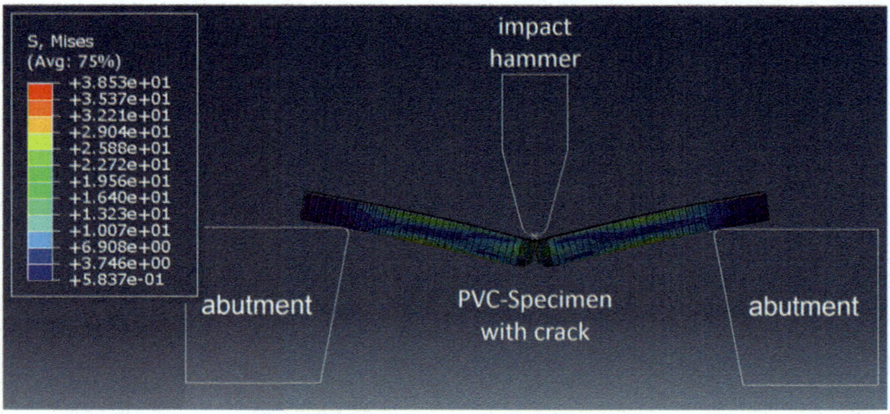

Fig. 14.9 Von Mises stress distribution in the simulation of a Charpy impact test of a flat 7 cm long PVC specimen

elastic/plastic simulation with the FE-software Abaqus includes 5747 elements of type C3D8R, whereas the hammer and the abutment are modelled as 3D analytic rigid shells. The simulation includes so-called "cohesive surfaces" to simulate the crack opening behaviour of the sample. Defining these cohesive surface properties is one of the greatest challenges in the modelling of material behaviour. Only with inverse methods will it be possible to find suitable material parameters requiring additional simulations that have to be compared with the experiments. Initial results show good agreement with the experimental data.

The simulation with flat PVC specimen is the first step; the second step is the analysis of arch-shaped PVC samples differing only in their shape from the flat ones in the Charpy impact tests. Thus, reliable information about shape is necessary (the shape dependency is discussed in more detail at the end of this paragraph).

The third step includes the natural material, which shows not only an arch-shaped form, but also inhomogeneous material properties (compare Sect. 14.3). Therefore, structural information, such as cell forms and dimensions, plus the number and dimensions of embedded vascular bundles need to be included in the simulation. Moreover, the cohesive surface properties will be of special interest here and need to be determined by inverse methods. The results of each experiment will differ in detail, because the tests are performed on a natural material with structural and mechanical variations that are, for example, attributable to differentially distributed vascular bundles. Hence, this step needs to include information about the geometric artefacts found in the first steps of simulation and the relevant structural information in order to estimate the quality of the results obtained with the natural material. A comparison with artificial materials such as PVC specimens is needed and will help to provide an improved interpretation of the features of the natural material.

As already mentioned the shape needs to be defined exactly. This can be obtained by scanning or exact measuring of the samples or by computer tomography (CT). Revealing the exact shape from the CT-images through a defined process was previously carried out in the group of the IMWF. One example is shown in Fig. 14.10

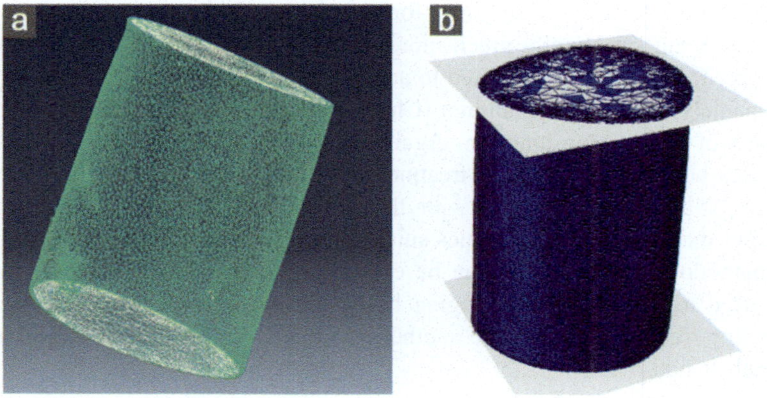

Fig. 14.10 Simulations of a PI sea urchin spine test specimen. (**a**) Mesh reconstructed from a CT scan and (**b**) screenshot of the FE simulation of the compression test with added pressure plates and filled with a bulk material with isotropic elastic properties as a first approximation

in which a sample of a PI sea urchin spine was analysed in a CT and converted into a 3D representation of the surface, which was then meshed and used in FE simulations. In this case, the resolution of the scan was high enough to reveal the macrostructure and the outer surface but not the microstructure such as the porous stereom.

14.6 Discussion

The challenge of scaling up natural constructions to sizes relevant for building construction is complicated for several reasons: natural biological materials show complex mechanical behaviour because of their complex hierarchical internal structure. This often creates a failure mode, which is neither ideally brittle nor ductile making it difficult to predict an upscaling based on classic reliability evaluation treatments such as Weibull algorithms.

In order to evaluate such theories, the main obstacle is the uniqueness of individual natural samples. A first step therefore involves the development of methods suitable for these unique materials in order to build up a reliable data base.

In our studies, promising model candidates are spines from large sea urchins, because they have an internal structure that can be characterised by quantitative parameters such as local densities, pore size and shape, while still being made of a single material (calcium carbonate) with little additions. This also makes them attractive for simulation treatment, which should be of great help in the evaluation of properties.

An unusual experimental treatment, namely the penetration of the quasi-brittle highly porous stereom of sea urchins, was chosen to yield quantitative strength and energy dissipation values. We have shown that the treatment represents a local probing method that allows structural evaluation before tests by computer tomography and/or by sectioning the partially destroyed samples after testing and that gives easily understood numbers for materials engineering.

Another, but structurally much more complex candidate is the tough coconut endocarp. Here, we have also tried to adapt a conventional material testing procedure, namely the impact pendulum (Charpy) test, to evaluate dynamic behaviour in terms of energy consumption. Again, we have the chance to investigate the structures both before and after treatment. The non-ideality of samples cut from fruits is still a challenge but, with the aid of computer simulations, we are confident that we can obtain values suitable for use in a scaling up evaluation. If this methodological approach can be evaluated for coconut endocarp, it will be transferred to even more challenging botanical materials such as highly energy dissipating tree barks, which are another role model for bio-inspired construction materials (Chap. 7).

Acknowledgements This work has been funded by the German Research Foundation (DFG) as part of the Transregional Collaborative Research Centre (SFB/Transregio) 141 'Biological Design and Integrative Structures'/project B01 'Scaling of Properties of Highly Porous Biological and Biomimetic Constructions'. The Plant Biomechanics Group Freiburg also thanks E. Heizmann and UNIVEG Freiburg, Germany, for providing the coconuts.

References

Andrews EW, Gioux G, Onck P, Gibson LJ (2001) Size effects in ductile cellular solids. Part II: experimental results. Int J Mech Sci 43(3):701–713

Bauer G, Schmier S, Thielen M, Speck T (2015) Energy dissipation in plants – from puncture resistant seed coats to impact resistant tree barks. In: Yamamoto H, Morita M, Gril J (eds) Proceedings of the 8th plant biomechanics conference, Nagoya, Japan, pp 190–195

Bažant ZP (2000) Size effect. Int J Solids Struct 37(1):69–80

Bažant ZP (2004) Scaling theory for quasibrittle structural failure. Proc Natl Acad Sci U S A 101(37):13400–13407

Chan E, Elevitch CR (2006) *Cocos nucifera* (coconut). Species Profiles for Pacific Island Agroforestry 2:1–27

Chan YL, Ngan AHW, King NM (2009) Use of focused ion beam milling for investigating the mechanical properties of biological tissues: a study of human primary molars. J Mech Behav Biomed 2(4):375–383

Danzer R (2014) On the relationship between ceramic strength and the requirements for mechanical design. J Eur Ceram Soc 34(15):3435–3460

Danzer R, Supancic P, Pascual J, Lube T (2007) Fracture statistics of ceramics–Weibull statistics and deviations from Weibull statistics. Eng Fract Mech 74(18):2919–2932

Franke E, Lieberei R, Reisdorff C (2012) Nutzpflanzen. Georg Thieme Verlag, Stuttgart

Griffith AA (1921) The phenomena of rupture and flow in solids. Phil Trans R Soc A 221:163–198

Grossmann JN, Nebelsick JH (2013) Comparative morphological and structural analysis of selected cidaroid and camarodont sea urchin spines. Zoomorphology 132(3):301–315

Krumbholz M, Hieronymus CF, Burchardt S, Troll VR, Tanner DC, Friese N (2014) Weibull-distributed dyke thickness reflects probabilistic character of host-rock strength. Nat Commun 5:3272

Kumar PS, Ramachandra S, Ramamurty U (2003) Effect of displacement-rate on the indentation behavior of an aluminum foam. Mater Sci Eng A 347(1):330–337

Lawn B (1993) Fracture of brittle solids. Cambridge University Press, Cambridge

Łysiak G (2007) Fracture toughness of pea: Weibull analysis. J Food Eng 83(3):436–443

Menig R, Meyers MH, Meyers MA, Vecchio KS (2000) Quasi-static and dynamic mechanical response of *Haliotis rufescens* (abalone) shells. Acta Mater 48(9):2383–2398

Mouginot R, Maugis D (1985) Fracture indentation beneath flat and spherical punches. J Mater Sci 20(12):4354–4376

Moureaux C, Pérez-Huerta A, Compère P, Zhu W, Leloup T, Cusack M, Dubois P (2010) Structure, composition and mechanical relations to function in sea urchin spine. J Struct Biol 170(1):41–49

Olurin OB, Fleck NA, Ashby MF (2000) Indentation resistance of an aluminium foam. Scr Mater 43(11):983–989

Presser V, Schultheiß S, Berthold C, Nickel KG (2009) Sea urchin spines as a model-system for permeable, light-weight ceramics with graceful failure behavior. Part I. Mechanical behavior of sea urchin spines under compression. J Bionic Eng 6(3):203–213

Seto J, Ma Y, Davis SA, Meldrum F, Gourrier A, Kim YY, Schilde U, Sztucki M, Burghammer M, Maltsev S, Jäger C, Cölfen H (2012) Structure-property relationships of a biological mesocrystal in the adult sea urchin spine. Proc Natl Acad Sci U S A 109(10):3699–3704

Su X, Kamat S, Heuer AH (2000) The structure of sea urchin spines, large biogenic single crystals of calcite. J Mater Sci 35(22):5545–5551

Taylor D (2000) Scaling effects in the fatigue strength of bones from different animals. J Theor Biol 206(2):299–306

Wagermaier W, Klaushofer K, Fratzl P (2015) Fragility of bone material controlled by internal interfaces. Calcif Tissue Int 97(3):201–212

Weibull W (1939) A statistical theory of the strength of materials. Generalstabens litografiska anstalts förlag, Stockholm

Weibull W (1951) A statistical distribution function of wide applicability. J Appl Mech 18:293–297

Winton AL (1901) Anatomy of the fruit of *Cocos nucifera*. Am J Sci 70:265–280

Yang F, Li JC (2013) Impression test—a review. Mat Sci Eng R 74(8):233–253

Yu HY, Imam MA, Rath BB (1985) Study of the deformation behaviour of homogeneous materials by impression tests. J Mater Sci 20(2):636–642

Chapter 15
Evolutionary Processes as Models for Exploratory Design

Long Nguyen*, Daniel Lang*, Nico van Gessel, Anna K. Beike, Achim Menges, Ralf Reski, and Anita Roth-Nebelsick

Abstract Biological evolution drives morphological diversity via genetic variation and results in a high level of adaptation, performance and resource efficiency. However, "biological design" arising from evolution is often counterintuitive and unexpected in a non-linear way. Evolutionary processes are undirected and very good at exploring novel design possibilities in an open-ended manner. Biological evolution thus differs profoundly from the gradualistic and constantly converging character of technical optimization with defined and static fitness functions. Evolutionary algorithms based on Darwinian principles are mainly developed for solving multi-criteria problems in technology. Technical goals are defined as fitness functions and the evolutionary mechanisms of selection, heredity, reproduction and mutation are employed as stochastic optimization processes. These metaheuristic algorithms do not include recent insights into micro- and macroevolutionary mechanisms derived from genomics, phylogenomics and population genomics. Similar to natural evolution, the architectural design process is an open-ended process exploring possible solutions. However, in order to navigate this vast and dynamic design space, most design methodologies in architecture are based on a typological approach. The designers, based on their knowledge and understanding of the problems, usually limit the solution space to a particular structural, constructional, spatial or programmatic type that is iteratively adapted to the particular design requirements. The constraints inherent in typology-based

* Both authors contribute equally for this chapter.

L. Nguyen (✉) • A. Menges
Institute for Computational Design (ICD), University of Stuttgart, Keplerstraße 11, 70174 Stuttgart, Germany
e-mail: long.nguyen@icd.uni-stuttgart.de

D. Lang (✉) • N. van Gessel • R. Reski
Plant Biotechnology, University of Freiburg, Schänzlestraße 1, D-79104 Freiburg, Germany
e-mail: daniel.lang@biologie.uni-freiburg.de

A.K. Beike • A. Roth-Nebelsick
Stuttgart State Museum of Natural History, Rosensteintraße 1, D-70191 Stuttgart, Germany

J. Knippers et al. (eds.), *Biomimetic Research for Architecture and Building Construction*. Biologically-Inspired Systems 9, DOI 10.1007/978-3-319-46374-2_15

design methodologies exclude a vast range of potentially more effective and better design variants. In contrast, the dynamics of biological evolution suggest ways of continuously expanding the design space towards new and unexplored possibilities, that can potentially in a new set of typologies that still satisfy the constraints. Thus, in architecture, evolutionary processes are more relevant as exploratory processes than as optimization tools.

15.1 Introduction

Biological innovation and morphological diversity are the product of evolutionary processes that are ultimately driven by genetic variation and diversification resulting in a high level of adaptation, performance and resource efficiency. Early on in the history of computer science, ideas arose for the adoption of evolutionary principles for the implementation of digital optimization tools that gradually refine a "population" of solutions according to certain fitness criteria. In his essay "Computing machinery and intelligence", Alan Turing (Turing 1950) compared the programming and education of a "child" learning machine to the process of natural evolution. The subsequent transition from theory to practice was accomplished by the digital simulation of evolutionary processes in the mid-1950s to early 1960s of the twentieth century (Fraser 1958; Barricelli 1962). The field gained further momentum with the work carried out on evolutionary strategies (Rechenberg 1971; Schwefel 1974). Genetic algorithms (GA) became popular with the book *Adaptation in Natural and Artificial Systems* (Holland 1975). This research gave birth to a new field in computer science and artificial intelligence known as evolutionary computing. The algorithms emerging from this field are collectively known as "evolutionary algorithms" (EA) and usually maintain a population of potential solutions that are subjected to recombination, mutation and selection for retention in the population by using predefined "fitness" criteria.

The underlying view of natural evolution dates back to the beginning of the twentieth century when the founders of population genetics reconciled Darwin's ideas concerning natural selection with Mendelian genetics into what is now known as the Modern Synthesis or Neo-Darwinism. Although the work on GAs largely remained theoretical until the 1980s, once the first GA conferences were held and General Electric had sold the first software including a GA, EAs can be considered as very early examples of biomimetics (mimicking a process found nature). Since its origins, the field of evolutionary computing has substantially diversified and EAs have been applied to many problems, including computational design. Surprisingly, except for a few instances of adaptations of additional genetic or evolutionary concepts such as transposable elements (Simoes and Costa 2000), the field largely remained uncoupled from biological research. Furthermore, if we consider, for example, the insights gained from evolutionary developmental biology or genome sequencing, biological research undoubtedly has progressed beyond the view of the Modern Synthesis. However, limited to no exchange has occurred between the disciplines of biology, computer science and design over the last decades. EAs and evolutionary simulation have also not attracted much attention in the biological mainstream. EAs depend on the exact quantitative knowledge of evolutionary rates to implement these principles. In EAs, the employed rates are usually tuned based on experience and adjusted to the problem. To our knowledge, *in silico* and *in vivo* evolutionary rates have not been consolidated so far. As we will see in the later sections of this chapter, although the terminology often is adopted from biological concept generators (indicated by italics in the subsequent text), in many cases, the meaning, i.e. the implementation, differs considerably (and sometimes fundamentally) from the biological source.

Many of the "designs" shaped by biological evolutionary processes often appear counterintuitive or unexpected. On a global scale, evolutionary processes are undirected and excel in the open-ended exploration of novel design possibilities. Thus, natural evolution differs profoundly from the gradualist and constantly converging character of technical evolutionary optimization with fitness functions. Technical goals are defined and static fitness functions and the evolutionary mechanisms of selection, heredity, reproduction and mutation are employed as stochastic optimization processes. Implementations of these mechanisms reconstitute microevolutionary changes that largely correspond to the biological variability observed at the population level. This resembles an optimization process analogous to the dominant genotype of a population that is optimally adapted to the respective ecological niche. However, as we will see later, true biological innovation and diversity are observed at the species level or above and is the product of additional macroevolutionary mechanisms usually spanning larger time intervals. Mostly still relying on the concepts of the Modern Synthesis, the metaheuristic EAs do not include recent insights into micro- and macroevolutionary mechanisms derived from genomics, phylogenomics and population genomics.

Although a certain renaissance of EAs occurred when coupled with Deep Learning (Tirumala 2014; David and Greental 2014), many computational scientists and machine learners consider EAs and GAs as no longer relevant. Many perceive them as too slow, too inefficient and prone to being stuck at local maxima, suggesting that the approach of transferring evolutionary principles to optimization and design problems has failed. This raises the obvious questions as to whether this view is correct or whether just the chosen abstractions of evolutionary processes are too simple to mimic evolutionary processes effectively. Furthermore, we need to ask whether and in what way biology can benefit from a realignment of the two fields and whether we can use EAs in a reverse biomimetic approach to gain novel biological insights.

Similar to biological evolution, the architectural design process is an open-ended process exploring possible solutions that satisfy the design requirements, which quite often include aspects that are extremely difficult to quantify. In order to navigate and search in the vast, dynamic, highly varying and multi-dimensional solution space, most design methodologies in architecture are based on a typological approach. This limits the solution space to a particular structural, constructional, spatial or programmatic type, which is iteratively adapted to the particular design requirements. The constraints inherent in typology-based design methodologies exclude a vast range of potentially more effective and better design variants. In contrast, the dynamics of biological evolution suggest ways of continuously expanding the solution space towards new and unexplored regions, as in macroevolution. Thus, in architecture, the exploratory nature of macroevolutionary processes is more relevant than the microevolutionary processes implemented in current EA optimization tools.

This chapter reviews the current state of EAs with relevance to computational design, aligns them with the current state of biological research, gives an overview of the computational approach to architectural design and highlights some of the relevant biological insights bridging the gap between micro- and macroevolution. Not all of these questions can be addressed at the global scale; any algorithmic instantiation of these principles aiming at avoiding being too simplistic or too divergent from the biological concept generator needs detailed and quantitative data. Finally, the chapter offers a perspective with regard to the way that these principles can inform the computational approach to biomimetic architectural design. The achievement of this goal requires interdisciplinary collaboration between the fields of biology, architecture, engineering and computer science.

15.2 Evolutionary Computation

15.2.1 Overview

Evolutionary computation (EC), or *evolutionary computing* is a subfield of *artificial intelligence* that comprises a group of algorithms known as *evolutionary algorithms* (EA) (Bentley and Corne 2001; Bentley et al. 2008). These algorithms are based on principles that are inspired by Darwin's theory of evolution and that act on a "population" of potential solutions to the problem at hand. The three most fundamental principles are:

1. *Recombination*: Individuals from the current *population* are combined to produce new individuals. The main idea is to allow "good" features from two (or sometime more) individuals to be (re)combined and carried on by the resulting offspring. Although the algorithmic details vary between the different versions, implementations or applications of the algorithms, in most cases, the recombination method involves some means of randomization.
2. *Mutation*: Each individual in the *population* has a certain probability of be *mutated* (i.e. randomly modified). This allows the production individuals that are possibly divergent from those in the current population and introduces a further source of variability.
3. *Selection*: Only the "best" individuals are retained in the *population*; the rest will be discarded. This ensures that, over time, the fitness of the individuals will increase. The selection procedure can be implemented explicitly by the computation of a fitness value for each individual and the selection of those with the highest values or by implicitly letting the individuals compete with each other.

Because the principles above are applied to the *population* of solutions over many iterations, they result in the gradual maximization of the fitness of the solutions. The converging character of this class of algorithms obviates their applicability to optimization problems.

EAs are generally grouped into four categories (Bentley 1999): genetic algorithms, genetic programming, evolutionary strategies and evolutionary programming. Among these, only genetic algorithms and genetic programming clearly distinguish between genotype and phenotype representations. As this distinction is important for the translation of macroevolutionary principles, the subsequent sections will focus solely on these two categories.

15.2.2 Genetic Algorithms

Genetic algorithms (GA) were introduced in the 1970s by John. H. Holland (Holland 1975) as an attempt to demonstrate and explain the adaptive processes found in natural systems and as an approach to produce artificial systems that exhibit adaptive characteristics similar to their natural counterparts. However, the same algorithm can be used for optimization (Bentley 1999). The object to be optimized is termed the *phenotype*, which is controlled by an underlying presentation known as the *genotype*. Both the terms *phenotype* and *genotype* are adopted from genetics. In architectural design, the *phenotype* is usually some type of three-dimensional (3D) shape. For example, a *phenotype* that is a 3D box shape can be controlled by an underlying *genotype* of three numerical parameters for width, length and height.

A typical GA (Fig. 15.1) comprises the iterative execution of many generations (cycles), which usually consists of seven stages:

Initialization (executed once): A *population* of solutions together with their corresponding *genotypic* representations is generated. The ideal *population* size (the number of individuals in the population) varies between applications and is constrained by the computational resources. The most straightforward way to generate these initial *ancestral* solutions is by assigning random parameter values to the genotype. Alternatively, they can be pre-generated by other optimization algorithms in order to ensure a minimum level of quality. Careful pre-generation of *ancestral populations* can help to improve the speed of finding an optimal solution (Rahnamayan et al. 2007). A comprehensive review of population initialization techniques can be found in (Kazimipour et al. 2014).

Recombination: *Genotypes* for new solutions are generated by combining *genotypes* from existing solutions. Prior to the actual recombination is a preparation step called *mating* whereby the algorithm decides which of the existing *genotypes* will be paired with each other to produce the *offspring genotypes*. The default and simplest approach is random *mating*, i.e. random selection of *mating* partners. Restricted mating procedures represent algorithmic means to implement non-random or assortative *mating* patterns found in nature, i.e. consider *genotypic* or

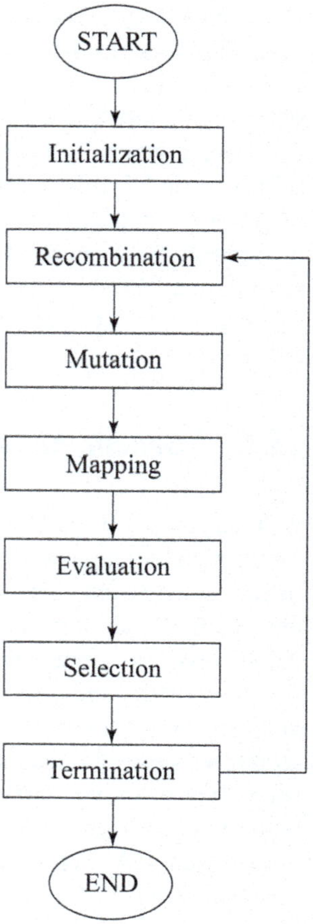

Fig. 15.1 Typical data flow of genetic algorithm

phenotypic compatibility or the similarity of the *mating* partners. For example, the two solutions in the same *mating* pair might be required to have a sufficiently similar *genotype* to avoid *offspring genotypes* that map to *phenotypes* with low *fitness* values (Deb and Goldberg 1989). Then again, the *mated genotypes* should not be too similar in order to prevent pre-mature convergence on a local maximum that might overlook a potentially better solution in another area of the search space; this is tackled by a procedure called incest prevention (Eshelman and Schaffer 1991). One popular version of *mating* allows the *fittest* solution (i.e. the one with the highest quality) to *mate* with every single other solution to allow its best features to combine with those from other solutions (Bentley 1999). Galán et al. (2013) generalized the traditional uniform *mating* method by attaching two extra parameters called the *mating* size and the *mating* index to each individual. The values of these parameters will decide whether an individual will *mate* with the most or least similar candidates in the *mating* pool or any in between. This allows the *mating* behaviour of each individual to vary between explorative (i.e. favouring variety) and exploitative (i.e. favouring convergence).

Furthermore, the values of the *mating* size and *mating* index do not have to be manually set by the user but are allowed to evolve together with other parameters in the *genotype*, resulting in a self-adaptive *mating* behaviour. The *genotype* of the *child* solution can be obtained in various ways, such as by using *crossover*, analogous to the way that homologous chromosomes recombine and exchange genetic material in meiosis during sexual reproduction. Whereas the biological recombination process involves a complex multi-step process most likely involving the tight regulation of the DNA damage machinery, artificial *crossover* in GA often represents a rather simple abstraction. In the original version described by Holland (1975), the parameters in a *genotype* are encoded as binary strings (i.e. genes). These gene strings are concatenated to form the *chromosome* binary string of that particular *genotype*. The *crossover* method will swap a region of the *father chromosome* string with the corresponding region of the *mother chromosome* string, resulting in two *children chromosome* strings. Alternatively to using binary representation, the real-value parameters can be combined into a *chromosome* vector and the *crossover* will act on two *parental chromosome* vectors (Wright 1991). In practice, the *crossover* operation can be applied many times to the same pair of *parental chromosomes* (but each time with a different *crossover* point) resulting in multiple distinct child solutions. Apart from the standard *crossover* procedure, real-value *chromosome* vectors can also be recombined by taking the weighted average of the parental parameter values by using Gaussian distribution as the basis for the weights (Ladkany and Trabia 2012). Apart from the oversimplification of the process of recombination, an obvious deviation from biological reality is present in the size and content of the *genome*. Eukaryotic genomes usually comprise multiple, independently inherited chromosomes that represent long-chained DNA polymers (i.e. vectors).

Mutation: Probabilistically, each individual solution in the current *population* can have their *genotype mutated* according to a certain probability value, often called the *mutation rate*. When *mutation* occurs, the *genotype* real-value parameters will be modified by adding a random offset to the current values based on some probability distribution function (Wright 1991) or, in the case of binary reorientation, a randomly chosen bit in the *chromosome* string will be altered (i.e. 0 becomes 1 or vice versa). *Mutation* allows unexplored regions of the search space in which potentially *fit* solutions can be found to be reached. Mühlenbein and Schlierkamp-Voosen (1993) have reported a *mutation rate* set inversely proportional to the number of *genes* (parameters) in the *genotypes* tends to give good results. GAs are usually tested in the range between 1E-03 and 0.5 (e.g. Kühn et al. 2013). *In vivo* mutation rates seem to be much lower and to depend on the outcome of the mutation. The spontaneous occurrence of indels (insertion or deletion of DNA stretches) has been recorded at a rate of 5E-12 per base per generation in yeasts (Zhu et al. 2014). Whereas selectively neutral changes, such as synonymous substitutions that affect the second and third codon position and that do not result in a change of the encoded amino acid, have been observed to occur at a rate of 1.9E-08 per site and year in mosses (Rensing et al. 2007).

Mapping: The *offspring genotypes* are mapped to their respective *phenotype* representation. For example, the simple 3D box *genotype* will map the parameters controlling the width, length and height to the actual geometry of the box. Not every region of the biological chromosomes directly encodes an attribute of the phenotype (i.e. gene) that is under selection. This phenomenon has resulted in the debate about junk DNA (Palazzo and Gregory 2014) that is tightly connected to the biological paradox that genome size is not correlated with organismic or morphological complexity (Gregory 2005; Greilhuber et al. 2005). Nevertheless, biological genomes are far more complex in gene content than their usual GA counterparts: unicellular bacteria usually harbour 1000–5000 genes, whereas multicellular eukaryotic genomes encode between 5000 and 10,0000 genes. Humans have about 20,000 genes and plants usually have around 30,000 genes (Sterck et al. 2007). Viruses can have similarly low gene numbers as those in common GA genomes but these strongly rely on their hosts for almost every aspect of their existence and are usually not considered as "alive" and coevolving with their respective host genomes. Only a few genes directly affect the morphology or behaviour of an organism. Furthermore, genes can be involved in multiple functions or affect multiple phenotypic traits (pleiotropy). In most cases, the products of multiple genes need to interact or act in sequence to implement a phenotype (polygenic inheritance). Thus, even if a gene has a direct phenotype, its activity might depend on the action of other genes (epistasis).

Evaluation: The *fitness* of each *phenotype* is evaluated via user-defined criteria. For example, the criterion can be the ratio of volume to surface area; this might serve as a *fitness* criterion, referring to the evaluation of a geometrical body. Alternatively, in case of a roof surface for a building, *fitness* might depend on the amount of solar energy that it receives, taking into account the occlusion caused by neighbouring buildings. The evaluation of the *phenotypical* consequences of a given *genotype in vivo* occurs on multiple levels and timescales. Not every change is evaluated or consequential directly after its occurrence. Organisms live in a changing environment that dynamically "evaluates" distinct aspects of the genotype.

Selection: A solution will be either retained or discarded based on its *fitness* value. Obviously, solutions with high *fitness* values need to be retained, whereas those with a lower *fitness* are discarded to ensure that the overall *fitness* of the solutions increases over time. The selection procedure can be deterministic, such that only a certain number of solutions with the highest *fitness* values are kept. This procedure is known as "truncation selection" (Mühlenbein and Schlierkamp-Voosen 1993). Stochastic methods provide an alternative to deterministic selection procedures. For example, in the "*fitness proportionate selection*" scheme, the probability that a solution will be retained is directly proportional to its *fitness* value. This means that solutions with low *fitness* scores still have a small non-zero chance of *survival* (rather than being certainly discarded as in the "*truncation selection*" scheme). In natural populations, not every trait is under strong purifying selection at all times. Thus, these methods introduce some aspects of balancing or relaxed selection and neutral evolution, which are seen as important drivers of diversification and biological innovation. The relaxed *selective pressure* on these suboptimal solutions in the current *generation* might eventually lead to even better *adapted* or more generalist solutions in later *generations*. It also tends to be computationally more efficient as it does not require the sorting of all individuals based on their *fitness* values. Another well-known stochastic selection scheme is the *tournament* scheme (Miller and Goldberg 1995), in which individuals are grouped randomly and only the *fittest individual* of each group will be retained in the *population*. In addition, *parents* from the current *generation* can also proceed to the next generation if they have the highest *fitness* values. This strategy is known as "*elitism*" and can help preserve solutions with very high *fitness*.

As we have seen above, mutational rates vary across sites of a genome. This is linked to distinctive and varying selective pressures associated with structural and functional constraints. Consequently, we can observe that the mutational rates of biological genomes and genes vary at the level of time and structure. Molecular phylogenetics document a high degree of heterogeneity of evolutionary rates among distinct sites of the genome or even within the same gene (Jia et al. 2014). These rates constantly change as a function of time. Furthermore, we can distinguish different directionalities of selection: (1) negative selective forces (purifying selection) result in a level of high sequence conservation of functionally or structurally important regions based on the effective elimination of altered alleles; (2) positive or "Darwinian" selection in contrast favours certain alleles as they positively influence fitness and is measurable by increased rates of change and diversification; (3) as we have seen earlier in the discussion of silent synonymous mutations, not every change is under selection. The relaxation of selection towards neutral evolution is of particular importance for the so-called junk DNA, such as transposable elements or integrated viroids.

Termination: The algorithm is stopped if the termination criterion is met. Otherwise, it will return to the recombination stage and continue iteratively until the termination criterion is met. Typically, the algorithm is terminated when a sufficiently good solution has been produced or the overall running time or number of *generations* has reached a certain threshold defined by the user.

15.2.3 Genetic Programming

Genetic programming (GP) can be regarded as a special case of GAs, whereby the objects being evolved are computer programs. This was originally developed by Koza (1992) as a strategy to create computer programs that can program themselves. The *genotype* of a program is typically represented as a tree structure and the *phenotype* is the program or function itself. In this regard, the *gene* representations of GP more closely resemble their biological counterparts, as they encode evolvable functions that can be cascaded. An example of such a scheme is shown in Fig. 15.2. The tree structure serves as a basis on which *recombination*

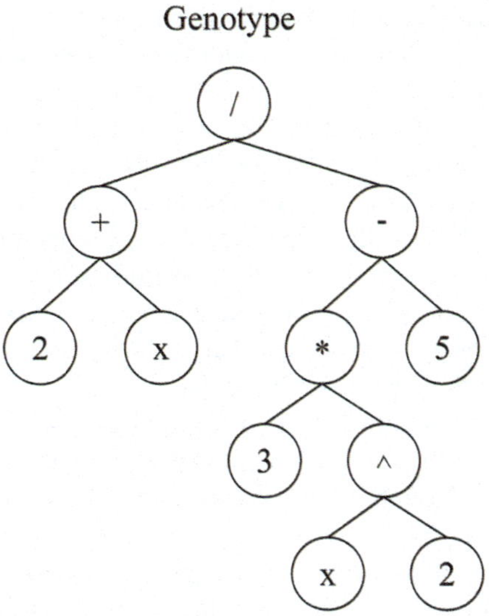

$$\frac{x+2}{3x^2-5}$$

Fig. 15.2 Genotype and phenotype of a mathematical formula being evolved using genetic programming

and *mutation* mechanisms such as *crossover* can act. As an example, *crossover* is achieved via the exchange of branches between two parental trees, resulting in two child trees that represent novel *genotypes* that subsequently can be mapped to their respective *phenotype* (Fig. 15.3).

15.3 Evolution and Computational Design in Architectural Exploration

15.3.1 Computational Approach to Architectural Design

The introduction of computation and programming as a design tool in architecture implies that a profound shift in attitude and thinking with regard to the very nature of the architectural design process itself is necessary in order to fully embrace the potential that computation can offer. This approach to design is referred to as *computational design* and is well-discussed in Menges and Ahlquist (2011).

Computational design entails new design methods that profoundly extend the range of possible architectural outcomes. This is because, rather than directly modelling the design by using standard geometric constructions such as points, lines, curves or flat and curved surfaces, a designer who adopts the computational approach models the steps that lead to the generation of the design solutions and lets them be executed by the computers. Typically, the steps here are founded upon many algorithms descending from various fields within computer science, mathematics, biology and physics, such as computer graphics, computational geometry, artificial intelligence, numerical computation, analysis and simulation. These algorithms are used to process large amounts of design and feedback data, transforming them into

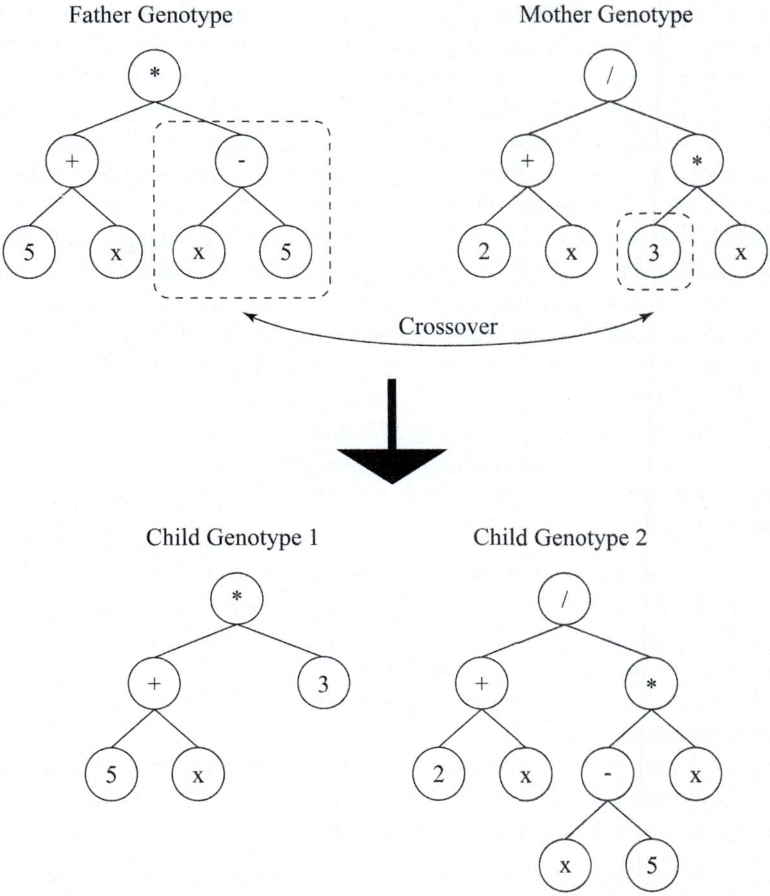

Fig. 15.3 Example of a crossover in the tree-structure representations of genotypes

design solutions that include both geometric and non-geometric information. Such features lead to the assumption that computational design will potentially lead to novel architectural outcomes that are simply not possible to achieve by using either manual techniques or their computerized version, which is often the found in current Computer-Aided Design workflows.

15.3.2 Evolutionary Computation in the Current Practice of Computational Design

The technology-affine part of the design community expressed interest in utilizing evolutionary computation very early (Frazer 1995). Among the broad range of approaches to evolutionary computation, GAs are the most widely used. On the one hand, this might be attributable to the prevalence of the concepts of Darwin and the Modern Synthesis in our society and, thus, ultimately might be linked to the "biomimetic promise" in which we perceive the outcome of natural evolution as optimal, efficient, sustainable and elegant (see Chap. 18). On the other hand, GAs are generic and flexible enough to be applied to a wide range of design problems. Most importantly, the genotype-phenotype mapping innate to GAs makes them particularly attractive for designers.

Overall, GAs are particularly suitable for architectural design problems in which various performance aspects and design criteria can be quantified and used as *fitness* criteria in order to guide the evolution of design solutions. Examples include lighting (Caldas and Norford 2002), acoustics (Sato et al. 2004; Spaeth and Menges 2011), view exposure (Menges 2012) or structural aspects (Dimčić 2011; Coelho et al. 2014). One advantage of GAs is their conceptual ease of use in a design environment and the lack of requirement for gradient information upon which many other optimization techniques depend, e.g. gradient descent.

The general approach to using GAs in architectural and structural design involves four major steps:

1. Defining a parametric model that captures the design solution space. The number of parameters is carefully picked in advance and, usually, their value ranges are also determined.
2. Defining the *fitness* criteria for the design.
3. Defining the way to compute the *fitness* value for a design generated by the parametric model in step 1.
4. Using GAs to explore the parameter space and the design solution space, so that eventually design solutions with high *fitness* values will be generated without having exhaustively to search the entire space.

Evolutionary computation in general and GAs in particular have been recurring features in the context of computational design. In many cases, GAs have been used as optimization tools. One of the most important and earliest pioneers in evolutionary architectural design is John Frazer. His work with evolutionary computation is documented in Frazer (1995). Examples of design problems that have been tackled by using evolutionary computation include, but are not limited to, free-form curved surface design (Hemberg et al. 2001, 2008), spatial layout (Jo and Gero 1998; Michalek et al. 2002), building envelope design (Tuhus-Dubrow and Krarti 2010), thermal and lighting performance (Caldas and Norford 2002; Wright et al. 2002) and urban design (Finucane et al. 2006). Evolutionary computation has also been applied in the related field of structural engineering; for example, Dimčić (2011) and Coelho et al. (2014) have shown the way that GAs can be used to optimize the design of shell structures.

Menges (2012) presented two case studies in which design solutions were generated by using an evolutionary approach. In the first case study, the form of a building envelope was evolved by using a GA guided by multiple *fitness* criteria, including floor area, envelope height, envelope slope, unobstructed view exposure and environmental criteria such as incident solar radiation in interior thermal loading (Fig. 15.4).

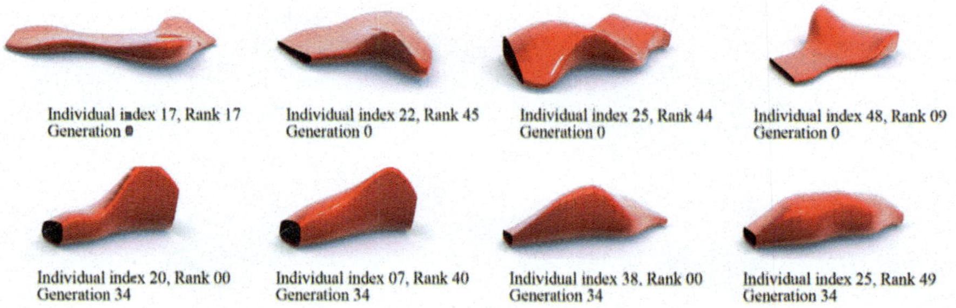

Fig. 15.4 Spatial envelope design of a structure on the Island of Boracay (Philippines) using GA, guided by multiple fitness criteria such as floor areas, height, slopes, incident solar radiation and interior thermal loading

The second case study dealt with the design of an urban block (Fig. 15.5). In this particular study, the designer attempted to overcome the typological design approach by devising a genotypic structure that actually consisted of five sets of genes, coupled with a so-called "*embryology* operator" (an advanced form of the genotype-phenotype mapping operator) that took each set of *genes* into account at various phases during the mapping (Fig. 15.6). In this regard, the *embryology* operator is a rough abstraction of a simple developmental program in biology. Similar to a developmental pathway in multicellular organisms, it establishes an expression hierarchy and implements a decision tree guiding the *developmental program*. In contrast to most eukaryotic developmental programs, the order of execution is encoded in the order of the *genes* on the *chromosomes*. An analogous counterpart of this can be seen in the operons found in unicellular bacteria in which genes encoding a specific pathway are co-localized on the bacterial genome and controlled by a single promoter. As we will see in the following section, developmental programs of multicellular eukaryotes are not coordinated via gene ordering and are much more complex.

15.3.3 Limitations of Current EAs for Computational Design

As we compare and contrast current approaches in evolutionary computation with advances in evolutionary biology, the great potential for enhancement seems to be worth investigating. In a biological context, evolutionary processes are undirected and open-ended and natural evolution has constantly produced novel designs that have not previously been seen. In this sense, the architectural design process itself is very similar to evolution. Thus, evolutionary principles seem to be more relevant to architectural exploration than design optimization with respect to a clearly defined set of goals. Indeed, as optimization tools, EAs are arguably falling out of favour as they tend to be quite inefficient compared with other superior techniques. This inefficiency is attributable to the rather "blind" searching mechanism of EAs,

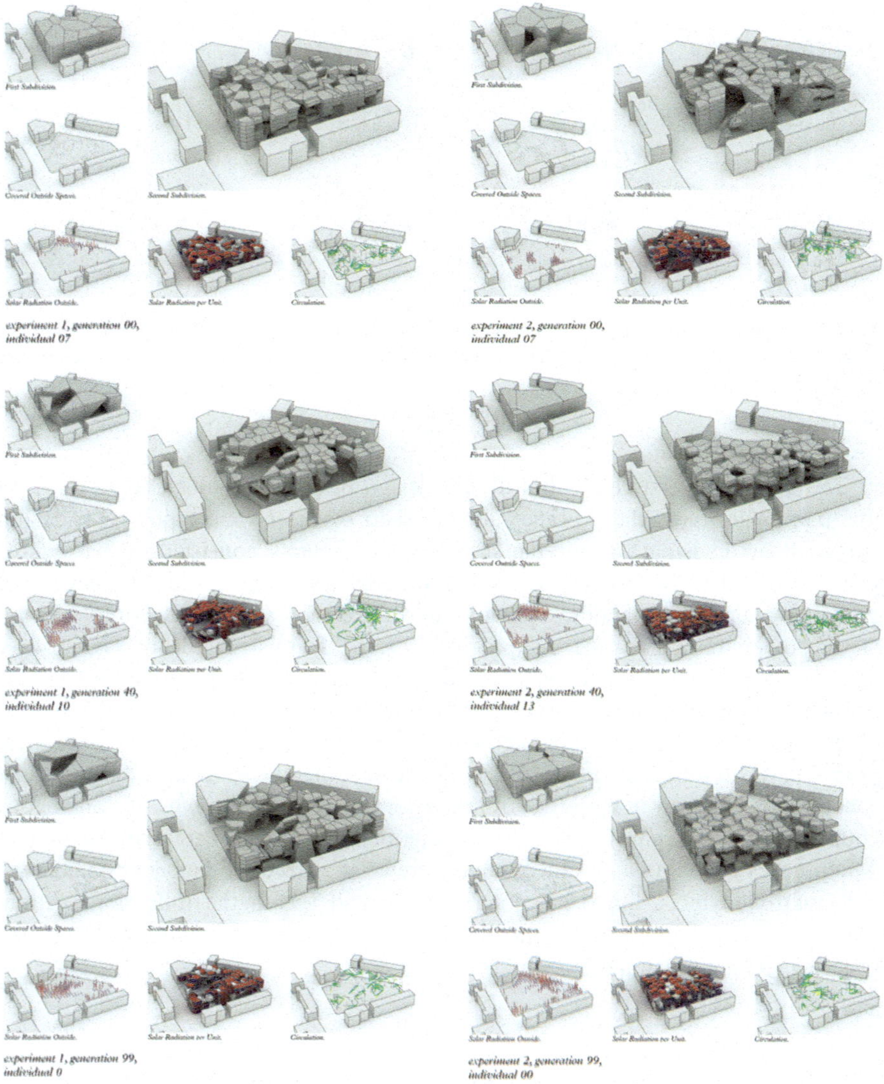

Fig. 15.5 Evolution of an urban block morphology driven by a GA, guided by five climatic fitness criteria (the amount of covered outside space, the cross-ventilation and solar radiation on both the overall morphology level and on the unit level) and two spatial fitness criteria (the number of units and the connectivity between them)

which generate hundreds or thousands of possible solutions in a manner of trial and error and do not take into account additional problem-specific information. However, a recent revival of interest has occurred in evolutionary computation, not for pure optimization purposes, but for more creative application domains, such as architecture, art and music (Bentley and Corne 2001).

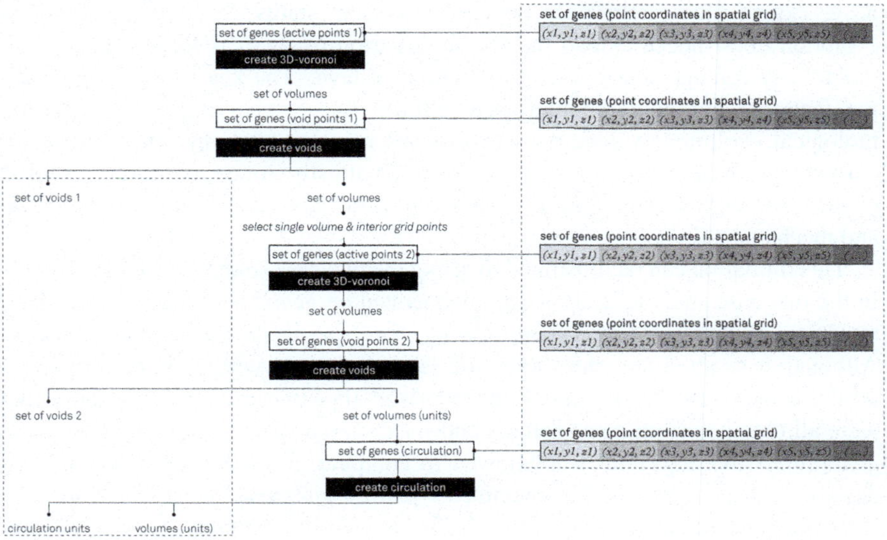

Fig. 15.6 The embryology operator that maps the genotype to phenotype via a multi-step hierarchical spatial subdivision scheme that drives the design shown in Fig. 15.5

To improve our understanding of an important limitation of the current approach in evolutionary computation, we should discuss the general workflow that involves four main steps as described in Sect. 18.3.2. The level of success in achieving good designs from the evolution process heavily depends on step 1. This is the point at which the designer must set up some sort of generative process that will transform a set of input parameters into a particular design solution. The designer must decide what aspects of the design can be changed and explored in order to maximize fitness criteria and then capture these aspects in these parameters. If the designer is overly ambitious and aims to capture as many aspects as possible, not only will he/she have to invest more time and effort in setting up the necessary parameters and the way that they drive the geometric relationships in the models, but this may also result in a very heavy parametric model that is slow to execute. This can seriously affect the optimization process as GAs work by generating hundreds or thousands of potential design solutions per generation, over many generations. Furthermore, the higher number of parameters results in a larger search space, i.e. an increased effort in order to find a good solution. This phenomenon is informally referred to as "the curse of dimensionality" and directly affects the algorithmic complexity of the problem (Bellman 1966). At the opposite extreme, if the designer only parametrizes a few aspects of the design, the search space will be very limited and the GA might not be able to find an acceptable solution at all.

The problem with the above approach is that, irrespective of the way that the designer sets up the parametric frameworks, the search space will be fixed and bounded. EAs will navigate the search space to find optimal designs but will never extend the search space beyond its current status. In contrast, biological evolution does indeed extend the search space on its own, as long as it does not violate any fundamental physical or biological laws. The key to this is probably the fact that not only the biological entities (e.g. organisms) evolve, but the biological evolutionary processes themselves are also evolving! In contrast, the current approach to evolutionary computation mostly concerns the evolution of objects (the digital genotypes and phenotypes) but not the evolutionary processes and mechanisms.

This limitation can be illustrated by using the two case studies in Menges (2012). In the first case study (Fig. 15.4), a GA operated on *genotypes* with the same fixed structure, by using a linear non-evolving *genotype-phenotype* mapping process. Although it resulted in a nice variety of design solutions with good performances, all these solutions shared similar characteristic appearances. This is because they were all obtained from a pre-defined and fixed search space. No matter how many *generations* the algorithm was allowed to compute, it would never produce any design that was outside of the predefined typology described by the search space.

In the second case study (Fig. 15.5), the designer employed a multi-step *embryology* mapping operator in order to diversify the search space (Fig. 15.6). This helped the design process to start breaking away from the current limitation. However, as the *embryology* operator itself was not evolving, ultimately the search space could not expand beyond that allowed by the *embryology* operator.

15.4 Molecular Motors of Micro- and Macroevolution

15.4.1 Introduction

Although its principle action and importance for the generation of biological diversity is scientifically undisputed, our understanding of the evolutionary mechanisms driving species diversification and organismic complexity has evolved substantially since the initial conception of EAs and the divergence of the field of evolutionary computation. The availability of entire genomes from an increasing number of model organisms and the accessibility of population dynamics at the genomic scale enabled by next-generation sequencing (NGS) technologies continuously inform comparative and evolutionary biology (Lang et al. 2008).

So far, out review of the role of evolutionary computation in architectural design, illustrated its current limitations and provocatively insinuated that these limitations are attributable to a combination of the level of abstraction (i.e. oversimplification) and a certain detachment of the evolutionary computation field from biological research beyond the Modern Synthesis. Furthermore, if aligned with the biological concept generators, EAs mostly correspond to a crude simulation of microevolutionary processes in viruses or very simple unicellular bacteria. This is certainly not the goal that engineers, designers or architects have in mind when they employ EAs as biomimetic products to utilize the exploratory powers of evolution. We think instead of the processes that have shaped the organismic and morphological complexity that is found in extant multicellular eukaryotes, such as that in plants ranging from unicellular photosynthetic eukaryotes (e.g. picoplankton *Ostreococcus*; Fig. 15.8c), multicellular algae (e.g. *Volvox*; Fig. 15.8b), mosses (e.g. *Physcomitrella* Fig. 15.8d) to, grasses (e.g. maize; Fig. 15.8e), shrubs (e.g. grape vine; Fig. 15.8f) and trees (e.g. poplar; Fig. 15.8g). Thus, as potential designers of evolution-inspired exploratory design tools, we are more interested in macro- than in microevolution.

From the perspective of the Modern Synthesis and in congruence with Darwin's model of the gradual acquisition and selection of changes, the distinction between micro- and macroevolution is not a fundamental one but is solely dependent on time and scale. Accordingly, the accumulation of small-scale random genetic changes over time (favouring changes that are advantageous for the survival of the organisms) eventually results in the rise of new species and higher-order taxa. Over decades, the Modern Synthesis has been constantly aligned to scientific progress by, for example, introducing single factors not necessarily qualifying as small changes, such as whole genome duplication events, and the various mechanisms of speciation, each in their own way resulting in reproductive isolation. However, the relative importance of these mechanisms in the light of species evolution largely remains unclear. In 1972, the gradualist model was challenged by palaeontologists Eldredge and Gould and their theory of punctuated equilibrium (Eldredge and Gould 1972). Derived from morphological observations in the fossil record, they proposed rapid bursts of evolutionary changes during relatively short timeframes as being the origin of new species punctuating longer periods of morphological stasis. Today, both models co-exist in the scientific community. Instead of two opposing views, they are rather seen as two distinct modes of evolution dynamically acting on multiple taxonomic levels and time scales.

As in biology, the variability-generating mechanisms in EAs ultimately need to be implemented and resolved down to the genotype level. Thus, any work on the improvement of EAs in this respect needs to target molecular processes. Nevertheless, what are the candidate molecular processes and mechanisms that affect macroevolution? What are the concepts determining morphological or organismic complexity? What are the molecular motors of diversification?

Comprehensively answering these questions would certainly require a whole series of books. Although evolutionary research has made tremendous progress since the dawn of the Modern Synthesis, it certainly still cannot provide the definitive answers. Indeed, because of the time scales and complexity of macroevolutionary processes, definite experimental proof for the causality of certain phenomena is extremely difficult to come by within a human lifetime. Darwin himself (1859) was very aware of this limitation: "We see nothing of these slow changes in progress, until the hand of time has marked the long lapses of ages, and then so imperfect is our view into long past geological ages, that we only see that the forms of life are now different from what they formerly were." This is the point at which biological research can benefit from a realignment with evolutionary computation and can use it in a reverse biomimetic approach in order to assess the macroevolutionary potential of candidate mechanisms and concepts.

For the remainder of this chapter, we will use plants as concept generators to highlight some of the candidate mechanisms and concepts driving macroevolution. Genomes of multicellular plants are complex being huge with so-called junk DNA and inherited on multiple chromosomes with usually around 30,000–40,000 genes (Sterck et al. 2007). In most instances, no simple 1:1 mapping occurs between genotype and phenotype. Biological processes are more like acyclic graphs than decision trees, and polygenetic inheritance, epistasis, multiple alleles and pleiotropy are common phenomena. Redundancy is the rule, not the exception. Genes are expressed and the central dogma "from gene to mRNA to protein" is being continuously revised and expanded (every step is regulated, i.e. dynamic). The dosage of gene products shapes the phenotype either via gene regulation (expression level, alternative splicing and degradation) and/or gene copy number (gene duplication) (Rensing 2014). From an engineering standpoint, all this seems counter-intuitive and it is difficult to believe that anyone would buy a machine or a building designed in this way. However, biological organisms are not intelligently designed; they are the result of a multi-scale, dynamic, stochastic and chaotic process. This is most evident in their genomes and encoded molecular processes.

Key concepts for our understanding of the way in which this apparent chaos is orchestrated and has evolved to form trees or humans from unicellular organisms are illustrated in Figs. 15.7 and 15.8. Gene products (proteins, rRNAs, tRNA, miRNAs and other non-protein-coding RNAs) interact in a spatio-temporal fashion to form a complex regulatory and enzymatic network. Multi-layered interaction networks and gene regulation are fundamental to an understanding of morphological evolution at the molecular level. Figure 15.7a depicts a randomly chosen subset of a protein-protein interaction network of a flowering plant. This interaction can be physical in the form of a multi-protein complex such as the dynamic protein complexes that control and initiate transcription (i.e. gene expression; Figs. 15.7c) or it can be indirect and causal (in trans) as in the regulatory network between a transcription factor and its target genes and their gene products (Fig. 15.7b).

Obvious differences can be found in the morphological complexity in the different plants listed earlier (Figs. 15.8b–g). One determinant feature of this complexity is the number of distinct cell types or tissues that these plants express in their life cycle.

In particular, gene regulatory networks and, consequently, the complements of gene regulators (GR: transcription factors [TF], transcriptional regulators [TR] and miRNAs) seem to be tightly linked to the evolution of organismic complexity (Levine and Tjian 2003; De Mendoza 2013). If we consider the phylogenetically inferred gains, losses and expansions of GR gene families along the lineages leading to plants (Fig. 15.8a), a significant correlation can be seen in the complements of GRs with complexity (Lang et al. 2010). The expansion of GR complements, and thus complexity, was achieved via multiple rounds of whole genome or large-scale duplication events (e.g. Lang et al. 2010; De Smet and Van de Peer 2012).

By sequencing the biological diversity ranging from the taxon to the population level, we can construct phylogenomics and population genomics frameworks to

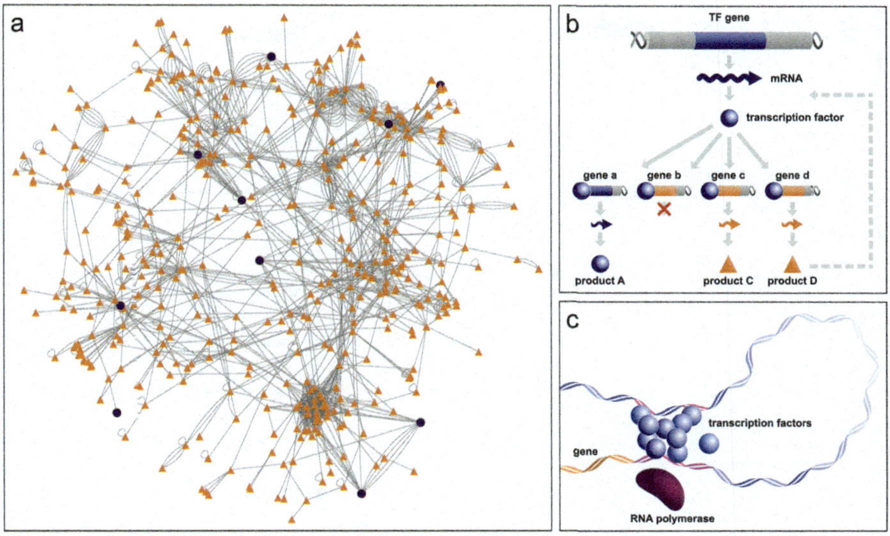

Fig. 15.7 Gene regulatory complexity by means of transcriptional regulation. (**a**) Depicts a complex network formed by a randomly chosen subset of interacting *Arabidopsis thaliana* proteins (Brandão et al. 2009; comprising 6905 nodes and 19938 edges in total). The nodes (451 of 6905) in this network represent gene products whose physical interactions, depicted as edges (1593 of 19938), were experimentally demonstrated. Transcription factors (TF) are highlighted as *blue circles*. Multiple edges between the same products represent the formation of complexes with more than one instance of each product and looped edges on single products represent multimers. A simple gene regulatory network is shown in (**b**). A TF itself is the product of its corresponding gene and physically binds to the promoter region of specific target genes. In that way the TF causally yet indirectly influences the expression of gene products, by either promoting (gene a, c, d) or suppressing (gene b) target gene expression. Such a target gene in turn might either encode for another TF (product A) or for any other gene product that either functions independently (product C) or forms a feedback loop (product D) that controls the expression of the original transcription factor, e.g. a microRNA. (**c**) illustrates a case of target gene regulation via complex formation of several transcription factors. The binding of transcription factors at the promoter region (*pink*) of the gene (*orange*) leads to a conformational change of the DNA structure and enables the binding of additional factors. Only by forming this complex, the RNA polymerase is recruited to the transcription start site and transcription is initiated

identify evolutionary processes (e.g. Wright and Andolfatto 2008). Combined analyses with further experimental data allows the bridging of the gap from genotype to phenotype and helps to explain the evolution of complex traits and organismic diversity (e.g. Freeling and Thomas 2006; Rensing et al. 2008; Lang et al. 2010; De Smet and Van de Peer 2012).

The use of the moss *Physcomitrella patens* and related species as evolutionary models and representatives of an early diverging lineage of land plants has provided important insights into land plant evolution including molecular adaptations related to the conquest of land (Rensing et al. 2008) and the evolution of organismic complexity (Lang et al. 2010).

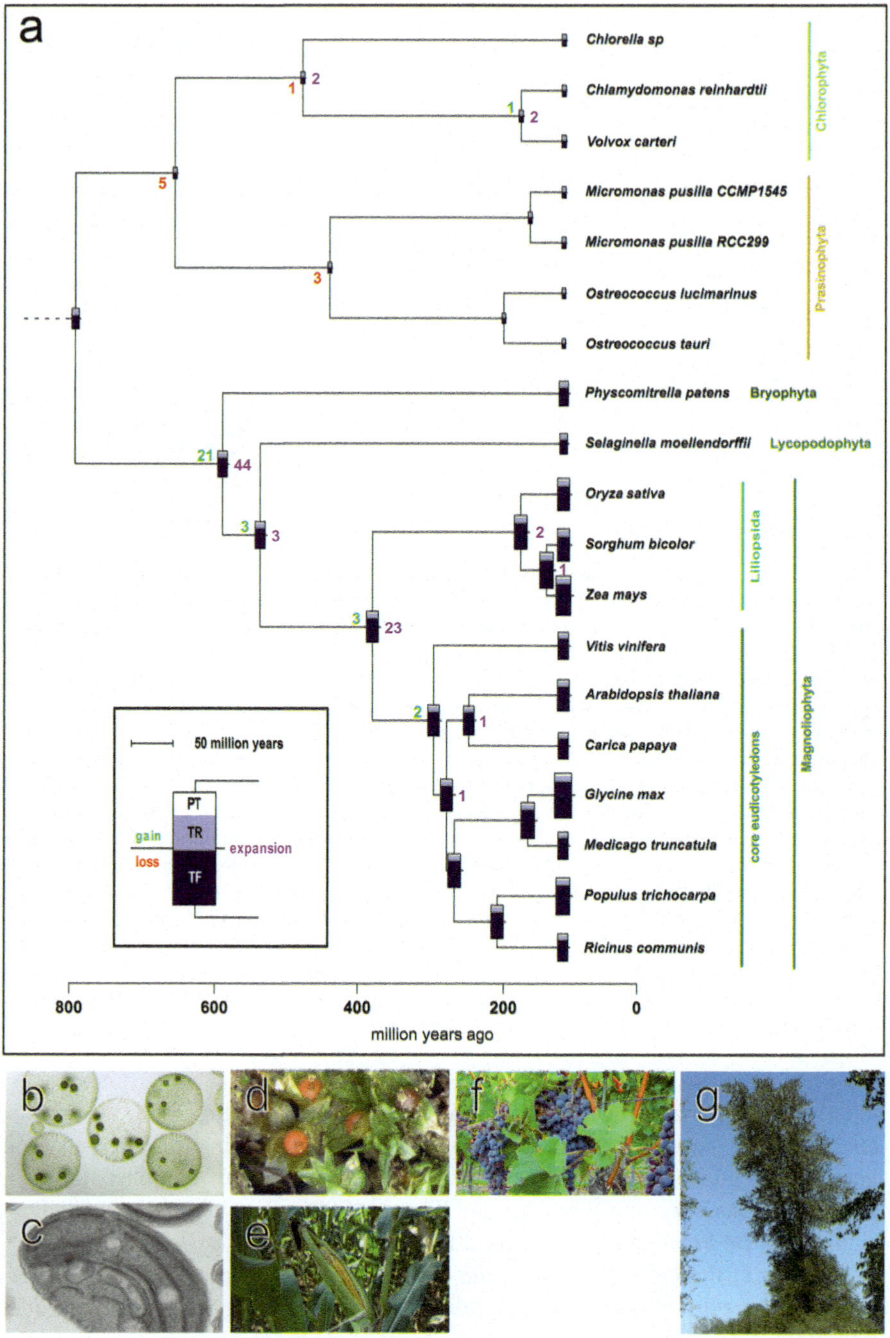

Fig. 15.8 Organismal complexity via expansion of gene regulator families in the evolution of plants. (**a**) shows the extant and reconstructed ancestral numbers of gene regulators (GR), namely transcription factors (TF, *dark blue*), transcriptional regulators (TR, *light blue*) and putative transcription-associated proteins (PT, *white*) along the phylogenetic tree of plant model species

The evolution and radiation of *P. patens* and its close relatives is thought to be shaped by whole-genome duplication events, i.e. polyploidization, and subsequent aneuploidization (Rensing et al. 2007; Beike et al. 2014).

The reconstruction of the polyploidization history of *P. patens* and related taxa enable tracing micro- and macro-evolutionary mechanisms behind speciation and morphological diversification. Furthermore, genomic analysis of these mosses provides an ideal opportunity to study the gene-phene evolution in the scope to serve as a concept generator that can inform the development of exploratory design tools.

15.5 Towards an Exploratory Design Framework Based on Evolutionary Principles

Concluding this chapter, we propose a re-evaluation of the abstractions and heuristics implemented in current EA approaches in order to achieve the transition from micro- to macro-evolutionary processes. We need to increase the complexity of our digital *organisms* and their *environment*. To reach this goal, we suggest the establishment of a flexible algorithmic framework that can be dynamically coupled to the results and hypotheses from current biological research encompassing a greater extent of the complexity of multicellular life. Because of its very nature, evolutionary biology does not usually deliver absolute ground truths, but instead candidate mechanisms and processes. Thus, the proposed framework needs to be dynamic and modular, allowing for experimental execution at low developmental cost to allow the direct integration of quantitative biological data such as evolutionary rates from multiple species or distinct scenarios.

From the perspective of computational design, this framework should enable the dynamic alteration of the *genotype-phenotype* mapping that accommodates both the evolution of 3D structures and more operational outputs, such as the behaviour of agent systems. Given this capacity, a truly explorative design tool that expands current design methods should be achievable. However, we also envisage obvious computational challenges in this approach. If traditional EAs with limited organismic complexity are currently computationally expensive, the challenge will be to find suitable abstractions and heuristics that increase organismic and operational complexity but nevertheless keep algorithmic costs in check.

Fig. 15.8 (continued) (Modified from Lang et al. 2010). At each node in the tree, i.e. for each common ancestor and extant species, the numbers represent the gain (*green*), loss (*red*) and expansion (*pink*) of GRs. Respective boxes depict the individual composition of GRs while their size corresponds to their absolute numbers. (**b**)–(**g**) illustrate the increase in organismic complexity coinciding with the expansion of GRs via a selection of extant species from (**a**), ranging from the multicellular and unicellular green algae *Volvox carteri* (**b**) and *Ostreococcus tauri* (**c**), respectively, over the moss *Physcomitrella patens* (**d**), the grass *Zea mays* (**e**) and the shrub *Vitis vinifera* (**f**) to the tree *Populus trichocarpa* (**g**)

Acknowledgements This work has been funded by the German Research Foundation (DFG) as part of the Transregional Collaborative Research Centre (SFB/Transregio) 141 'Biological Design and Integrative Structures'/project B02.

References

Barricelli NA (1962) Numerical testing of evolution theories. Acta Biotheor 16:69–98. doi:10.1007/BF01556771

Beike AK, von Stackelberg M, Schallenberg-Rüdinger M et al (2014) Molecular evidence for convergent evolution and allopolyploid speciation within the Physcomitrium-Physcomitrella species complex. BMC Evol Biol 14:158. doi:10.1186/1471-2148-14-158

Bellman R (1966) Dynamic programming. Science (80-) 153:34–37. doi: 10.1126/science.153.3731.34

Bentley PJ (1999) Evolutionary design by computers. Morgan Kaufmann Publishers, San Francisco

Bentley PJ, Corne DW (eds) (2001) Introduction to creative evolutionary systems. Morgan Kaufmann Publishers, San Francisco

Bentley DR, Balasubramanian S, Swerdlow HP et al (2008) Accurate whole human genome sequencing using reversible terminator chemistry. Nature 456:53–59. doi:10.1038/nature07517

Brandão MM, Dantas LL, Silva-Filho MC (2009) AtPIN: *Arabidopsis thaliana* protein interaction network. BMC Bioinf 10:454. doi:10.1186/1471-2105-10-454

Caldas LG, Norford LK (2002) A design optimization tool based on a genetic algorithm. In: Automation in construction, pp 173–184

Coelho RF, Echenagucia TM, Pugnale A, Richardson JN (2014) Genetic algorithms for structural design. In: Adriaenssens S, Block P, Veenendaal D, Williams C (eds) Shell structures for architecture: form finding and optimization. Routledge, London, pp 290–294

Darwin C (1859) On the origin of species. John Murray, London

David OE, Greental I (2014) Genetic algorithms for evolving deep neural networks. In: Proceedings of the 2014 conference companion on genetic and evolutionary computation. ACM Press, New York, pp 1451–1452

De Mendoza A (2013) Transcription factor evolution in eukaryotes and the assembly of the regulatory toolkit in multicellular lineages. PNAS 110(50):E4858–E4866

De Smet R, Van de Peer Y (2012) Redundancy and rewiring of genetic networks following genome-wide duplication events. Curr Opin Plant Biol 15:168–176

Deb K, Goldberg D (1989) An investigation of niche and species formation in genetic function optimization. In: Proceedings of the 3rd international conference on genetic algorithms, pp 42–50

Dimčić M (2011) Structural optimisation of grid shells based on genetic algorithms. University of Stuttgart, Stuttgart

Eldredge N, Gould SJ (1972) Punctuated equilibria: an alternative to phyletic gradualism. In: Schopf TJM (ed) Models in paleobiology. Freeman Cooper, San Francisco, pp 82–115

Eshelman L, Schaffer J (1991) Preventing premature convergence in genetic algorithms by preventing incest. In: Proceedings of the 3rd international conference on genetic algorithms, pp 115–122

Finucane E, Derix C, Coates P (2006) Evolving urban structures using computer optimisation techniques. In: Generative Art 2006 GA2006, IX Generative Art conference, 13–15 December 2006

Fraser AS (1958) Monte Carlo analyses of genetic models. Nature 181:208–209

Frazer J (1995) An evolutionary architecture. Architectural Association, London

Freeling M, Thomas BC (2006) Gene-balanced duplications, like tetraploidy, provide predictable drive to increase morphological complexity. Genome Res 16:805–814. doi:10.1101/gr.3681406

Galán S, Mengshoel O, Pinter R (2013) A novel mating approach for genetic algorithms. Evol Comput 21:197–229

Gregory TR (2005) The C-value enigma in plants and animals: a review of parallels and an appeal for partnership. Ann Bot 95:133–146. doi:10.1093/aob/mci009

Greilhuber J, Dolezel J, Lysák MA, Bennett MD (2005) The origin, evolution and proposed stabilization of the terms "genome size" and "C-value" to describe nuclear DNA contents. Ann Bot 95:255–260. doi:10.1093/aob/mci019

Hemberg M, O'Reilly U-M, Nordin P (2001) GENR8 – a design tool for surface generation. In: Goodman ED (ed) Genetic and evolutionary computation conference late breaking papers, pp 413–416

Hemberg M, O'Reilly U-M, Menges A et al (2008) Genr8: architects' experience with an emergent design tool. In: The art of artificial evolution. Springer, Berlin, pp 167–188

Holland JH (1975) Adaptation in natural and artificial systems: an introductory analysis with applications to biology, control, and artificial intelligence. University of Michigan Press, Ann Arbor

Jia F, Lo N, Ho SYW (2014) The impact of modelling rate heterogeneity among sites on phylogenetic estimates of intraspecific evolutionary rates and timescales. PLoS One 9, e95722. doi:10.1371/journal.pone.0095722

Jo J, Gero J (1998) Space layout planning using an evolutionary approach. Artif Intell Eng 12:149–162

Kazimipour B, Li X, Qin A (2014) A review of population initialization techniques for evolutionary algorithms. In: Proceedings of the IEEE congress on evolutionary computation, pp 2585–2592

Koza JR (1992) Genetic programming as a means for programming computers by natural selection. Stat Comput 4:87–112

Kühn M, Severin T, Salzwedel H (2013) Variable mutation rate at genetic algorithms: introduction of chromosome fitness in connection with multi-chromosome representation. Int J Comput Appl 72:31–38

Ladkany G, Trabia M (2012) A genetic algorithm with weighted average normally-distributed arithmetic crossover and twinkling. Appl Math 3:1220–1235

Lang D, Zimmer AD, Rensing SA, Reski R (2008) Exploring plant biodiversity: the Physcomitrella genome and beyond. Trends Plant Sci 13:542–549. doi:10.1016/j.tplants.2008.07.002

Lang D, Weiche B, Timmerhaus G et al (2010) Genome-wide phylogenetic comparative analysis of plant transcriptional regulation: a timeline of loss, gain, expansion, and correlation with complexity. Genome Biol Evol 2:488–503. doi:10.1093/gbe/evq032

Levine M, Tjian R (2003) Transcription regulation and animal diversity. Nature 424:147–151

Menges A (2012) Biomimetic design processes in architecture: morphogenetic and evolutionary computational design. Bioinspir Biomim 7:015003. doi:10.1088/1748-3182/7/1/015003

Menges A, Ahlquist S (eds) (2011) Computational design thinking. Wiley, Chichester

Michalek J, Choudhary R, Papalambros P (2002) Architectural layout design optimization. Eng Optim 34:461–484

Miller B, Goldberg DE (1995) Genetic algorithms, tournament selection, and the effects of noise. Complex Syst 9(3):193–212

Mühlenbein H, Schlierkamp-Voosen D (1993) Predictive models for the breeder genetic algorithm I. Continuous parameter optimization. Evol Comput 1:25–49

Palazzo AF, Gregory TR (2014) The case for junk DNA. PLoS Genet 10, e1004351. doi:10.1371/journal.pgen.1004351

Rahnamayan S, Tizhoosh H, Salama M (2007) A novel population initialization method for accelerating evolutionary algorithms. Comput Math Appl 53:1605–1614. doi:10.1016/j.camwa.2006.07.013

Rechenberg I (1971) Evolutionsstrategie – Optimierung technischer Systeme nach Prinzipien der biologischen Evolution. TU, Berlin

Rensing SA (2014) Gene duplication as a driver of plant morphogenetic evolution. Curr Opin Plant Biol 17:43–48. doi:10.1016/j.pbi.2013.11.002

Rensing SA, Ick J, Fawcett JA et al (2007) An ancient genome duplication contributed to the abundance of metabolic genes in the moss *Physcomitrella patens*. BMC Evol Biol 7:130. doi:10.1186/1471-2148-7-130

Rensing SA, Lang D, Zimmer AD et al (2008) The *Physcomitrella* genome reveals evolutionary insights into the conquest of land by plants. Science 319:64–69. doi:10.1126/science.1150646

Sato S, Hayashi T, Takizawa A et al (2004) Acoustic design of theatres applying genetic algorithms. J Temporal Des Archit Environ 4:41–51

Schwefel H-P (1974) Numerische Optimierung von Computer-Modellen. TU, Berlin

Simoes A, Costa E (2000) Using genetic algorithms with asexual transposition. In: Genetic and evolutionary computation conference

Spaeth AB, Menges A (2011) Performative design for spatial acoustics: concept for an evolutionary design algorithm based on acoustics as design driver. In: Respecting fragile places, pp 461–468

Sterck L, Rombauts S, Vandepoele K et al (2007) How many genes are there in plants (... why are they there)? Curr Opin Plant Biol 10:199–203. doi:10.1016/j.pbi.2007.01.004

Tirumala SS (2014) Implementation of evolutionary algorithms for deep architectures. In: Proceedings of the second international workshop on artificial intelligence and cognition, pp 164–171

Tuhus-Dubrow D, Krarti M (2010) Genetic-algorithm based approach to optimize building envelope design for residential buildings. Build Environ 45:1574–1581. doi:10.1016/j.buildenv.2010.01.005

Turing A (1950) Computing machinery and intelligence. Mind LIX:433–460

Wright A (1991) Genetic algorithms for real parameter optimization. In: Foundations of genetic algorithms, pp 205–220

Wright S, Andolfatto P (2008) The impact of natural selection on the genome: emerging patterns in *Drosophila* and *Arabidopsis*. Annu Rev Ecol Evol Syst 39:199–213

Wright JA, Loosemore HA, Farmani R (2002) Optimization of building thermal design and control by multi-criterion genetic algorithm. Energy Build 34:959–972. doi:10.1016/S0378-7788(02)00071-3

Zhu YO, Siegal ML, Hall DW, Petrov DA (2014) Precise estimates of mutation rate and spectrum in yeast. Proc Natl Acad Sci U S A 111:E2310–E2318. doi:10.1073/pnas.1323011111

Chapter 16
Fabrication of Biomimetic and Biologically Inspired (Modular) Structures for Use in the Construction Industry

Daniel Coupek, Daria Kovaleva, Hans Christof, Karl-Heinz Wurst, Alexander Verl, Werner Sobek, Walter Haase, Götz T. Gresser, and Armin Lechler

Abstract The transformation of biological paradigms into building construction involves the transfer of structure and system-defining properties from biological role models to construction-specific and innovative non-construction-specific systems and processes. The challenge of manufacturing biomimetic and bio-inspired structures includes the provision of methods and procedures that allow the mapping of biological features on a production-related description. The methodological approach requires the validation and verification of existing production methods at the small scale (model, elementary cell) in order to transfer findings to the production of components at the construction scale. Additionally, the biological features that cannot be reproduced by existing methods require further adjustment or the development of new methods for appropriate transfer. A basic condition for the further development of such production procedures is the possibility of manufacturing complex structures based on biological strategies concerning resource and energy consumption, waste production and greenhouse gas emissions.

D. Coupek • K.-H. Wurst • A. Verl • A. Lechler (✉)
Institute for Control Engineering of Machine Tools and Manufacturing Units, University of Stuttgart, Seidenstr. 36, 70174 Stuttgart, Germany
e-mail: armin.lechler@isw.uni-stuttgart.de

D. Kovaleva • W. Sobek • W. Haase
Institute for Lightweight Structures and Conceptual Design (ILEK), University of Stuttgart, Pfaffenwaldring 7 + 14, 70569 Stuttgart, Germany
e-mail: werner.sobek@ilek.uni-stuttgart.de

H. Christof • G.T. Gresser
Institute for Textile Technology, Fibre Based Materials and Textile Machinery (ITFT), University of Stuttgart, Pfaffenwaldring 9, 70569 Stuttgart, Germany
e-mail: goetz.gresser@itv-denkendorf.de

J. Knippers et al. (eds.), *Biomimetic Research for Architecture and Building Construction*, Biologically-Inspired Systems 9, DOI 10.1007/978-3-319-46374-2_16

16.1 Strategies for Fabrication of Biomimetic Structures

Generally, the family of layer-based production processes is suitable for the development of complex geometries and structures, the functional integration of components and, consequently, the generation of lightweight structures based on natural models. In addition to the complexity of the geometries of such models, the variety of employed materials and their resulting behaviour has to be considered. Since a basic principle of load-bearing living matter is the transfer of the load via matrix-embedded fibres, various macroscopic features of construction materials will be achieved by the targeted placement of diverse fibrous materials. Thus, the production of fibre-reinforced structures with a graded behaviour should be possible by layer-based processes and by spatial pultrusion. Hence, known layer processes and pultrusion methods will serve as a starting point, although their suitability must be evaluated and further developed with respect to the requirements of biomimetic construction components. Three process classes have been identified to enable the fabrication of biomimetic structures:

- reinforced Fused Deposition Modelling (FDM) with endless fibres,
- (braid-) pultrusion and fibre placing processes,
- spraying and casting of functionally graded concrete.

The extent to which these production methods can complement each other or can be used as a hybrid production process in the same plant will be examined and considered.

This chapter fulfils a cross-sectional role, as for all investigated biological structures the developed results have to be verified in a first step by means of a technical reproduction. Based on these results, the methodology and future production methods for structures on a construction scale have to be developed. The gradation of materials, porosities and textures allow the spatial differentiation of the structural and the physical properties as they appear in many biomimetic models, resulting in multifunctional solutions. The investigation and the design of guiding systems for print and extrusion heads are also essential, as are pultrusion facilities to produce biomimetic parts on a construction scale.

16.1.1 Requirements for the Fabrication of Biomimetic Structures

From the analysis of biological models, various principles of material and structural organisation need to be transferred to reproduce biomimetic structures at an elementary cell size and, later, on a construction scale. For this purpose, production parameters need to be extracted from biological principles by abstraction and simulations. Thus, initial boundary conditions and requirements for process sequences and, consequently, for the modification of the production processes can be derived.

The first group of characteristic features can be derived from functionally graded porous or cellular natural structures that have a unique combination of low weight and high performance. Many biological systems use lightweight porous structures not only for mechanical support, but also as functional components for heat and mass transfer. Since plants have developed individual strategies to adapt to various circumstances, one goal is to transfer some of these properties to industrially feasible, porous construction materials. Based on the biological role model, requirements are derived for the fabrication of porous workpieces. This will lead to optimal physical properties with regard to frost-resistance, thermal insulation and moisture transport in heterogeneous porous structures. Pores can be created either by internal material properties (e.g. foam) or by designed workpiece features. In FDM, pores and hollow structures can be printed by using extremely fine nozzle diameters (<< 1 mm). Thermal insulation in concrete workpieces can be created by developing functionally graded closed- or open-pore structures. Another source of inspiration from natural porous and layered graded materials is their ability to combine strength and ductility, which result in high energy dissipation, in order to avoid overcritical loads. The transfer of such properties into construction-specific materials and structures is of particular interest for buildings in regions with high seismic activity or exposed to dynamic loads. Methodologies for transfer include the evaluation and development of graded porous and layered fibre-reinforced structures with the ability to locally adapt mechanical properties, such as strength and toughness, through the controlled variation of material and structural distribution. By using functionally graded concrete or fibre-reinforced FDM, such structures can be created with variations of density, porosity and fibre orientation in accordance to a specific layout.

The second group of characteristic features is related to fibre-based and fibre-reinforced materials and structures. Scientific investigations of natural role models have shown that all plants consist of natural fibre-reinforced materials. In particular, when it comes to branched load-bearing structures, nature delivers interesting concepts regarding the orientation of the fibres within the branching structures (Milwich 2015). For example, use of the principle of fibre orientation in *Shefflera* plants has led to a new braiding technique. This allows the fabrication of a lightweight branched structure made of fibre-reinforced polymers with load-adapted fibre orientation that serves as the outer formwork for concrete filling. Branched column structures, produced with this technology, are able to bear torsional, flexural and tensile forces.

Fibre-based materials can also be used for the development of self-adjusting structures in various architectural applications. A broad variety of the movable structures in nature are flat or shell-shaped and, therefore, the technical adaption will consist of fibre-reinforced plastic components with fibre orientation. In particular, the special hingeless moving mechanism of plants can be technically transcribed. The active movement of traps in carnivorous plants such as *Aldrovanda vesiculosa* deliver principles for adjustable structures that can be used as shading elements, which serve the demand for a more energy efficient and sustainable style of architecture.

Natural systems can also serve as a source of inspiration for complex composite structures and even processes. For example, the shells of sand dollars, highly adapted sea urchins, have been studied as a biological concept generator for segmented shells in buildings. Sand dollars show great geometric variety with respect to overall shape and plate arrangements. The high load-bearing capacity of the sand dollar skeleton is well adapted to its environment and it shows morphological features that are also required of shells in building construction (mainly flat curvature, apertures and column-like connections). The shells of sand dollars consist of modular polygonal plates linked at their edges by finger-like calcite protrusions and organic fibres and can thus serve as a suitable model for shells in building construction. The physical representation of the skeleton and joints by additive fabrication processes and the influence of additive manufacturing on the joints are being studied. Strategies for fibre-reinforced joint mechanisms will be derived in order to reach uniform stiffness and stability, even in segmented structures.

Based on the shell formation of molluscs, we aim to develop a new additive process for architectural envelopes. One promising biological model is shell formation in the land snail, which produces complex shapes with highly structured, functional surfaces. In contrast to most other shell-producing animals, snails produce building material extruded at a single groove of productive tissue, which is similar to current additive fabrication processes. Land snails extrude a thin polymer film (periostracum) that is reinforced with calcium carbonate. The periostracum is the outer-most layer of the shell providing the formwork for calcification and a functional surface with a specific structure (water repellent, self-cleaning, etc.). The load-bearing material of the shell is the thick calcium carbonate layer. The basic principles that can be transferred into technical fabrication processes are a multi-scale workpiece geometric representation and multi-scale production processes of surface layer, formwork and load-bearing structure.

16.1.2 Workpiece Optimisation

An optimised workpiece can be characterised by the synthesis of design methods, material properties and fabrication techniques towards a maximum performance within predefined design constraints. The formation of natural systems can serve as a source of inspiration in an approach to such problems, since the interdependencies between shape, microstructure and fabrication are inherent to the process of growth, which appears as a long-term adaptation of an organism to ever-changing external and internal conditions. Organisms develop performative structures by using limited resources in a deliberate combination of material constituents with highly differentiated placement of them under fabrication constraints. Thus, for the fabrication of bioinspired structures, the interdependencies between design methods, material properties and fabrication techniques need to be considered in order to reproduce such material and structural distributions. This becomes possible by the adjustment of process parameters (trajectories, feed-rates, temperatures) or process components

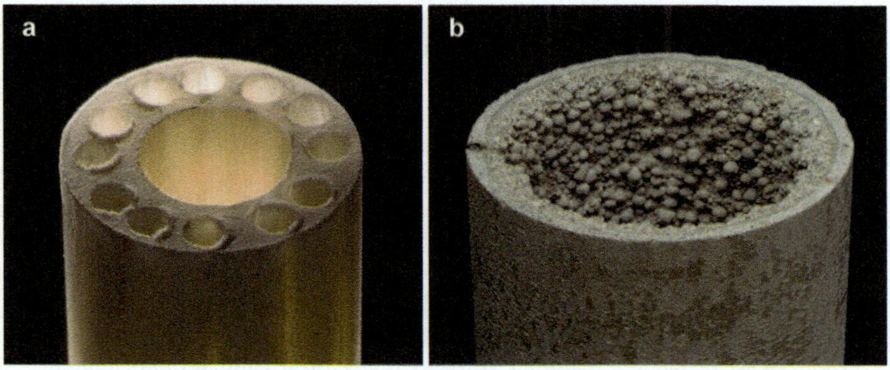

Fig. 16.1 Geometrically and topologically optimised structures; (**a**) 'technical plant stem' produced with pultrusion process, ITV, Denkendorf; (**b**) functionally graded column, ILEK, University of Stuttgart

(development of printing heads, nozzle design, material feeding systems) or the development of new processes (functionally graded concrete spraying, new additive processes).

Various strategies have been developed for fibres, matrix materials and their combinations. For example, for fibre-based materials in such fabrication processes as pultrusion, fibre tape-laying or fibre-reinforced FDM, the arrangement of the direction of fibre placement results in improved material layout and thus higher performance (Fig. 16.1a). In this case, the fibre layout is developed in accordance with material properties and under fabrication constraints (continuous deposition, collision check, multi-axes trajectories).

For matrix materials (e.g. polymers and minerals), another way for the optimisation of the workpiece, which has proved to be particularly suitable, is the controlled variation of material properties (density, material volume fraction, etc.) within the design volume. For example, in concrete casting and spraying technology, variation of density can be achieved by the layer-wise casting of graded concrete mixtures, by the use of a controlled demixing process (Fig. 16.1b) or, for complex three-dimensional gradients, by the application of guided gradient mixture deposition based on density distribution layout. In all the cases described above, the integration of material properties, the design of the layout and the transfer of layout parameters to the fabrication parameters is required.

16.2 Fast Fabrication of Lightweight Porous Structures by Using Fibre-Reinforced FDM

Research work with respect to additive manufacturing focuses on fibre-reinforced FDM, as FDM is a popular production method for the generation of structures, both at the model size and, later, at construction scale. Selective Laser Sintering

(SLS) and Laminated Object Modelling (LOM) can be used in specific applications as complementary processes. Building materials need to be produced with a wide variety of macroscopic properties, especially by reinforcing the matrix material with diverse fibre materials and by variation of the fibre directions in different areas. The use of endless fibres together with an additive processes has not been investigated in detail yet and so the basic feasibility of this approach needs to be examined first. Current limitations that have to be overcome are small working areas, uniform infill strategies and low building rates. The working area can be increased by new kinematic concepts such as cable robots, mobile robots or portal systems. Another approach is to fabricate small segments that have to be assembled on site, requiring new joining strategies in order to ensure the performance of optimised structures. Current FDM trajectory planning software tools offer only limited amounts of infill strategies (e.g. diagonal, honeycomb, solid, etc.) that are used for the whole workpiece. Topologically optimised workpieces require new and locally variable infill patterns, especially for fibre-reinforced parts, as the fibres must be placed in the direction of stresses. The low building rates, which make it difficult to fabricate large parts, can be increased by parallelisation of the extrusion and by increasing the nozzle diameter, influencing the path planning and trajectory controlling algorithms.

16.2.1 Workpiece Optimisation for Additive Manufacturing

Workpiece orientation in standard FDM processes is of great importance concerning mechanical properties and overhanging surfaces. New constraints arise when fibres are placed in the workpiece with a specific orientation. Both two- and three-dimensional trajectories require systems with more than three degrees of freedom for fibre and print head orientation, exceeding the state of the art of FDM machines. In addition, three-dimensional trajectories might generate collisions between the printing head and previously printed zones of the workpiece; this has to be avoided by the path planning tool. The goal is to determine, for each area of the workpiece, a possible series of workpiece/head orientations according to required fibre trajectories, while taking into account previously deposited material, potential collisions and overhangs. A mandatory requirement is the accessibility of all printing positions and therefore the feasibility for producing the workpiece in its current position. If several positions and orientations of the workpiece are feasible, optimisation towards printing speed, accuracy and minimisation of support material can be performed. In current FDM construction processes, the infill is built at the same time as the perimeters. The workpiece is defined only by the surface and the whole part is filled by using the same infill pattern (diagonal infill, honeycomb, etc.), neglecting internal stress and geometry.

The objective is to define new three-dimensional infill strategies compatible with fibre deposition trajectories and topological optimisation of the workpiece, as fibre reinforcements are also needed inside the workpiece. The infill strategies support and guarantee the internal fibre placement derived from finite elements simulations.

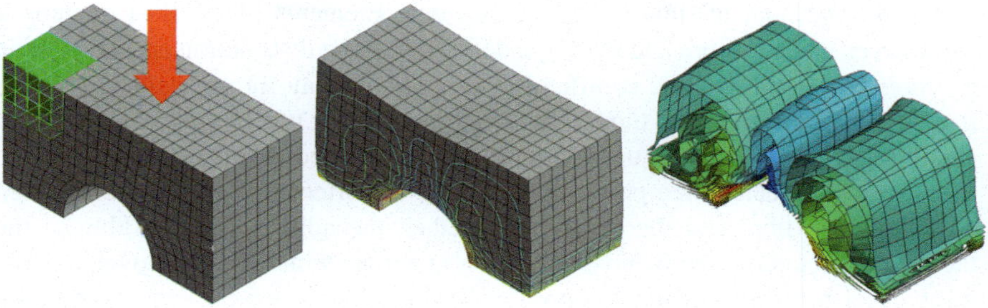

Fig. 16.2 Topological optimisation of an assembled structure consisting of sub-modules (e.g. cubes)

Large architectural structures can be built by assembling smaller segments printed on conventional machines. Thus, multi-scalar topological optimisation has to be carried out for both the single segments and the assembled structure (Dadalau et al. 2009). This results in micro-scale optimisation for single segments and macro-scale optimisation for the entire structure (Fig. 16.2).

16.2.2 Fibre-Reinforced FDM with Use of Endless Fibres

Currently, the mechanical strength of conventional FDM parts is approximately 80–85 % of the corresponding injection moulded parts (Fischer et al. 2013a). Integration of fibres into workpieces in the direction of stresses increases the strength. The thermoplastic filament that is used in conventional FDM systems can handle tensile and compression forces allowing the trivial material supply that can be found in state of the art printers. The filament is pulled by a motor from the filament roll and is pushed into a heater and finally out of the extrusion nozzle. In contrast, fibres can handle only tensile forces requiring new material supply strategies. One concept available on the market uses two nozzles, one for the filament and the second for a carbon fibre. This machine is limited to three axes and therefore reinforcement is only feasible inside the layers and not between them. Another concept of an extended FDM process has been developed by using a six-axis industrial robot, including the possibility of extruding fibre-reinforced polymer materials (Fischer et al. 2013b). However, only flat workpieces have been built by using conventional slicing software. As online cutting of the fibre is not possible in this system, the workpieces are limited to very simple shapes.

The two major drawbacks of the presented solution are: the low infill of fibres (<20 %) and the absence of the possibility for cutting the fibre during the extrusion process. This results in a continuous path that never leaves the workpiece and to limitations in the geometry. Therefore, a new printing head must be developed to overcome these drawbacks together with trajectory planning strategies.

For the mechanical simulation (by the Finite Elements Method) of composite workpieces made of fibre and matrix, a definition of the distribution and orientation of fibres within the volume is difficult. One possible way for mechanical analysis is to build a finite elements model based on the extruder trajectories, whereby each single trajectory element is used to build a single finite element with well-known characteristics, such as the presence of a fibre, the orientation of fibre, etc. When depositing the fibre and thermoplastic together, the adhesion of the fibre to the thermoplastic matrix, the behaviour of both materials while cooling down, and the mechanical behaviour of the composite filament while being pushed or pulled on the workpiece over curved trajectories are all critical but unknown parameters.

A critical bending angle must not be exceeded as this will damage the fibre bundles. Therefore, a possible set of parameters can be determined related to the fibre and thermoplastic characteristics, extrusion temperature, radii, tilt angle, overhang angle, etc. The extrusion angle can be minimised by tilting the extrusion head in reverse to the trajectory direction so that it pulls the filament backwards instead of bending it. This has to be planned in the CAM tool. For applying these trajectories to the process, two additional axes in the printing head are needed. In addition to those two axes, the nozzle has to be designed in a new way to minimise the extrusion angle (Fig. 16.3). Current nozzles do not allow fibre integration into the thermoplastic melt. In addition, short nozzles reduce the possible extrusion angle. The goal is to develop a new nozzle that allows the introduction of the fibre into the melt without causing air bubbles. The nozzle design must be chosen to reduce the restrictions imposed on the extrusion angle. The nozzle mechanics will be designed to integrate the fibre into the melted thermoplastic, so that it is pulled from its material supply system. A curved shape of the nozzle adapter should allow the extrusion angle to be increased easily in order to avoid collisions of the printing head with the built part or machine.

An uncontrolled fibre feeding system without feedback can lead to loose fibre segments in the system, causing bending or ravelling, which might damage or even break the fibre bundles. If the fibre and thermoplastic filament feed rate are not

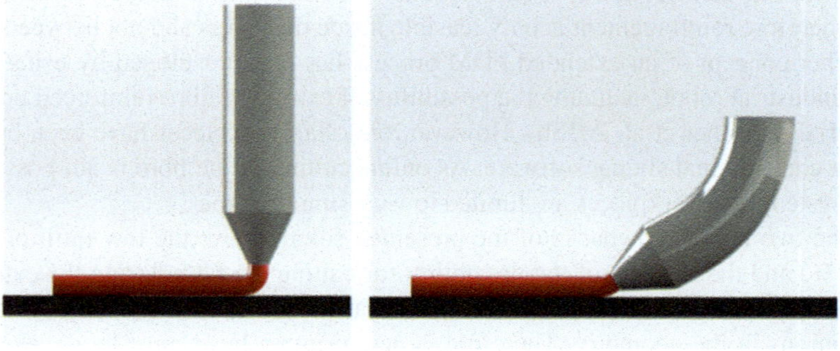

Fig. 16.3 Constrained angle for fibre integration requires new nozzle design

synchronised, the workpiece geometry and internal stresses will be affected. In the worst case, distortion of the workpiece might occur during the building process. By the integration of sensor systems, the fibre tension and extrusion speed must be controlled so that a defined fibre behaviour is guaranteed. Fibre tension can be controlled by dancer control with an integrated force sensor. The dancer can be either a passive or an active spring, e.g. electric drive. Controlling the fibre tension in this way guarantees that the fibre remains in the defined feed path under defined tension. An idler pulley with integrated resolver can be used for measuring the actual fibre feed rate. As fibres are able to receive only tensile forces and as the main feed tensile force is induced by the extruded thermoplastic, the control system can only reduce the current velocity by an electro-dynamic brake at the fibre roll or with an additional actuator.

16.2.3 Nozzle Array for Fast Material Deposition

Large-scale structures require higher building rates than current FDM machines can offer. For increasing the extruded material volume per time, new concepts with nozzle arrays are being developed, enhancing the possibilities of today's systems. Nozzle arrays allow the parallelisation of the extrusion process, thus increasing the material feed (Fig. 16.4). Depending on the rotating angle of the array with regard to the print head feed direction, different structures can be produced (several parallel lines, closed area or one thin wall). If the printing head moves in the direction of the array, a thin wall can rapidly be built up; however, z-axes in the nozzles are required to avoid collision of a nozzle with extruded material from its antecessor. FDM processes with nozzle arrays require new strategies for path planning and controlling. The main goal of these methods is to avoid collisions with previously printed areas while reducing building time by parallel extrusion. Additional axes can

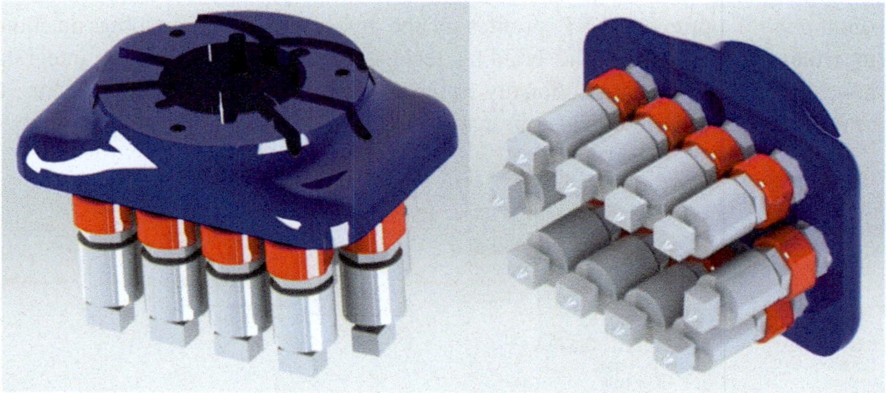

Fig. 16.4 Concept of a printing head with a nozzle array consisting of ten nozzles

be included in the array to increase the degree of freedom of the nozzles. However, it becomes more complex to control the system, as the additional axes must be included into the kinematic transformation inside the numeric control system of the machine in order to fabricate workpieces, as described by G-code that is used in conventional machining.

16.3 Pultrusion and Braiding Processes for Fibre-Reinforced Lightweight Structures

Pultrusion is one of the very few processes available for manufacturing endless fibre-reinforced composites continuously. It allows the production of profiles, such as rods, beams or hollow tubes in a wide range of possible constant cross-sections with different matrix materials (thermoset and thermoplastic resins) and fibre materials (e.g. carbon, glass, basalt etc.). Because of their advantageous cost, pultruded composite profiles are widely used in the automotive and aviation industries. High production output also makes the pultrusion process particularly suitable for manufacturing composite parts for the construction industry in high quantities. With the traditional pultrusion process shown in Fig. 16.5, only unidirectional reinforced profiles can be produced. These profiles can bear high tensile and flexural loads. Because of the lack of fibres in the transverse direction, the capability to bear torsional loads is low.

Various projects aim at enhancing the potential of the pultrusion process in the production of more complex profiles and shapes. For example, pultrusion lines for the production of curved profiles have recently been developed (Mallick 2008). However, to date, no pultrusion line is capable of producing complex hierarchical or gradient three-dimensional structures, observed in natural concept generators. Other manufacturing methods are also not fully suitable for building structures such as those found in nature.

A first step towards a cost-effective pultrusion process for manufacturing (biomimetic) optimised FRP profiles is the integration of a braiding machine into the pultrusion line. The braiding technique makes it possible to integrate helically oriented fibres additionally to the unidirectional-oriented fibres. Thus, a braid-pultruded profile can also bear torsional loads. The braiding process allows

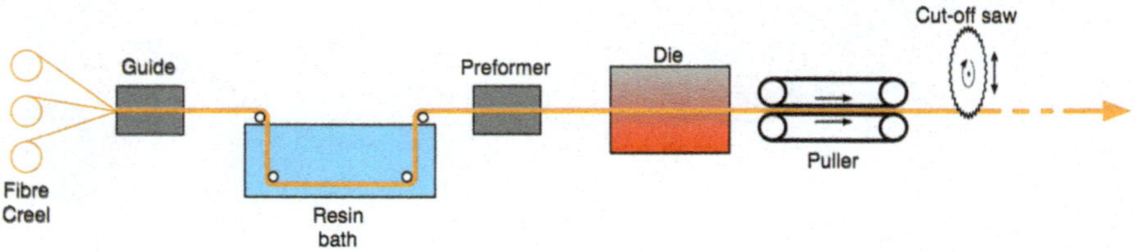

Fig. 16.5 Standard pultrusion process for thermoset resins

the adjustment of the fibre angles between 10° and 70° along the pultrusion axis. Therefore, depending on the braiding angle, the ratio between bending stiffness and torsional stiffness can be adjusted.

In close cooperation, biologists and engineers examined diverse plant stems, such as the giant reed (*Arundo donax*) and horsetail (*Equisetum hyemale*), to find possibilities for enhancing the mechanical properties of pultruded profiles. The research work finally led to the development of the technical plant stem profile with significantly higher structural damping capabilities compared with a tubular-shaped profile with the same cross-sectional area (Ghomeshi 2015). This offers the possibility of mass and information transport through the composite structure. In this case, the biomimetic optimisation of the structure not only contributes to the mechanical properties, but also adds new functions such as sensor integration possibilities to the profiles. This is a step towards mass production of intelligent composite structures.

16.3.1 Adjustment of the Pultrusion Process for the Production of Spatial Bioinspired Lightweight Components

Braid-pultrusion allows variation of the mechanical properties of the profile without changes in the type of reinforcement fibres or the matrix by adjusting the fibre angles. Nevertheless, in the present state, the pultrusion process is limited to profiles with a constant cross section.

To overcome the current limitations of the pultrusion process, a new approach to producing three-dimensional (spatial) fibre-reinforced parts is being investigated. The aim is to develop further the pultrusion process, which then would be capable of aligning fibres spatially without the need of a layup tool or core. A second aim is to develop a bioinspired topologically optimised structure by a gradual change in fibre orientation and fibre density within very small areas. A highly sophisticated topology at various hierarchical levels is observed in plankton species. In the plankton-tech project, optimised natural topology has been analysed and transferred into various composite structures as seen in Fig. 16.6 (Milwich 2015).

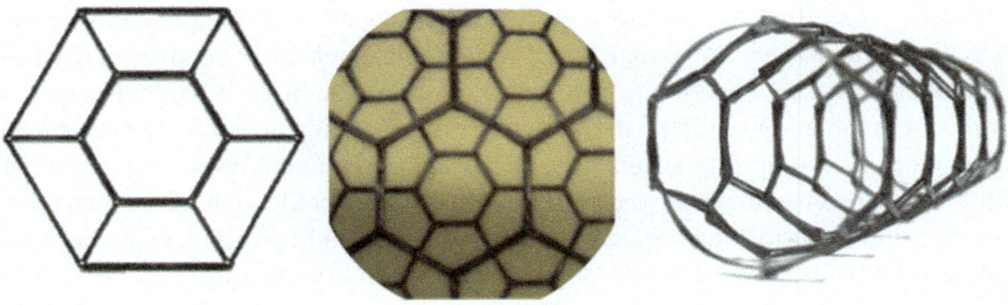

Fig. 16.6 Biomimetic technical structures inspired by diatoms

The placement of fibres in a spatial manner can be achieved by mounting the pultrusion die onto an industrial robot. This allows the pultrusion die to reach almost every coordinate within the working space and basically makes spatial layup possible. As the fibres need to be pulled through the die, a pulling device is mounted in front of the die. The second aim (changing fibre orientation/density) can be attained by using multiple pultrusion dies with different fibre materials or different fibre thicknesses of the same fibre material.

For the production of spatial structures through a moving pultrusion die, a fast curing thermoset matrix system is needed to achieve high production speeds. The investigation of various matrix systems has led to the conclusion that thermoplastic matrix systems are only partly suitable for the above-mentioned pultrusion process. The high viscosity of the thermoplastic resins can cause an insufficient fibre wet out and therefore good mechanical properties will not easily be achieved (Mallick 2008). An option to improve the wet out is the use of reinforcing fibres, which are pre-impregnated with thermoplastic matrix (thermoplastic prepregs). These prepregs are relatively stiff and not suited for a fast change of the composite properties.

In the case of thermoset matrix systems, epoxy resins are a standard matrix material for many FRP parts. A disadvantage of epoxy resins is that they need time-consuming curing/post-curing cycles during production (Mallick 2008). Especially for the complex plankton-tech hierarchical parts as seen in Fig. 16.6, each hierarchical level has to be fully cured before the next level can be produced. The use of epoxies would take too much time to produce those composite structures. Therefore, epoxies will not be suitable initially.

Since the curing time is crucial for the spatial pultrusion process, a fast curing thermoset matrix system needs to be identified. The most promising option is the use of a matrix system that does not cure through heat but through a reaction initiated by ultraviolet light. Such so-called photocuring or UV-curing resins are mostly urethane acrylates, polyester acrylates, amino acrylates and epoxy acrylates (Duan et al. 2010). Tests with a radiation-curable mixture of acrylated resin and acrylated monomer have shown very good results with regard to curing time. Depending on the match between the photocuring agent and the UV-light source, curing times of less than 1 min can be achieved. Evaluated by curing time, UV-curing resins seem to be the ideal matrix material for a spatial pultrusion process.

Not only curing time, but also the cross-linking degree of the resin can be adjusted by shifting the wavelength or the intensity of the UV-light source. Low-intensity irradiation of the photocuring agent slows down the cross-linking reaction and the resin can be brought into a gel state where it is highly viscous but still deformable. This offers the possibility of changing locally the strength of the pultruded composite component for later deformation. With this potential, interlocking connections between different rovings or profiles can be formed. Once the non-fully interlocking segment is brought into the right position, it can be cured to completion by spotlighting the segment again with UV radiation.

This new spatial pultrusion process is still in the concept phase and has not yet been validated. Several technical parameters such as the maximal horizontal span without deflection, minimal redirection radius without buckling, possible fibre volume fraction, resulting pulling forces and possible speed of production need to be examined.

16.3.2 Automated Tape Laying – Fibres Following the Flux of Force

Automated fibre placement technologies offer the possibility of placing reinforcement fibres in almost any desired angle with only a minimum amount of waste fibres. Three different fibre placement technologies are mainly available nowadays (Neitzel et al. 2014). Dry-fibre placement technology is capable of placing dry rovings of any reinforcement fibre material on a substrate. Either the fixation of the fibre on the substrates is achieved by a wet binder or the fibre can be stitched onto a textile substrate with a sewing thread. The infiltration of the fibres with resin occurs after all required fibres for one component have been laid up. This process is limited to a low thickness of entire layers. Because of the compact fibre layup, the fibre wet out becomes difficult and the risk of dry spots rises.

The difficult fibre wet out for parts with large dimensions can be avoided by using fibres that are pre-impregnated with thermoset resins (prepreg-tapes). After placing the tapes in the desired orientation on a surface, curing will take place in an autoclave. Another possibility is to impregnate reinforcement fibre rovings on-line in a resin bath; this allows the precise placement of delicate structures as seen in Fig. 16.6b (Milwich 2015).

In contrast to pultrusion, tape laying is much better suited for the use of reinforcement fibres that are pre-impregnated with thermoplastic matrix material. Preconfigured thermoplastic tapes are used that are consolidated either on-line through thermal heaters or lasers or after layup in a heated tool or press. Prepreg-tapes, which have a minimum width of 5 mm, can be placed on a curved plane but cannot follow a curved path on a flat surface. Therefore, prepreg-tapes can follow the flux of force only within certain limitations. This can be compensated within certain limits by an optimised layup planning with multiple, differently oriented layers of tape. Figure 16.7 shows the layup of thermoplastic tapes on a plane surface.

The efficient use of thermoplastic tape laying technologies will be investigated with the goal of manufacturing biological and topological optimised lightweight components with hierarchical structures. In addition, the link to other fabrication technologies will be investigated and evaluated.

Fig. 16.7 Thermoplastic carbon fibre tape being laid on metal sheet

16.4 Spraying and Casting of Functionally Graded Concrete Structures

Fabrication of bio-inspired functionally graded concrete structures and components opens the possibility to reconsider the way that such widely used construction material as concrete is applied in the building industry. Investigations on functionally graded concrete technology developed in Institute for Lightweight Structures and Conceptual Design at University of Stuttgart have shown that the functional gradation of material properties in accordance with structural and thermal requirements leads to a higher performance and the multi-functionality of a structure while reducing the mass, embodied energy and CO_2 emissions (Heinz et al. 2011). The fabrication process has been specifically developed for large-scale construction components with the resolution of the material gradation of down to five millimetres, depending on material constituents. Current research is focusing on methods for the transfer of biological principles from structural biomaterials into the fabrication of bio-inspired lightweight, multifunctional and recyclable functionally graded concrete components and structures at the construction scale. In order to transfer these principles, investigations into the gradient layout, material properties of mineral matrix and reinforcement materials, appropriate formwork and joining techniques with regard to fabrication technologies need to be carried out. The research aims to develop methods for the integrative design and fabrication of such structures on the way towards efficient construction for the built environment of tomorrow.

16.4.1 *Materials for Development of Functional Concrete Gradients*

In order to enable the transfer of principles of biological gradients to the production of functionally graded concrete, the properties of the materials and their variations need to be evaluated and adjusted in accordance with production-based criteria, as described in the first section of this chapter.

The current approach to the gradation of concrete properties is based on lightweight aggregate concrete technology and is characterised by the manipulation of bulk density, substituting the volumetric portion of the fine aggregate by the lightweight aggregate in the required proportion. With the objective of developing pure mineral components that are suitable for recycling, expanded glass has been preferred over other lightweight aggregates, such as expanded clay or perlite. Lightweight aggregate concrete has proven better workability and controllability of gradients, compared with other types of lightweight concrete processing, such as foams of aerated lightweight concrete.

Two concrete mixtures with the highest and lowest density values have been developed as a basis for the production of gradients by combining them in required proportions (Herrmann et al. 2015). In order to facilitate handling and analysis at the development stage, the reference mixtures are composed of cement (CEM I 52,5 R), water and fine aggregate (sand with aggregate size up to 2 mm) for the heavy mixture and additionally of the lightweight expanded glass aggregate for the low-density mixture (Heinz et al. 2011).

The lower density limit depends on the size of the lightweight aggregates and the use of admixtures and pore-forming agents and currently reaches 800 kg/m^3 when aggregates with a grain size of 1–2 mm are used. Thus, the bulk density value can be reduced by 65 % compared with the heavy mixture with a density of 2300 kg/m^3. For comparison, highly mineralised biological tissues, such as corals and echinoids, exhibit similar density ranges, varying from fully dense calcite areas with a density of 2700–3000 kg/m^3 to highly porous areas of 900–1000 kg/m^3.

In order to expand the range of density gradients in the direction of the lower limits, the adjustment of such parameters as size, shape or arrangement of lightweight aggregates has been investigated (Fig. 16.8). The increase of the grain size from 1 to 4 mm contributes to the reduction of density by ca. 10 %. However, the conveyance of the mixture with larger aggregates becomes problematic in the case of spraying application technology with the use of warm pump spraying machines; because of the unequal weight and insufficient adhesiveness of mixture components, lightweight aggregates fail to pass through the pump to the hose. In this case, a production-specific adjustment of the mixture should be provided. Another approach is to combine the appropriate grain sizes in required proportion in order to increase the packing density of aggregate particles, resulting in an increase of volume fraction of lightweight aggregate and, at the same time, the reduction of the amount of binder that results in a decrease of the bulk density value.

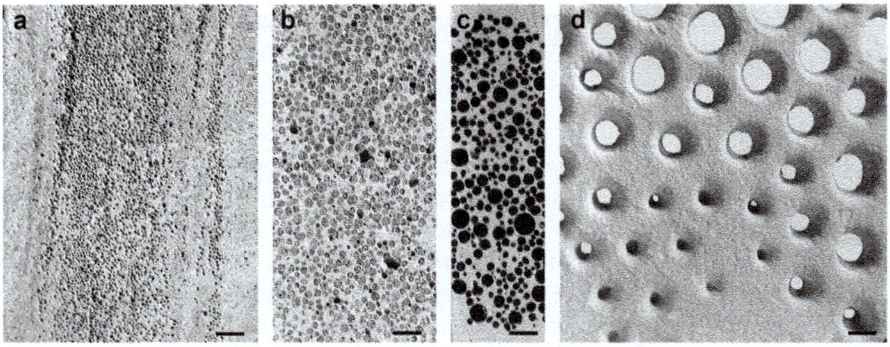

Fig. 16.8 Various sizes and methods for porosity gradient. (**a**, **b**, **c**) – lightweight aggregate concrete with grain size of 1–2 mm, 2–4 mm, 4–8 mm, accordingly; (**d**) – open porous concrete; all sample are at one scale, scale mark = 1 cm

Other approaches to the production of concrete with graded densities are currently being investigated, including the development of mineral hollow spheres with diameters from 10 mm up to 200 mm that are placed on a predetermined position during concrete application procedure (casting or spraying). Moreover, methods for the production of open porous concrete structures are being developed by using removable or dissoluble formwork materials, such as frozen sand or other two-phase materials.

16.4.2 Casting and Spraying of Functionally Graded Concrete

The essential part of functionally graded concrete technology that affects mixture development and puts constraints on layout design is the concrete application technique. Two different layer-based techniques of casting and spraying have been investigated at Institute for Lightweight Structures and Conceptual Design at University of Stuttgart for application of gradient concrete and are currently in use (Woerner et al. 2015).

For both techniques, two concrete mixtures with minimum and maximum density values need to be adjusted under production-specific constraints in terms of aggregate grading curves, water/cement ratio, suitable admixtures, etc. Casting technology proves to be easier in handling and is particularity suitable for layered layouts with a unidirectional gradation of material densities. In order to achieve a certain desired gradient, two previously prepared mixtures are blended in the required proportions and applied layer by layer in the formwork. Of importance, mixtures have to be developed with regard to their compatibility by blending gradients, as otherwise, undesirable effects such as impaired packing density can be observed. The density and the thickness of the layer when using lightweight aggregate concrete can be easily predicted and controlled. However, several drawbacks are experienced in casting technology with regard to its use in gradient production,

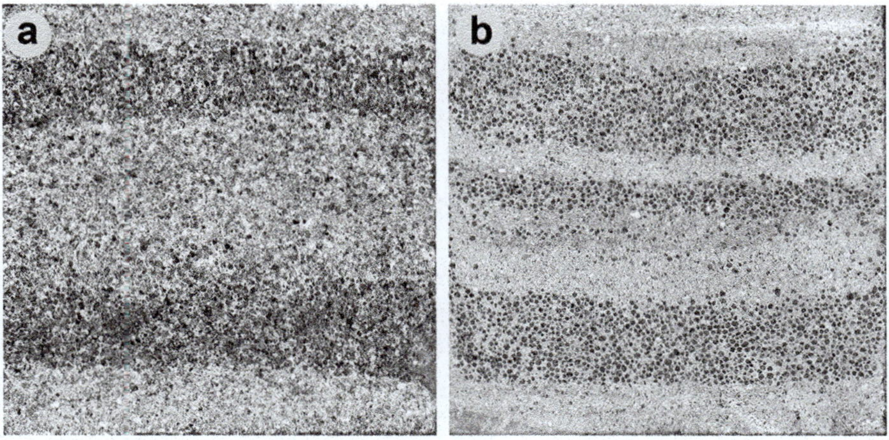

Fig. 16.9 Gradient samples produced by: (**a**) casting and (**b**) spraying technology

such as an inability to compact concrete after casting, because of the presence of the different concrete densities, and limitations in the achievement of spatially complex gradients.

In the case of more complex two- and even three-dimensional gradients, spraying technology proves to be more suitable. By spraying, two concrete mixtures are preblended and pumped through the hoses to the nozzle. The variation of material properties is achieved by manipulation of the amount of conveyed material in each hose through the regulation of pump rotation speed. With a continuous increase of flow rate of one mixture and a decrease of the other, a seamless unidirectional gradient between two mixtures is achievable. The application of material on the surface under air pressure results in appropriate compaction, which enables the application of material in three dimensions, even in overhead areas.

Currently, the use of the two nozzles, installed at a distance of 10 cm from each other and 40–60 cm from the surface, results in inhomogeneous material distribution. The mixing of material at the intersection of flows (Fig. 16.9b), and the diameter of applied material of 25–30 cm, which is suitable for unidirectional and two-dimensional gradients, still puts several constraints onto the resolution of three-dimensional gradients. Both dry and wet spraying processes are being currently optimised in terms of material premix in one nozzle and adjustment of resolution of deposited material in order to ensure the precise and repetitive fabrication of complex three-dimensional gradients.

16.4.3 Case Studies on Gradient Layout Fabrication

Biological systems build their structures under fabrication constraints, adopting the physical and chemical properties of materials that they are using. Similarly, the

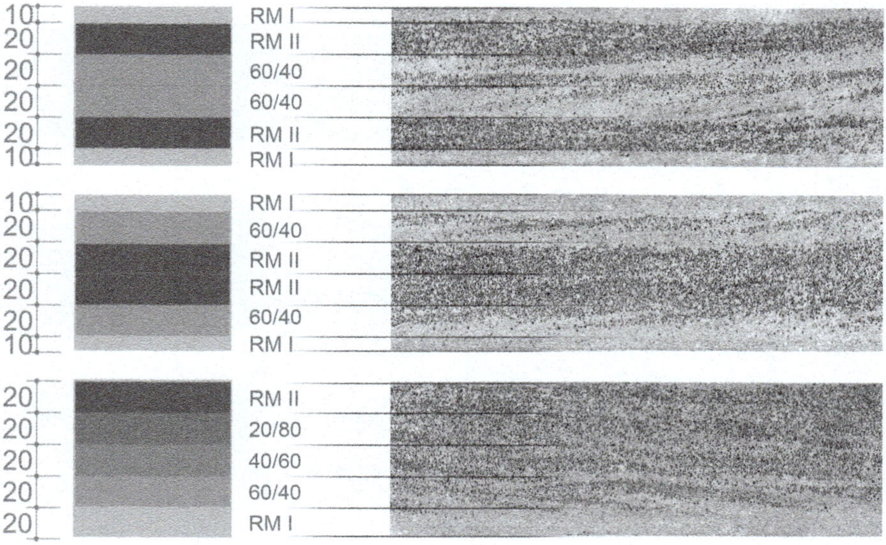

Fig. 16.10 Unidirectional gradients produced by casting and spraying technologies

gradient layout for the production of functionally graded concrete structures represents the distribution of microstructural and/or mesostructural properties based on material parameters and production constraints, integrating material, form, structure and function of the component. Two case studies are presented to demonstrate the way that bio-inspired unidirectional and two-dimensional gradients can be achieved by the functional gradation of concrete.

Functionally Graded Sandwich Structures Several types of layouts have been developed and produced, both by casting and by dry spraying techniques, in order to test the reproducibility of unidirectional gradients that can be used when a combination of structural performance, such as shock absorbance with the integration of additional function, such as thermal insulation, is needed. This type of structure is widespread in natural systems, for example, in animal dermal armours. Layouts vary in the distribution and thicknesses of the gradient layers. Figure 16.10 shows the precision of reproduction and the impact of the fabrication technique on the layout outlook.

Functionally Graded Shell Segment An open-porosity shell segment was produced to test its potential for the fabrication of mesostructural two-dimensional concrete gradients that can be used when optimised structural behaviour needs to be combined with contradicting function, such as light penetration (Fig. 16.11b). The Stress pattern, derived from the structural analysis model, was translated to a mesostructural distribution of concrete porosity, taking into account material constraints such as strut stability. Such open porous structures can be observed in many natural systems, for example, highly mineralised porous calcite structures in echinoids and other marine organisms.

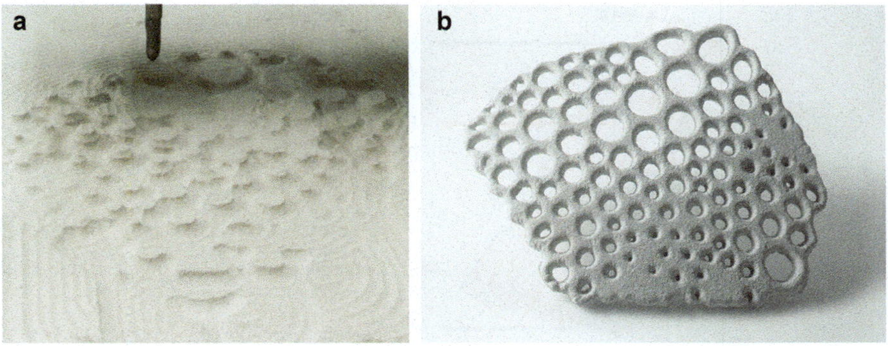

Fig. 16.11 Fabrication of porous structures with use of frozen sand formwork. (**a**) Frozen sand milling; (**b**) casted concrete workpiece with graded porosity

Despite the fact that the shape and size of pores are constrained by material properties and the process of biomineralisation, considerable variation in the porosity of the meshwork from fully dense areas up to 65 % can still be observed, ensuring the permeability of structure for nutritional transport within the organism.

In this particular case, the fabrication of such structures becomes possible with the use of the frozen sand formwork technique (Fig. 16.11a), which is waste-free and allows the preparation of complex three-dimensional moulds by milling a frozen mixture of water and sand (Gericke et al. 2015). It also has potentials for the development of complex spatial moulds, since it has no constraints, such as undercuts or bottleneck problems. As soon as the concrete is cast into the formwork the cooling of the sand is stopped, it unfreezes and, after the required period of concrete hardening, it can be freely removed and reused.

16.5 Outlook for Combination of Fabrication Processes in Construction Industry

The combination of fabrication processes exhibits a high potential for the production of complex bioinspired structures, mainly because natural systems are multiscale hierarchical structures and multi-material composites, despite being problematic to fabricate within one single process.

Three process classes, described in this chapter, offer the processing of various material groups (fibres, polymer and mineral matrices) and operate on different scales. This allows the use of the advantages of one process to overcome the constraints of another, such as large-scale concrete component production, detailed and complex structures by FDM and endless fibre reinforcement of various matrix materials.

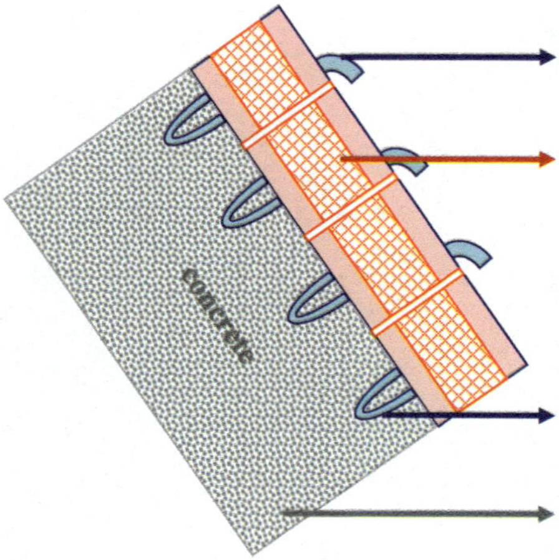

Fig. 16.12 Schematic section of multilayered prototype of shell structure

One of the case studies for such a combination of fabrication processes with the aim of achieving structural and functional performance is, for example, the concept of the large-scale prototype of a structural envelope inspired by the shells of land snails and consisting of several layers (Fig. 16.12). Here, the high-level integration of the formation processes and functions that complement each other are observed. A functional outer layer with complex geometry also acting as the formwork for the load-bearing structure is produced by fibre-reinforced FDM. The characteristics of this process allow the printing of topologically optimised workpieces that have an adapted internal structure and additional functions, such as insulation, shock absorbing and heat and mass transport. The formwork either can be printed as one piece or can be assembled from smaller segments, the latter of which requires strategies for multi-material joints. The outmost surface layer of this formwork can provide various functions, depending on the surface structure (water repellent, self-cleaning, etc.). Either this surface layer can be printed in the same process step or it can be created by a separate coating process. Once the required volume of the formwork is built up, concrete spraying is initiated to reinforce the functional layer and finally to create the load-bearing structure. The fibre-reinforced connecting layer is produced to compensate the concrete shrinking effect and to guarantee an appropriate bond between the layers.

Such an approach of combining these three classes of processes into one multi-material and multi-scale fabrication process opens up the possibility of producing new types of components and structures that can be employed in architecture and are beyond the current state of the art fabrication techniques.

Acknowledgements This work has been funded by the German Research Foundation (DFG) as part of the Transregional Collaborative Research Centre (SFB/Transregio) 141'Biological Design and Integrative Structures'/project B04.

References

Dadalau A, Hafla A, Verl A (2009) A new adaptive penalization scheme for topology optimization. Prod Eng Res Dev (WGP) 3(4):427

Duan Y, Wang Y, Tang Y et al (2010) Fabrication and mechanical properties of UV-curable glass fiber-reinforced polymer–matrix composite. J Comp Mat 45(5):565–572. doi:10.1177/0021998310376107

Fischer A, Rommel S, Bauernhansl T (2013a) New fiber matrix process with 3D fiber printer – a strategic in-process integration of endless fibers using fused deposition modeling (FDM). Digital product and process development systems, Bd. 411, Springer, Berlin Heidelberg, pp 167–175

Fischer A, Rommel S, Verl A (2013b) 3D fibre printer – generativ gefertigte thermoplastische Kunststoff Bauteile mit Endlosfaser Integration. Tagungsband Anwenderforum

Gericke O, Haase W, Sobek W (2015) Herstellung von Freiform-Betonbauteilen mittels einer gefrorenen Schalung aus Wasser und Sand, DAfStb-Jahrestagung mit 56. Forschungskolloqium, Stuttgart

Ghomeshi R (2015) Entwicklung von biomimmetisch optimierten, pultrudierten Faserverbundprofilen mit verbesserten dynamischen Eigenschaften. Dissertation, University of Stuttgart

Heinz P, Herrmann M, Sobek W (2011) Herstellungsverfahren und Anwendungsbereiche für Funktional Gradierte Bauteile im Bauwesen. Final report Research Initiative Future Building, University of Stuttgart

Herrmann M, Wolf C, Sobek W (2015) Design and manufacturing of optimal structures made from functionally graded concrete. Proceedings of the International Association for Shell and Spatial Structures (IASS) 2015, Amsterdam

Mallick PK (2008) Fiber-reinforced composites: materials manufacturing and design. CRC Press, Boca Raton

Milwich M (2015) Biomimetic engineering of tailored, ultra-lightweight fibrous composites. In: Hamm C (ed) Evolution of lightweight structures – analyses and technical applications. Springer, Dordrecht

Neitzel M, Mitschang P, Breuer U (2014) Handbuch Verbundwerkstoffe: Werkstoffe, Verarbeitung, Anwendung

Woerner M, Sippel S, Schmeer D, Garrecht H, Sobek W, Sawodny O (2015) Automated spraying of functionally graded concrete components – analysis of the process parameters. Proceedings of the International Association for Shell and Spatial Structures (IASS) 2015, Amsterdam

Chapter 17
Structural Design with Biological Methods: Optimality, Multi-functionality and Robustness

Debdas Paul, Layla Koohi Fayegh Dehkordi, Malte von Scheven, Manfred Bischoff, and Nicole Radde

Abstract We present ideas and concepts towards defining a common framework unifying abstract metrics in order to quantify key features of technical load-bearing structures and biological systems. Our aim is to transfer biological concepts to technical systems at this abstract level rather than on the basis of their outward appearance or actual functionality. This means that the biological concept generators for load-bearing structures do not have to be load-bearing structures themselves but may instead achieve rather different functionalities. We intend to carry out this transference by generalizing graph-based abstractions of both technical and biological worlds to allow comparisons to be made at an abstract level. We focus in particular on the intrinsically competing aims of optimality versus multi-functionality and robustness. In this review, we present initial attempts towards defining suitable quantitative measures for robustness to serve as a common ground for studying technical systems and biological systems simultaneously. We discuss generic properties of a ubiquitous signalling network motif and potential relationships to a minimal model for a robust truss structure. These case studies suggest that topological complexity can serve as a common source for a design that is insensitive to perturbations and thus robust in the measures of both worlds.

17.1 Introduction

In this study we focus on the analysis of biological concept generators for load-bearing structures in view of the intrinsically competing aims of optimality versus multi-functionality and robustness. Our overall goal is to obtain a deeper understanding of the principle underlying mechanisms at an abstract, formal level. This comprises, in particular, the investigation of two partly complementary

D. Paul • N. Radde (✉)
Institute for Systems Theory and Automatic Control, University of Stuttgart, Pfaffenwaldring 9, 70550, Stuttgart, Germany

L. Koohi Fayegh Dehkordi • M. von Scheven • M. Bischoff
Institute for Structural Mechanics, University of Stuttgart, Pfaffenwaldring 7, 70550, Stuttgart, Germany

J. Knippers et al. (eds.), *Biomimetic Research for Architecture and Building Construction*, Biologically-Inspired Systems 9, DOI 10.1007/978-3-319-46374-2_17

hypotheses derived from the same basic consideration: the design process of biological systems has to maintain multiple functions simultaneously in order to ensure survival under various environmental conditions. Robustness against perturbations and imperfections is a crucial property that systems optimized towards one single goal typically do not possess.

Hypothesis 1: Assuming that multi-functionality always constitutes a compromise between opposing optimization criteria, we hypothesize that the performance of technical load-bearing structures, for which only some of these criteria are important, can in principle even outperform the biological role model. This can only be achieved by approaches that go beyond a pure copying of the example and requires an understanding of basic natural design principles.

Hypothesis 2: Systems emerging from evolutionary processes are intrinsically more complex and more robust than "minimal" solutions optimized towards a single specific function. This hypothesis was originally motivated by recent biological literature related to a concept termed *functional robustness*, which so far has mainly been used in connection with intracellular regulation processes.

One of the methodological concerns of this project is the performance of abstraction in a formal approach, e.g. based on graph theory or matrix representations, rather than the phenomenological or geometrical appearance of the biological and technical systems. This should lead to the transfer of biological principles beyond the original biological functionality. More precisely, a biological example for a load-bearing structure does not have to be a load-bearing structure itself but can be any biological entity exhibiting a similar formal structure. Thus, our investigations do not focus on particular biological systems as role models but, rather, search for general network motifs that are ubiquitous in nature and that are known robustly to facilitate certain functions.

In particular, intracellular signal transduction networks are known to function in an extremely robust way. This robustness is often tightly related to their inherently complex network structure (Batchelor and Goulian 2003; Caicedo-Casso et al. 2015; Song et al. 2007; Wagner 2005). Likewise, for the case of load-bearing structures, topological complexity is well known to be correlated with structural robustness, which is demonstrated via a simple toy problem.

This study presents preliminary results providing such abstract representations of biological concept generators on the one hand and load-bearing structures on the other. Section 17.2 elaborates on the meaning of optimality and robustness in the context of intracellular networks and technically achieved load-bearing structures. In Sects. 17.3 and 17.4, we present two case studies from both the biological and technical worlds, in which we quantify robustness on concrete minimal example models. Relationships between both studies and a preview of future steps within the project are discussed in Sect. 17.5.

17.2 Optimality, Multi-functionality and Robustness in Biological Systems and Technical Load-Bearing Structures

Unlike technical systems, biological systems are living objects that seem to be optimized for survival. In particular, bacteria are extremely flexible and able to maintain multiple functions essential for survival under quite a variety of different conditions (Hart et al. 2011). Living individuals are open systems that sense their environment and have developed various ways to react to signals during evolution. Signalling pathways, on which we will focus, take over a key role in these processes. Signal transduction is usually initiated via the activation of membrane receptors, leading to the recruitment and activation of proteins further downstream. These transmit the signal to the cytoplasm and the nucleus, ultimately leading to changes in the gene expression pattern.

A properly working signaling pathway shows incredibly reliable functioning. It can cope with environmental changes, large internal variations e.g. in the amounts of proteins, and large stochastic fluctuations in the process of transcription or attributable to interactions involving low copy numbers. Signalling pathways can tolerate all of these variations and are robust to many kinds of perturbations. Furthermore, signalling pathways are often able to distinguish between real signals, which are enhanced during propagation, and spurious signals, which are filtered out at the same time. A reliable signalling is indeed essential for survival, since malfunctioning compromises fitness for bacteria or may perturb tissue homeostasis and lead to severe diseases.

Load-bearing structures, i.e. engineering structures that are predominantly designed to carry loads, do not seem to have anything in common with biological signalling networks at first glance. However, the outer appearance of simple truss structures, like the one shown in Fig. 17.3 below, indicates that formal similarities exist. Indeed, this is also true for structures composed of surfaces or three-dimensional bulky solids that do not geometrically appear as networks of nodes connected with one-dimensional members.

On an abstract level, load-bearing structures – like signalling networks – are expected to perform certain functions: they are potentially subject to environmental changes, uncertainties and imperfections. Loads (external “signals”) are supposed to be carried safely without structural collapse or excessive deformation. From an extremely simplified point of view, one can say that, if the structure is *optimal*, the functions are achieved at minimal cost (e.g. use of material or structural weight), whereas if it is *robust*, it has a minimal vulnerability to perturbations. Such perturbations can, for instance, be deviations of the applied load from the expected load (the design load).

From the point of view of structural mechanics, structural complexity can be associated with the degree of *statical indeterminacy* n_s of a structure. Without delving into technical details, this is a measure for the degree of redundancy of a structure. A *statically determinate* structure has a zero degree of statical indeterminacy $n_s = 0$. For a given external load, the "path" of the loads to the supports and the individual member forces are determined by an equilibrium, independent of the stiffness or load-carrying capacity of the individual members. If one member fails, the entire structure collapses, as in a chain.

Statically indeterminate structures with $n_s > 0$ offer various different possibilities for safely transferring the load to the bearings (one might think of parallel chains or a network of chains). If one member fails, a different member might compensate this by increased internal forces. As such, statically indeterminate structures are more robust. Topologically, they are more complex. Compared with statically determinate structures, they are also more expensive in terms of material and weight. Thus, structures that are optimized towards a specific goal (minimal weight for a given load) are typically statically determinate and thus possess zero robustness.

In both disciplines, questions concerning mechanisms underlying this robustness have kept researchers busy for many years. For biological systems, research in this direction has currently reached a new dimension, mainly because of the enormous advances in experimental techniques in the last few decades. These measurements now allow intracellular processes to be observed in a way that is fundamentally different from standard methods that were in use 30 years ago. Such development, of course, facilitates a quantitative and mechanistic understanding of processes that regulate the occurrence of diverse phenotypes. It also poses interesting new challenges, such as meaningful interpretations within a unifying framework. Hence, unsurprisingly, mathematical concepts and models enter the scene.

Indeed, mathematical models have highly facilitated our understanding of general patterns and processes underlying reliable functioning, which is also frequently named functional robustness (Barkai and Shilo 2007; Batchelor and Goulian 2003; Blanchini and Franco 2011; Blüthgen and Legewie 2013; Clodong et al. 2007; Emmert-Streib and Dehmer 2009; Lapidus et al. 2008; Stelling et al. 2004). The number of studies that focus on this term has exploded in the last 10 years or so. From these studies, one essential conclusion (which is of particular importance also in the context of our Collaborative Research Center SFB-TRR 141, in which our project is embedded) is that functional robustness is tightly related to the structure of the underlying interaction network and its complex regulation. Signalling pathways are by no means chains of proteins in which the signal is simply propagated from one candidate to the next in, say, a linear way. They instead form a complex network of interacting biomolecules, with multiple, frequently occurring motifs that act on different time scales. Moreover, signalling pathways are not independent in vivo but communicate with each other. In fact, this context-dependency and non-modularity complicates the transfer of synthetic networks from cell-free environments to in vivo systems considerably, which is a major problem for the design of synthetic functional networks (Klavins 2014; Mishra et al. 2014). Overall, as it is the case for load-bearing structures, this leads to highly complex and redundant structures. An interesting and challenging question is whether any structural properties of signalling networks that are related to robust functioning can, in an abstract way, serve as concept generators for the design of load-bearing structures.

In structural engineering, the opposition of optimality and robustness is well understood. During the design process, one seeks a certain compromise of both extremes. When using mathematical optimization methods, one has to be aware of the fact that the tendency of high performance, but less robust systems, crucially depends on the formulation of the optimization problem itself and, in particular, the objective function. The research field of *robust design optimization* systematically addresses this issue and offers feasible solutions and powerful tools for design engineers.

In the spirit of this research project, however, we try to pursue a different strategy, namely directly to analyse the structure in view of its redundancy. Starting from a traditionally optimal design (i.e. high performance but low robustness) considerations on the basis of *redundancy matrices* help to deduce design changes that significantly increase robustness while not compromising optimality. Interpretation of these redundancy matrices as an abstract indicator of robustness (or a robust design) is research in progress on the path to the overall aim of the research project outlined above.

17.3 Robustness Versus Optimality Based on the Example of a Ubiquitous Signalling Network Motif

As a case study to investigate robustness, we compare a ubiquitous signalling network motif, a three-tired phosphorylation cascade, with simpler activation motifs. Model structures are shown in Fig. 17.1 on the left.

Figure 17.1c shows the reaction network of a three-tired phosphorylation cascade, in which a signal, modelled as a dynamic input, $u(t)$, describes the activity of a protein upstream of the cascade. A first protein X is activated upon dual phosphorylation in a distributive manner, i.e. phosphorylation of the first and the second site occurs via two independent binding events. Both phosphorylations are necessary for the protein to become fully active. Unphosphorylated protein is denoted by X, whereas single and double phosphorylations are denoted by X_1 and X_2, respectively. This module triggers phosphorylation of a second protein Y in the same distributive manner. We assume, for simplicity, that these phosphorylations are only triggered by X_2, i.e. only X_2 acts as a kinase on Y. We are primarily interested in the behaviour of Y_2, since this readout usually has many substrates and thus regulates the overall cellular response to the signal.

Prominent examples for the appearance of such phosphorylation cascades are the mitogen-activated protein kinase pathway (MAPK) family and Akt signalling (see e.g. Brightman and Fell 2000; Fritsche-Guenther et al. 2011; O'Shaughnessy et al. 2011; Santos et al. 2007). Both pathways are well-investigated experimentally and from a modelling perspective. They can show a rich and cell-type-specific variety of dynamic behaviours such as oscillations, bistability, bimodality, graded

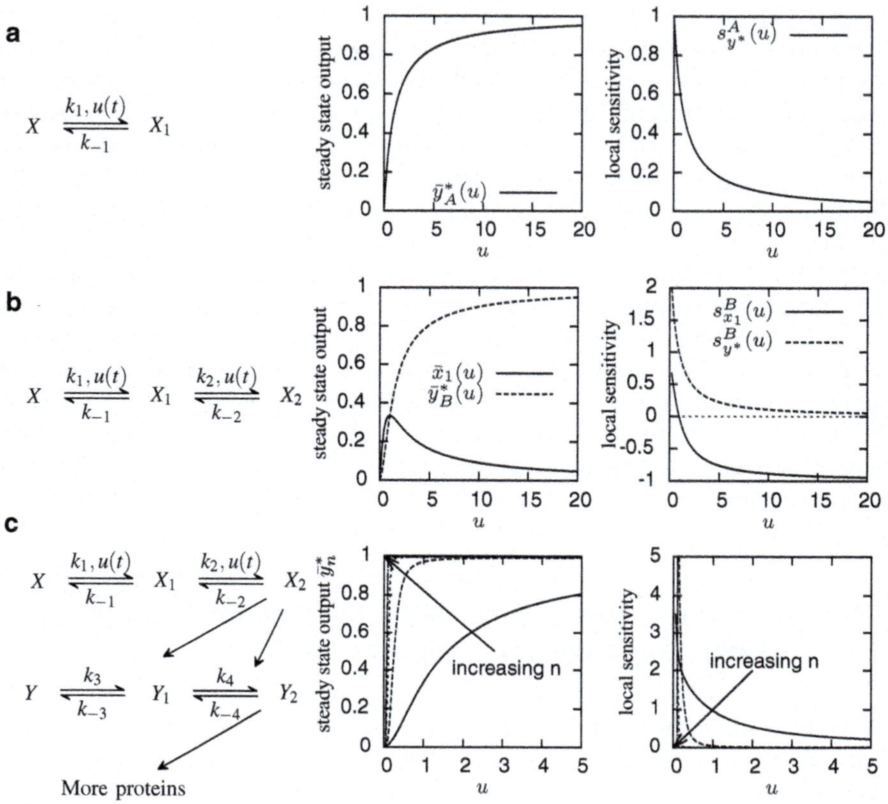

Fig. 17.1 Steady state sensitivity analysis of different model versions (A–C), as described in the main text (Eqs. (17.1), (17.2), and (17.3), respectively), and reaction rate constants $k_1, k_{-1}, k_2, k_{-2}, k_{-3}, k_{-4} = 1, k_3, k_4 = 20$. These rates were chosen such that the signal is reliably propagated throughout the network

or switch like responses, signal amplification and ultrasensitivity (Birtwistle et al. 2012; Kholodenko 2000; Legewie et al. 2007; Markevich et al. 2004; Qiao et al. 2007; Xiong and Ferrell 2003). Furthermore, both pathways are deregulated in several diseases and are prominent drug targets in the clinic (Grieco et al. 2013; Kolch 2005; Kolch et al. 2005). Here, we aim to investigate the following questions:

1. In which way does the hierarchical structure of this motif (i.e. the sequence of activation of the different proteins) contribute to a robust functioning of the signalling pathway?
2. What is the particular role of the intermediate single phosphorylated protein, which is usually much less active than the double phosphorylated form?

In order to answer these questions, we compare this motif with two simpler network structures. The simplest network to propagate a signal consists of an input that directly activates a single protein X, which then determines the cellular response by regulating downstream substrates (Fig. 17.1a). This network is optimal in terms of energy consumption, since each phosphorylation reaction consumes energy in the form of an ATP molecule and since only one single protein is required to propagate the signal. In a second model, we include the distributive activation mechanisms (Fig. 17.1b). We assume conservation of total protein masses in all three model variants, and neglect slow turnover rates, a common and reasonable assumption for signalling pathways. Mass conservation relationships are used in order to substitute unphosphorylated protein amounts. All variables are normalized to these total masses. Reaction rates are described by mass action kinetics, i.e. rates are proportional to the product of concentrations of all educt species. Michaelis-Menten kinetics for kinase and phosphatase-driven phosphorylation and dephosphorylation reactions would be an alternative, more complex description (Gunawardena 2005; Markevich et al. 2004). We denote the relevant model outputs by y_i^*, where $i = A, B, C$.

Model variant A consists of a single protein that becomes fully active when phosphorylated at a single site. Its' activity is described by

$$\dot{x}_1 = -(k_1 u(t) + k_{-1})\, x_1 + k_1 u(t), \quad y_A^*(t) = x_1(t), \tag{17.1}$$

In model variant B, a single protein is activated upon dual phosphorylation in a distributive manner. We assume that both phosphorylations are initiated by the input $u(t)$, which corresponds to the assumption that both reactions are triggered by the same kinase. The model reads

$$\dot{x}_1 = k_1 u(t)(1 - x_1 - x_2) - k_2 u(t) x_1 - k_{-1} x_1 + k_{-2} x_2 \tag{17.2a}$$

$$\dot{x}_2 = k_2 u(t) x_1 - k_{-2} x_2, \quad y_B^*(t) = x_2(t). \tag{17.2b}$$

Model variant C consists of a phosphorylation cascade of two proteins X and Y, where the double phosphorylated protein X_2 acts as a kinase that triggers the phosphorylation of protein Y,

$$\dot{x}_1 = k_1 u(t)(1 - x_1 - x_2) - k_2 u(t) x_1 - k_{-1} x_1 + k_{-2} x_2 \tag{17.3a}$$

$$\dot{x}_2 = k_2 u(t) x_1 - k_{-2} x_2 \tag{17.3b}$$

$$\dot{y}_1 = k_3 x_2 (1 - y_1 - y_2) - k_4 x_2 y_1 - k_{-3} y_1 + k_{-4} y_2 \tag{17.3c}$$

$$\dot{y}_2 = k_4 x_2 y_1 - k_{-4} y_2, \quad y_C^*(t) = y_2(t). \tag{17.3d}$$

After specifying the biological models, we have to quantify the terms perturbations and robustness in this context. Here, we consider the following settings:

First, we investigate the input/output behaviour to variations in the strengths of the input signals. Therefore, the steady state of the output is analysed as a function of a constant input u. The normalized local sensitivity coefficient

$$s^i_{x_j}(u) = \frac{\partial \ln \bar{x}_i(u)}{\partial \ln u} = \frac{u}{\bar{x}_i(u)} \frac{\partial \bar{x}_i(u)}{\partial u} \tag{17.4}$$

is used as an inverse measure for robustness in this setting. It quantifies the relative change in the steady states upon a relative change in the input u and is frequently used as a measure for robustness in systems biology (Batchelor and Goulian 2003; Blüthgen and Legewie 2013; Caicedo-Casso et al. 2015; Dexter and Gunawardena 2012; Dexter et al. 2015; Ouldridge and ten Wolde 2014). Expressions can partly be derived analytically, allowing an investigation without specifying exact parameter values for all concentrations and rate constants. Thus, this approach can provide insights that are tightly related to the network structure itself and that do not rely on particular parameter sets. We note here that the term absolute concentration robustness was introduced for pure mass-action reaction networks by Feinberg and coworkers (Shinar and Feinberg 2010, 2011) and denotes perfect robustness of steady state concentrations. Experimentally verified examples for such a robustness are adaptation processes in bacterial chemotaxis and the robustness of metabolic functions to changes caused by mutations (see Shinar and Feinberg (2011) and references therein).

Results are also shown in Fig. 17.1. Figures in the centre show the outputs $\bar{y}_i^*(u)$ of the systems as functions of constant inputs u. Respective sensitivity coefficients are illustrated on the right.

For model variant A, $\bar{y}_A^*(u)$ and $s^A_{y*}(u)$ are given by

$$\bar{y}_A^*(u) = \bar{x}_1(u) = \frac{u}{u + \frac{k_{-1}}{k_1}} \quad \text{and} \quad s^A_{y*}(u) = \frac{\partial \ln \bar{y}_A^*(u)}{\partial \ln u} = \frac{1}{1 + \frac{k_1}{k_{-1}} u}. \tag{17.5}$$

Thus, the fraction of active protein increases hyperbolically with increasing input strengths u. The sensitivity decreases monotonically to zero, with decay constant given by $u = {}^{k_{-1}}/_{k_1}$.

For model variant B, the steady state concentrations and sensitivities are given by

$$\bar{x}_1(u) = \frac{k_1 u}{\frac{k_1 k_2}{k_{-2}} u^2 + k_1 u + k_{-1}}, \quad s^B_{x_1}(u) = \frac{k_1 - \frac{k_1 k_2}{k_{-2}} u^2}{\frac{k_1 k_2}{k_{-2}} u^2 + k_1 u + k_{-1}} \tag{17.6}$$

$$\bar{y}_B^*(u) = \bar{x}_2(u) = \frac{k_2}{k_{-2}} \bar{x}_1(u) u, \quad s^B_{y*}(u) = \frac{\partial \log \bar{x}_1(u)}{\partial \log u} + 1. \tag{17.7}$$

The steady state coordinate $\bar{x}_1(u)$ first increases for small u and approaches zero for larger inputs. The output $\bar{y}_B^*(u)$ shows a monotonic and sigmoidal increase and approaches 1 for large inputs. The output sensitivities for models A and B a rather similar, except that the output sensitivity is higher for small u compared with model A.

The steady states of model variant C are given by

$$\bar{x}_1(u) = \frac{k_1 u}{\frac{k_1 k_2}{k_{-2}} u^2 + k_1 u + k_{-1}}, \quad \bar{x}_2(u) = \frac{k_2}{k_{-2}} \bar{x}_1(u) u \tag{17.8a}$$

$$\bar{y}_1(u) = \frac{k_3 \bar{x}_2(u)}{\frac{k_3 k_4}{k_{-4}} \bar{x}_2(u)^2 + k_3 \bar{x}_2(u) + k_{-3}}, \quad \bar{y}_2(u) = \bar{y}^*_C(u) = \frac{k_4}{k_{-4}} \bar{y}_1(u) u. \tag{17.8b}$$

Sensitivities are calculated numerically for this example.

To show the effect of the hierarchical structure, Fig. 17.1c presents a comparison of the output behaviours and sensitivities for model B ($n = 1$ protein), model C ($n = 2$ proteins) and models with increasing numbers of proteins. The steady state curve becomes steeply sigmoidal with increasing number of proteins. Thus, long protein cascades show ultrasensitive behaviors and can act as efficient switches. The threshold value is very low here, but generally depends on the parameters and can be tuned otherwise. As a consequence, sensitivity is no longer graded, but becomes very high near the threshold and rapidly decreases to zero thereafter. Thus, once the signal reaches the threshold, the signalling cascade is reliably turned on and is no longer sensitive to variations in the input strengths. At the same time, signals that are too low and might be spurious, are filtered out.

As a second approach, we investigate the ability of all network structures to filter out noise. Of course, in reality, no clear separation occurs between a real signal and noise. However, we approach this concept by assuming that high frequencies in the input signal are more likely to mimic stochastic fluctuations and perturbations than slow and consistent changes. Signalling pathways should, to some extent, be able to filter out these spurious signals, since unintended activation can be severely deleterious and unnecessarily alters transcription and metabolism (Dexter et al. 2015; Krantz et al. 2009). At the same time, the system should reliably react to real signals. To address these filtering properties in our study, we follow ideas described in Hersen et al. (2008), who have shown experimentally that the bandwidth of the HOG MAP kinase pathway acts as a low-pass filter, and investigate the behaviour of the system outputs when stimulated with an oscillating input

$$u(t) = \begin{cases} 0 & t \leq 0 \\ 20 + 5\sin(\omega t) & t > 0 \quad \omega = 2\pi f \end{cases} \tag{17.9}$$

For this constant shift, steady state outputs have similar values for all models and output amplitudes can directly be compared. We simulated the models until a stationary oscillating response $y^*_{i,s}(t)$ was reached, which was then used to determine the peak-amplitude of the output as

$$A^*_i(f) = 0.5(\max(y^*_{i,s}(t)) - \min(y^*_{i,s}(t))), \tag{17.10}$$

where the term $\max(y^*_{i,s}(t)) - \min(y^*_{i,s}(t))$ is the semi-amplitude (Zhou 2013) over the time window in which the steady state response is perturbed with an input of frequency f in Hz. Results are shown in Fig. 17.2. Figure 17.2a shows stationary response amplitudes $A^*_C(f)$ for different input frequencies. As expected, $A^*_C(f)$ decreases with increasing frequencies. Figure 17.2b summarizes output amplitudes for all three model variants. For low frequency inputs, model responses accurately follow the course of the input signal. For higher frequencies, the systems increasingly act as integrators that average over fast input changes. Models A and B behave very similarly in this analysis, whereas the output amplitude of model C is much smaller. Thus, this network structure partly acts as an integrator even for lower frequencies. Hence, it senses and responds to consistent average input changes rather than to short and rapid fluctuations. Respective sensitivities can be seen in Fig. 17.2c.

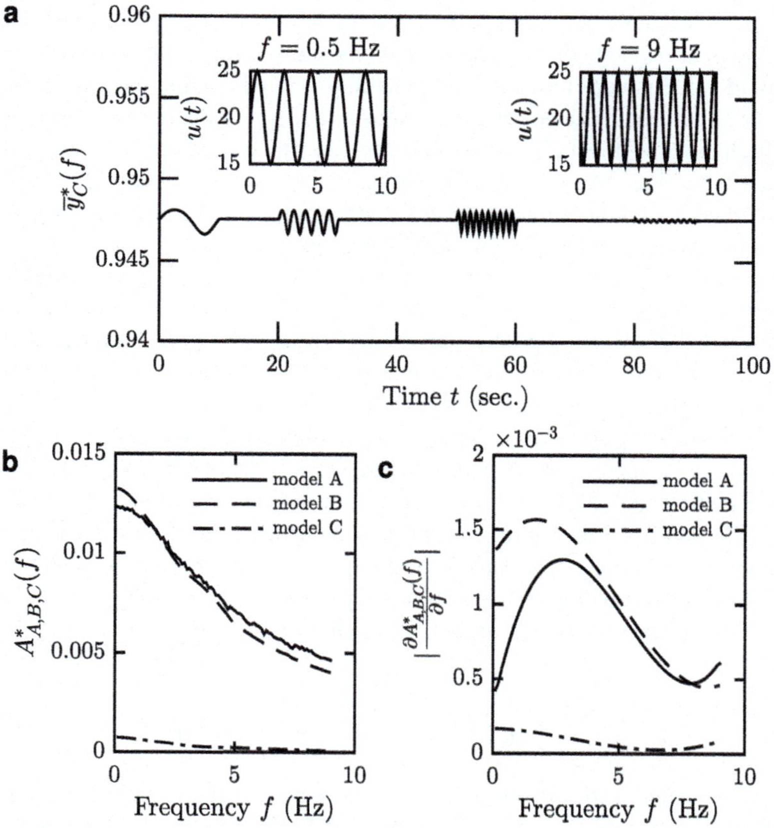

Fig. 17.2 Response of all model variants (A–C) to oscillating input signals, (**a**) Stationary responses $y^*_{C,s}(t)$ of model C to different input frequencies: 0.1, 0.5, 5 and 9 Hz (0.5 and 9 Hz inputs are shown as examples) at different time intervals, (**b**) Output amplitudes $A^*_i(f)$ $i = A, B, C$, (**c**) Local output sensitivities, which have been approximated by polynomial regression. Model parameters are the same as in Fig. 17.1

In summary, frequently occurring distributive phosphorylation cascades can act as efficient switches that do not respond to low signals but function in an insensitive and robust way once a signal has reached a threshold. Moreover, they are able to filter out high frequencies, which are likely to be caused by random fluctuations, and act as integrators that process slow and consistent changes.

The questions arise as to how definitions and measurements for robustness in biological systems compare with those for technical structures and whether some of the concepts introduced here for biological networks can be transferred to load-bearing structures. Some answers will be provided in the following section.

17.4 Optimality and Robustness of a Simple Truss System

To take a look at the notions of optimality and robustness for structures, a simple two-dimensional example will be introduced. Many simplifications are of course involved compared with real engineering structures but the overall ideas of optimal and robust structures can be examined without loss of generality.

Consider the truss structure depicted in Fig. 17.3. The ten members are connected by hinges to each other and have the elastic moduli of $E = 2.1 \cdot 10^8\,\mathrm{kN/m^3}$ and density of $\rho = 7850\,\mathrm{kg/m^3}$. Throughout this section, the total mass of the structure will be kept constant as $m_{\mathrm{tot}} = 500\,\mathrm{kg}$. The structure is supported at the left hand side and carries a point load of $P = 200\,\mathrm{kN}$. To evaluate the robustness, a second load P_{imp} in the middle of the structure will be introduced later.

For the initial design the cross sections of all members are assumed to be identical. Considering the weight ρ of the material and the allowable total mass m_{tot}, the cross section area is calculated to $A = 1.5368 \cdot 10^{-3}\,\mathrm{m}^2$.

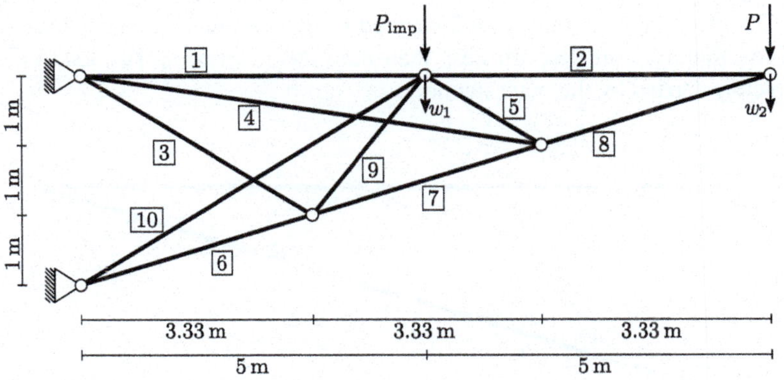

Fig. 17.3 Initial design of the truss structure consisting of ten members with the load P applied. The cross sections of all members are initially identical. The imperfection load P_{imp} is later introduced to evaluate the robustness

To evaluate the performance of the structure, the mean displacement of the two nodes at the top of the structure is defined as $w = \frac{1}{2}(|w_1| + |w_2|)$. In general, a smaller mean displacement is associated with a stiffer and better structure. The initial design yields a mean displacement of $w = 8.1359 \cdot 10^{-2}$ m.

17.4.1 Optimality

In structural engineering optimization is a widely used tool to improve the performance of structures. In many cases the aim is to increase the stiffness of a structure for a given mass or vice versa in order to reduce the mass without decreasing the stiffness. A good overview and examples of structural optimization can be found in MacBain and Spillers (2009), Bendsœand Sigmund (2004), and Bletzinger and Ramm (2014).

In this section a better structure compared with the truss example introduced above will be found by cross section optimization. For this purpose, the cross section of the inner members A_i is considered to be different from the cross section of the outer members A_o. This means that, for the optimization, there are two design variables A_i and A_o. The aim of the optimization is to increase the stiffness of the structure, i.e. to reduce the mean displacement w, keeping the mass constant. In mathematical terms, this means the objective function $w = \frac{1}{2}(|w_1| + |w_2|)$ is to be minimized with the constraint $m_{tot} = 500$ kg. To avoid non-physical results for the cross sections A_i and A_o a second constrained $A \geq 4.0 \cdot 10^{-6}$ m^2 is introduced.

The problem is solved analytically and the optimal design of the structure is found to be $A_o = 3.1120 \cdot 10^{-3}$ m^2 and $A_i = 4.0 \cdot 10^{-6}$ m^2, which corresponds to the constraint for the cross section area. This means the optimal design is found on a boundary of the admissible region in design space, which is common for the optimization of structures. A visualization of the optimal design can be seen in Fig. 17.4. The inner members are observed to have a much smaller cross section area and therefore smaller stiffness than the outer members. The load is almost completely carried by the outer members.

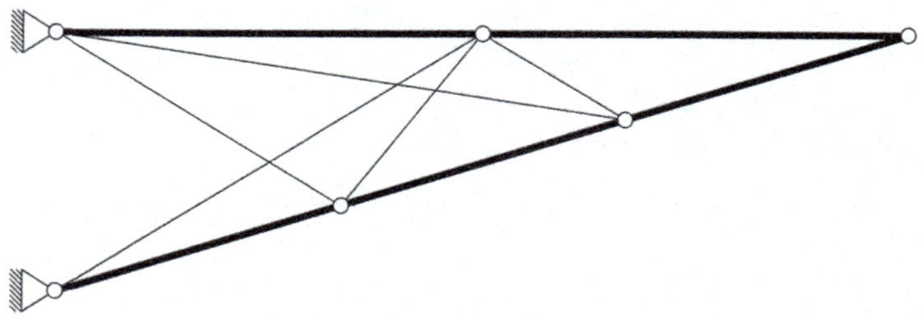

Fig. 17.4 Optimal design of the truss structure with the thickness of the lines proportional to the cross-sectional area of the members

The design is close to a statically determinate structure, exclusively consisting of two outer members (1–2 and 6–7–8, without hinges in between). It is typical for the applied optimization strategy to provide structures with almost no redundancy (cf. Sect. 17.4.2).

The value of the objective function, i.e. the mean displacement for the optimal design, is calculated to be $w = 4.1215 \cdot 10^{-2}$ m, which is a reduction by approx. 50 % in comparison with the initial design. This means that, with the same amount of mass, the optimal structure is twice as stiff as the initial design.

17.4.2 Robustness

Mechanically optimal structures tend to be highly sensitive to imperfections if they are neglected in the optimization process, as in our example. Small perturbations in the load case and material properties or geometrical imperfections can severely affect the overall behaviour of the structure. The ability of a structure to maintain its performance in spite of perturbations is called robustness. As perturbations cannot be avoided in engineering structures, the sensitivity to these uncertainties in geometry, materials and loads is also addressed in the field of optimization. Many publications deal with robust design optimization, e.g. Robinson et al. (2004) and Beyer and Sendhoff (2007).

For the problem defined above, robustness is investigated for one exemplary perturbation in the load case by applying a a second load P_{imp} to the system. The magnitude of this perturbation is measured as the ratio between the imperfection load and the original load $\frac{P_{imp}}{P}$.

In Fig. 17.5, the value of the objective function is plotted with respect to the change in the imperfection load, for both the initial design and the optimal design. As can be seen, for no imperfection load, the optimal design shows 50 % less mean displacement than the initial design. However, the mean displacement of the optimal design strongly depends on the imperfection load and, even for small imperfections

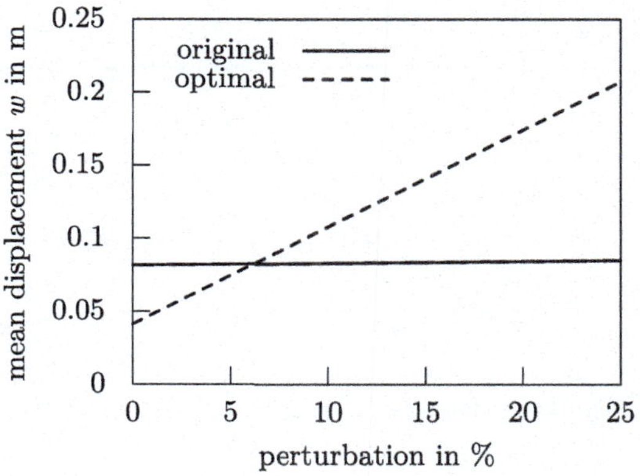

Fig. 17.5 Mean displacement w in dependence of the load perturbation for the initial design and optimal design

of above 6 %, the mean displacement becomes larger than for the initial design. The initial design maintains its performance, even for large perturbations.

Like for biological systems, the sensitivity is the change in the output (here: mean displacement) upon a change in the input (here: perturbation). Thus, the slope of the curves in Fig. 17.5 corresponds to the sensitivity of the two designs towards the load imperfection and is an inverse measure for the robustness (cf. Eq. (17.4)).

17.4.3 Redundancy in Truss Systems

As an alternative to robust design optimization algorithms, the redundancy contribution introduced in Ströbel (1995) can be applied to increase the robustness of the optimal design. The redundancy contribution describes the distribution of the static indeterminacy in a truss system and can be collected in the redundancy matrix **R**.

The redundancy contribution R_{ij} for a truss member i is calculated from the change in length Δl^i attributable to a unit elongation Δl_0^j of any member in the system,

$$R_{ij} = \frac{\Delta l_0^i - \Delta l^i}{\Delta l_0^j} = \frac{-N^i \frac{l^i}{E^i A^i}}{\Delta l_0^j}. \tag{17.11}$$

Considering the stiffness $\frac{EA}{l}$ of element i this can also be expressed in terms of the normal member force N^i in the truss member i. The redundancy contributions are a measure of the internal constrain in a system. The trace of the redundancy matrix **R**, i.e. the sum of all R_{ii} is equal to the degree of statical indeterminacy of the entire structure.

For the truss system defined above, the main diagonal entries of the redundancy matrix for the initial design are visualized in Fig. 17.6. As can be observed, the sum of all contributions is equal to 2, which is the degree of statical indeterminacy for this system. In addition, one can see that the two members at the tip of the truss have

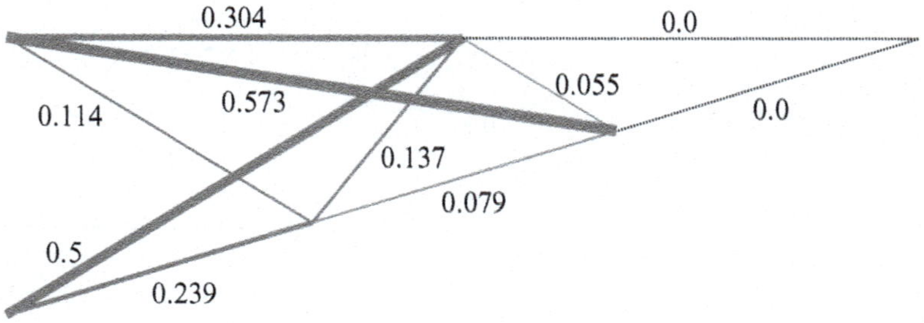

Fig. 17.6 Redundancy contributions for the initial design

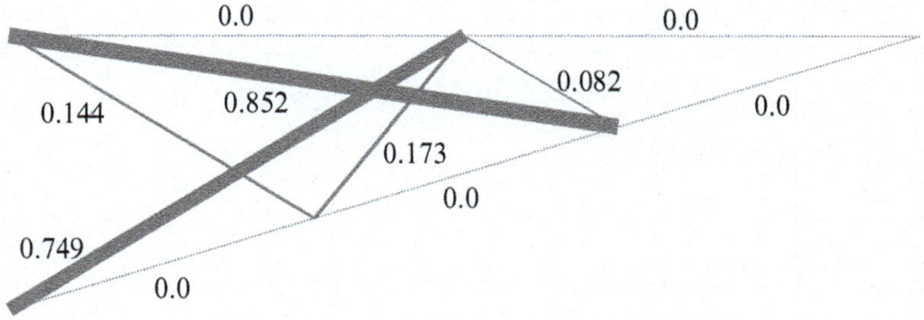

Fig. 17.7 Redundancy contributions for the optimal design

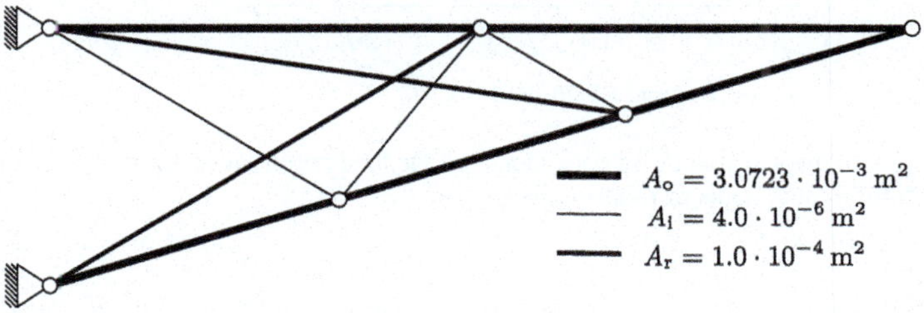

Fig. 17.8 Robust design of the truss structure

zero redundancy. This part of the structure is statically determinate, i.e. removal of one of these members will lead to the collapse of (this part of) the structure.

In comparison with this, the redundancy contributions for the optimal design are shown in Fig. 17.7. Here, all the outer members have zero redundancy. As explained in Sect. 17.4.1, this means that this part of the system acts as a statically determinate system. For the optimal design, the load is almost completely carried by the outer members and removal of one of these will lead to a collapse.

The analogy with a statically determinate system implies that changes in stiffness to the outer members of the structure will not change the load bearing behaviour. If we want to increase the robustness and change the load-bearing behaviour for the case of perturbations, we need to change the properties of redundant members.

Considering the redundancy distribution of the optimal design shown in Fig. 17.7, the two members that possess the highest redundancy, namely member 4 and 10, are taken into account for modification. To increase the robustness of the complete system, the stiffness (i.e. the cross sectional area A_r) of these two members is increased by a factor of 25. To keep the total mass of the system constant, the cross sectional area of the outer members A_o is slightly reduced. The cross section of the other inner members A_i is kept the same as that for the optimal design. The exact numbers for the cross sectional areas are given in Fig. 17.8.

The robustness of the new system is then studied by applying the load perturbation again. The result is shown in Fig. 17.9 with the curve labeled "robust". In order to have a better comparison with other possibilities of making the system more

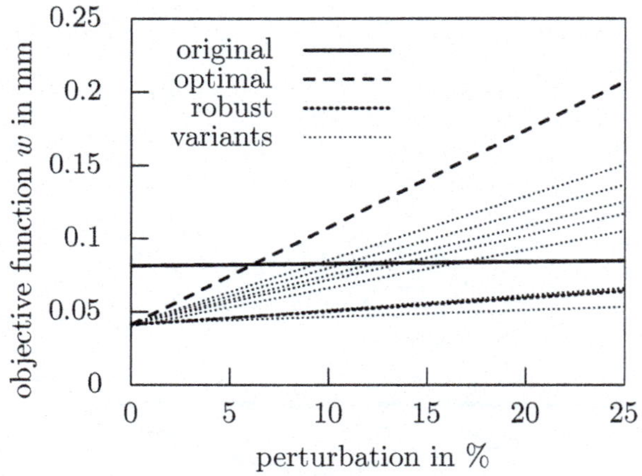

Fig. 17.9 Mean displacement w in dependence of the load perturbation for the initial and optimal design and different variants of robust designs

robust, the same study is performed for all other combinations of the two inner members with a higher stiffness. The results of all these variants are also shown in Fig. 17.9.

As can be seen, the variants of a robust design show highly different sensitivity towards the applied load imperfection. Some of the curves have a slope almost similar to the optimal design, whereas others are robust with small slopes. The robust design chosen on the basis of the redundancy contributions (label "robust") is not the variant with the smallest sensitivity towards the exemplary load imperfection. However, it shows the same optimal behaviour for no imperfection as the optimal design and a relative high robustness for the tested perturbation. To obtain a representative and more objective measure for the robustness of the different variants, other imperfections need to be considered. For now, this small example merely serves as a proof of concept.

In the spirit of the overall aim of this project, the redundancy distribution can be seen as an abstract measure for the robustness of a structure. A better understanding of this matrix-based property might lead to a "natural design" of robust structures as an alternative to robust design optimization algorithms. Furthermore, information from redundancy matrices might also be an ingredient for the objective function in a robust design optimization.

In summary, although robustness, perturbations and outputs have been defined in a completely different way in our two examples, both the biological and the mechanical example show that the network or graph structures play an important role for a robust functioning. Furthermore, in both examples complexity in topology and, in particular, redundancy, oppose optimality but are tightly related to an increased robustness. Moreover, the inverses of sensitivities of model outputs to perturbations serve as suitable measures for robustness in both worlds.

17.5 Conclusions

In this study we describe first attempts to bring two rather complementary research disciplines together, namely structural mechanics and intracellular networks. We have approached this aim by focusing on suitable examples for trade-offs between optimality, robustness and complexity from both the biological and the mechanical side. We have furthermore started to define graph-based and matrix-based abstractions for biological regulation networks, in particular intracellular signalling pathways and technical load-bearing structures, both of which allow the investigation of relationships between robustness and complexity at an abstract meta-level.

In the field of systems biology we have focused on ubiquitous signalling network motifs and their role for noise filtering and reliable signal propagation. For engineering structures, an in-depth study has been initiated concerning the interplay between statical indeterminacy, redundancy and robustness of load bearing structures, plus their quantification on the basis of abstract matrix-based measures.

It is difficult to find completely unifying representations of biological networks and load-bearing structures, even on an abstract mathematical level. Although several elegant robustness measures defined on pure mathematical graphs, such as Ricci curvatures (Sandhu et al. 2015), have been studied, which looked promising for our purposes at a first glance, we see several issues with these theories regarding our goals. This is related to the fact that some conceptual differences between molecular networks and load-bearing structures are important for a robust functioning in both worlds and hence impede a completely unifying framework. Importantly, literature about robustness in biological systems clearly indicates that the dynamics of interacting components are essential for a robust functioning. Thus, a purely static abstraction seems unable fully to capture the functional robustness of biological systems. Similarly, and in contrast to biological networks, properties of load-bearing structures are determined by material properties and the exact geometry of the structure, including angles and hinges. The direct reduction of these objects to the types of graph that have proved to be useful for dynamic biological systems is thus similarly impossible. Furthermore, signaling networks and load-bearing structures are subject to quite different perturbations: While the functionality of signaling networks has to persist despite stochastic fluctuations and intrinsic and extrinsic noise, load-bearing structures have to cope with material inhomogeneities or imperfection loads.

Although a considerable gap towards unifying descriptions of biological networks and load-bearing structures thus still exists, the discovery of common ground to describe optimality, complexity and robustness has been quite successful so far in our opinion. A suitable definition of robustness in terms of local sensitivities seems to be a useful and universal concept for both worlds and works independently of the way that the perturbations and outputs are precisely defined. This study presents the first results concerning these commonalities and suitable phenomenological representations.

In future, we aim to transfer general concepts from biological systems to technical load-bearing structures and vice versa. Of course, here the question arises as to why the kinds of signalling pathway modules investigated here could serve as appropriate subjects for this purpose. First, this motif is well-investigated experimentally and quantitative models have been introduced that are, to some extent, also predictive. This provides us with a good starting point to construct representative toy models. Furthermore, this motif seems to be a convenient illustrative example for our presumed hypothesis of a positive correlation between network complexity and robustness. Moreover, we believe that our approach is sufficiently general to transmit ideas to mechanical load-bearing structures in the future.

A challenging new direction is for example to systematically screen for network motifs that are capable of providing robust solutions in both worlds. A suitable and nice approach in this direction for the task of perfect adaptation of enzyme networks is presented in Ma et al. (2009), who revealed that the simultaneous realization of precision and sensitivity can only be achieved by networks including negative feedback loops or incoherent feed forward loops. These motifs are also known to play a central role in the regulation of signaling pathways. However, as discussed above, this is only possible if we are able to find a suitable mapping between biological networks and load-bearing structures in terms of abstract measures for complexity and robustness.

Acknowledgements This work has been funded by the German Research Foundation (DFG) as part of the Transregional Collaborative Research Centre (SFB/Transregio) 141 "Biological Design and Integrative Structures"/project B05.

References

Barkai N, Shilo B-Z (2007) Variability and robustness in biomolecular systems. Cell 28:755–760

Batchelor E, Goulian M (2003) Robustness and the cycle of phosphorylation and dephosphorylation in a two-component regulatory system. Proc Natl Acad Sci U S A 100(2):691–96

BendsœMP, Sigmund O (2004) Topology optimization. Springer, Berlin/Heidelberg

Beyer H-G, Sendhoff B (2007) Robust optimization – a comprehensive survey. Comput Method Appl M 196:3190–3218

Birtwistle MR, Rauch J, Kiyatkin A, Aksamitiene E, Dobrzynski M, Hoek JB, Kolch W, Ogunnaike BA, Kholodenko BN (2012) Emergence of bimodal cell population responses from the interplay between analog single-cell signaling and protein expression noise. BMC Syst Biol 6(109):1–12

Blanchini F, Franco E (2011) Multistability and robustness of the MAPK pathway. In: Proceedings of 50st IEEE conference on decision and ctrl and European control conference, Orlando, pp 2214–19

Bletzinger K-U, Ramm E (2014) Computational form finding and optimization. In: Adriaenssens S, Block P, Veenendaal D, Williams C (eds) Shell structures for architecture: form finding and optimization. Routledge, London/New York

Blüthgen N, Legewie S (2013) Robustness of signal transduction pathways. Cell Mol Life Sci 70:2259–2269

Brightman FA, Fell DA (2000) Differential feedback regulation of the MAPK cascade underlies the quantitative differences in EGF and NGF signalling in PC12 cells. FEBS Lett 482:169–174
Caicedo-Casso A, Kang H-W, Lim S, Hong CI (2015) Robustness and period sensitivity analysis of minimal models for biochemical oscillators. Sci Rep 5(13161):1–13
Clodong S, Dühring U, Kronk L, Wilde A, Axmann I, Herzel H, Kollmann M (2007) Functioning and robustness of a bacterial circadian clock. Mol Syst Biol 3(90):1–9
Dexter JP, Gunawardena J (2012) Dimerization and bifunctionality confer robustness to the isocitrate dehydrogenase regulatory system in Escherichia coli. J Biol Chem 288(8):5770–5778
Dexter JP, Xu P, Gunawardena J, McClean M (2015) Robust network structure of the Sln1-Ypd1-Ssk1 three-component phospho-relay prevents unintended activation of the HOG MAPK pathway in Saccharomyces cerevisiae. BMC Syst Biol 9(17):1–15
Emmert-Streib F, Dehmer M (2009) Information processing in the transcriptional regulatory network of yeast: functional robustness. BMC Syst Biol 3(35). doi: 10.1186/1752-0509-3-35
Fritsche-Guenther R, Witzel F, Sieber A, Herr R, Schmidt N, Braun S, Brummer T, Sers C, Blüthgen N (2011) Strong negative feedback from Erk to Raf confers robustness to MAPK signalling. Mol Syst Biol 7(489). doi: 10.1038/msb.2011.27
Grieco L, Calzone L, Bernard-Pierrot I, Radvanyi F, Kahn-Perlés B, Thieffry D (2013) Integrative modelling of the influence of MAPK network on cancer cell fate decision. PLoS Comput Biol 9(10):e1003286. doi: 10.1371/journal.pcbi.1003286
Gunawardena J (2005) Multisite protein phosphorylation makes a good threshold but can be a poor switch. Proc Natl Acad Sci U S A 102(41):14617–14622
Hart Y, Madar D, Yuan J, Bren A, Mayo AE, Rabinowitz JD, Alon U (2011) Robust control of nitrogen assimilation by a bifunctional enzyme in E. Coli. Mol Cell 41:117–127
Hersen P, McClean MN, Mahadevan L, Ramanathan S (2008) Signal processing by the HOG MAP kinase pathway. Proc Natl Acad Sci U S A 105(20):7165–7170
Kholodenko B (2000) Negative feedback and ultrasensitivity can bring about oscillations in the mitogen-activated protein kinase casacde. Eur J Biochem 267:1583–1588
Klavins E (2014) Lightening the load in synthetic biology. Nat Biotechnol 32(12):1198–1200
Kolch W (2005) Coordinating ERK/MAPK signalling through scaffolds and inhibitors. Nat Rev Mol Cell Biol 6:827–838
Kolch W, Calder M, Gilbert D (2005) When kinases meet mathematics: the systems biology of MAPK signalling. FEBS Lett 579:1891–1895
Krantz M, Ahmadpour D, Ottosson L-G, Warringer J, Waltermann C, Nordlander B, Klipp E, Blomberg A, Hohmann S, Kitano H (2009) Robustness and fragility in the yeast high osmolarity glycerol (HOG) signal transduction pathway. Mol Syst Biol 5(281):1–7
Lapidus S, Han B, Wang J (2008) Intrinsic noise, dissipation cost, and robustness of cellular networks: the underlying energy landscape of MAPK signal transduction. Proc Natl Acad Sci U S A 105(16):6039–6044
Legewie S, Schoeberl B, Blüthgen N, Herzel H (2007) Competing docking interactions can bring about bistability in the MAPK cascade. Biophys J 93:2279–2288
Ma W, Trusina A, El-Samad H, Lim WA, Tang C (2009) Defining network topologies that can achieve biochemical adaptation. Cell 138:760–773
MacBain KM, William RS (2009) Structural Optimization. Springer, Boston
Markevich NI, Hoek JB, Kholodenko BN (2004) Signaling switches and bistability arising from multisite phosphorylation in protein kinase cascades. J Cell Biol 164(3):353–359
Mishra D, Phillip MR, Allen L, Del Vecchio D, Weiss R (2014) A load driver device for engineering modularity in biological networks. Nat Biotechnol 32(12):1268–1275
O'Shaughnessy E, Palani S, Collins JJ, Sarkar CA (2011) Tunable signal processing in synthetic MAP kinase cascades. Cell 144(1):119–131
Ouldridge TE, ten Wolde PR (2014) The robustness of proofreading to crowding-induced pseudo-processivity in the MAPK pathway. Biophys J 107:2425–2435
Qiao L, Nachbar RB, Kevrekidis IG, Shvartsman SY (2007) Bistability and oscillations in the Huang-Ferrell model of MAPK signaling. PLoS Comput Biol 3(9):1819–1826

Robinson TJ, Borror CM, Myers RH (2004) Robust parameter design: a review. Qual Reliab Eng Int 20(1):81–101. http://onlinelibrary.wiley.com/doi/10.1002/qre.602/abstract

Sandhu R, Georgiou T, Reznik E, Zhu L, Kolesov I, Senbabaoglu Y, Tannenbaum A (2015) Graph curvature for differentiating cancer networks. Sci Rep 5(12323). doi: 10.1038/srep12323

Santos SDM, Verveer PJ, Bastiaens PIH (2007) Growth factor-induced MAPK network topology shapes Erk response determining PC-12 cell fate. Nat Cell Biol 9(3):324–330

Shinar G, Feinberg M (2010) Structural sources of robustness in biochemical reaction networks. Science 327:1389–1391

Shinar G, Feinberg M (2011) Design principles for robust biochemical reactiion networks: what works, what cannot work, and what might almost work. Math Biosci 231:39–48

Song H, Smolen P, Av-Ron E, Baxter DA, Byrne JH (2007) Dynamics of a minimal model of interlocked positive and negative feedback loops of transcriptional regulation by cAMP-response element binding proteins. Biophys J 92:3407–3424

Stelling J, Sauer U, Szallasi Z, Doyle FJ, Doyle J (2004) Robustness of cellular functions. Cell 118:675–685

Ströbel D (1995) Die Anwendung der Ausgleichungsrechnung auf elastomechanische Systeme, Ph.D. thesis Institut für Anwendungen der Geodäsie im Bauwesen, Universität Stuttgart

Wagner A (2005) Circuit topology and the evolution of robustness in two-gene circadian oscillators. Proc Natl Acad Sci U S A 102(33):11775–11780

Xiong W, Ferrell JE (2003) A positive-feedback-based bistable memory module that governs cell fate decision. Nature 426:460–465

Zhou T (2013) Encyclopedia of systems biology. Springer New York chapter Oscillation Amplitude, New York, pp 1616–1616

Chapter 18
Bio-inspired Sustainability Assessment – A Conceptual Framework

Rafael Horn, Johannes Gantner, Ludmila Widmer, Klaus Peter Sedlbauer, and Olga Speck

Abstract Because of the tremendous challenges of the impacts caused by the globally growing economy, the targeted development of sustainable innovation is an inevitable social responsibility. Despite some advances, however, sustainability has not yet been integrated into product development on a broad scale. Although bio-inspired innovations seem to offer solutions, the transfer of sustainability through the bio-inspiration process is only conducted implicitly and the possible fulfilment of the 'promise of bio-inspiration' is only assessed retrospectively.

In view of this situation, a bio-inspired sustainability concept is defined by conflating sustainability and bio-inspiration and is made concrete by framing an integrated assessment approach. The concept links current sustainability assessment practice, exemplified by sustainability in construction and aspects of sustainability in biological systems. The basic assessment structure is derived from biological systems, which provide necessary functions through the efficient use of scarce resources. Its application covers the complete development process of bio-inspired innovations, providing feedback and thus decision support with a focus on sustainability. Hence, the implicit sustainability transfer of bio-inspiration is enhanced by targeted transfer and by a 'commitment of bio-inspiration' to create both sustainable and bio-inspired innovations.

As the assessment method itself is expected to be bio-inspired, it is constructed based on characteristics of biological systems such as effectiveness, adaptivity, multifunctionality and resilience.

R. Horn (✉) • J. Gantner • K.P. Sedlbauer
Fraunhofer Institute for Building Physics, Wankelstraße 5, D-70563 Stuttgart, Germany
e-mail: rafael.horn@ibp.fraunhofer.de

L. Widmer
Plant Biomechanics Group Freiburg, Botanic Garden, Faculty of Biology, University of Freiburg, Schänzlestraße 1, 79104 Freiburg, Germany

O. Speck
Plant Biomechanics Group Freiburg, Botanic Garden, Faculty of Biology, University of Freiburg, Schänzlestraße 1, 79104 Freiburg, Germany

Freiburg Centre for Interactive Materials and Bioinspired Technologies (FIT), University of Freiburg, Georges-Köhler-Allee 105, 79110 Freiburg, Germany

J. Knippers et al. (eds.), *Biomimetic Research for Architecture and Building Construction*, Biologically-Inspired Systems 9, DOI 10.1007/978-3-319-46374-2_18

18.1 Introduction

Since the first human civilizations arose, Mankind has tried to subject its surrounding environment to its own needs. As long as societal metabolisms primarily based on biogenic resources and human populations remained low, environmental impacts were of a mainly local or regional character and only rarely affected the functionality of the ecosystem. From the time of the industrial revolution and the triumph of the global market economy which followed, these impacts have grown tremendously. Human activities nowadays interfere with the functions of the global ecosystem to an extent that is visible on a geographical scale, leading to a new era mainly influenced by human activities, the Anthropocene (Cruzen 2002). Although the consequences of human impacts on soil, water, the atmosphere, in particular, and ecosystems in general cannot be predicted precisely, the ongoing and rising activities of Mankind hold risks for the fundamentals of life. To limit these risks and to facilitate liveable conditions for future generations, the paradigm of sustainability has been developed. It aims to balance social and economic aspects of human activities without undermining the quality of the surrounding environment. Its implementation is directly addressed by the United Nations within its Agenda, introducing the Sustainable Development Goals and being committed to fulfil them by 2030 (United Nations 2015). However, in the spotlight of the continuous and unrestricted growth of population and economy, human activities are still strongly linked to a growing resource demand and depletion of the environment. Consequently, the development of systems that are in agreement with the Sustainable Development Goals is an urgent necessity.

Biologically inspired[1] systems promise to provide some of these solutions. Learning from nature is linked with the hope of learning from biological solutions that seem to be evolutionarily optimised, ecologically adapted and low risk. Von Gleich coined the term ‘biomimetic promise’, which implies that, because of their inspiratory flow and knowledge transfer from nature to technical products, biomimetic solutions have the intrinsic potential to contribute to sustainable technological development (von Gleich et al. 2007). Although this is identified by von Gleich as being a significant normative content of biomimetics, it has not been assessed at a general level. Moreover, we need to clarify whether living nature provides aspects that can be explicitly transferred to bio-inspired technical innovations with the aim of making them more sustainable. This is followed by the question as to whether the methods of sustainability assessment can be improved by biological systems as concept generators.

[1]To improve readability, we use the umbrella terms ‘biologically inspired’, ‘bio-inspired’, ‘biological inspiration’ and ‘bio-inspiration’ for all innovations inspired by a biological role model that is not directly involved in the production. If the terms biomimetic or biomimetics are used, we should point out that, in these cases, a development according to the VDI definitions of biomimetics is achieved and a knowledge transfer of ‘inspiration’ and ‘functional principle’ has taken place (Speck et al. 2016).

In the framework of the CRC-Transregio 141, all three objectives outlined above are embedded within a dedicated project in close cooperation with the other subprojects. It aims at developing a bio-inspired sustainability concept including a model for the assessment of sustainability potentials of biologically inspired systems. Therefore, the retrospective assessments currently used will be enhanced towards an *ex ante* assessment, accompanying the development of biologically inspired systems and, thus, actively supporting the achievement of bio-inspired and sustainable solutions within the development process. Future case studies will deal with examples of application and evaluation. To facilitate this, a basic assessment framework will be developed that requires the definition of the desired characteristics of the assessed systems within the area of application, including the identification of goals. As the concepts of bio-inspiration and sustainability are claimed to be compatible, their interoperability and their discrepancies have to be identified. The sustainability assessment of technical systems offers a number of comprehensive assessment schemes, particularly in the construction sector. As sustainability is a manmade concept derived from domesticated nature, we can determine to what extent biological systems are able to act as concept generators for sustainable bio-inspired innovations and bio-inspired sustainability assessment. The assessment concept based on the investigation of the underlying technical system is then complemented through the integration of sustainability aspects given by biological systems. It acts as a basis from which to describe the desired system properties, at least in terms of social, economic and ecological aspects, but also with regard to functionality. In the following, the conceptual framework is presented based on current sustainability assessment practice exemplified by sustainability in construction and aspects of sustainability in biological systems. Furthermore, the rules for the evolution of the assessment structure are specified.

18.2 Sustainability of Technical and Biological Systems – Common Elements and Delimitations

To facilitate a conjoint sustainability assessment model including both biological and artificial systems, the two paradigms are evaluated regarding their interoperability in order to draw conclusions within a structured approach. Therefore, the historical background and the state of the art of addressing sustainability are depicted and assessed in a structured approach to identify strengths, weaknesses and inconsistencies. This assessment is then complemented by elements that can be transferred from biological concept generators to sustainability assessment.

18.2.1 Sustainability of Technical Systems

Before fossil resources were discovered, revolutionising Mankind's energy supply system, humans were dependent on wood as their main source of energy and building material. At the beginnings of the industrial revolution in the eighteenth century, the invention of the steam engine required increasing amounts of fire-wood that could not be satisfied adequately (Grober 2013). To constrain the threatened shrinking of forest areas, a method of proactive forestry was developed based on the premise of a persistent utilisation (Carlowitz 1713). Forestry practice, subsequently defined as sustainability, was inspired by living nature in terms of the preservation of the natural equilibrium. This was basically an attempt to mimic the apparently stable conditions of untouched forests, enhanced by the idea of enduring anthropogenic utilisation.

When rediscovered in the twentieth century, the paradigm of sustainability was still based on a concept of preservation. Its reference to the principle of nature conservation was maintained, although the premise of the balance of nature was outdated even within the scientific community (Leach et al. 2010). While becoming relevant through the problems arising together with economic global growth, the paradigmatic concept was extended to include social justice and shifted towards an anthropogenic principle. These principles of anthropocentric equity and preservation of nature seem to act as antagonists and are complemented and further complicated by the paradigm of economic growth (Atkinson et al. 2007). Numerous approaches have been developed to address and facilitate sustainability but recent societal developments on a global scale point towards an ongoing unsustainable development (Moore 2011).

Although the paradigm of sustainability and the utopia behind them have been commonly acknowledged since the UN conference in Rio de Janeiro in 1990, the transfer from theory to practice is often missing. One of the main obstacles to the implementation of sustainability in concrete actions lies in the constructive ambiguity that was necessary to facilitate the development of a generally accepted definition (Moore 2011). This includes the fact that the goals are frequently conflicting at the operational base. These goals are known as three pillars or circles of sustainability and include environmental impacts, social acceptability and economic viability. As far as today's society and recent technologies are concerned, a sustainable solution is not possible without accepting a trade-off (Robinson 2004).

In order to face the lack of concrete recommendations and the onward march of unsustainable development, numerous attempts have been developed to interpret sustainability and to implement the paradigm (Pesqueux 2009). These interpretations not only address the quantitative sustainability assessment of systems, but also provide guidance on a more generalised scale. Although no consistent segmentation of these general sustainability strategies is possible, they can be associated into three basic approaches. Efficiency, or eco-efficiency, is the most popular strategy, promoting sustainable development by increasing efficiency in terms of material and energy. Although the related methods are not explicitly limited to technical system efficiency, they mainly focus on increasing the energy efficiency of technological solutions (von Weizsäcker et al. 2010; Schmidt-Bleek and Bierter 1998). This strategy is applied frequently and is deeply entrenched within technical development processes but bears several potential limitations. One of these is the potential rebound effects that include increasing consumption patterns compensating the efficiency-induced gains. Sufficiency strategies are premised on this assumption and claim that a real decoupling of growth and environmental impact is not feasible. On this account, these strategies underlie a general change in human behaviour, as the individual is expected to set limits to consumption based on self-limitation and to focus on relevant items (Daly 1991). Instead of the potentially negative and prohibiting picture of the sufficiency strategy, consistency models aspire to positive impulses. Consistency is thereby understood as an eco-effectivity strategy striving for complete integration of ecological and social aspects with economic considerations (Hawken et al. 1999; Braungart and McDonough 2014). The difference between consistency and sufficiency can also be described as weak and strong sustainability. While strong sustainability considers both the ecosystem and the earth's natural resources as irreplaceable and therefore demands that this so-called natural capital is preserved, in weak sustainability, natural and socio-economic goods are fully interchangeable (Dietz and Neumayer 2007). This includes the outstanding problems as to the extent that anthropogenic impacts have on these ecosystems until they collapse and, especially, the consequences for the ecosystem services on which Mankind is dependent. Although extensive progress has been made on the topic of ecosystem behaviour under anthropogenic influence, we do not as yet know the precise way that ecosystems react to disruptions of global relevance and high dynamics. As for other highly uncertain system responses of essential impact on human existence, the risk of unpredictable hazards is given. In order to address this adequately, the concept of resilience is considered to complement the concept of sustainability (Bloesch et al. 2015).

Whatever strategy is chosen to develop sustainable systems, the assessment of solutions in terms of their contribution to sustainability is fundamental. Concerning the three pillars solely for the assessment of environmental impacts, the method of Life Cycle Assessment (LCA) has prevailed (Klöpffer and Grahl 2009). For the assessment of economic or social implications, several approaches are available but are not as yet fully developed (Finkbeiner et al. 2010). All assessment methods that are being used today are based on the principle of damage prevention and thus are impact-focused. Furthermore, most assessments are based on the simplifying premise of static systems and do not take into account interdependencies and the dynamic nature of technical, natural, social, economic and ecological systems. As an example of a number of sustainability assessment approaches, sustainability certification schemes within the construction sector are depicted in the following section (for further approaches, see Grießhammer et al. (2007); Blok et al. (2013) or UNEP/SETAC Life Cycle Initiative (2011)).

Within the last few decades, several institutions have worked on the subject of sustainable buildings and have developed methods systematically to assess and review the sustainability of buildings. Starting with and based on the British certification system BREEAM (Building Research Establishment's Environmental Assessment Method), several country-specific certification systems have been developed. They are mainly based on the existing systems while taking national standards and structural conditions into consideration. Building on the experiences of the previously existing systems and with regard to further methodological

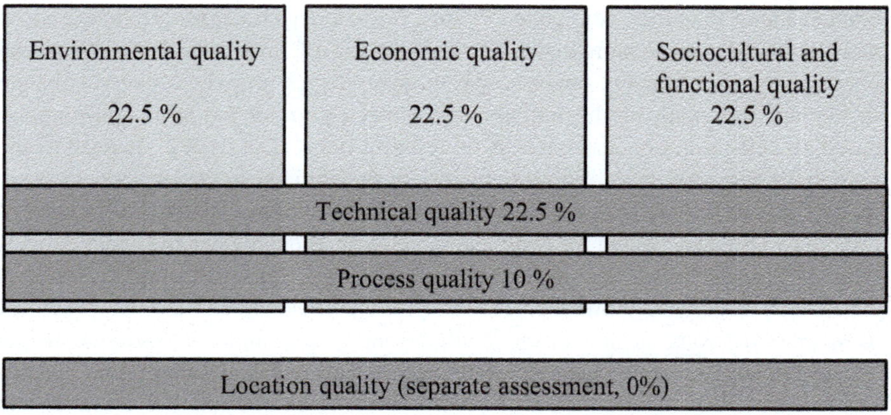

Fig. 18.1 Core criteria and associated weighting factors showing the assessment structure of the DGNB system using the example of new construction of buildings

developments in the sustainability assessment sector, some systems such as the German DGNB (Deutsche Gesellschaft für Nachhaltiges Bauen) system have been developed more recently and can be regarded as second generation systems. Their approach is based on life cycle thinking and aspires to address all relevant fields of sustainability. Therefore, not only energetic and environmental aspects, but also sociocultural, technological and economic indicators are contemplated. Furthermore, the scoring system is not action-oriented but provides, where available, quantifiable target values based on scientifically sound models (Ebert et al. 2010).

The DGNB system stands exemplary for the certification schemes of the second generation. Its basic assessment structure follows the principle of *protection areas* constituting natural sociocultural and economic values. These are structured as depicted in Fig. 18.1, whereas each of these core criteria is again broken down to several indicators. For each indicator, the requirements concerning calculation rules, data quality and documentation are stated in a profile. The calculated value is compared with a reference value given by a reference building, and points are provided, if defined quality levels are exceeded. The given points are then weighted within a two-stage process to analyse the overall building performance according to the weighting key shown in Fig. 18.1 (Ebert et al. 2010).

The upcoming certification schemes within the construction sector facilitate a comprehensive sustainability assessment but demand high expenditures and thus cannot yet be feasibly applied on a broad scale (Ebert et al. 2010). Moreover, the schemes are focused on buildings and do not aim at sustainable product development. If the performance of a product is strongly influenced by the hierarchical structure of its surrounding system, as is the case in building products and buildings, this interdependence should be considered within product development. Despite their comprehensive structure, building certification schemes are not directly applicable when conclusions have to be drawn for building products based on building sustainability (Wittstock 2012). As the building products constitute the building's sustainability, however, the contribution of a single product can be improved with the aim of optimising the sustainability of the buildings in which it might be applied. Furthermore, the impact-focused approach of sustainability assessment models is based on functional equivalence. This frequently seems to oppose improvements in functionality, as they are often associated with an increase in resource demand (Jung et al. 2014).

One of the key aspects when striving for sustainability is the targeted development of products focusing on their life cycle impact on society and environment. When it comes to developing a sustainability assessment that accompanies product development, several aspects are crucial. The applicability throughout all development phases constitutes a major requirement that has to be joined with the principle of scientifically sound assessment methods. In previous concepts, the assessment either is focused on the design guideline level (see Crul and Diehl (2007); Reap (2009) or Pedersen Zari (2014)) or is a comprehensive, but retrospective, assessment of existing products (Grießhammer et al. (2007); FAO (2014); Blok et al. (2013)). For consistent development support, a framework is required that adapts accordingly to the development phase and the data availability.

18.2.2 Sustainability in Biological Systems

Learning from biological concept generators is linked to the hope of the gaining of knowledge of sustainability with all its facets such as sustainable innovations, sustainability strategies and suitable sustainability assessments. On an initial impulse, nature or biological systems seem to be, for a number of reasons, the appropriate reference systems for comparisons. One reason might be that sustainability concepts originate from the biological system of a (commercial) forest as developed by von Carlowitz (Carlowitz 1713). Another reason might be that humans are part of the natural system that has evolved over the last 3.8 billion years. The anthropocentric viewpoint is clearly mirrored by the term 'surrounding environment' (in German 'Um-Welt'). In contrast to the man-made paradigm of sustainability in all its diversity, evolution is neither anthropocentric, nor teleological (Speck et al. 2016) and does not preserve a status quo. On the contrary, biological systems are characterised by the dynamics of evolutionary adaptations. In contrast to technical systems, biological systems underlie evolutionary processes of mutation, recombination and selection in an ever-changing environment with the result of multifunctional and optimised structures or processes after several generations. However, whereas biological systems do not act in a goal-oriented or conscious manner as a whole, most artificial systems do. In addition, the term 'ecological sustainability' suggests the subject Ecology, a field in the biological sciences involving studies of the interaction between living organisms and their environment. Even though 'ecological sustainability' and 'environmental sustainability' are often used as synonyms, ecology and the environment are not synonymous terms in the natural sciences.

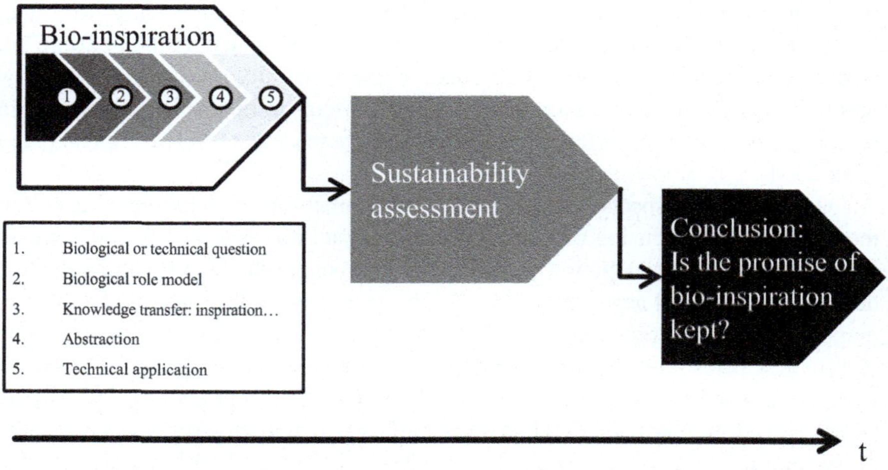

Fig. 18.2 Three-stage validation procedure as to whether the 'promise of bio-inspiration' in terms of an implicit sustainability potential of bio-inspired innovations (products, process engineering, methods) is kept. The first stage depicts the process of bio-inspiration based on advanced VDI criteria. In the second, the sustainability assessment is conducted retrospectively for the three pillars of sustainability. In the third step, a conclusion is drawn as to whether the 'promise of bio-inspiration' is fulfilled

In the framework of CRC 141, we need to clarify principally whether learning from nature per se leads to sustainable solutions in technology and to what extent we can make the development of bio-inspired technical solutions sustainable. With regard to the 'biomimetic promise' (von Gleich et al. 2007); Antony et al. (2014) have suggested a structured approach for the quantification and operationalisation of the intrinsic sustainability potential of biomimetic innovations. The suggested approach can be adapted to bio-inspired solutions. As depicted in Fig. 18.2, the implementation of the 'promise of bio-inspiration' consists of three stages: the verification that a technical innovation is bio-inspired, the assessment of its sustainability and a final conclusion as to whether the 'promise of bio-inspiration' is kept.

In general, the transfer of biological insights into technical innovations is a systematic approach that can be characterised by at least five objectively variable attributes such as model, knowledge transfer, natural appearance, same function and status of development. On the basis of this characterisation, a semi-quantitative analysis can be carried out that ranges from 0 (no impact) to 1 (maximum impact) describing the impact of natural models on technical products and methods. Two results are of utmost importance in the context of this paper: the attribute 'function' plays an important role, whereas currently sustainability plays no role at all (Speck et al. 2016).

With respect to the knowledge transfer of a 'functional principle', the meaning of the word 'function' in the context of technical and biological systems needs to be defined. The German zoologist Günther Osche referred to the meeting of (at times contradictory) requirements of functionality and evolutionary adaptation: "Die Natur kann kein Schild heraushängen, auf dem steht: Wegen Umbau geschlossen" ("Nature can't display a sign saying: closed for reconstruction"; translation by the authors, personal communication of G. Osche during lectures at the University of Freiburg). Over the course of evolution, adaptations are possible but always need to build on existing structures and functions, even if new structures and functions arise. Moreover, another difference in terms of functionality is the role of growth as a central driver: biological systems are based on the principle of growth and decay resulting in evolution, whereas artificial systems, especially economic systems in their present form, are built on the paradigmatic concept of growth (Purdey 2012).

Following von Gleich, learning from nature in general can take place in terms of learning from (i) form-function relationships, (ii) processes of evolution and (iii) successful concepts of evolution (von Gleich et al. 2007). These all differ from each other with regard to the quality of knowledge transfer (e.g. inspiration, functional principle or algorithm, Speck et al. 2016), with respect to the hierarchical level (e.g. molecules, cells, tissues, organs, organisms, populations, biocoenosis, ecosystem and biosphere) and to the respective time scales (e.g. from milliseconds to millions of years). Focusing on sustainability to be transferred, difficulties arise because the well-known sustainability indicators (e.g. primary energy consumption, greenhouse gas emissions) and sustainability strategies (efficiency, sufficiency and consistency) originate from a technology-driven viewpoint. From a biologist's perspective, indicators such as robustness, resilience, adaptivity, self-X-properties, multifunctionality and optimisation are more relevant. Further studies will concentrate on biological examples in order to determine consistent key indicators for application in the field of sustainable development and sustainable living.

In terms of bio-inspired sustainability assessments of artificial systems, these differences between biological and technical systems indicate both transfer options and limitations. As human activities influence the global ecosystem to an obvious extent, the responsibility for preventing dramatic changes is evident. However, in contrast to biological systems, artificial systems are developed consciously and thus their impacts can be considered in a structured approach (von Gleich et al. 2007). As biological systems are highly hierarchically organised and as their subsystems show strong interdependencies, their functional relationships may offer methodological approaches for addressing these issues. Furthermore, specific aspects based on the rules of civilization, such as aspects of morality and civil rights, do not have a biological counterpart, although they may have an evolutionary justification (FitzPatrick 2016).

The paradigm of sustainability does not fully apply within living nature. Nevertheless, the transfer of knowledge from biological concept generators in general and, in particular, the potential of knowledge transfer involving sustainability offer potential improvements to current sustainability assessment models. Even though several elements concerning artificial systems are also found in biological systems, as stated above, they exceed these by mainly two aspects, namely, by aspects of culture in the sense of the philosophy of arts and humanity including equity and human rights, which have no direct counterparts in biological systems.

18.3 Biologically Inspired Sustainability

To assess bio-inspired concepts and products with regard to their sustainability, the assessment structure itself should be designed inspired by biology. In the following, a bio-inspired sustainability concept is presented and applied in order to form a basis for an assessment model, namely an assessment and system structure. It is based on currently applied sustainability assessment concepts in the construction sector and on other disciplines and is complemented by bio-inspired aspects. Based on existing case studies, the concept will be developed and will further evolve accompanied by the developments presented within the previous chapters. Although the concept has been developed specifically for the given context, i.e. bio-inspired structures within constructional application, it is not restricted to this view.

18.3.1 Basic Structure

In terms of their level of abstraction, both bio-inspiration and sustainability are paradigms depicting ideal states and are thus, first and foremost, of a utopian nature. To transfer these vague statements conjointly to a feasible level, they are merged within a concept of bio-inspired sustainability. As depicted in Fig. 18.3, several approaches can be derived from this: the retrospective assessment as described by Antony et al. (2014) and the integrated assessment as presented herein. In terms of their basic approach, the main distinction lies in the implementation of the sustainability assessment. Within previous assessments, the creation of sustainable solutions was expected to be implicitly transferred through bio-inspired innovations and came in the form of the promise of bio-inspiration. For the current approach, this implicit transfer is expanded towards an integrated assessment, ensuring the creation of a sustainable solution through a targeted approach with the objective of accessing the sustainability potentials that arise from bio-inspiration. This development is mainly motivated by the fact that an explicit knowledge transfer of sustainability is both possible and desirable. It simply requires additional attributes in combination with a modified developmental approach. The contribution of bio-inspired innovations (products, process engineering, methods) to sustainability should not be a mere accidental event measured in a subsequent step after having previously finished the development (Antony et al. 2014) but should rather be an explicit target from the very beginning of the development (Fig. 18.3).

This shift is again evident in the methods that facilitate the application of the approaches. Whereas the method of Antony et al. (2014) complies to the

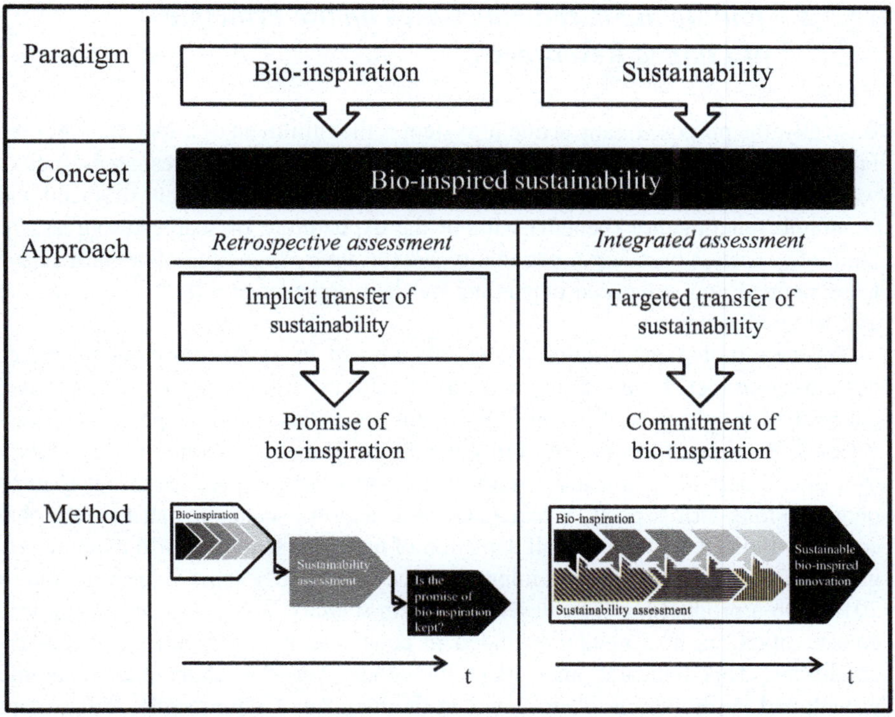

Fig. 18.3 Classification and structuring of the present integrated assessment for bio-inspired sustainability compared with the recently applied approach of a retrospective assessment described by Antony et al. (2014). The paradigms and the underlying concept form the basis for both approaches. Whereas the retrospective method uses an *ex ante* approach to assess the sustainability of bio-inspired innovations, the integrated assessment includes continuous feedback loops of the bio-inspiration process and the sustainability assessment in order to facilitate the commitment of bio-inspiration to sustainable innovations

currently applied *ex ante* assessment that cannot provide guidance within the development process of bio-inspired innovations, the integrated assessment includes a continuous feedback mechanism. As the bio-inspiration process evolves towards a bio-inspired product, the sustainability assessment adapts likewise. Because only little information is available at the beginning, the assessment starts from a screening assessment. As information availability increases, the assessment structure is refined and, as such, facilitates a consistent accompaniment for the whole process of bio-inspiration. For fully developed innovation, a comprehensive assessment is provided for both sustainability and bio-inspiration. Practitioners can integrate their sustainability targets into the development process and thus make informed decisions.

18.3.2 Joining the Paradigms Based on the Principle of Limited Resources

Basically, the purpose of any artificial system is the fulfilment of a specific function. For a sustainable product, this has to be obtained preferably in a resource-efficient way. Hence, a sustainable system is specified through the fulfilment of social and economic functionality (benefit) without the expenditure of scarce resources. To facilitate a system based on functionality and resource expenditure, both terms have to be defined and structured depending on their appearance within the biological and artificial systems.

Whereas biological systems are mainly limited by competition for resources depending on their local scarcity and fulfil functions with the objective of survival (survival of the cheapest, Vincent (2002)), this restriction does not apply to artificial systems. Because of resource mobility within the globalised economy, the scarcity of artificial systems is not restricted to local and short-term relevance but focuses on global long-term aspects. When assessed on a global scale, the patterns of global economy indicate that the overall depletion of natural resources is still dynamically growing, despite its tremendous impacts on the world ecosystem (Valero and Valero 2010). On a product level, however, long-term scarcity is not presently taken into consideration, as it is neither included in economic nor in regulatory boundary conditions. This indicates an evident dichotomy between short-term economic growth and the limitation of resources in the long run. Consequently, the concept of scarce resources is not considered when developing products with state of the art methods but will be dealt with as a role model for developing and assessing sustainable systems. Within a sustainable system, all resources should be handled in a way that keeps the surrounding system intact. Consequently, the exploitation of scarce resources in general should be avoided, except for the possibility of circular resource flows (Meadows et al. 1972). As long as the cause-and-effect chains and correlated tipping points cannot be predicted precisely, the overarching goal is to minimise the utilisation of resources by means of their scarcity. Controversially addressed in recent scientific discussions, scarcity is basically structured in terms of renewability, recyclability and criticality (Klinglmair et al. 2014). Furthermore, the topics of spatial relevance, dynamics and system resilience need to be considered (Leach et al. 2010). Socio-economic aspects, such as criticality, are omitted with regard to as yet unresolved questions and ongoing scientific discussions (Schneider et al. 2015; Die Partner des Begleitprojekts MaRKT 2015). Resources are defined according to the Kosmol et al. (2012) and thus cover both sources and sinks. This implies that the expenditure of resources is not only restricted to the sourcing and consumption of scarce resources, but also covers the pollution of the environment and the receptivity depletion of sinks. Materials that can be completely obtained from and are fully reusable by ecosystems are not regarded as scarce resources. For biological resources, the regeneration rate is considered as suggested by Klinglmair et al. (2014).

In terms of functionality, biological and artificial systems show certain overlaps. Biological systems are subjected to evolution and therefore mainly optimise their functional properties when evolutionary pressure is present and forces the system to evolve or become extinct. This implies the restriction of functions to aspects of relevance in relation to a selective regime, constituted as relational functions (Walsh 1996). Artificial systems prevail based on similar criteria, as their implementation is mainly determined through the concurrence situation and the market demand. Both systems thus handle functionality as an aspect of competition and have no intension of optimising functionality as an end in itself. Functionality can therefore be regarded as a saturation function focused on exceeding the (unsaturated) performance of reference systems. Within recent assessments, the aspect of functionality is mainly addressed through the introduction of a functional unit (within general LCA studies) or a functional equivalent if buildings are assessed. Both approaches mainly apply to the target function that a system has to fulfil. However, although this approach might be adequate for technical singular functionalities, multiple or fuzzy functions cannot be addressed by these means. Moreover, the scattering of the degree of functional performance within technical functions cannot be addressed through the currently applied functional unit. To include various functional aspects within sustainability assessment, a function fulfilment assessment method has to be chosen. The function fulfilment is depicted as a proportion of the actual functions of a system and its target functions. Furthermore, system functionality is, in this context, restricted to the design function according to Achinstein (1977) and does not cover any unintended use. For buildings and building products, the design functions are exhaustively addressed by Moro (2009) in terms of building physical functions.

18.3.3 Assessment Structure

The sustainability assessment framework is inspired by biology, both for structural concept based on the principle of limited resources and on a meta-level following several functional requirements. These constitute the framework of the assessment model and simultaneously define its success criteria as given in the subsequent description of the assessment model. The framework itself addresses the system functions and resource depletion. They conjointly depict the sustainability of any assessed system in terms of its resource efficient function fulfilment. Defined according to Eq. 18.1, the sustainability of a system rises with an increase in the degree of function fulfilment and a decrease in scarce resource expenditure.

$$S_{sys} = \frac{F_{sys}}{R_{sys}} \tag{18.1}$$

S_{sys}: Sustainability of a given system $[0 \ldots. 1]$
F_{sys}: (Multi-) functionality of a given system $[0 \ldots. 1]$
R_{sys}: Resource expenditure function of a given system $[1 \ldots. \infty)$

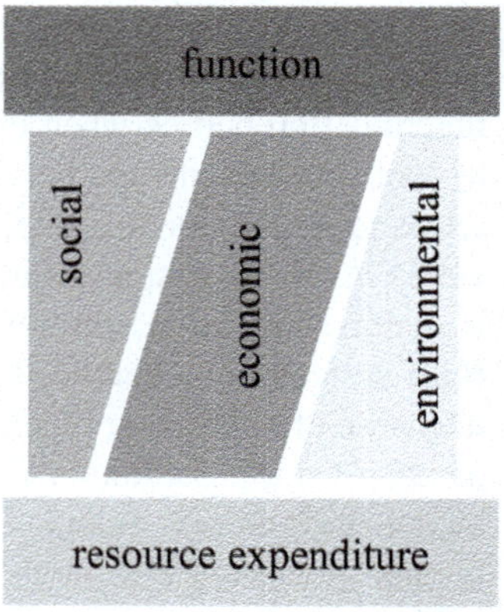

Fig. 18.4 Graphic representation of bio-inspired sustainability

Resource expenditure is addressed conjointly for all required resources including sinks and sources and is depicted on a normalised scale with a lower boundary of 1, constituting the minimal amount of resources required to fulfil the (multi-) functionality. As resource expenditure rises, the resource effort required for the actual functionality increases. Functionality is defined as the saturation function in accordance with the distinctions outlined above, constituting the degree of system functionality ranging from 0 to 1. System sustainability ranges from 0 to 1 accordingly, if addressed without a reference, on an absolute level. Although the expectation of absolute sustainability according to this specification is neither possible nor expedient, absolute sustainability leads to a quantifiable counterpart of the sustainability paradigm on system and product scale. Nevertheless, the improvement potential of any system can be assessed straightforwardly by comparing it with a reference system. Following the principle of relational functions, functionality is thus enhanced through the relationship to its target functionality, constituting the functional requirements of the reference system. This method is furthermore transferred to the resource expenditure function, including the resource expenditure of the reference system as a normalisation benchmark.

A representation of the concept is depicted in Fig. 18.4, where sustainability is shown as a ratio of revenues and expenses connected by three pillars. Whereas, in classical representations, the pillars stand for themselves and build the abstract foundation of sustainability, they are here representing instruments to describe the relationship of functionality and expended resources. Recently, artificial systems have mainly fulfilled social and economic functions through the utilisation of environmental and economic resources. This is shown by the different widths of the pillars: environmental resources are exploited to fulfil social functions, when neither scarce social resources nor environmental functions are prevailing. However, these connections can also be applied in future artificial systems and are depicted accordingly.

The concept agrees with the fundamental assumptions of the Brundtland report, as these are constructed as a societal guiding principle and need to be transformed in order to be applied to specific systems. To apply the concept to a specific system, the three pillars are completed through the implementation of functionality and resource expenditure as biological aspects. As within nature, resources are only chosen with regard to their properties and their scarcity and, thus, the concept can be identified as being based on an extended weak sustainability principle. Furthermore, an assessment of the eligible resources against each other is also possible, if their scarcity is considered accordingly.

18.4 Evolution of the Assessment Model

To operationalise the bio-inspired sustainability concept, elements of natural systems such as multifunctionality and the minimum use of resources are transferred in order to assess artificial systems. Therefore, both indicators and weighting schemes have to be developed into a method that meets the following requirement criteria constituting the model to be bio-inspired: consistency, effectivity, adaptivity and resilience (derived from Vester 2011; von Gleich et al. 2007; Reap 2009). Consistency is achieved, if the same assessment results, within a reasonable deviation, are found for different stages within a project development and levels of detail known to or given by the practitioner. Effectivity is addressed through the amount information yield compared with the required effort, stated as 'practitioner satisfaction'. The flexibility and expandability in an indicator and weighting scheme constitutes the adaptivity of the model. If the assessment model is resistant to failures and system deflection, it is regarded as being resilient.

Guided by this quality criteria, an assessment model can be developed based on existing indicators and weighting schemes, starting with the systems available within building sustainability assessment (Eberl 2013; Moro 2009; Wittstock 2012) and evolving them by using data from existing building products focusing on bio-inspired systems (Antony et al. 2014; Lienhard et al. 2011; Reap 2009). Within this phase, the success criteria will be made concrete in order to facilitate a validation process. As the model aims to develop both bio-inspired and sustainable solutions, the application will also include an assessment of the impact from natural concept generators. Thus, the developed model paves a pathway towards an integrated adaptive sustainability assessment and facilitates the achievement of the commitment of bio-inspiration.

Acknowledgements This work has been funded by the German Research Foundation (DFG) as part of the Transregional Collaborative Research Centre (SFB/Transregio) 141 'Biological Design and Integrative Structures'/project C01 'The biomimetic promise: natural solutions as concept generators for sustainable technology development in the construction sector'.

References

Achinstein P (1977) Function statements. Philos Sci 44(3):341–367

Antony F, Grießhammer R, Speck T, Speck O (2014) Sustainability assessment of a lightweight biomimetic ceiling structure. Bioinspir Biomim 9(1):16013. doi:10.1088/1748-3182/9/1/016013

Atkinson G, Dietz S, Neumayer E (2007) Handbook of sustainable development. Edward Elgar, Cheltenham/Northampton

Bloesch J, von Hauff M, Mainzer K et al (2015) Sustainable development integrated in the concept of resilence. Problemy Ekorozwoju – Probl Sustain Dev 10(1):7–14

Blok K, Huijbregts M, Roes L et al (2013) A novel methodology for the sustainability impact assessment of new technologies. http://www.prosuite.org/c/document_library/get_file?uuid=bdbb04e9-1a34-434b-85a8-44bafb28155b&groupId=10136. Accessed 1 Jun 2016

Braungart M, McDonough W (2014) Cradle to Cradle: Einfach intelligent produzieren, Ungekürzte Taschenbuchausg. Piper, München, Zürich

Carlowitz HCV (1713) Sylvicultura oeconomica oder haußwirthliche Nachricht und Naturgemäßige Anweisung zur Wilden Baum-Zucht. Johan Friedrich Braun, Leipzig

Crul M, Diehl J (2007) Design for sustainability: a practical approach for developing economies. http://www.d4s-de.org/manual/d4stotalmanual.pdf. Accessed 1 Jun 2016

Cruzen PJ (2002) Geology of mankind: the Anthropocene. Nature 415:23

Daly HE (1991) Steady-state economics, 2nd edn, with new essays. Island Press, Washington, DC

Die Partner des Begleitprojekts MaRKT (2015) Leitfaden zur Bewertung von Ressourceneffizienz in Projekten der BMBF-Fördermaßnahme MatRessource. http://www.matressource.de/fileadmin/redakteure/pdf/Leitfaden_Bewertung_von_Ressourceneffizienz_V4.pdf. Accessed 1 Jun 2016

Dietz S, Neumayer E (2007) Weak and strong sustainability in the SEEA: concepts and measurement. Ecol Econ 61(4):617–626. doi:10.1016/j.ecolecon.2006.09.007

Eberl S (2013) OPEN HOUSE: Assessment guideline. http://www.openhouse-fp7.eu/assets/files/OPEN_HOUSE_AG1.2.pdf. Accessed 1 Jun 2016

Ebert T, Eßig N, Hauser G (2010) Zertifizierungssysteme für Gebäude: Nachhaltigkeit bewerten; internationaler Systemvergleich; Zertifizierung und Ökonomie. Inst. f. internat. Architektur-Dokumentation, München

FAO (2014) The Sustainability Assessment of Food and Agriculture systems (SAFA) guidelines: version 3.0. Accessed 1 Jun 2016

Finkbeiner M, Schau EM, Lehmann A, Traverso M (2010) Towards life cycle sustainability assessment. Sustainability 2(10):3309–3322. doi:10.3390/su2103309

FitzPatrick W (2016) Morality and evolutionary biology. The Stanford encyclopedia of philosophy, Spring 2016 Edition

Grießhammer R, Buchert M, Gensch C-O et al (2007) PROSA – Product Sustainability Assessment. http://www.prosa.org/fileadmin/user_upload/pdf/PROSA-gesamt_Finalversion_0407_red.pdf. Accessed 1 Jun 2016

Grober U (2013) Die Entdeckung der Nachhaltigkeit: Kulturgeschichte eines Begriffs. Kunstmann, München

Hawken P, Lovins AB, Lovins LH (1999) Natural capitalism: creating the next industrial revolution, 1st edn. Little, Brown and Co., Boston

Jung J, von der Assen N, Bardow A (2014) Sensitivity coefficient-based uncertainty analysis for multi-functionality in LCA. Int J Life Cycle Assess 19(3):661–676. doi:10.1007/s11367-013-0655-4

Klinglmair M, Sala S, Brandão M (2014) Assessing resource depletion in LCA: a review of methods and methodological issues. Int J Life Cycle Assess 19(3):580–592. doi:10.1007/s11367-013-0650-9

Klöpffer W, Grahl B (2009) Ökobilanz (LCA): Ein Leitfaden für Ausbildung und Beruf. Wiley, Weinheim

Kosmol J, Kanthak J, Herrman F et al (2012) Glossar zum Ressourcenschutz. https://www.umweltbundesamt.de/sites/default/files/medien/publikation/long/4242.pdf. Accessed 1 Jun 2016

Leach M, Stirling A, Scoones I (2010) Dynamic sustainabilities: technology, environment, social justice, Pathways to Sustainability Series. Earthscan, London

Lienhard J, Schleicher S, Poppinga S et al (2011) Flectofin: a hingeless flapping mechanism inspired by nature. Bioinspir Biomim 6(4):45001. doi:10.1088/1748-3182/6/4/045001

Meadows DL, Randers J, Behrens III., William W (1972) The limits to growth: a report for the club of Rome's project on the predicament of mankind, 1. print. Universe Books, New York

Moore FC (2011) Toppling the tripod: sustainable development, constructive ambiguity, and the environmental challenge. Consilience: J Sustain Dev 1(5):141–150

Moro JL (2009) Baukonstruktion: vom Prinzip zum Detail. In: Grundlagen, vol 1. Springer, Berlin, Heidelberg

Pedersen Zari M (2014) Ecosystem processes for biomimetic architectural and urban design. Archit Sci Rev 58(2):106–119. doi:10.1080/00038628.2014.968086

Pesqueux Y (2009) Sustainable development: a vague and ambiguous "theory". Soc Bus Rev 4(3):231–245. doi:10.1108/17465680910994227

Purdey SJ (2012) Economic growth, the environment and international relations: the growth paradigm, vol 17, 2nd edn. Routledge, London/New York

Reap J (2009) Holistic biomimicry: a biologically inspired approach to environmentally benign engineering. Dissertation, Georgia Institute of Technology

Robinson J (2004) Squaring the circle? Some thoughts on the idea of sustainable development. Ecol Econ 48(4):369–384. doi:10.1016/j.ecolecon.2003.10.017

Schmidt-Bleek F, Bierter W (1998) Das MIPS-Konzept: Weniger Naturverbrauch–mehr Lebensqualität durch Faktor 10. Droemer, München

Schneider L, Berger M, Finkbeiner M (2015) Abiotic resource depletion in LCA—background and update of the anthropogenic stock extended abiotic depletion potential (AADP) model. Int J Life Cycle Assess 20(5):709–721. doi:10.1007/s11367-015-0864-0

Speck O, Speck D, Horn R et al (2016) Biomimetic – bio-inspired – biomorph – sustainable? An attempt to classify and clarify biology-derived technical developments. Bioinsp. Biomim. (in press)

UNEP/SETAC Life Cycle Initiative (2011) Towards a life cycle sustainability assessment: making informed choices on products. http://www.unep.org/pdf/UNEP_LifecycleInit_Dec_FINAL.pdf. Accessed 1 Jun 2016

United Nations (2015) Transforming our world: the 2030 agenda for sustainable development: A/RES/70/1

Valero A, Valero A (2010) Physical geonomics: combining the exergy and Hubbert peak analysis for predicting mineral resources depletion. Resour Conserv Recycl 54(12):1074–1083. doi:10.1016/j.resconrec.2010.02.010

Vester F (2011) Die Kunst vernetzt zu denken: Ideen und Werkzeuge für einen neuen Umgang mit Komplexität; ein Bericht an den Club of Rome, 8th edn. Dt. Taschenbuch-Verl, München

Vincent JF (2002) Survival of the cheapest. Mater Today 5(12):28–41. doi:10.1016/S1369-7021(02)01237-3

von Gleich A, Pade C, Petschow U, Pissarskoi E (2007) Bionik: Aktuelle Trends und zukünftige Potenziale. Institut für ökologische Wirtschaftsforschung, Berlin

von Weizsäcker, Ernst Ulrich, Desha C (2010) Faktor Fünf: Die Formel für nachhaltiges Wachstum. Droemer, München

Walsh DM (1996) Fitness and function. Br J Philos Sci 47:553–574

Wittstock B (2012) Methode zur Analyse und Beurteilung des Einflusses von Bauprodukteigenschaften auf die Nachhaltigkeitsbewertung im Rahmen der Zertifizierung von Gebäuden. Dissertation, Universität Stuttgart

Chapter 19
Making Life "Visible": Organism Concepts in Biology and Architecture as the Basis for an Interdisciplinary Synopsis of Constructional Biomimetics

Gerd de Bruyn, Oliver Betz, James H. Nebelsick, Manfred Drack, and Mirco Limpinsel

Abstract Within biomimetics, scientists are challenged by the interdisciplinary exchange of knowledge and concepts, which include functional principles in complex systems of biological organisms, buildings and machines. One concept, that is used in biology as well as in architecture and in engineering, is the concept of the "organism". Despite representing the primary hierarchical level on which morphological form and functionality interact, the individual organism, as a functional unit, has been increasingly neglected within modern biology. A similar trend can be recognized within modern architecture: as an integral concept, the built form has been lost from view. This article raises the question as to how the term "organism" and its function in the discourse of architecture can be conceptualized and possibly used as a unifying concept in interdisciplinary biomimetic research. While in biology, the "organism" is a more or less well defined concept to denote living entities, in an architectural sense, it functions as a model or topos, i.e. a commonly plausible semantic form that is usually not explicitly stated, but still becomes operative in establishing form decisions. As a case example, in this contribution, the focus is on the use of the "organism" in the German Romantic discourse of architecture and aesthetics, namely in the writings of Schelling and in Schinkel's architectural designs. Thereby, it becomes apparent how a scientific term can be transferred into a model for designing buildings.

G. de Bruyn (✉) • M. Limpinsel
Institute for Architectural Theory (IGMA), University of Stuttgart, Keplerstr. 11, 70174 Stuttgart, Germany
e-mail: g.debruyn@igma.uni-stuttgart.de

O. Betz • M. Drack
Evolutionary Biology of Invertebrates, Institute of Evolution and Ecology, University of Tübingen, Auf der Morgenstelle 28, 72076 Tübingen, Germany

J.H. Nebelsick
Department of Geosciences, University of Tübingen, Hölderlinstraße 12, 72074 Tübingen, Germany

J. Knippers et al. (eds.), *Biomimetic Research for Architecture and Building Construction*, Biologically-Inspired Systems 9, DOI 10.1007/978-3-319-46374-2_19

19.1 Introduction

Within biomimetics, scientists are challenged by the interdisciplinary exchange of knowledge and concepts, which include functional principles in complex systems represented not only by biological organisms, but also by buildings and machines. Despite epitomizing the primary hierarchical level on which morphological form and functionality become conjoined, the individual organism, as a functional unit, has been increasingly neglected within modern biological agendas. A similar trend can be recognized within modern architecture: as an integral concept, the *built form* as such has been lost from view. Not only architecture, in its encyclopedic approach, but also the life sciences struggle with the modern phenomenon of specialization within the arts and sciences. It is thus of prime importance to focus once again on the integrative nature of organisms, in the fields of biology, architecture and engineering.

Biologists and architects obviously have completely different things in mind when the word "organism" is used. The main difference is the status of the word itself. In biology, the organism is considered as a *concept*, a well defined language tool to denominate certain real things (Toepfer 2011). An architect will instead use the word as a *metaphor* to describe distinct properties of objects that are not actually organisms as such, but buildings, towns and so on. Nevertheless, both the biological and the architectural usages of the term "organism" refer to the notion of integrating various functional parts into a single complete system. Investigation of these basic semantic structures is instrumental in extending the field of biomimetics into a full scientific discipline that also integrates basic epistemological theories, instead of solely dealing with exploitable knowledge as an auxiliary field. A basic step in this direction is the recapitulation and further development of the theory of the organism and the evolution of a common system and vocabulary of concepts and structures for the field of architectural and constructional biomimetics. Basic approaches, principles and concepts by which complex biological constructions and architectural artifacts are characterized and explained with respect to their functionalities need to be explored. In defining a common vocabulary of basic terms and concepts regarding the systemic level of the organism, an approach is needed that also facilitates communication between biologists, architects and engineers.

This approach entails the documentation and evaluation of historical and existing ideas including concepts and classifications of the organism in biology, architecture and engineering. Furthermore, functional aspects resulting from biological, architectural and technical organismal constructions and their interlinked subsystems have to be compiled, formalized and systematized. This has to be carried out in order to establish an interdisciplinary and formalized vocabulary for architectural and constructional biomimetics documenting both congruencies and differences between the use of the word "organism" and systems of concepts of the three disciplines.

19.2 The Organism Concept

Biological organisms, architectural buildings and machines are highly complex entities whose analytic study has necessitated an increasingly reductionist stance in modern biology, contemporary architecture and building technology (Caplan 1987; Riedl 2000; Nönnig 2007). As a consequence, both biology and architecture have been subdivided into numerous sub-disciplines and have developed beyond and lost track of their primary manifestation, namely that of the living organism or the building, respectively (Laubichler 2005; Nachtigall 2010).

In biology, this development is expressed in scientific endeavours that increasingly follow approaches that, on the one hand, reduce biological processes to their molecular basis and, on the other, focus on supra-organismic levels such as populations and ecosystems. Extremely reductionistic or superordinate research programs, however, must not be taken as marking the endpoints of biological research, but should be reappraised as the beginning of a new integrative approach encompassing the entire organism (Sauer 1992). This view has been promoted since the second half of the twentieth century with the rise of new disciplines such as eco-physiology, ethoecology, or ecomorphology. In morphology, an integrative approach with regard to the form and function of organisms in their relationship to the external environment has been becoming increasingly relevant (e.g. Karr and James 1975; Motta and Kotrschal 1992; Reilly and Wainwright 1994). Whereas functional morphology investigates the exact relationship between form and function (e.g. Homberger 1988), ecological morphologists focus on the function of morphological structures in an ecological (environmental) and evolutionary (historical) framework in order to promote the general understanding of the ecological and evolutionary consequences of animal constructions and organismal functions. In order to understand complex morphological structures, separate organs and entire organisms, the emergent qualities of the interlinked subsystems of organisms must be considered, whereby the properties and functions of distinct units cannot be entirely ascribed to the properties of their subsystems (Campbell and Reece 2009). A combined analytic and synthetic approach linking, as far as possible, the various levels of description and explanation is thus required in order to describe the multiple networked and interacting complex chains of cause and effect within living organisms (Zoglauer 1994). Biological organisms are conceived as functionally integrated systems that are hierarchically composed of single organs ("tools") that have been formed by both ontogenetic adaptations and evolutionary transformations (cf. Gegenbaur 1870 as modified by Maier 1999; Fig. 19.1). In contrast to technical systems, biological organisms are, among others, characterized by metabolism, reproduction and evolution as based on genetic alterations.

With regard to engineering, concepts from biology were made prominent by the development of cybernetics. Norbert Wiener states in *Cybernetics, or Control and Communication in the Animal and the Machine* (Wiener 1948) that, by utilizing

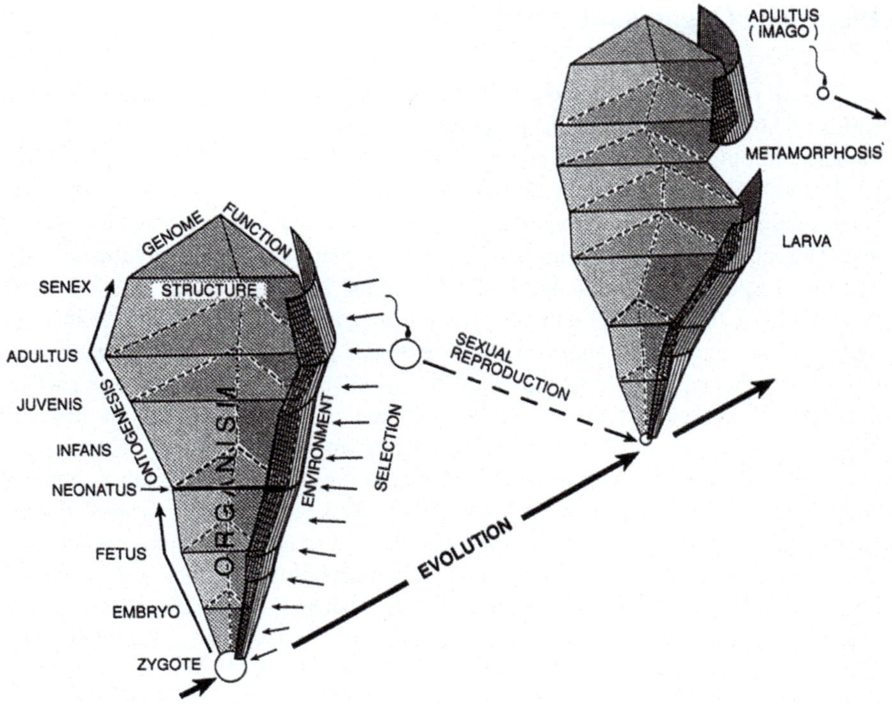

Fig. 19.1 Representation of a biological organism stressing its historical dimension. The organism is considered as the result of ontogenetic adaptation (*vertical axis*) and evolutionary transformation (*horizontal axis*). From Maier (1999)

feedbacks in its operation, a machine becomes similar in its behaviour to a living organism. The mere explanation of phenomena in the living by means of feedback was too narrow for the organismic biologist Ludwig von Bertalanffy (Drack and Pouvreau 2015). Cybernetics provided a common language and method for biology and technology, as applied in many fields of technology today. The biological organism as a functional entity is used as a role model for robotics, system control engineering and artificial intelligence. In today's construction engineering and building technology, however, the notion of the *machine as an organism* is of minor relevance. In this regard, an historical assessment of the notion of the organism in engineering has to be developed.

Classical architectural theory has always called for the organic unity of the work of art. In this respect, nature serves as a model in two different ways: its outward beauty is imitated on the one hand and its organic structure is reconstructed on the other. Plato's ideas of *organic unity* have spread to all fields of art including architecture (Plato 1981, Phaedrus: 264c). Since the nineteenth century, these ideas have also become a strategy of legitimization to modernize architecture itself. Noticeably, architects thereby refer to biology and engineering without differentiating between the organism and the machine (for examples see van Eck 1994; Blümle and Schäfer 2007).

In architecture and engineering, however, a classification of "organic" aspects in such a stringent manner as used in biology does not exist – although metaphorical identifications of organism and building can be found thoughout the history of architecture. For example, one could consider buildings as being analogous to individual biological organisms. In this sense, architectural units such as the office or the apartment and engineering structures such as balloon framing might be compared with organic systems such as the cardiovascular system; smaller units such as the galley kitchen are then comparable with single organs and so on. Parts of a single organ can be regarded as functional elements, such as a door or a window, whereas architectural fabric and material correspond to biological tissues. An important contrast to biology is that architecture and engineering keep and vary the systematization of the "organic" at the level of the example that is being analysed.

Technically speaking, such analogical use of the term "organism" does not deal with the *concept* of the organism in the same manner as biologists do when describing their objects of research. Instead, they seem to be mere metaphors. This, of course, is also because biological organisms are objects of study that have evolved by natural processes and whose structures and functional properties need to be disentangled by explicit scientific research. In contrast, architects and engineers deal with man-made artefacts, whose configurations and functions are known from the beginning. Nevertheless, the term "organism" fulfils a positive function in discourses related to art and design. In order to understand the organism concept and to facilitate interdisciplinary communication, a closer look at its origin and use is needed.

In the following section, the question is raised as to how the word "organism" and its function in the discourse of architecture can be conceptualized. Rather than a scientific concept, the word functions as a *model* or *topos*, i.e. a commonly used semantic term that is usually not explicitly stated as such, but still operates as a functional concept. As a case in point, the use of the "organism" in the German Romantic discourse of architecture and aesthetics is discussed. In the writings of philosophers such as Schelling, and its appropriation by architects including Schinkel, the transformation of a concept into a design model becomes perceptible. The German Romantic is chosen for this case study, because the concept of the organism was fundamentally re-interpreted in philosophy, aesthetics and the arts. The emphatic use of this concept in the nineteenth century might seem slightly mannered today as it is both conceptually ambiguous and normative. For this very reason, however, the implications of the organism model have remained latently effective in both architectural practice and notions about form and design (Ripley and Bhushan 2016:30).

19.3 The Organism Model in Architecture

When thinking about a definition of the organism concept, the major epistemological obstacle is the notion of the "concept" itself, as the classical theory of the structure of concepts always implies that necessary and sufficient conditions exist under which something can be subsumed under a given concept. This definition means that no dissent is allowed about whether a concept characterizes a certain thing or not. Whereas this is largely unproblematic in the sciences, the situation is completely different in other fields such as the humanities and arts. It quickly becomes clear that the definition of a concept of the organism cannot be attained if the various disciplinary approaches are simply applied in an equal manner and if the "concept" in the objectifying scientific sense is used.

At present, a general tenure is that only science is capable of solving the riddle of life and of explaining the irreducibility of living organisms. In Schelling's time, however, hope remained that philosophy, namely *Naturphilosophie,* would contribute to a basic understanding of nature. At the height of the German Romantic *Naturphilosophie*, Schelling claimed that nature and the spirit had to be identical in order to overcome idealism and to restore what he called the original unity of ideality and reality. This idea, namely that desire and reality would coincide and that the cosmos would turn out to be meaningful, continues to be a tempting idea and is certainly thriving in the superficial ideas within esoterics. It can also be found in the thinking of the eminent scholar and cybernetician Gregory Bateson who, like Schelling, rejected the dualism of body and mind. He argued that the phenomenon of life can only be adequately described by using metaphoric language (Bateson 1972). Thus, even at present, poetry can indeed play a role in explaining nature.

This note on the imagery of language is, at any rate, an incentive to reconsider the aspects that the humanities can contribute to the notion of the organism. Luckily, more can be said about those things that interest and affect us than rigorous concepts can express. Concepts as such are not only explanatory, but they also always distort the things that they attempt to explain. Concepts transform something that is highly individual into a mere platitude and thus miss everything that is unique about it. Moreover, general aspects usually affect us less than specific ones. What strikes us most about a person or a work of art is its originality. That is why many words both prescind and substantiate at one and the same time. In other words, words fulfil a "hard" conceptual and a "soft" intuitive function.

Such an idea also applies to the word organism. Like all words, it is not that which it designates. This is trivial, but it still indicates something important, namely, that words open associative spaces and allow us to develop them in different directions and to give them new horizons. This also leads to misunderstandings, even in the sciences. In addition to the word itself, few commonalities exist when a zoologist describes the "organism" of a beetle and an architect calls a house a "functioning organism". Whereas for the zoologist, all living entities are organisms, the architect will list a number of buildings that he does not consider organic, although the first colleague that comes along might well contradict this definition.

Aside from natural scientists, anyone who speaks of organisms without referring to actual living entities merely deals with analogies. Things that do not exist in nature are described by applying a natural role model as suggested by Auguste Perret, the French pioneer of modern concrete construction, who wrote in 1952:

> The great buildings of our time consist of a skeleton, a structure in steel or reinforced concrete. The framework is to the building what the skeleton is to the animal; as the animal's skeleton, measured, balanced, symmetrical, contains and supports the most various and most variously situated organs, so must the structure of a building be composite, measured, balanced and also symmetrical. It must be capable of containing various and variously situated organs, organisms, that are required by the function and purpose. (Perret 1952, translation from Guillén 2006:53–54, modified)

For two reasons, the "organic" serves well both as a model and a pattern. First, it can demonstrate, as pictorially and ostensively as possible, the aspects that one aims to understand: the order of the cosmos, the state and the city and of the work of art. Shapes and structures that we assume to be natural immediately make sense, whereas artificial shapes and structures are always in need of explanation. Second, outside of academic discourse, the measures of value that we extract from nature can lead to debates that might have political implications and are usually dominated by lay persons.

Although this might be difficult to comprehend as a biologist, public debates offer the opportunity whereby complex issues can be discussed not only by experts, but by society as a whole. Of course, a broader audience cannot be reached by means of detailed explanations of scientific terms. Instead, the terms are replaced by moral judgments related to environmental conservation, animal experiments, our consumption of meat and so on. Unlike architects and artists, scientists generally abstain from a moral and aesthetic assessment of their observations and findings, or at least they try to. Architectural theorists, who are more concerned with the way that buildings appear to the audience and to critics do not even assume a scientific basis when describing a building or a sculpture as organic.

A scientific language as such is not to be found in the midst of architectural and art theory, it being instead associative, intuitive, and impulsive. Instead, architectural and art theoreticians use highly artificial languages such as those used by artists. Characteristic of their scientific approach, the humanities concentrate on the imagery of language. How else can this be? Science always deals with objects that are, as such, not predefined as being scientific until done so. Thus, biologists first determine an object as being scientific by studying them in the first place. Similarly, literary scholars reflect on and define their objects of interests which, per se, do not belong to predefined categories.

Scholars who, for example, investigated what was referred to as organic or pertaining to organisms in nineteenth and twentieth century music, did not assume that they would actually deal with scientific terms. Instead, they assumed these to be part of a scientifically relevant *model* (Thaler 1984). Thus, music and music theory are not concerned with the organism as a *concept*. Instead, composers use a *model* of the organism to solve formal problems. In architecture and arts, the organism is therefore not used as a scientific term per se but as a model. The "organism" is, in the ancient Greek meaning of this word, a *topos,* a categorization of a motif that generally has positive connotations and is understood by almost everyone (Perelman and Olbrechts-Tyteca 2004; Limpinsel 2016). In this respect, the organism corresponds to ethical and aesthetic value judgments.

The questions are: how has the organism model been used in architecture and what function does it serve? These questions, for which we do not have definite answers at this state of the project, are directly linked to biomimetics in which the organism is maintained as a *topos*. Why? First, because biomimetics tries to fulfil wishes that go beyond a mere optimism of progress connected to industry and technology. In this respect, Schelling's desire to reconcile "spirit and nature", so to speak, lingers on. Second, biomimetics becomes aesthetic as soon as it allies itself with architecture and design. As long as biomimetics interacts not only with active technological fields, but also with ethical and aesthetic fields, the word "organism" will be used in both its scientific and topical connotations.

Long before the nineteenth century, architects sought for natural role models of what we call aesthetic coherence. Compared with natural organisms, the evaluation of a building's aesthetic organization seemed much easier. The building ought to look and function as pleasingly as natural organisms – in fact, even better! Only the modern world has changed this notion. At the moment, increasingly detailed research has led to the insight that in many cases natural products are far superior to the complexity of human inventions and artefacts. In addition, the tendency to take nature as a role model has been constantly increasing during modernization. This applies to both lay persons and experts. Whereas an expert might consider the unification of micro- and macrobiological knowledge into one indivisible organism to be utopian, the lay person who only has a vague idea of this dilemma is even more convinced of the organism as a *topos*.

Karl Friedrich Schinkel could not have known of Darwin's scientific explanation of the diversity of life through variation and selection, but he would have hardly accepted it, since from Fichte's lectures, he had learned that only those disciplines guided by philosophy could be considered scientific. One can characterize Schinkel (1781–1841) as a young cousin of the early Romanticists to all of whom Fichte had been a teacher. The example of Novalis shows, however, that these early romanticists were not only interested in *Naturphilosophie* but also in the emerging science of nature. For this reason, Novalis enrolled at the renowned Bergakademie in Freiberg, which offered the highest standard of scientific education in Germany at the time.

Novalis' move to Freiberg was due to his wish to be near his friend Johann Wilhelm Ritter who was one of the most important naturalist of the early Romanticism. Ritter developed a prototype of an accumulator that was known as a "Rittersäule" or "Ladungssäule". Indeed, experiments using his invention to explore galvanism, which was a fashionable branch of research at the time, would eventually cause Ritter's death.

After having experimented with prepared frogs' legs, the Italian Luigi Galvani came to the conclusion that a distinct form of electricity must have been inherent in these organisms (Fig. 19.2). "Animal electricity" occurring in nature was considered to be a special type of electricity differing from that of man-made origin. Ritter was

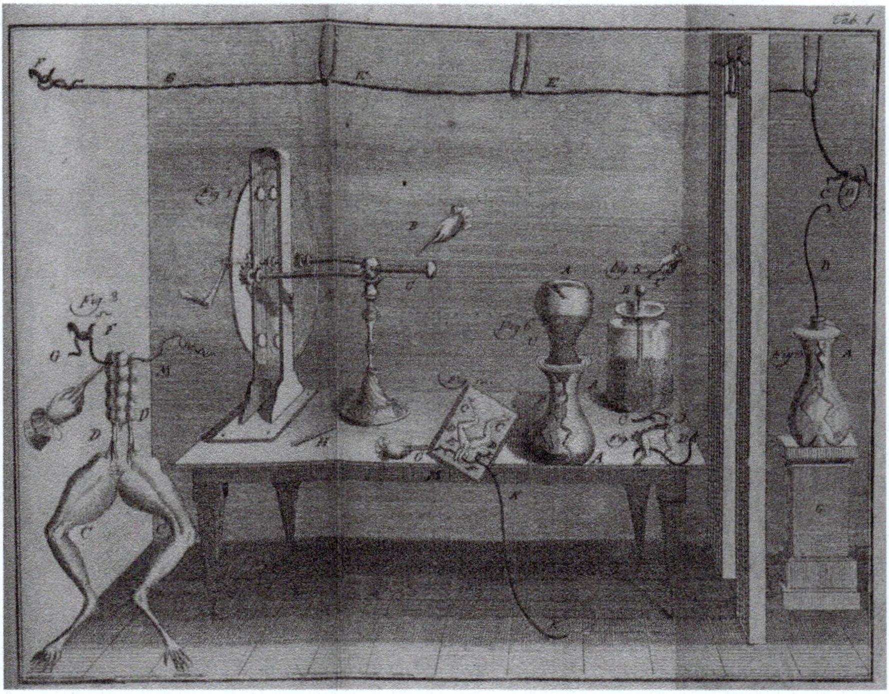

Fig. 19.2 Apparatus used for experiments on frogs' legs. From *Wikimedia Commons*

convinced that the key to life was to be found in this specific type of electricity. As we all know from "Frankenstein", this could only lead to one result. In fact, Ritter discovered ultraviolet radiation and, interestingly, Mary Shelley's novel is full of flashing lightning strikes. Even 18 years before "Frankenstein" was published in 1818, Novalis tried to reveal the life-giving principles of electricity through several sequences of awakening in his novel "Heinrich von Ofterdingen". The fact that many biologists, towards the end of the eighteenth and the beginning of the nineteenth century, thought that electricity was the major distinguishing factor between biologically active and inactive matter demonstrates that Ritter was not alone in this opinion.

Schinkel was also impressed by a self-contained power having to be assumed as a source of life and was convinced of another basic idea of romanticism, namely that art had to seek an "energetic" role model in nature. Art was supposed to define itself as a process rather than represent a finished product and should draw its liveliness from nature. In the excerpts of his "Architektonisches Lehrbuch" that are entitled "Romantische Skripte", Schinkel explains that architecture, like any other art, should make life visible and that the work of architecture, the building, should never stand alone:

Fig. 19.3 Schinkel, Stage design for Undine 1816. Staatliche Museen zu Berlin, Kupferstichkabinett. From *Kupferstichkabinett, Berlin*

> In der Baukunst muss wie in jeder Kunst *Leben* sichtbar werden, man muss die Handlung des gestaltens der Idee sehn, und wie die ganze bildliche Natur ihr zu Gebothe steht und sich hervordrängt herandrängt um ihrem Willen zu genügen. Das Werk der Bau*kunst* muss nicht dastehn als ein abgeschlossener Gegenstand [. . .]. (Schinkel 1979:32)

What, however, should architecture represent? Schinkel's answer to this question is revealed in his stage designs for romantic operas (Fig. 19.3). He also specifies all the attributes that dead (immobile) architecture shares with living (mobile) organisms:

> Ein Streben, ein Sprossen ein Crystallisieren, ein Aufschießen, ein Drängen ein Spalten, ein Fügen, ein Drücken, Biegen, Tragen, Setzen, Schmiegen, Verbinden, Halten ein Liegen und Ruhn welches letztere aber hier [. . .] als lebendiges Handeln gedacht werden muss, dies sind die Leben andeutenden Erfordernisse in der Architectur. (Schinkel 1979:32)

Later, as the chief construction officer of Prussia and a leading neoclassicist, Schinkel did show the same fervor in this regard. He, nevertheless, never abandoned the ideas from his *Sturm und Drang* period. We can even retrace these ideas when regarding the *Altes Museum* in Berlin (Fig. 19.4). Schinkel was so proud of this building that he made it accessible to the public immediately after its completion. In doing so, Schinkel developed a new typology as previously existing museums were restricted to castles or similar buildings. Instead, Schinkel confronted the Berlin City Palace, which lies opposite to the museum to the south, with a magnificent but nonetheless bourgeois architecture. On three sides, Schinkel's Museum is an unadorned cubic structure, which in its interior conceals a spectacular rotunda modeled on the Pantheon (Fig. 19.5). Schinkel's claim to classicism is

Fig. 19.4 Altes Museum, historical photograph, before 1854. From *Wikimedia Commons*

primarily expressed by the Greek Stoa-inspired portico formed of Ionic columns in monumental order.

In order to satisfy the romantic intentions that Schinkel had earlier so emphatically embraced, the *Alte Museum* had not only to demonstrate typological autonomy, but also had to make life "visible". To some extent, this is achieved by the persons situated within it. The building itself, however, is intended to be a "*Streben und Sprossen*" ("Aspiring and Sprouting") and to obey the model of an organism. More precisely, the museum conforms to two organism models (1) the neo-classical model, emanating from the harmonious organization of a divine world order that seeks to reflect the unification of disjunct parts into a perfect whole; (2) the romantic model intending to reflect vitality. Schinkel had considered both *topoi*, obviously preferring calmness over motion. Nevertheless, on entering the entrance hall, one can feel the expression of life in the promenade character of the hallways and in the staircase and the ambiguous visual axes connecting the inside with the outside (Fig. 19.6).

Why was Schinkel, as an architect, so restrained with respect to the demands of the organism model, of which he had been so enthusiastic as a young painter? Was it perhaps because he believed that construction was about harmonizing the antipodal forces of load and bearing, instead of inciting them? In his designs, however, Schinkel aimed for a life of calmness and tranquility as opposed to that

Fig. 19.5 Altes Museum, Rotunda (painting by Carl Emanuel Conrad). From *Wikimedia Commons*

of turbulence. He was correct, after all, in as much as architecture can only fulfil the requirements for flexibility by waiving its artistic values.

In modernism, the gain in liveliness promised by the organism model of art caused architecture to lose its aspirations to being an art form. Building "organically" meant becoming functional and building spaces that no longer constrained the vitality of the "new man".

Only because Schinkel cautiously "galvanized" the *Alte Museum,* as electric lighting had not yet been invented, could it be modern and nevertheless a work of art. Architecture could only meet the organism model more effectively when it stopped pretending to be a form of art. In 1926, Adolf Behne stated, with good

Fig. 19.6 Altes Museum, staircase (drawing by Schinkel). From *Wikimedia Commons*

reason, that functionalism and organic architecture were identical (Behne 1926). The idea behind this was that a complete building could be adapted to functionality just like a door handle is adapted to our hands.

As the artist, be it a painter or an industrial designer, will always be involved in design decisions, the portion of subjectivity in artistic production can only be reduced, not eliminated. In the past, the organism model established a connection between the objectivity of nature and the subjectivity of art. In the same manner, biomimetics can connect biology, which is reminiscent in its name, with the mimetic impulses that incited people to mimicry and, in the later stages of development, to the Arts. This can only succeed if biomimetics retains the emphatic concept of the organism, a concept that modern art inherited from *Naturphilosophie*.

19.4 Conclusions and Future Prospects

The example presented in this contribution shows the manner in which the organism model can be used not only to generate concrete architectural designs, but also to legitimate a certain way of speaking and thinking about the actual production of a "good" design. It also integrates arts and architecture into a holistic model of seeing the world, whereby nature, technology and arts form one unit. In Schelling's time, this was mainly a philosophical issue. The influence of modern natural sciences was, however, even then becoming noticeable. Thus, the "organism" provides both a practical model and its theoretical foundation.

Based on this idea, the current potential of the organism concept to integrate biological, technical and aesthetic approaches is investigated. Whereas Schelling's contemporaries could resort to a more or less universal terminology, the field of knowledge applied in Schinkel's designs has today fragmented into independent disciplines. How can such disparity be attacked? First a mutually understandable language needs to be established. Although the organism is more than a mere example, as it also conceptually conveys the idea of such integration itself, many (technical) terms and concepts as used in the daily practical work of biologists, engineers and architects need to be considered in order to facilitate interdisciplinary communication. Therefore, important concepts used in the particular disciplines are being collected and defined in a way in which the members of other fields can understand them. Second, the practical and the conceptual effects of the use of these terms need to be investigated, as was shown above for the "organism". Such implications can be called the *scope* of a concept, stressing not its denomination but its more associative implications in external fields. Thus, it is intended to establish an interdisciplinary and formalized vocabulary for the field of architectural and constructional biomimetics documenting both congruencies and differences between the organism concepts and systems of concepts of the three disciplines and to compile significant (technical) terms and their definition in the form of a theoretically founded and interdisciplinarily controlled vocabulary for the field of architectural and constructional biomimetics. Experience so far has shown that, for instance, terms such as form, function, adaptation, structure, material or principle are used in different ways by the different disciplines involved in joint biomimetic projects. Such inconsistencies might even constrain an unimpeded knowledge transfer from biology to technical applications and *vice versa*.

The analysis of the biomimetic use of language and concepts will eventually form the basis for a future evaluation to the extent to which computer-assisted system-theoretical approaches and (bio-)ontologies can help to comprehend complex organisms thereby enabling relationships of cause and effect to be drawn between various structures and structural levels. By means of defined and formalized vocabularies, (bio-)ontologies will become modern (bio-)informatical instruments used to represent and exchange the knowledge of certain disciplines in the form of digital networks. These networks are currently being developed in many scientific disciplines. For the field of biomimetics, an ontology will largely facilitate the interdisciplinary exchange of knowledge by setting a common data standard interconnecting databases from different disciplines and systemic levels (cf. Vincent 2014) and scientifically enabling the comprehension of complex organisms, i.e. biological individuals, engineering scientific machines and architectural buildings. This will help to mould (architectural and constructional) biomimetics into a science that also integrates basic theories, instead of solely extracting exploitable knowledge from biological role models.

Starting from the concept of the organism used in the three disciplines, the biological organism approach can be expanded by making use of technomorphous organism models (Zoglauer 1994) and, at the same time, by reflecting the irreducible emergent character of organismal sub-systems coupled to one another. As a specification of natural organisms, such reflections need to consider their capacity for self-regulation and their embedment into an ontogenetic and evolutionary formative process that is shaped by interaction with their environment. In order to develop the organism concept in accordance with the scopes of the “organism” dealt with in architecture (buildings) and engineering (machines), for the field of architectural biomimetics, the strong contrast between the organism as a real scientific concept on the one hand and the organism as a model or topos on the other, could potentially be resolved by developing a biomimetical interpretation of the organism topos, in which the topos retains its philosophical and speculative connotations, but at the same time keeps up with the insights of modern biology. The example of Schinkel showed, how properties ascribed to living nature can motivate design decisions and, more generally, aesthetic judgement. Another property, which is derived from general assumptions about biological organisms and then becomes a feature of technical inventions is the idea of “sustainability” (see Chap. 18).

The analysis of the organism model in the history of architecture shows that this can only succeed when the complexity and ambiguity of the “organism” is taken seriously. This also describes the challenge of providing a useful biomimetic terminology. How can a common vocabulary be found that unites “hard” concepts with “soft” models or, in other words, “hard” significance with “soft” scope? This finally can be used to develop a theoretically based, interdisciplinary system of concepts for the field of architectural and constructional biomimetics. The notion of “unification” can become a major outcome of defining the organism model within the auspices of biomimetics, thus serving to unite the natural sciences on the one hand and arts and humanities on the other.

Acknowledgements This work has been funded by the German Research Foundation (DFG) as part of the Transregional Collaborative Research Centre (SFB/Transregio) 141 ‘Biological Design and Integrative Structures’, project C02 ‘Organism concepts in biology and architecture as the basis for an interdisciplinary synopsis of constructional biomimetics.’

References

Bateson G (1972) Steps to an ecology of mind. Collected essays in anthropology, psychiatry, evolution, and epistemology. Jason Aronson, Northvale/London

Behne A (1926) Der moderne Zweckbau. Drei Masken Verlag, Berlin

Blümle C, Schäfer A (2007) Organismus und Kunstwerk. Zur Einführung. In: Blümle C, Schäfer A (eds) Struktur, Figur, Kontur. Abstraktion in Kunst und Lebenswissenschaften. Diaphanes, Berlin, pp 9–25

Campbell NA, Reece JB (2009) Biologie. Pearson Studium, München

Caplan AL (1987) Why the problem of reductionism in biological science will not go away. Growth 51:22–34

Drack M, Pouvreau D (2015) On the history of Ludwig von Bertalanffy's "General systemology", and on its relationship to cybernetics – Part III: convergences and divergences. Int J Gen Syst 44:523–571

Gegenbaur C (1870) Grundzüge der vergleichenden Anatomie. W. Engelmann, Leipzig

Guillén MF (2006) The taylorized beauty of the mechanical: scientific management and the rise of modernist architecture. Princeton University Press, Princeton/Oxford

Homberger DG (1988) Models and tests in functional morphology: the significance of description and integration. Am Zool 28:217–229

Karr JR, James FC (1975) Eco-morphological configurations and convergent evolution in species and communities. In: Cody ML, Diamond JM (eds) Ecology and evolution of communities. Belknap, Cambridge, MA, pp 258–291

Laubichler MD (2005) Systemtheoretische Organismuskonzeptionen. In: Krohs U, Toepfer G (eds) Philosophie der Biologie. Suhrkamp, Frankfurt am Main, pp 109–124

Limpinsel M (2016) Was macht Wissen plausibel? Topische Objektkonstitution in den Geisteswissenschaften. In: Kämper H, Warnke IH, Schmidt-Brücken D (eds) Textuelle Historizität. Interdisziplinäre Perspektiven auf das historische Apriori (Discourse Patterns 12). De Gruyter, Berlin/Boston, pp 105–121

Maier W (1999) On the evolutionary biology of early mammals – with methodological remarks on the interaction between ontogenetic adaption and phylogenetic transformation. Zool Anz 238:55–74

Motta PJ, Kotrschal KM (1992) Correlative, experimental, and comparative evolutionary approaches in ecomorphology. Neth J Zool 43(2–3):400–415

Nachtigall W (2010) Bionik als Wissenschaft: Erkennen – Abstrahieren – Umsetzen. Springer, Berlin

Nönnig JR (2007) Architektur. Sprache. Komplexität. Acht Essays zur Architekturepistemologie. Bauhaus Universität, Weimar

Perelman C, Olbrechts-Tyteca L (2004) Die neue Rhetorik. Eine Abhandlung über das Argumentieren. Ed. Josef Kopperschmidt, problemata, Bd. 149, Stuttgart-Bad Cannstatt

Perret A (1952) Contribution à une théorie de l'architecture. Cercle d'études architecturales, Paris

Plato (1981) Werke in acht Bänden. Ed. Gunther Eigler. Bd. 5, Phaidros. Parmenides. Briefe. Wissenschaftliche Buchgesellschaft, Darmstadt.

Reilly SM, Wainwright PC (1994) Conclusion: ecological morphology and the power of integration. In: Wainwright PC, Reilly SM (eds) Ecological morphology. The University of Chicago Press, Chicago/London, pp 339–354

Riedl R (2000) Strukturen der Komplexität. Eine Morphologie des Erkennens und Erklärens. Springer, Berlin

Ripley RL, Bhushan B (2016) Bioarchitecture: bioinspired art and architecture—a perspective. Phil Trans R Soc A 374

Sauer KP (1992) Morphologie und Evolution. Verh Dtsch Zool Ges 85(2):349–357

Schinkel KF (1979) Das architektonische Lehrbuch. Ed. Goerd Peschken. Deutscher Kunstverlag. Berlin

Thaler L (1984) Organische Form in der Musiktheorie des 19. und beginnenden 20. Jahrhunderts. Musikverlag Emil Katzbichler, München.

Toepfer G (2011) Organismus. In: Toepfer G (ed) Historisches Wörterbuch der Biologie. Geschichte und Theorie der biologischen Grundbegriffe. Metzler, Stuttgart

van Eck C (1994) Organicism in nineteenth-century architecture. An inquiry into its theoretical and philosophical background. Architectura & Natura Press, Amsterdam

Vincent JFC (2014) An ontology of biomimetics. In: Goel AK et al (eds) Biologically inspired design. Springer, London

Wiener N (1948) Cybernetics, or control and communication in the animal and the machine. MIT Press, Cambridge, MA

Zoglauer, T (1994) Modellübertagungen als Mittel interdisziplinärer Forschung. In: Maier W, Zoglauer T (eds) Technomorphe Organismuskonzepte. Modellübertragungen zwischen Biologie und Technik. problemata 128, frommann-holzboog, Stuttgart, pp 12–24

Chapter 20
Perceptual Orientation and Spatial Navigation in Dense Urban Environments

Patrik Schumacher

Abstract This paper poses the problem of perceptual orientation and spatial navigation in dense urban environments and proposes a scenario and methodology of urbanization in analogy to natural ecologies in which land formation, streams and rivers, flora and fauna come together according to rules of interdependency that build up a complex variegated order. Despite their complexity, such natural environments are eminently legible and navigable. The vision of a perceptually tractable complex urban order contrasts with the urban disarticulation that has resulted from the urbanization processes of the last 40 years, since the demise of modernism and modernist urban planning.

20.1 Historical Socio-economic Condition of Urban Planning's Demise

We are witnessing a sustained drive towards urban concentration in global hub cities such as London, New York, Tokyo, Shanghai and Sao Paulo. Within the contemporary network of society, an individual's productivity depends on being plugged into urban professional and cultural networks that exist only in the big cities. What each of us is doing needs to be continuously re-calibrated with what everybody else is doing. That is what all further productivity gains will depend upon. A new level of communicative density is thus required that is only available in the metropolis. This underlies what economists measure as 'agglomeration economies'. In the provinces, you are cut off and thus unproductive. We all experience this feeling and that is why we rightly pile into the city; the more centrally we can locate, the better. Since the neat division into work and leisure has disappeared and we feel the vital urge to remain plugged into the network 24/7, it is as important for us to live in the city as it is inevitable for us to work in the city. Everything presses into the centre, the more the merrier. This spells a new desire for an unprecedented degree of urban intensification and mixity, a desire that is

P. Schumacher (✉)
Zaha Hadid Architects, 10 Bowling Green Lane, London EC1R 0BQ, UK
e-mail: Patrik.Schumacher@zaha-hadid.com

J. Knippers et al. (eds.), *Biomimetic Research for Architecture and Building Construction*, Biologically-Inspired Systems 9, DOI 10.1007/978-3-319-46374-2_20

currently frustrated by outmoded planning restrictions. The new urban dynamic is not only a fascinating challenge and task for architects, but also requires new degrees of freedom for urban entrepreneurs who need this liberty to experiment, discover and create the best ways to weave the new urban texture and to garner the potential synergies through new intricate programmatic juxtapositions. Only an unhampered market process can enable such a path of discovery and has the information-processing capacity and agility to generate a viable complex variegated urban programmatic order for this new dynamic societal context. The old planning brakes had to be released in terms of land use and density restrictions. The power of urban planning and control has been eroded accordingly. If we take a global glance at our big cities, it is amazing that any planning systems are in place at all. The amorphous urban agglomerations that our cities have become certainly do not allow us to see any trace of these regulatory efforts. Planning is fighting a losing rear-guard battle. The thesis elaborated in this paper not only accepts this fact, but moreover extrapolates it to municipal urban planning's final demise as the logical and desirable culmination of the current socio-economic trend towards what we might call the Post-Fordist Network Society. Cities are our civilisation's 'super brains' continuously searching for new ideas and elaborating innovations that lead to the continuous re-programming of the manufacturing and agricultural production robots that churn out our means of life in emptied-out production landscapes, after which the agricultural products are sorted in robotic warehouses and shipped back to us city dwellers. The most crucial information processing and creation of ideas in the 'super brain' are happening in face-to-face communications. The myriad ordered spaces of the city operate as an integrated system of communication interfaces that need to be browsed and navigated by all to maximize their interaction density and relevancy, generating more productivity enhancing 'brain power'. This network cannot be planned. It must be given plenty of freedom to evolve via negative and positive feedback loops. Plasticity is here as much a precondition of learning and intelligence upgrading as in the case of our brains proper. (It's only the big cities that evolve brain power and brain-drain all else.)

Large-scale city planning first started to recede during the 1970s and, since then, urbanism as a discourse, discipline and profession has all but vanished. The disappearance of urbanism coincides with the crisis of Modernism, which can be interpreted as the way in which the crisis of the Fordist planned economy manifested itself within architecture. The 50 core years of architectural modernism (1925–1975) were also the golden era of urbanism. During this period, the advanced industrial nations urbanized on a massive scale. This was also the era of Fordism, i.e. the era of mechanical mass production and the era of the planned/mixed economy. The state dominated much of the city building via extensive public investment in infrastructure, social housing, schools, hospitals, universities, etc. This made large-scale long-term physical planning possible. In Western Europe, energy, utilities, broadcasting, railways and many heavy industries had been nationalized. This further enhanced the feasibility of large-scale long-term urban planning. The most congenial societal context for modernist urbanism existed within the socialist block with its centrally planned economy. Socialism delivered the logical conclusion of the tendencies of the era, rolling out the technological achievements of the era in a predictable, centrally planned manner, literally delivering the uniform consumption standard made possible by Fordist mass production to every member of society. Consequently, we find the fullest expression of modernist urbanism in the Eastern Block. Civilization evolved further. The crisis of Fordism, Post-Fordist restructuring, the neo-liberal turn in economic policy (privatization, deregulation) and the collapse of the Eastern Block system all coincided with the crisis of modernism in architecture and urbanism. The bankruptcy of Modernist planning gave way everywhere to the same visual chaos of laissez faire urban expansion and agglomeration under the auspices of stylistic pluralism and the anti-method of collage. Our contemporary cities exude vitality but they are no longer symbols of order but are instead poised to tip over into a menacing disorder. Like all urban agglomerations, London is expanding without bounds and without shape. The only features that give otherwise amorphous mega-cities a recogniseable shape are natural landscape features such as rivers, hills and valleys (Figs. 20.1 and 20.2).

20.2 Concepts of Order: The Dialectic of Urban and Conceptual Order

Architecture and the city used to be the ultimate incarnations and indeed models of all concepts of order: ancient cities such as Babylon, ancient Roman cities, the ideal and built cities of the Renaissance and the Baroque cities. Architectural figures and urban-ordering patterns offer the archetypical paradigm of any concept or order: a place for everything and everything in its place, within a bounded realm, subdivided by an orderly geometric arrangement.

The built environment does indeed provide the historical/material conditions for the emergence and transformation of those deep-seated conceptual ‘architectures’ that are constitutive of our abstract thinking in general. The Classical architectural canon that developed following the rebirth of architecture in the Italian Renaissance served as the source domain for the transference of analogies of order within all domains of abstract conceptual thinking, including the most sophisticated philosophies of the Enlightenment.

Diagrammatic conceptualizations became prevalent during the Renaissance and continued to be prominent until the end of the eighteenth century. However, there was but one schema for such conceptualizations: the organizational schema of the ideal city, with a concentric and radially segmented order, representing the classificatory pyramid as the only tool of abstract thinking of this era (and still prevalent today). This schema of an ideal city was itself abstracted from the prior material reality of the unplanned medieval walled city that evolved by a quasi-material (economic) process with the ring as most efficient boundary and with growth via concentric (ring upon ring) expansion. Thus, at the root of this

universal conceptual structure lies the spontaneous evolutionary morphogenesis of the concentrically growing walled city.

The spatial order of the built environment constitutes one of the primary analogical source domains for schemata of thinking. This phenomenon is not restricted to prehistoric times but has remained an active force all along. The permanent possibility and immediacy of architectural/spatial analogies as resource of our thinking are reinforced and confirmed through our daily navigation of architectural orders. Order might still be defined as 'everything in its place and a place for everything'. 'Architecture' is still at the root of most of the conceptual schemata that we rely on today: sequences, branching graphs, grids, concentric nesting, etc. This deep-seated power of architectural tropes, embedded in the most general operations of logic and language, poses a formidable barrier to radical innovation. The transcendence of these tropes is so difficult because the very patterns of conception (clear and distinct thinking) are locked within these tropes. An architecture that today is self-conscious of its formative role within the domain of conceptual order should be able to challenge deeply ingrained patterns of thought by effective spatial intervention.

Renaissance, Baroque and planned modernist cities such as Brasilia delivered further recognisable paradigms of order. The Baroque used primary and secondary axes and ever-ramifying symmetries as new ordering structures. Modernism added the principles of separation, specialization and repetition. Brasilia was perhaps the last sizeable city or urban district that projects a clear visual urban order: the last beautiful city development.

Symmetry has been made the norm in Classical architecture. Conceptual symmetries inform all Classical conceptual formalisms within science and philosophy up to the end of the eighteenth century: for instance, the Kantian table of categories is marked by an insistence upon symmetric order, signifying completeness. This insistence on symmetry, which was for Kant an unreflected a priori of his theory, strikes us today as an irrational formalism. Modernism established the possibility of designing architectural orders without symmetry but was still based on orthogonal grids and seriality. Relationships of exclusion, inclusion, subsumption and subdivision, plus lists and sequences, operate in all modernist architectural and urban orders. Such ordering devices are ubiquitous in modern civilization. These abstracted architectural tropes have been recycled back from thinking to building, i.e. to concrete architectural construction, thus perpetuating their hold over our

◄

Figs. 20.1 and 20.2 City of London: Communication density becomes physically manifest. London is paradigmatic example of the urban concentration process in global hub cities. As more and more large iconic structures pile into the financial district, the urban scape becomes increasingly chaotic, an unintentional bricolage. The planning process is evidently failing to stem the visual chaos and is unable to establish any semblance of urban order. Like all urban agglomerations, London is expanding without bounds and without shape. The only features that give otherwise amorphous mega-cities a recogniseable shape are natural landscape features such as rivers. (Photos: Patrik Schumacher) conceptual, social and spatial order. Deconstructivism tried to cut this loop. It was the vehicle by which philosophy returned to its roots in architecture in an effort to break this circle of repetition through direct action in space. Deconstructivism did indeed violate longstanding conceptions of clear and distinct order. Initially, this work was limited to the creative destruction of order, producing gestures of disruption and disorder. However, a new repertoire of ordering principles emerged that was able to increase the capacity of architecture to organize and articulate the more complex life-processes and social institutions that had started to emerge. For instance, Deconstructivism elaborated a capacity for spatial overlap and interpenetration of domains. This capacity recognizes a salient trend in contemporary social institutions whereby conditions of multivalency become more and more widespread. The main point here is to increase the repertoire of both conceptual and spatial ordering principles and to upgrade their capacity to structural complexity. The contemporary style of Parametricism (for an elaborate exposition of Parametricism consult Schumacher 2009 and also Schumacher 2012) is well prepared to continue the Deconstructivist project of expanding architecture's repertoire of conceptual and spatial ordering way beyond the capacity of Deconstructivism or any prior style.

All urbanization since 1980 has produced "ugly" amorphous urban agglomerations without recognisable order and identity. However, if real estate and rental markets are able to deliver synergetic programme mixes, i.e. programmatic order, then the hypothesis might be ventured that the disorder that we perceive in our contemporary urban agglomerations is only apparent, only visual. The urban morphological cacophony obscures the underlying programmatic order, i.e. the evolved urban system of interaction offerings. This obscurity is attributable not only to the disarticulation produced by the prevalent pluralism of styles, but also partly to our limited conceptual repertoire of recognizing more complex systems of order.

Le Corbusier insists that: 'the house, the street, the town … should be ordered; … if they are not ordered, they oppose themselves to us (Le Corbusier 1925, p. 15). Le Corbusier's limitation is not his insistence upon order but his limited concept of order in terms of Classical geometry. Complexity theory (or chaos theory) in general, and the research of Frei Otto in particular, have since taught us to recognize, measure and simulate the complex patterns of order that emerge from processes of self-organization. Phenomena such as the 'donkey's path' and the urban patterns resulting from unplanned settlement processes can now be analysed and appreciated in terms of their underlying logic and rationality, i.e. in terms of their hidden regularity and related performative power that result from the consistent constraining pressures that have underpinned their process of formation. Le Corbusier realized that, although 'nature presents itself to us as a chaos … the spirit which animates nature is a spirit of order' (Le Corbusier 1925, p. 18). However, his understanding of nature's order was limited by the science of his day. He lacked the concepts and computational tools that can now reveal the complex order of those apparently chaotic patterns by means of simulating their lawful 'material computation'. Parametricist sensibility gives more credit to the 'pack-donkey's path' as a form of adaptive material computation than to the simplicity of clear geometries that can be imposed in one sweeping move. Frei

Otto's pioneering work on natural structures included work on settlement patterns. His starting point was the distinction and interplay of occupation and connection as the two fundamental processes that are involved in all processes of urbanization (Otto 2009). His mapping of existing patterns and their geometric analyses were paralleled by physical experiments that were conceived as analogues modelling crucial features of the settlement process. In terms of occupation, he distinguished distancing and attractive occupations. For distancing occupation, he used magnets floating in water and, for attractive occupation, he used floating polystyrene chips. A more complex model integrates both distancing and attractive occupation, whereby the polystyrene chips cluster around the floating magnetic needles that maintain distance among themselves (Otto 2009, p. 45). The result closely resembles the typical settlement patterns found in real urban landscapes.

This suggests the existence of potentially discernible patterns in the apparent visual chaos of contemporary urban agglomerations. The assumption is that these patterns can be clarified and accentuated by architectural articulation. Underlying rules – economic rules – guide individual decisions that form the (so far mostly obscure) patterns. The proposition put forward here is that these economic-programmatic rules should be aligned with rules of architectural translation that make the intricately ordered complexity of the urban life processes visually legible and avoid the visual pollution and obfuscation that stems from the current unprincipled cacophony of disparate architectural translations. The processes of architectural translation do not need to follow a uniform script as if conducted by a single hand but can be delivered on the basis of multiple authors working within a shared language, with the shared ethos of making and maintaining connections, resonances and continuities across a field of diverse urban riches.

20.3 Beauty and the Evolution of Concepts of Order

What is beauty, including urban beauty? Whatever appeals at first sight. Being impressed by beauty is a gut reaction, triggered by a perceptual encounter. This immediate gut reaction operates according to an underlying rationality. The recognition of beauty within a built environment is the recognition of the vitality of this environment on the basis of its mere appearance, prior to a more in-depth experience and verification of its functionality. This works because of the extent to which subjects are conditioned by prior experience. However, as society evolves, what was once vital might have become dysfunctional. Aesthetic sensibilities have to be adapted via aesthetic revolutions. New vital societal processes might be unduly constrained by the established order of beauty. They break out of this order and the environments that they find or bring forth appear ugly. Their aesthetic rejection becomes a fetter on their further progress. A contradiction develops that can only be solved by an aesthetic revolution. Sensibilities need to be (periodically) brought into line with the morphological conditions of the most vital social life-processes. In this sense beauty keeps changing its physiognomy. However, is the category of beauty really devoid of any features that persist across its different concrete historical manifestations? If this were so, we would not be able to see the beauty of earlier styles. However, contemporary society – inclusive of contemporary architects – is still touched by the beauty (filigree order) of the Gothic, by the beauty (simple elegance) of the Renaissance, by the beauty (intense plasticity) of the Baroque, etc. Contemporary architects recognize the beauty of past eras, although they would not find it appropriate to use any of these older styles to frame contemporary institutions. Is it possible to identify an invariant characteristic, a universally applicable condition, that must be met by all environments, and even by all phenomena, and that is recognized as beautiful? Yes, there is an invariant aspect that guides all discriminations of beauty versus ugliness: the sensation of beauty is always bound to a sense of order as distinct from chaos. Order as the universal and invariant aspect of beauty has been alluded to by many classical definitions of beauty. For instance, Leon Battista Alberti's famous definition reads as follows: 'Beauty is that reasoned harmony of all the parts within a body, so that nothing may be added, taken away, or altered, but for the worse (Alberti 1988, p. 156). The positive principle of harmony/order is emphasized by reference to an 'integral body' and contrasted with a mere agglomeration: 'The harmony is such that the building appears a single, integral, and well-composed body, rather than a collection of extraneous and unrelated parts' (Alberti 1988, p. 24). The same point is further explicated by negating its opposite, which might thus be taken as the implicit definition of the ugly: a composition should be 'neither jumpy, nor confused, nor disorganized, nor disconnected, nor composed of incongruous elements, ... nor too disjointed or distant from the rest of the body' (Alberti 1988, p. 163). Alberti references order via the phrase 'harmony of all the parts'. However, his insistence on completeness, i.e., that nothing may be added, taken away or altered, is specific to Classical architecture and can no longer be considered a universal and invariant feature of beauty. Alberti's concept of an organic whole, with symmetry and strict rules of proportion, with a state of completeness or perfection that tolerates neither additions nor subtractions, describes a general ideal of beauty that remained in force from the Renaissance until the Historicism of the nineteenth century. The restrictions of symmetry, proportion and wholeness/completeness were abandoned within twentieth-century modernism. Instead, order was maintained via the order of the module and the grid and via the order of dynamic equilibrium. In addition, features such as simplicity and lightness were pursued, further specifying the modernist sense of beauty. The formal heuristics of parametricism call for order via lawful differentiation and correlation. These concepts are implemented via rule-based (algorithmic) design processes. A sense of order as distinct from chaos is maintained in all historical concretizations of the code of beauty. Order vs chaos is thus the invariant criterion of beauty. However, the criterion of order vs chaos is insufficient to give an operational definition of beauty that could fully guide the concrete application of the code values of beautiful vs ugly. The order vs chaos criterion is still too abstract and leaves too many possibilities open. There can be many different forms of ordering, of relating non-arbitrarily. Order is a necessary but not a sufficient condition of beauty. Being attracted to order and repulsed by chaos might be a biologically hardwired response, i.e. the rationality of this response might be based on biological evolution rather than on cultural evolution or on conditioning on the basis of individual experience. Chaos, the absence of any perceived order, is disorienting and thus threatening, especially if the whole environment lacks order. If the environment is partially ordered and partially configured randomly, then it makes sense that attention is drawn towards the ordered aspects, ignoring the less ordered or accidental configurations. The probability that a random configuration of entities constitutes an interrelated functioning assemblage is very low. Where entities are configured into an order, the presumption is justified that these entities somehow add up to a unit of interaction. Ordered configurations are thus more likely to constitute a force than random configurations, a force that should be reckoned with. Complex order inspires curiosity and awe, random configurations, such as a heap of garbage or the disarticulated agglomerations of suburbia, are usually taken no notice of, except negatively for their ugliness and thus absence of interest. All natural systems are ordered in some way. However, the complexity of many natural phenomena prevented the recognition of their order and beauty in earlier times. Animal forms (and animal formations such as flocks) are more organized than plant forms. Attention to animals is of higher evolutionary importance than attention to plants. Cultural evolution further confirmed the privileging of order over disorder. The more ordered appearance of the early city-based civilizations (Babylon, Maya Civilization etc.), compared with village-based clan societies, correlates with the superiority of these civilizations. The effort to give order to the built environment has been a constant feature of the process of civilization.

20.4 Evolving Multi-species Ecologies as a Model for an Evolving Multi-author Urban Order

The market process is an evolutionary process that operates via mutation (trial and error), selection and reproduction. It is self-correcting and self-regulating, leading to a self-organized order. Thus, we might assume that the land use allocation and thus the programmatic dimension of the urban and architectural order are determined by architecture's private clients within a market process that allocates land resources to the most valued uses. However, in the absence of stylistic and methodological coherence, we cannot expect the underlying programmatic order to become legible as a spatio-morphological order. For this to happen, we must presume a hegemonic stylistic and methodological paradigm that has the versatility and ordering capacity to translate the social order into a complex variegated spatial order. A shared paradigm offers the prospect of coherence across multiple authors working for multiple clients. No controlling hand needs to be presupposed.

The thesis presented here proposes the analogy of an unplanned multi-author parametric urbanism with a multi-species ecology. Consider the way that the various features and creatures within a natural environment coalesce to create a complex variegated order on the basis of rules (in turn, based on the complex interaction of multiple laws of nature) that establish systematic correlations between the various organic and inorganic subsystems that make up a natural landscape. The topography correlates with the path of the river, but the river together with the topography and sun orientation differentiates the flora, whereas the differentiation of the flora together with the river and topography shapes the differentiation and distribution of the fauna, which in turn impacts back on the fauna and thus often also on rivers and even the topography. Whereas this causality is complex and not easy to unravel, correlations and thus inference potentials are established in all directions and provide information to those who wish to navigate such a landscape. The key here is the build-up of correlations and associations (irrespective of the underlying causality). Each new species of plant or animal proliferates according to its own rules of adaptation and survival. For instance, moss grows differentially on the terraced rock surface of certain shaded slopes, i.e. depending on the surface pattern, sun orientation, self-shading rock formation, etc. A population of a certain species of bird might subsequently settle on these slopes accordingly. In the same way, parametricism envisages the build-up of a densely layered urban environment via differentiated rule-based architectural interventions that are designed via scripts that form the new architectural sub-systems, just like a new species settles into a natural environment. This process delivers rich diversity, which is nevertheless fully correlated, when designed according to the heuristics of parametricism. Each new architect/author can be uniquely creative in inventing and designing the rules/scripts of his/her project and participate in their own unique way in the build-up of a variegated information-rich urban order. The analogy also extends to the navigation in rule-based environments: the urbanite's intuitive orientation within a parametric urban environment functions analogously to animal cognition/navigation in a natural environment (Fig. 20.3).

20.5 Parametricism Delivers Urban Neg-Entropy

The only viable candidate for the next hegemonic epochal style is parametricism. Neither a hegemonic postmodernism, nor a hegemonic deconstructivism can overcome the visual chaos that allows the proliferation of differences to collapse into global sameness (white noise). Both postmodernism and deconstructivism operate via collage, i.e. via the unconstrained agglomeration of differences. Deconstructivism can be looked at as the aesthetic sublimation of the urban process of 'garbage spill' collage. Only parametricism has the capacity to combine an increase in complexity with a simultaneous increase in order, via the principles of lawful differentiation and multi-system correlation. Only parametricism can overcome the visual chaos and white noise sameness that laissez faire urbanization produces everywhere. Parametricism holds out the possibility of a free market urbanism that produces an emergent order and local identity in a bottom up process, i.e. without relying on political or bureaucratic power. The values and methodological principles of parametricism are prone to produce path-dependent self-amplifying

Fig. 20.3 Complex variegated order via multi-author coherence, Istanbul Cultural District, Studio Hadid/Schumacher, Yale University, 2013. This design experiment in un-planned multi-author urban order demonstrates the way that coherence, interarticulation and resonance can emerge if independent authors work from within the shared paradigm of parametricism that enables and calls for mutually adaptive, affiliative and resonating design moves

local identities, starting with the given natural features and settlements. Its ethos of contextual affiliation and ambition to establish or reinforce continuities allows for the development of unique urban identities on the basis of local contexts, topography, climate, etc.

Parametricist order does not rely on the uniform repetition of patterns as Modernist urbanism does. In contrast to Baroque or Beaux Arts master-plans, parametricist compositions are inherently open ended (incomplete) compositions. Their order is relational rather than geometric. They establish order and orientation via the lawful differentiation of fields, via vectors of transformation, and via contextual affiliations and subsystem correlations. This neither requires the completion of a figure, nor – (in contrast to Modernist master-plans) the uniform repetition of a pattern. In principle, infinitely many creative ways are possible to transform, to affiliate, to correlate. A unique, unpredictable, but recognisable and legible order (which allows for orienting inferences) will emerge as long as all architects acquire the required skills and are able to create, within the parametrist paradigm and ethos that calls for continuities and affiliations, under the critical eye and peer pressure of each other. A hegemonic parametricism thus holds out the prospect of a free market urban order (Fig. 20.4).

If we look at the historical progression of styles, we find that the last 300 years have established what we might call architecture's entropy law: all gains in terms of design freedom and versatility have been achieved at the expense of urban and architectural order, i.e. increases in versatility had to be bought by a progressive degeneration of architecture's ordering capacity. The increase of the designer's

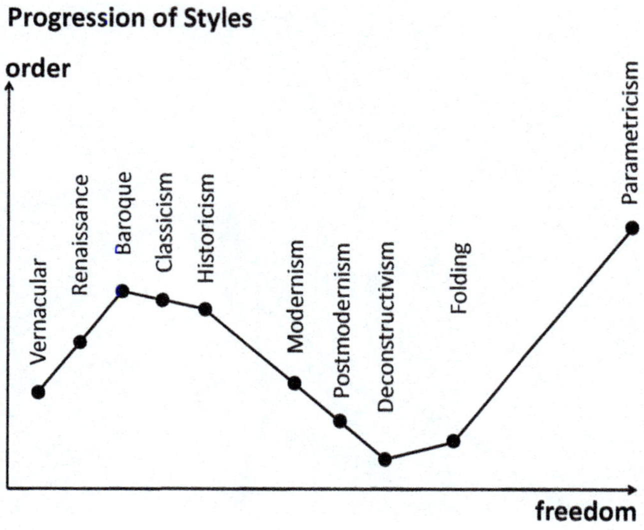

Fig. 20.4 Progression of styles: Freedom vs order, graph by Patrik Schumacher. Parametricism achieves an inversion of architecture's entropy law. Freedom must be bought by giving up order until the techniques of parametricism give a new powerful ordering capacity to the discipline of architecture, a capacity that delivers a simultaneous enhancement of freedom and order

degrees of freedom was established via the enrichment of architecture's formal-compositional repertoire. This increase in freedom/versatility was the paramount criterion of progress in architecture's pursuit of matching the requisite variety of societal complexity. Like the move from classical architecture to Modernism, the move from Modernism via Postmodernism to Deconstructivism delivered an expansion of degrees of freedom and versatility (to accommodate a more complex society) that was paid for by a relaxation or rejection of rules of composition, i.e. of means of ordering, and thus a resultant degeneration of the visual order.

Order was progressively eroded. This long trend of a negative correlation of freedom and order can be reversed under the auspices of parametricism. Parametricism offers the simultaneous increase in freedom and order and thus inaugurates a new phase of architectural neg-entropy. Parametricism's radical ontological and methodological innovation translates into a massive leap in both the dimensions of architectural progress considered here, i.e. it entails an unprecedented expansion of architecture's compositional freedom and an unprecedented leap in architecture's ordering capacity through the deployment of algorithms and associative logics.

Parametricism is the first style that delivers further degrees of freedom and versatility in conjunction with a simultaneous increase in its ordering capacity via new compositional rules such as affiliations, gradients and associative logics. In principle, all design moves are now rule-based and thus have the potential to enhance the visual order and hence the legibility of the built environment in the face of increased complexity (Fig. 20.5 and 20.6).

Fig. 20.5 Deconstructivism: Further degrees of freedom – Further degeneration of order, Zaha Hadid 1986. Deconstructivism abandons orthogonality and all historical motifs to regain the freedom of abstraction and intensifies the principle of collage by allowing superimposition and interpenetration as much as juxtaposition

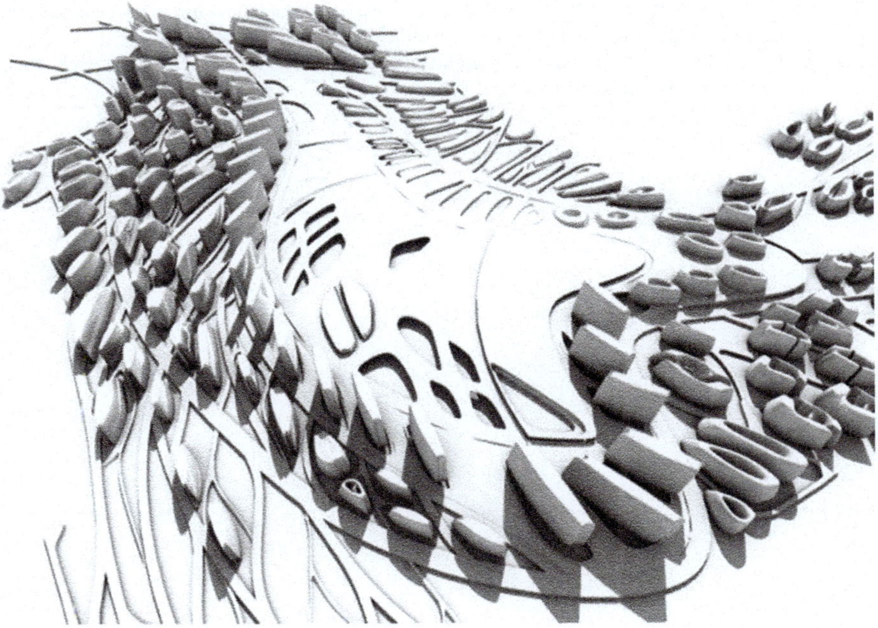

Fig. 20.6 Parametricism: Pronounced increase in freedom – sharp increase in order, Zaha Hadid Architects, Masterplans for Appur. Parametricism expands the repertoire and thus freedom with spline/nurb-based curve-linearity and gradient swarm formations. Parametricism hugely extends architecture's ordering capacity via the scripting or agent-based emergence of associative logics

Parametricism can now be considered manifestly superior to all other styles that are still pandered to and pursued. This implies that parametricism should sweep the market and put an end to the current pluralism of styles that has resulted from Modernism's crisis and that has been allowed to go on for far too long because of ideological inertia. The plurality of styles must make way for a sweeping, universal, i.e. hegemonic parametricism, to enable architecture finally to exert, once more, a vital decisive transformative impact on the built environment, in the way that Modernism did in the twentieth century.

References

Alberti LB (1988) On the art of building in ten books. (trans. Rykwert J, Leach N, Tavernor R). MIT Press, Cambridge, MA/London

Le Corbusier (1987) The city of tomorrow and its planning. Dover Publications, New York (Translated from French original Urbanisme, Editions Crès & Cie (Paris), 1925)

Otto F (2009) Occupying and connecting – thoughts on territories and spheres of influence with particular reference to human settlement. Edition Axel Menges, Stuttgart/London

Schumacher P (2009) Parametricism – a new global style for architecture and urban design. In: AD architectural design – digital cities, vol 79, No 4, July/August 2009

Schumacher P (2012) The autopoiesis of architecture, volume 2, a new agenda for architecture. Wiley, Hoboken, March 2012

The manufacturer's authorised representative in the EU is Springer Nature Customer Service Centre GmbH, Europaplatz 3, 69115 Heidelberg, Germany. If you have any concerns regarding our products, please contact ProductSafety@springernature.com

Printed and bound by CPI Group (UK) Ltd, Croydon, CR0 4YY
30/09/2026
02225501-0010